AF880501

Antibody Expression and Production

Cell Engineering

Volume 7

Series Editor

Professor Mohamed Al-Rubeai
UCD School of Chemical and Bioprocess Engineering
University College Dublin
Dublin, Ireland

Editorial Board

Dr Hansjorg Hauser
Helmholtz Centre for Infection Research
Braunschweig
Germany

Professor Michael Betenbaugh
Johns Hopkins University
Baltimore, USA

Professor Martin Fussenegger
Swiss Federal Institute of Technology
Zurich, Switzerland

Professor Nigel Jenkins
National Institute for Bioprocessing Research and Training
Dublin, Ireland

Dr Otto-Wilhelm Merten
A.F.M.-Genethon 11
Gene Therapy Program
Evry, France

For other titles published in this series, go to
http://www.springer.com/series/5728

CELL ENGINEERING

Vol. 7: Antibody Expression and Production

Edited by

Mohamed Al-Rubeai

University College Dublin, Ireland

Springer

Editor
Prof. Mohamed Al-Rubeai
School of Chemical and Bioprocess
Engineering
University College Dublin
Dublin
Ireland
m.al-rubeai@ucd.ie

ISSN 1389-6946
ISBN 978-94-007-1256-0 e-ISBN 978-94-007-1257-7
DOI 10.1007/978-94-007-1257-7
Springer Dordrecht Heidelberg London New York

Library of Congress Control Number: 2011928367

© Springer Science+Business Media B.V. 2011
No part of this work may be reproduced, stored in a retrieval system, or transmitted in any form or by any means, electronic, mechanical, photocopying, microfilming, recording or otherwise, without written permission from the Publisher, with the exception of any material supplied specifically for the purpose of being entered and executed on a computer system, for exclusive use by the purchaser of the work.

Printed on acid-free paper

Springer is part of Springer Science+Business Media (www.springer.com)

Contents

Contributors

Mohamed Al-Rubeai School of Chemical and Bioprocess Engineering, and Conway Institute of Biomolecular and Biomedical Research, University College Dublin, Dublin, Ireland, m.al-rubeai@ucd.ie

Silvia Biocca Department of Neuroscience, University of Rome "Tor Vergata", 00133 Rome, Italy, biocca@med.uniroma2.it

Michael Butler Department of Microbiology, University of Manitoba, Winnipeg, MB, Canada R3T 2N2, butler@cc.umanitoba.ca

Ulrike Dahlems PharmedArtis GmbH, Forckenbeckstr. 6, D-52074 Aachen, Germany, Ulrike.dahlems@pharmedartis.de

Ben Dionne Department of Microbiology, University of Manitoba, Winnipeg, MB, Canada R3T 2N2, umdionnb@cc.umanitoba.ca

Lisa E. Goldsmith Developmental Therapeutics Program, Fox Chase Cancer Center, Philadelphia, PA 19111, USA, lisa.goldsmith@fccc.edu

Antonio Granell Instituto de Biología Molecular y Celular de Plantas (IBMCP), UPV-CSIC, 46022 Valencia, Spain, agranell@ibmcp.upv.es

XueJun Han Amgen Inc., Seattle, WA 98119, USA, xhan@amgen.com

Grit Hehmann PharmedArtis GmbH, Forckenbeckstr. 6, D-52074 Aachen, Germany, Grit.hehman@pharmedartis.de

Stephan Hellwig PharmedArtis GmbH, Forckenbeckstr. 6, D-52074 Aachen, Germany, Stephan.hellwig@pharmedartis.de

Arthur Hewig Amgen Inc., Seattle, WA 98119, USA, hewiga@amgen.com

Nigel Jenkins National Institute for Bioprocessing Reasearch and Training, University College Dublin, Dublin, Ireland, nigel.jenkins@nibrt.ie

Thomas Jostock Novartis Pharma AG, CH-4002 Basel, Switzerland, Thomas.jostock@novartis.com

Paloma Juarez Instituto de Biología Molecular y Celular de Plantas (IBMCP), UPV-CSIC, 46022 Valencia, Spain, pjuarez@ibmcp.upv.es

Masamichi Kamihira Department of Chemical Engineering, Faculty of Engineering, Kyushu University, Fukuoka 819-0395, Japan, kamihira@chem-eng.kyushu-u.ac.jp

Yoshinori Kawabe Department of Chemical Engineering, Faculty of Engineering, Kyushu University, Fukuoka 819-0395, Japan, kawabe@chem-eng.kyushu-u.ac.jp

Roland E. Kontermann Institut für Zellbiologie und Immunologie, Universität Stuttgart, 70569 Stuttgart, Germany, roland.kontermann@izi.uni-stuttgart.de

Darrin Kuystermans School of Chemical and Bioprocess Engineering, and Conway Institute of Biomolecular and Biomedical Research, University College Dublin, Dublin, Ireland, dkuystermans@sanfordburnham.org

Thomas Linden Bioprocess Development, Merck & Co, Rahway, NJ 07065, USA, Thomas_Linden@Merck.com

Georg Melmer PharmedArtis GmbH, Forckenbeckstr. 6, D-52074 Aachen, Germany, georg.melmer@pharmedartis.de

Dafne Müller Institut für Zellbiologie und Immunologie, Universität Stuttgart, 70569 Stuttgart, Germany, dafne.mueller@izi.uni-stuttgart.de

Diego Orzaez Instituto de Biología Molecular y Celular de Plantas (IBMCP), UPV-CSIC, 46022 Valencia, Spain, dorzaez@ibmcp.upv.es

Thomas Potgieter Bioprocess Development, Merck & Co, Rahway, NJ 07065, USA, Thomas_Potgieter@Merck.com

Matthew K. Robinson Developmental Therapeutics Program, Fox Chase Cancer Center, Philadelphia, PA 19111, USA, matthew.robinson@fccc.edu

Alejandro Sarrion-Perdigones Instituto de Biología Molecular y Celular de Plantas (IBMCP), UPV-CSIC, 46022 Valencia, Spain, asarrion@ibmcp.upv.es

Maureen Spearman Department of Microbiology, University of Manitoba, Winnipeg, MB, Canada R3T 2N2, spearma1@cc.umanitoba.ca

Hiroshi Tachibana Department of Infectious Diseases, Tokai University School of Medicine, Isehara, Kanagawa 259-1193, Japan, htachiba@is.icc.u-tokai.ac.jp

Masataka Takekoshi Department of Molecular Life Science, Tokai University School of Medicine, Isehara, Kanagawa 259-1193, Japan, mtakekos@is.icc.u-tokai.ac.jp

Raymond Tyther National Institute for Bioprocessing Reasearch and Training, University College Dublin, Dublin, Ireland, r_tyther@hotmail.com

Ganesh Vedantham Amgen Inc., Seattle, WA 98119, USA, vedanthg@amgen.com

Hideki Yamaji Department of Chemical Science and Engineering, Graduate School of Engineering, Kobe University, Kobe 657-8501, Japan, yamaji@kobe-u.ac.jp

Dongxing Zha GlycoFi Inc. (A wholly-owned subsidiary of Merck & Co Inc.), Lebanon, NH 03766, USA, Dongxing_Zha@merck.com

Chapter 1
Expression of Antibody in Mammalian Cells

Thomas Jostock

Abstract Despite having a reputation to be costly, mammalian cell culture processes are used to produce the majority of currently marketed recombinant biopharmaceuticals, many of which are antibodies. Historically, mammalian cells were mainly chosen for whole IgG manufacturing because of product quality requirements like e.g. glycosylation, folding and assembly of the individual chains, which make microbial expression difficult. However, during the past 2 decades, significant progress has been made in both: Speed of mammalian cell line generation and yield of the manufacturing processes, which makes mammalian systems also from a commercial point of view more and more competitive to microbial expression. The following chapter provides an overview of state of the art mammalian cell line technologies for antibody manufacturing starting from vector and selection systems over host cell lines to screening methods.

1.1 Introduction

An increasing portion of marketed drugs and drugs under development are biopharmaceuticals, of which antibodies currently are the most common molecule class. Still, IgG is the preferred format for many indications because of high plasma half-life, long time stability, suitability for efficient affinity purification and, if anticipated, immunological effector functions. However, the tetrameric nature of an IgG molecule and the fact that glycosylation is essential for many antibody functions make it a challenging protein for microbial expression. With prokaryotic systems like *E.coli*, proper folding of the Ig domains, assembly of the four antibody chains and correct disulfide bond formation are only possible after extensive optimisation efforts and careful adjustment of expression levels (Simmons et al. 2002) and the produced is antibody aglycosylated, which limits the spectrum of therapeutic applications. In eukaryotic microbial systems like yeast and fungi secretion of proteins occurs via endoplasmic reticulum and Golgi apparatus and is aided by chaperones and cofactors which support folding and assembly of complex multimeric proteins like antibodies. The glycosylation patterns however, significantly differ from those

T. Jostock (✉)
Novartis Pharma AG, CH-4002 Basel, Switzerland
e-mail: Thomas.jostock@novartis.com

M. Al-Rubeai (ed.), *Antibody Expression and Production*, Cell Engineering 7,
DOI 10.1007/978-94-007-1257-7_1, © Springer Science+Business Media B.V. 2011

of human cells and are considered as a major immunogenicity risk factor in humans. Activities to humanize the glycosylation patterns of different microbial hosts are ongoing and have already given rise to genetically engineered *P. pastoris* strains capable of generating human-like N-glycans (Choi et al. 2003, Hamilton et al. 2003, Gerngross 2004).

Mammalian expression for manufacturing recombinant biopharmaceuticals was at first applied for t-PA (tissue plasminogen activator) which entered the market in 1986 (Wurm 2004). Monoclonal antibody therapeutics were initially manufactured using hybridoma technology, which can be applied for mouse antibodies but not for the less immunogenic chimeric and humanized antibodies and is difficult for fully human antibodies (human B-cell hybridomas tend to be not stable enough for manufacturing and bare significant viral and prion safety risks). These factors in conjunction with the upcoming of in vitro technologies for the isolation of fully human antibodies, like phage display, promoted stepwise replacement of hybridoma technologies for manufacturing purposes by recombinant DNA methods.

Production yields of recombinant mammalian cell lines quite drastically increased during the past 2 decades (Fig. 1.1a). While t-PA was produced with an upstream process yield of 50 mg/L, today for monoclonal antibodies yields of 5 g/L and more are not unusual (Wurm 2004, Birch and Racher 2006). Recently, yields of up to 10 g/L for fed-batch processes and more than 20 g/L for high cell density perfusion cultivations have been reported (Lee 2008). Much of this over 100-fold improvement in volumetric productivity is due to increased cell densities in modern cell culture processes (Birch and Racher 2006), but part of it also to improved cell

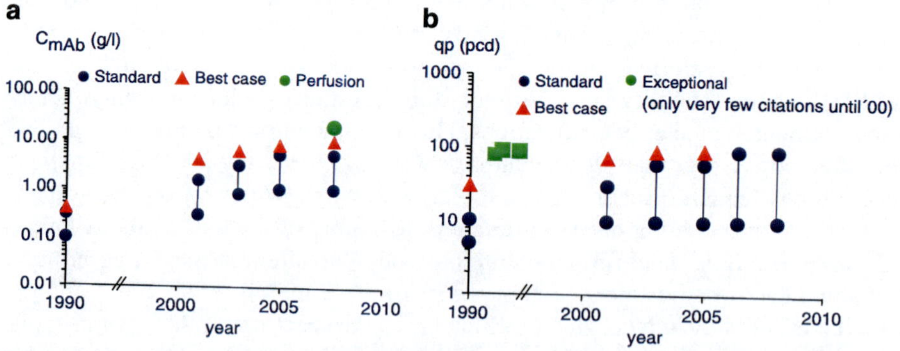

Fig. 1.1 Evolution of antibody production yields. **a** Evolution of volumetric productivity of antibody production processes in mammalian cells. An increase of about 100-fold was achieved in the past 20 years. The values and ranges shown in the graph are based on personal experience, literature and conference presentations without claiming completeness. The *Y-axis* is drawn in logarithmic scale. **b** Evolution of cell specific productivity (qp) of mammalian antibody producing cell lines. An increase of about tenfold was achieved during the past 20 years. The values and ranges shown in the graph are based on personal experience, literature and conference presentations without claiming completeness. The *Y-axis* is showing productivities in pg per cell and day (pcd) and drawn in logarithmic scale. (Modified from (Knopf 2008))

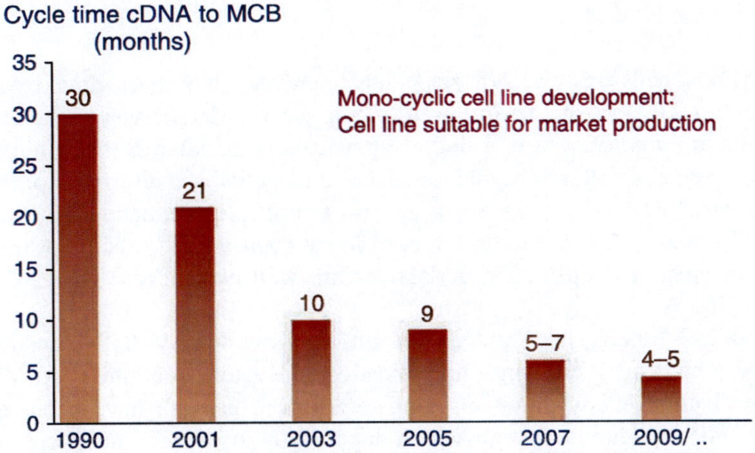

Fig. 1.2 Evolution of cycle times for cell line development. Time from transfection to preparation of a master cell bank is shown. An about sixfold reduction of cycle times has been achieved during the past 20 years. The values and ranges shown in the graph are based on personal experience, literature and conference presentations without claiming completeness. (Modified from (Knopf 2008))

lines with about tenfold higher cell specific productivities (Fig. 1.1b). Besides yield, also speed of recombinant cell line development has been substantially improved (Fig. 1.2). In the early 1990s it could easily take 2 years from vector construction to a master cell bank. Today, timelines of around 20 weeks are feasible to generate high performing cell lines and master cell banks (Jostock et al. 2008, Jostock 2009).

This huge progress in mammalian expression technologies is one of the factors that helped to make antibodies a dominant class of molecules in clinical and preclinical development since manufacturing of sufficient amounts is not considered to be a major obstacle anymore.

1.2 Expression Systems

1.2.1 Transient Expression Systems

Mammalian transient expression systems are well established and are commonly used for the production of smaller amounts of antibody drug candidates and other proteins (reviewed in (Geisse 2009, Geisse and Fux 2009)). This is of particular interest in early project phases, where many different candidates are screened to identify final lead candidates. Here, the high speed and throughput achieved with transient expression is of great advantage and compensates for lower yields compared to stable transfected cell lines.

1.2.2 Host Cell Lines

Typically, host cells that are well transfectable and capable of episomal replication of suitable expression vectors are used. Among those, derivatives of the HEK293 (human embryonic kidney) cell line are particularly popular. Besides high transfection efficiencies with affordable transfection reagents, constitutive expression of the adenoviral E1A transactivator, a general transcription enhancer (Fussenegger et al. 1999), make HEK293 cells a very efficient transient expression system. Two variants of episomal replicating derivatives are well established: HEK293-T and HEK293-EBNA.

In HEK293-T cells the T antigen of simian virus 40 (SV40) is constitutively expressed which facilitates high amplification of vectors containing a SV40 origin of replication. Copy numbers of up to 200 000 per cell have been reported for the SV40 replication system which supports high expression levels but also finally can culminate in cell death and limits the duration of the culture time after transfection (Van Craenenbroeck et al. 2000). For IgG antibodies, high throughput transient transfection (96-well) for screening applications and expression levels of 15–20 mg/L have been described with HEK293-T cells (Jostock et al. 2004).

The copy numbers after episomal replication in HEK293-EBNA cells is considerably lower (5–100 per cell) and driven by the EBNA1 transactivator and oriP from Epstein-Barr virus (EBV). Due to high retention rates (92–98% per cell generation) long retention times of plasmids are achieved resulting also in long transgene expression periods (Van Craenenbroeck et al. 2000). Transient expression yields of 40 mg/L for IgG antibodies (Meissner et al. 2001) in 1–3 L bioreactor scale and 0.5 g/L in 100-L bioreactor scale (Girard et al. 2002) have been reported for suspension-adapted HEK293-EBNA cells. Further optimization of transfection protocols, vectors and culture conditions in conjunction with co-expression of growth factors and cell cycle regulators have resulted in antibody productivities of over 1 g/L (Backliwal et al. 2008a, b).

Besides those two episomal replication competent HEK293 derivatives, HKB-11 (HEK293 fusion with lymphoma cell line) and 293 Freestyle (293-F) cells are commonly used for transient expression of antibodies and other proteins (Geisse 2009).

Recently, Chinese hamster ovary cells (CHO), which are the most widely used host cells for stable expression and manufacturing, gained popularity for transient expression, too, although typically lower yields are achieved (Suen et al. 2010, Geisse 2009, Li et al. 2007a). Especially, for early characterization of therapeutic protein candidates, transient CHO based expression systems are attractive since they are more close to the final large scale manufacturing process than HEK293 systems. This reduces the risk of differences in post-translational modifications of the protein with potential impact on pharmacokinetis and potency between early, transiently produced material and final material produced from stable CHO cell lines (Suen et al. 2010).

Apart from HEK293 and CHO, African green monkey kidney (COS) cells (Jostock et al. 1999, 2001, Evans et al. 1995) and baby hamster kidney (BHK) cells (Bi et al. 2003) are utilized for transient expression but to a lesser extend.

1.2.3 Expression Vectors

High producing transient expression systems are dependent on high transfection efficiencies. With many transfection methods, small vectors reach higher gene transfer rates than large ones and large vectors tend to result in instabilities and low yields from large scale plasmid preparations (Geisse 2009). Thus, for transient expression minimal vector backbones with only the most necessary elements are advantageous. As mentioned before, a viral origin of replication for episomal replication of the plasmid in suitable host cells is often included in the vector.

In order to supply the host cell with both, light and heavy chain of an antibody, different strategies can be followed (Fig. 1.3). Co-transfection of individual plasmids for the heavy and light chain has the advantage of comparably small vector sizes and the possibility to vary the ratio of heavy to light chain encoding plasmids (Schlatter et al. 2005). Alternatively, both chains can be expressed from one plasmid with a mono-cistronic "tandem" or "sandwich" setup, with an internal ribosomal

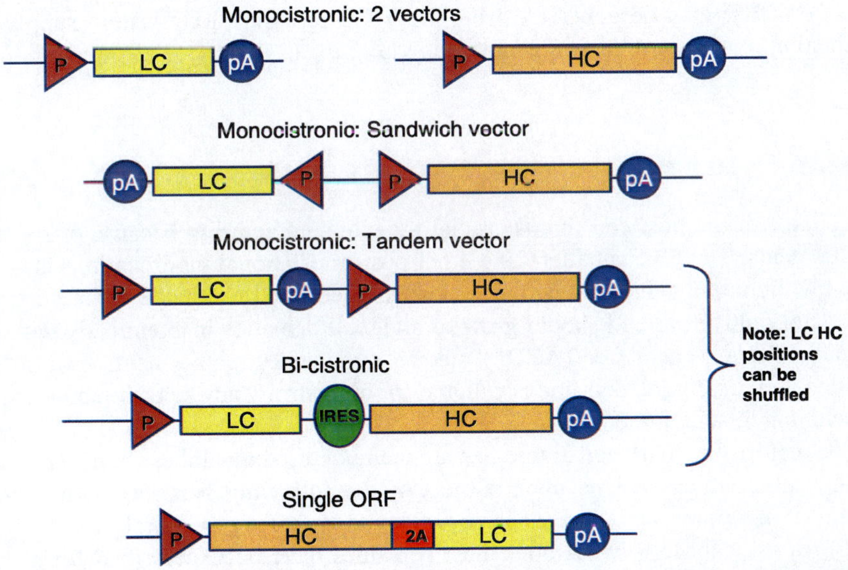

Fig. 1.3 Vector setups for antibody light and heavy chain co-expression. A schematic representation of possible vector strategies to drive expression of both chains of an antibody in recombinant mammalian cells is shown. Light and heavy chain can be encoded on separate plasmids or combined in single vector with a mono-cistronic, bi-cistronic or single ORF setup

entry site (IRES) based di-cistronic setup or with a mono-cistronic single open reading frame (ORF) setup (Fang et al. 2005, Jostock et al. 2010a, Li et al. 2007a).

1.3 Stable Expression

Manufacturing of biopharmaceutical antibodies usually is done with stable transfected recombinant cell lines. The main advantages compared to transient expression systems are the potential of using very large scale (up to 20,000 L) and a high degree of batch to batch consistency.

Stable cell line development generally follows a common scheme of sequential steps (Fig. 1.4) and ideally is done in a chemically defined and animal component free environment (Birch and Racher 2006). First, the expression vector(s) have to be transferred into the cells, typically via transfection. Then, one ore more selection steps are applied to enrich for cells that have integrated the expression vector into their genome and that are over-expressing the transgene (Fig. 1.4a). To assure monoclonality of the production cell lines, one or more single cell cloning steps are applied and followed by screening for high productivity. Further characterization steps of candidate clones usually include analysis of growth behavior, production stability, bioreactor suitability and product quality (Fig. 1.4b). Finally selected clones are conserved as master cell banks (MCBs) which are extensively tested for adventitious agents like viruses. Repeated generations of working cell banks (WCBs) may be generated from the MCB to assure long-lasting supply of production runs according to the material requirements.

1.3.1 Host Cell lines

The choice of the host cell line is crucial for biomanufacturing because of several requirements: The glycan pattern and other post-translational modifications have to have low immunogenicity in vivo. The host cell needs to be suitable for large scale bioreactor cultivation and should grow to high cell densities in chemically defined and animal component free (ACDF) media. Viral safety requirements have to be fulfilled and ideally the cell line is suitable for high stringency selection and single cell cloning in ACDF media.

The majority of marketed antibodies are manufactured in Chinese hamster ovary (CHO) cells, which will be main subject of the following paragraphs, followed by mouse myeloma cell lines such as NS0 and Sp2/0-Ag14 (Birch and Racher 2006). Besides those, several other host cell lines have been described for stable protein expression including Baby hamster kidney (BHK21) (Durocher and Butler 2009) and human cell lines like HEK293 (Graham et al. 1977, Durocher and Butler 2009, Shaw et al. 2002) and HKB11 (Cho et al. 2002). More recently, the human retinoblastoma cell line Per.C6 has been shown to be suitable for stable antibody expression (Jones et al. 2003). Other cell lines that have been described for stable

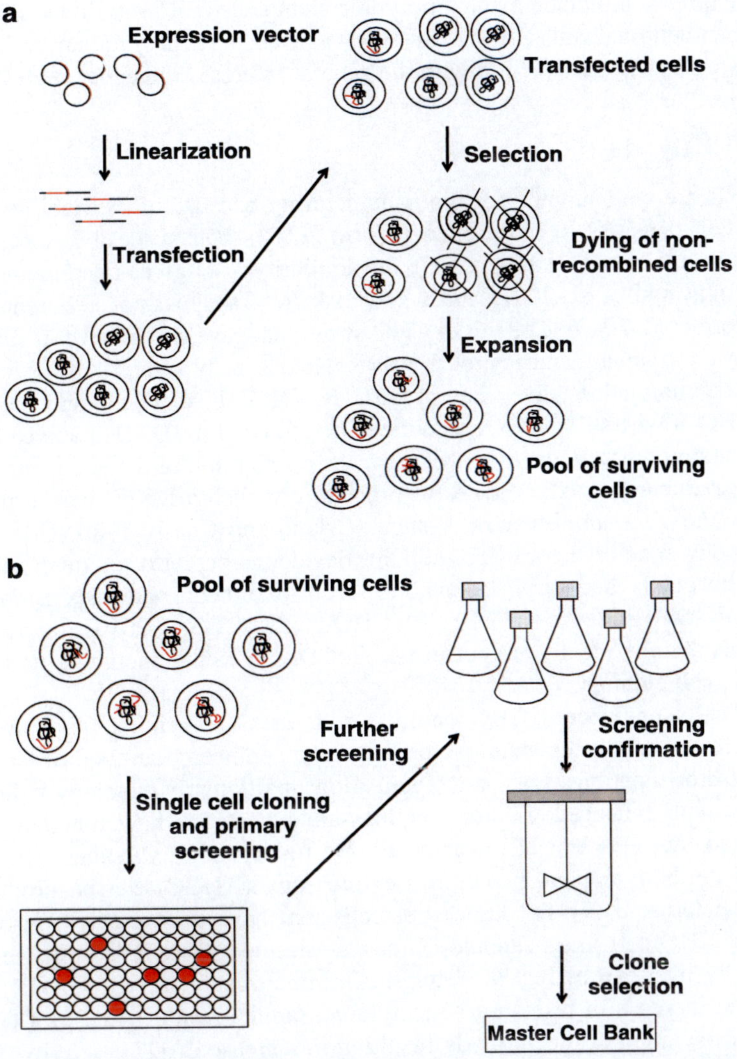

Fig. 1.4 Overview of a typical cell line development process. **a** Transfection and selection: Cell line development starts with an expression vector which, optionally after linearization, is transfected in the host cell line. Subsequently, one or more selection steps are applied to selectively kill cells that did not stably integrate the expression vector. Depending on the stringency of the selection system, the result is a more or less diverse pool of surviving cells. **b** Single cell cloning and screening: Monoclonal cell lines are generated from transfected and selected pools via single cell cloning, which can be selective or random. Several layers of screening are applied to identify high performing clone candidates. Screening results from multiwell plates and shake flasks may be confirmed in small scale bioreactors prior to selecting the final clone and manufacturing of the master cell bank

protein expression include avian embryonic stem cells (EB66) (Olivier et al. 2010) and human neuronal cells (Rose et al. 2005) as well as primary human cells that are immortalized upon transfection with the gene of interest (Schiedner et al. 2008).

1.3.1.1 CHO Cells

CHO cells are well suitable for biomanufacturing because of their ability to grow to high cell densities in bioreactors (Wurm 2004), their succesibility for genetic manipulation (Cacciatore et al. 2010), the similarity of N-glycan patterns to human proteins (Jayapal et al. 2007) and a low risk for transmission of human viruses (Cacciatore et al. 2010). The initial CHO strain that gave rise of several sub strains that are used for manufacturing today was isolated already in the 1950s (Puck 1958). The most common derivatives thereof are CHO-K1, CHO-DXB11 and CHO-DG44. While CHO-K1 are still quite close to the wild type cell line, DXB11 and DG44 have undergone random mutagenesis cycles for the elimination of the endogenous dihydrofolate reductase (DHFR) genes in order to establish a DHFR based metabolic selection and gene amplification system (Urlaub and Chasin 1980, Urlaub et al. 1983). Today, state-of-the-art CHO cell line development platforms are mostly using parental host cells that are pre-adapted to serum-free media and suspension culture like CHOK1SV (Porter et al. 2010a, b), CHO-S (Pichler et al. 2010) and others. All procedures thereby ideally are performed in CDACF media, including transfection and single cell cloning.

Genetically engineered CHO parental cell lines are gaining more and more importance, especially for manufacturing of glycoenineered antibodies with modulated effector functions. For many applications, particularly oncology indications, antibodies with enhanced effector functions are expected to have a higher efficacy in vivo and thus efficient at lower doses. De-fucosylated antibodies with highly increased antibody-dependent cellular cytotoxicity (ADCC) can be produced in fucosyltransferase-8 (FUT8) knock-out cells that have been created on basis of the DG44 CHO cell line (Yamane-Ohnuki et al. 2004, 2008). A similar effect is achieved by over-expressing N-acetylglucosamine transferase III (GnTIII) which leads to an increase in bisecting N-acetylglucosamine residues and to a decrease in core fucosylation of the antibody resulting in increased ADCC activity (Umana et al. 1999).

In addition to modifications of the product, also metabolic engineering of CHO cells to increase the productivity and process yields has been explored. Apoptosis engineering, secretion engineering and cell cycle engineering are three of the main strategies that have been followed (reviewed in (Fussenegger et al. 1999, Dinnis and James 2005) and described in Volumes 4 and 6 of this issue). Over-expression of different B-cell lymphoma-2 (Bcl-2) family members of anti-apoptotic genes has been tested in several cell lines, including CHO, and proven to protect cells from apoptosis under certain stress conditions like sodium butyrate addition to boost transcription (Kim and Lee 2000), nutrient/growth factor deprivation (Mastrangelo et al. 2000, Simpson et al. 1998, Ishaque and Al-Rubeai 2002, Meents et al.

2002, Goswami et al. 1999) or hyper osmotic pressure (Kim and Lee 2002b). However, state of the art cell culture processes have minimized such pro-apoptotic conditions to a degree that often no positive effect on productivity is achieved by over-expressing anti-apoptotic genes (Birch and Racher 2006, Dinnis and James 2005).

Inhibition of pro-apoptotic genes is another way to engineer more apoptosis resistant cell lines and has been done by over-expressing a dominant negative mutant of caspase-9 in CHO cells. As a result, such cells showed a delay in the onset of apoptosis in a bioreactor and consequently a prolongation of viability of several days (Van De Goor 2004). Antisense RNA of caspase-3 was shown to inhibit sodium butyrate induced apoptosis (Kim and Lee 2002a).

Attempts to expand the capabilities of the folding and secretion apparatus of production cell lines by secretion engineering have started already in the 1990s. Over-expression of the chaperone BiP or protein disulfide isomerase (PDI), have led to different results (reviewed in (Dinnis and James 2005)). While PDI over-expression had a negative effect on secretion of a Tumor necrosis factor alpha receptor (TNFR)-Fc fusion protein (Davis et al. 2000), increased antibody production was seen in another study (Borth et al. 2005). Recently, over-expression of X-box binding protein 1 was shown to increase endoplasmic reticulum (ER) content and antibody productivity in CHO-DG44 cells (Becker et al. 2008). However, this phenotype tends to be unstable over prolonged cultivation time, due to general survival disadvantages of XBP-1 over-expressing cells (Becker et al. 2010). Combining XBP-1 with caspase inhibitor XIAP (x-linked inhibitor of apoptosis) over-expression led to inhibition of XBP-1 induced apoptosis and improved cell survival and productivities (Becker et al. 2010). Besides ER engineering also Golgi complex enhancement has been explored. Increased antibody secretion in CHO-DG44 cells over-expressing ceramid transfer protein (CERT) or a hyperactive mutant thereof was demonstrated (Florin et al. 2009).

The concept behind cell cycle engineering is that growth arrested cells reach higher cell specific productivities. Thus, a system where proliferation can be regulated by inducing growth arrest at high cell densities can offer superior yields (reviewed (Fussenegger et al. 1999)). Regulated expression of cyclin-dependent kinase inhibitors (CDIs) was shown to allow control of proliferation and to achieve high productivities in growth arrested cells (Fussenegger et al. 1998, Mazur et al. 1999). The various strategies are described in more detail in Volume 6 of this issue.

1.3.2 Expression Vectors

Compared to transient expression approaches, vectors for stable expression usually contain additional features like selectable marker genes or chromatin opening elements. Commonly used vector technologies are described in the following paragraph.

1.3.2.1 Expression Cassette Setup

As already described above different possibilities to facilitate co-expression of antibody heavy and light chains exist (Fig. 1.3).

Classical vector strategies include co-transfection of individual plasmids for the heavy and the light chain (Kaloff and Haas 1995, Montano and Morrison 2002, Schlatter et al. 2005). The genes thereby can have either a genome like intron-exon structure (Kalwy et al. 2006) or a cDNA structure (Li et al. 2007a). Advantages of such a strategy are comparably small vectors and the option to vary the ratio of heavy chain to light chain encoding plasmids. Disadvantages are the lack of control over the ratio of heavy to light chain insertion into the host genome and the risk of multiple insertion sites for the different plasmids.

Vectors combining complete expression cassettes for heavy and light chain on a single plasmid are also commonly used and are proven to be suitable to generate high producing cell lines (Kalwy et al. 2006, Schlatter et al. 2005). In "tandem" vectors, both cassettes have the same orientation and are positioned sequentially, while "sandwich" vectors contain both cassettes in opposed orientation. Such "double gene" vectors offer a high probability of having equal copy numbers for both cassettes after integration and even optional gene amplification steps. However, such vectors can reach a significant size, especially if genomic sequences for the transgenes are used which can make handling and production of the plasmid difficult and might be a limiting factor for gene transfer efficiencies during transfection.

One alternative possibility is to combine heavy and light chain in a single expression cassette with a bi-cistronic setup containing an internal ribosomal entry site (IRES) (Jostock et al. 2004, Li et al. 2007a, b). IRES sequences are genetic elements, often of viral origin, that can drive cap-independent translation initiation of cistrons located downstream of the element (Borman et al. 1994). With bi-cistronic vectors, both chains are encoded on a single mRNA in the expressing cell, which leads to balanced expression of both polypeptides. However, the translation initiation efficiency of IRES elements usually is significantly lower as that of the 5′-cap of the mRNA, leading to an excess of translation product from the upstream cistron (Kaufman et al. 1991). Studies using mutant IRES elements having different translation initiation efficiencies showed that this is strongly effecting overall expression levels (Li et al. 2007b).

Recently, a different technology has been described, that allows co-expression of the heavy and the light chain from a single open reading frame (ORF) (Fang et al. 2005, Jostock et al. 2010a). In such a setup, heavy and light chain are encoded as a single polypeptide with a self-processing 2A sequence motive of viral origin and a furin cleavage site in between both antibody chains. Upon translation and secretion, self-processing of the 2A sequence occurs via a yet undefined mechanism which separates the heavy and light chain peptides. Furin catalysed proteolysis and the activity of carboxy-peptidase remove the remaining amino acids of the motive and the heavy and the light chain are assembled to fully functional tetrameric IgG molecules (Fang et al. 2005, Jostock et al. 2010a). One of the possible advantages of the single ORF approach is the forced equimolar expression of the heavy and the light chain genes.

1.3.2.2 Regulatory Elements

Usually, strong viral promoter/enhancer combinations like cytomegalovirus (CMV) immediate early gene region promoter/enhancer (Imhof et al. Volume 6 of this issue) or strong constitutively active cellular promoters like those of translation elongation factors (EFs) (Li et al. 2007a, Cacciatore et al. 2010) are used to drive antibody expression in mammalian cells. Besides high transcription levels and 5′ un-translated regions (UTR) with high translation initiation activity, also susceptibility to silencing of the transgene expression by e.g. methylation is an important criterion for suitable promoters. Gene silencing can be influenced by surrounding endogenous condensed chromatin and is correlated with histone modifications and CpG DNA methylation at the promoter. The histone modifications thereby seem to be the primary events in gene silencing (Mutskov and Felsenfeld 2004). Inclusion of an intron in the expressed RNA is believed to be beneficial for transgene expression and RNA stabilisation.

On the 3′-end, a suitable polyadenylation signal is needed to complete the expression cassettes.

Use of chromatin modulating motifs such as S/MARs (scaffold/matrix attachment regions), UCOEs (ubiquitous chromatin opening elements) or STAR® (stabilizing and anti-repressor) elements (Fig. 1.5a) to flank the expression cassettes of the antibody chains has been described to increase the ratio of high producing clones after random integration of the vector and to support long term stability of transgene expression (Gorman et al. Volume 6 of this issue, Cacciatore et al. 2010, Benton et al. 2002, Kwaks et al. 2003, Otte et al. 2007, Jostock et al. 2008). The upper limit of productivities achieved with such vector elements however is similar to vectors without chromatin opening motifs, but, less clone screening is necessary to identify high producers (Jostock et al. 2008). Such vectors therefore can be useful in cases where only limited clone screening efforts are possible. Additionally, the probability of high production stability may be elevated due to anti-repressor activities of such elements (Kwaks et al. 2003).

1.3.2.3 Targeted Integration Systems

With a random integration approach, the generation of stable transfectants with high expression levels and high stability of transgene expression often requires screening of a high number of clones. Alternatively, different recombination mechanisms can be used, which allow targeted integration of the transgene at genomic sites that are known to allow high and stable expression (L.Garma-Norton et al. Volume 6 of this issue, Cacciatore et al. 2010). Acceptor cell lines can be generated using reporter or selection genes or a combination of both to find and label genomic integration sites with high expression activity and long term stability.

In the mammalian cell lines that are usually used for recombinant protein expression, homologous recombination is a much less frequent event than in yeast cells or embryonic stem cells, which makes this mechanism quite inefficient for biotechnological processes. One possibility to boost homologous recombination in such cell lines is to introduce targeted double strand breaks at the desired site of integration.

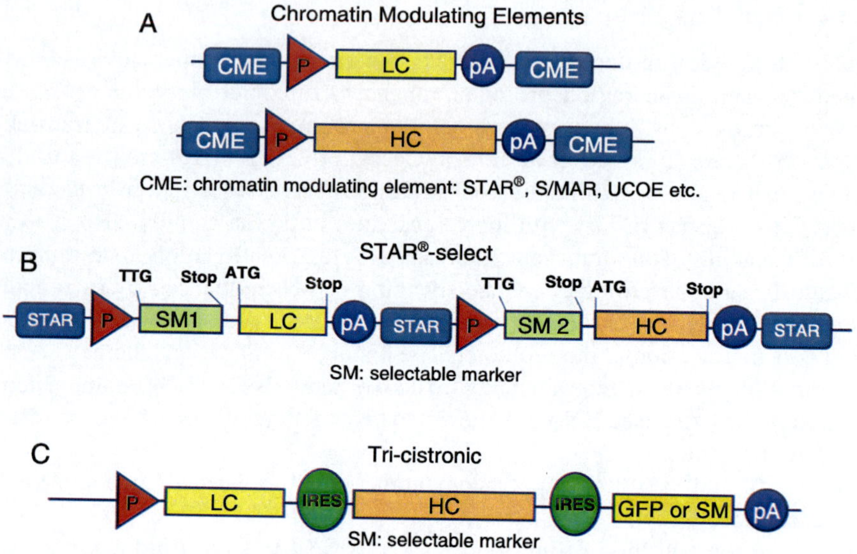

Fig. 1.5 Examples of special vector setups for stable cell line generation. **a** Chromatin modulating elements: Flanking the antibody expression cassettes with chromatin modulating elements such as S/MAR, UCOE, STAR® or others can increase the abundance of producing cells after transfection and selection and may convey clonal production stability. **b** STAR®-select: Selection marker and antibody coding regions are positioned in single expression cassettes with the selection marker upfront. The selection markers have attenuated start-codons and all ATG codons in the coding region have been abolished. Thus, antibody chains are translated with high efficiency, while selection markers are only weakly translated from the same mRNA. This leads to a very tight linkage of selection marker and gene of interest and to such high selection stringency that STAR® elements are included in the vector to increase the number of surviving cells after selection. **c** Tri-cistronic vectors: IRES elements are used to combine antibody chains and selection marker a reporter gene like GFP in a single expression cassette and to drive translation initiation of the downstream cistrons. Antibody chains and selection marker or reporter genes are expressed from the same mRNA

This can be achieved by using highly specific nucleases with large recognition sequences like wild-type or engineered meganucleaes (Arnould et al. 2010) or by using artificial hybrid nucleases consisting of a zinc finger DNA recognition moiety and a non-sequence specific cleavage domain of a restriction endonuclease (zinc finger nucleases ZFNs) (Porteus and Carroll 2005).

Alternatively, targeted integration can be mediated by recombinases that catalyze DNA strand exchange reactions between short target sequences (Bode et al. 2000). Among the first described site specific recombination systems are those based on the recombinases Cre from the phage P1 and Flp from the yeast *Saccaromyces cerevisae* with their recognition sites LoxP and FRT respectively (O'Gorman et al. 1991, Fukushige and Sauer 1992). Targeted integration can be driven by recognition sites localized on a circular targeting vector and on a pre-engineered locus in the genomic DNA. However, the efficiency of the integration reaction is limited by the competition with the thermodynamically favoured re-excision reaction (Bode et al. 2000). Pairs of mutant recognition sites were developed that allow advanced gene exchange procedures (RMCE, recombinase mediated cassette exchange) for Cre (Araki et al. 1997) and Flp (Seibler et al. 1998, Oumard et al. 2006).

1.3.2.4 Artificial Chromosomes

As an alternative approach to classical plasmid vector technologies that rely on integration in the host cell genome, an artificial chromosome based system (artificial chromosome engineering ACE) has been described (Lindenbaum et al. 2004). An artificial chromosome has been generated from an existing mouse chromosome and engineered to carry multiple recombination acceptor sites for targeted insertion of transgenes. Catalyzed by an engineered lambda phage integrase multiple copies of the gene of interest can be inserted in transiently transfected mammalian cells acting as a host for the recombination system. Recombined artificial chromosomes can be isolated by flow cytometry and then introduced to the expression host, where it is replicated and segregated among daughter cells. Antibody expression levels of up to > 1 g/L in CHO cells and production stabilities of up to 70 days have been achieved with this approach (Kennard et al. 2009b). A nice feature of this technology is that the chromosomes can be re-isolated from expressing clones and introduced to different parental host cell in order to compare their suitability to express a certain protein of interest without having a bias due to random integration mediated position effects (Kennard et al. 2009a).

1.3.3 Selection Systems

Selectable marker genes are a key feature of most vectors for stable expression. A broad range of different markers is available and the performance of the selection system strongly depends on the host cell properties.

One group of selection systems is based on marker genes that mediate resistancy to antibiotic substances like G418/Neomycine, Puromycine, Hygromycine and Zeocine. The co-responding marker genes often derive from microbial organisms and are enzymes that can de-activate the antibiotic reagent. With suitable vectors such systems are applicable for CHO cells and particularly suitable to rapidly kill non-transfected and non-expressing cells. Disadvantages of such markers are that they are regarded as non-amplifiable and that high expression of the resistancy marker by some cells of a mixed population may lead to fast inactivation of the selection reagent, allowing also non-expressing cells to survive.

Another group are metabolic selection systems such as dihydrofolate reductase (DHFR) or glutamine synthetase (GS) that are using enzymes or other components of metabolic key pathways like nucleotide or amino acid synthesis as marker genes (Cacciatore et al. 2010, Birch and Racher 2006). In a host cell background, where the endogenous counterparts of the selectable marker genes are not expressed, like for example the DHFR negative CHO cell lines DG44 and DXB11, transfected cells can be selected in a medium lacking a vital metabolite. Only cells expressing the ectopic metabolic selectable marker are able to synthesize this metabolite from precursors present in the medium. DHFR for example catalyzes generation of reduced folates which are crucial metabolites for nucleotide synthesis, while GS synthesises the essential amino acid glutamine from glutamate (Birch and Racher 2006, Cacciatore et al. 2010). The selection stringency can be further increased by adding inhibitors of the metabolic markers like methotrexate to inhibit DHFR or methionine sulphoximine to inhibit GS. Using high concentrations of such inhibitors also host cells expressing some endogenous marker enzyme can be used with this selectable marker. Methotrexate is a folate analogue which inhibits the enzymatic activity of DHFR. By adding methotrexate to the cells and by stepwise increasing the concentration of methotrexate, cells can be enriched that have undergone gene duplication events leading to a higher transgene copy number (Birch and Racher 2006, Cacciatore et al. 2010). Such multistep gene amplification procedures can significantly increase the productivity of cell lines but are time consuming and thus are less and less used in the pharmaceutical industry because of the lack of time in modern drug development processes.

A rather new approach is to use folate receptor as a dominant metabolic selectable marker. Here, cells over-expressing folate receptor are selected under folate deprivation conditions to enrich high producing transfectants (Jostock et al. 2010b).

Another example for a high stringency selection system is STAR®-select (Fig. 1.5b). Here, due to a very tight linkage of selection marker and gene of interest combined with a strong attenuation of the selection marker expression, a high proportion of high producing cells are found after selection (van Blokland et al. 2007, Otte et al. 2007, Jostock et al. 2008). This is achieved by placing the coding region of a selectable marker gene upstream of the gene of interest in a bicistronic setup, whereby the marker gene contains a strongly attenuated start codon. Hence, both genes are expressed from a single mRNA with a huge excess of translated product from the gene of interest with its optimal start codon compared to the selectable marker. This leads to such high selection stringencies that chromatin opening STAR®-elements are included in the vectors in order to get a sufficient numbers of surviving cells.

IRES elements offer another opportunity to combine gene of interest and selection marker or reporter gene on a single mRNA (Fig. 1.5c) (DeMaria et al. 2007, Li et al. 2007a, Liu et al. 2000). Typically, the gene of interest is placed upstream of the IRES element to obtain maximal translation levels driven by the mRNA 5′-cap, while selection markers are placed downstream of the IRES elements. Here, attenuated IRES variants with reduced translation initiation efficiency are of special interest in order to increase selection stringency (Li et al. 2007b).

1.3.4 Screening of Cell Line for High Productivity and Stability

Following selection, single cell cloning and screening are the next steps in cell line development (Fig. 1.4b). Although, after high stringency selection, quite high productivities can be obtained with non-clonal populations, manufacturing of clinical material is done with monoclonal cell lines to ensure consistent product quality and yield after upscaling. Also, monoclonality of the producer cell line is requested by the regulatory authorities (e.g. EMEA notes for guidance ICHQ5B and ICHQ5D). However, production stability and product quality consistency are not always given for all clones and can also be influenced by the antibody sequence. Thus, production stability and product quality are analyzed to finally confirm the suitability of a recombinant cell line for bio-manufacturing.

1.3.4.1 Cloning Technologies

Methods for single cell cloning of CHO cells include limiting dilution, colony picking by hand or using an automated system (e.g. ClonepixFL), and flow cytometry. While limiting dilution cloning is non-selective, flow cytometry and automated colony picking systems allow selective cloning of high producing cells. Several commonly used technologies are described below, a comprehensive and more detailed overview is given in Volume 6 of this issue by Browne and Al-Rubeai.

Limiting Dilution Cloning

This rather classical method works by seeding cells in a suspension with low cell density (e.g. calculated to be < 0.5 cells/well) into multiwell plates which leads to a high statistical probability of monoclonality for the resulting cell lines. The lower the calculated number of cells per well, the higher is the probability of monoclonality. However, since there is still a certain proportion of non-monoclonal cell lines arising from wells with more than one cell after seeding, a second round of limiting dilution cloning of the candidate cell line may need to be done to assure monoclonality. Alternatively, visual observation or imaging of the seeded cells in the multiwell plates right after cloning can be done to monitor, which of the candidate cell lines arise from a single cell.

While a clear advantage of this approach is the independency of special equipment, disadvantages are a comparably low number of clones per multiwell plate and the lack of selectivity of the cloning procedure, which can make excessive screening efforts necessary to identify high performing cell lines.

Flow Cytometry Cloning

Flow cytometers are capable of analyzing and sorting cells (Fluorescence activated cell sorting, FACS) at very high throughput. In combination with the ability to seed individual cells in multiwell plates this makes flow cytometry a very powerful technology for selective single cell cloning. Pre-requisite for selective FACS-cloning is that a fluorescence read-out is possible where the measured fluorescence intensity correlates with the production level of the gene of interest (reviewed in (Browne and Al-Rubeai 2007, Carroll and Al-Rubeai 2004)).

Technologies that lead to some degree of cell surface fixation of the protein to be produced, like affinity matrix capture display techniques (Manz et al. 1995, Holmes and Al-Rubeai 1999, Borth et al. 2000), are suitable to drastically reduce screening efforts by selective enrichment of high producing cells prior to cloning or by selective flow cytometry based direct cloning of high producing cells. The cells thereby are stained with suitable reagents to detect the surface displayed protein of interest.

Also, certain staining procedures have been developed that allow cell surface staining of the secreted product without fixation (Cacciatore et al. 2010, Borth et al. 2000, Brezinsky et al. 2003, Carroll and Al-Rubeai 2004).

Alternative strategies include co-expression of reporter genes (DeMaria et al. 2007, Sleiman et al. 2008, Li et al. 2007a) and microdroplet encapsulation (Powell and Weaver 1990, Kenney et al. 1995).

Automated Colony Picking

Similar to the well established colony picking systems for microbial organisms, since recent years also systems for mammalian cells, like the ClonePix FL™, from Genetix are available. In order to obtain colony formation with suspension growing cells in serum free conditions, cells are plated in dishes using semi-solid medium. Detection of high producing colonies is facilitated by embedding a suitable fluorescently labeled staining reagent in the semi-solid medium. According to the principle of an Ochterlony double immunodiffusion assay, diffusion of high amounts of produced antibody into the medium leads to formation of "halo"-shaped fluorescent immunorprecipitates with the (polyvalent) staining reagents in the surrounding of the colony. Based on imaging data generated by the system, colonies can be ranked and selected for picking taking parameters like fluorescence intensity and colony size into account.

Laser-Enabled Analysis and Processing

Cyntellects Laser-Enabled Analysis and Processing (LEAP™) system is a high throughput instrument with the ability to image cells, to quantify secreted antibody and to eliminate individual cells with a targeted laser (Hanania et al. 2005). Secreted antibody is captured by an affinity matrix on the bottom of the plates and can be detected via fluorescent staining. Clonal populations are generated by "purifying" the well from all unwanted cells using the targeted laser. The technology is described in detail in Volume 6 of this issue.

1.3.4.2 Screening Formats

Screening for High Performing Clones

The parameter that is usually analyzed to identify candidate clones for stable production in a primary screening is productivity. In later screening stages additional parameters include growth behaviour and product quality attributes.

Initial screening rounds are typically performed in multi-well formats such as 96- or 24-well plates. In multiwell culture systems, normalisation of the starting cell density is difficult; therefore usually overgrown cultures are prepared to compensate for differences in the number of inoculated cells for comparing the performance of individual clones. The volumetric productivity (product titers) determined from supernatants of such overgrown cultures are used for a first ranking of clones. High volumetric productivity can be a result of high cell specific productivity of a poor growing clone or high growth rates of a clone with low cell specific productivity. For optimal performance in fed-batch processes, however, clones need to have the right balance between growth rate and cell specific productivity (Birch and Racher 2006). This balance cannot be reliably assessed in such a basic primary small scale screen so that there is only limited correlation between multiwell screening results and bioreactor performance (Porter et al. 2010a, b).

Therefore, further screening layers, which offer better control over the seeding cell density and culture conditions, are applied. In shake flasks, suspension growing cells can reach cell densities similar to those in bioreactors. Further, shake flask cultures offer the option to monitor growth and metabolites and to add feeds, which made shake flask models in batch and fed-batch mode a widely used screening method. If optimized protocols for a given host cell line are in place, fed-batch production with suspension growing cells in shake flasks can result in very high yields already with antibody product concentrations of over 4 g/L in the culture medium (Jostock et al. 2008). Initial product quality profiles can be assessed from shake flask cultures as well by purifying the antibody from the harvested supernatant using protein-A or -G chromatography. Semi-automated small to mid-scale culture systems with some on-line analysis capabilities like pH and/or turbidity as a measure of cell density are getting more and more prevalent. By combining online analysis with automated pH control and feeding, such systems aim to better simulate large scale bioreactor processes in order to screen the candidate clones for their fit to large scale production conditions.

The screening results are usually confirmed for the one or more candidate clones with the most preferred properties in small to mid-scale bioreactor experiments to support the final clone selection decision prior to generating a master cell bank.

Production stability testing

Besides productivity and growth behaviour also production stability is a key parameter especially if up-scaling to large culture volumes is anticipated. Production stability is assessed experimentally by sub-cultivating the candidate cell lines for the period required to expand the cells from a frozen vial to production scale until harvest (typically 10–14 weeks), while regularly checking productivity in standardized assays. Depending on the selection system and the application of the product, sub-cultivation may be done in presence or absence of selection reagents. Ideally, no drop of productivity and no change in product quality are observed during the stability assessment, but with most expression systems, besides stable clones also some unstable clones may be obtained (Jostock et al. 2008, Jostock et al. 2010a) The vector and selection system can have some influence, but, the site of integration and the surrounding chromatin structure is expected to have a major impact.

Several different mechanisms may lead to a loss of productivity over time for a given cell line. Gene amplified clones with very high copy numbers may loose copies during prolonged cultivation, leading to a reduced transgene dosis which can finaly also translate into reduced expression levels.

Gene silencing is another mechanism known to influence production stability and can be driven by DNA methlyation, histone modifications and changes in the chromatin structure. Although the primary gene sequences remain unchanged, such epigenetic changes can be passed on to doughter cells during cell divisions so that sub-popuations of cells having a gene silenced phenotype are forming within a monoclonal cell line. In case cells with the antibody expression silenced phenotype are growing somewhat faster than the non-silenced cells, high producing clones are overgrown and stepwise displaced by the lower- or non-producing subpopulations.

Chromatin modulating elements like STAR®, S/MAR or UCOE as well as targeted integration systems with "silencing resistant" integration sites are possible ways to minimize the probabiliy of unstable expression in cell line development.

1.4 Conclusions

Mammalian expression of antibodies is well established in research, development and bio-manufacturing. CHO cells thereby are the dominating host cell for stable expression and also gaining importance for transient expression. Recent improvements in host cell lines, culture media, vector and screening technologies led to a drastic increase of volumetric and specific productivities while at the same time cycle times for cell line generation have been massively reduced. Cell line development platforms, as they are currently used by the biopharmaceutical industry and contract manufacturers, have reached a high level of robustness, time- and cost-eficiency. However, further improvements are possible and necessary in order to reach a level of optimisation that is more comparable to the production of mass products. With biosimilars entering the markets, more competition for cost efficient manufacturing can be expected. With new genome engineering tools like targeted nucleases and the gaining knowledge from omics approaches (Kantardjieff et al. 2009, Nissom et al. 2006, Santiago et al. 2008, Wlaschin et al. 2006, Yee et al. 2008), host cell engineering in order to improve cell line perfomance or product attributes are abolishing some of the natural limits of classical host cells.

Acknowledgements I would like to thank Hans-Peter Knopf and Burkhard Wilms for their help and for proof-reading of the manuscript.

References

Araki, K., Araki, M. & Yamamura, K. (1997) Targeted integration of DNA using mutant lox sites in embryonic stem cells. *Nucleic Acids Res*, 25, 868–72.

Arnould, S., Delenda, C., Grizot, S., Desseaux, C., Paques, F., Silva, G. H. & Smith, J. (2010) The I-CreI meganuclease and its engineered derivatives: applications from cell modification to gene therapy. *Protein Eng Des Sel*, 2011 Jan, 24(1–2), 27–31.

Backliwal, G., Hildinger, M., Chenuet, S., Dejesus, M. & Wurm, F. M. (2008a) Coexpression of acidic fibroblast growth factor enhances specific productivity and antibody titers in transiently transfected HEK293 cells. *N Biotechnol*, 25, 162–6.

Backliwal, G., Hildinger, M., Chenuet, S., Wulhfard, S., De Jesus, M. & Wurm, F. M. (2008b) Rational vector design and multi-pathway modulation of HEK 293E cells yield recombinant antibody titers exceeding 1 g/l by transient transfection under serum-free conditions. *Nucleic Acids Res*, 36, e96.

Becker, E., Florin, L., Pfizenmaier, K. & Kaufmann, H. (2008) An XBP-1 dependent bottle-neck in production of IgG subtype antibodies in chemically defined serum-free Chinese hamster ovary (CHO) fed-batch processes. *J Biotechnol*, 135, 217–23.

Becker, E., Florin, L., Pfizenmaier, K. & Kaufmann, H. (2010) Evaluation of a combinatorial cell engineering approach to overcome apoptotic effects in XBP-1(s) expressing cells. *J Biotechnol*, 146, 198–206.

Benton, T., Chen, T., Mcentee, M., Fox, B., King, D., Crombie, R., Thomas, T. C. & Bebbington, C. (2002) The use of UCOE vectors in combination with a preadapted serum free, suspension cell line allows for rapid production of large quantities of protein. *Cytotechnology*, 38, 43–6.

Bi, J. X., Buhr, P., Zeng, A. P. & Wirth, M. (2003) Human c-fos promoter mediates high-level, inducible expression in various mammalian cell lines. *Biotechnol Bioeng*, 81, 848–54.

Birch, J. R. & Racher, A. J. (2006) Antibody production. *Adv Drug Deliv Rev*, 58, 671–85.

Bode, J., Schlake, T., Iber, M., Schubeler, D., Seibler, J., Snezhkov, E. & Nikolaev, L. (2000) The transgeneticist's toolbox: novel methods for the targeted modification of eukaryotic genomes. *Biol Chem*, 381, 801–13.

Borman, A. M., Deliat, F. G. & Kean, K. M. (1994) Sequences within the poliovirus internal ribosome entry segment control viral RNA synthesis. *Embo J*, 13, 3149–57.

Borth, N., Mattanovich, D., Kunert, R. & Katinger, H. (2005) Effect of increased expression of protein disulfide isomerase and heavy chain binding protein on antibody secretion in a recombinant CHO cell line. *Biotechnol Prog*, 21, 106–11.

Borth, N., Zeyda, M., Kunert, R. & Katinger, H. (2000) Efficient selection of high-producing subclones during gene amplification of recombinant Chinese hamster ovary cells by flow cytometry and cell sorting. *Biotechnol Bioeng*, 71, 266–73.

Brezinsky, S. C., Chiang, G. G., Szilvasi, A., Mohan, S., Shapiro, R. I., Maclean, A., Sisk, W. & Thill, G. (2003) A simple method for enriching populations of transfected CHO cells for cells of higher specific productivity. *J Immunol Methods*, 277, 141–55.

Browne, S. M. & Al-Rubeai, M. (2007) Selection methods for high-producing mammalian cell lines. *Trends Biotechnol*, 25, 425–32.

Cacciatore, J. J., Chasin, L. A. & Leonard, E. F. (2010) Gene amplification and vector engineering to achieve rapid and high-level therapeutic protein production using the Dhfr-based CHO cell selection system. *Biotechnol Adv*, 28, 673–81.

Carroll, S. & Al-Rubeai, M. (2004) The selection of high-producing cell lines using flow cytometry and cell sorting. *Expert Opin Biol Ther*, 4, 1821–9.

Cho, M. S., Yee, H. & Chan, S. (2002) Establishment of a human somatic hybrid cell line for recombinant protein production. *J Biomed Sci*, 9, 631–8.

Choi, B. K., Bobrowicz, P., Davidson, R. C., Hamilton, S. R., Kung, D. H., Li, H., Miele, R. G., Nett, J. H., Wildt, S. & Gerngross, T. U. (2003) Use of combinatorial genetic libraries to humanize N-linked glycosylation in the yeast Pichia pastoris. *Proc Natl Acad Sci USA*, 100, 5022–7.

Davis, R., Schooley, K., Rasmussen, B., Thomas, J. & Reddy, P. (2000) Effect of PDI overexpression on recombinant protein secretion in CHO cells. *Biotechnol Prog*, 16, 736–43.

Demaria, C. T., Cairns, V., Schwarz, C., Zhang, J., Guerin, M., Zuena, E., Estes, S. & Karey, K. P. (2007) Accelerated clone selection for recombinant CHO CELLS using a FACS-based high-throughput screen. *Biotechnol Prog*, 23, 465–72.

Dinnis, D. M. & James, D. C. (2005) Engineering mammalian cell factories for improved recombinant monoclonal antibody production: lessons from nature? *Biotechnol Bioeng*, 91, 180–9.

Durocher, Y. & Butler, M. (2009) Expression systems for therapeutic glycoprotein production. *Curr Opin Biotechnol*, 20, 700–7.

Evans, M. J., Hartman, S. L., Wolff, D. W., Rollins, S. A. & Squinto, S. P. (1995) Rapid expression of an anti-human C5 chimeric Fab utilizing a vector that replicates in COS and 293 cells. *J Immunol Methods*, 184, 123–38.

Fang, J., Qian, J. J., Yi, S., Harding, T. C., Tu, G. H., Vanroey, M. & Jooss, K. (2005) Stable antibody expression at therapeutic levels using the 2A peptide. *Nat Biotechnol*, 23, 584–90.

Florin, L., Pegel, A., Becker, E., Hausser, A., Olayioye, M. A. & Kaufmann, H. (2009) Heterologous expression of the lipid transfer protein CERT increases therapeutic protein productivity of mammalian cells. *J Biotechnol*, 141, 84–90.

Fukushige, S. & Sauer, B. (1992) Genomic targeting with a positive-selection lox integration vector allows highly reproducible gene expression in mammalian cells. *Proc Natl Acad Sci USA*, 89, 7905–9.

Fussenegger, M., Bailey, J. E., Hauser, H. & Mueller, P. P. (1999) Genetic optimization of recombinant glycoprotein production by mammalian cells. *Trends Biotechnol*, 17, 35–42.

Fussenegger, M., Schlatter, S., Datwyler, D., Mazur, X. & Bailey, J. E. (1998) Controlled proliferation by multigene metabolic engineering enhances the productivity of Chinese hamster ovary cells. *Nat Biotechnol*, 16, 468–72.

Geisse, S. (2009) Reflections on more than 10 years of TGE approaches. *Protein Expr Purif*, 64, 99–107.

Geisse, S. & Fux, C. (2009) Recombinant protein production by transient gene transfer into Mammalian cells. *Methods Enzymol*, 463, 223–38.

Gerngross, T. U. (2004) Advances in the production of human therapeutic proteins in yeasts and filamentous fungi. *Nat Biotechnol*, 22, 1409–14.

Girard, P., Derouazi, M., Baumgartner, G., Bourgeois, M., Jordan, M., Jacko, B. & Wurm, F. M. (2002) 100-liter transient transfection. *Cytotechnology*, 38, 15–21.

Goswami, J., Sinskey, A. J., Steller, H., Stephanopoulos, G. N. & Wang, D. I. (1999) Apoptosis in batch cultures of Chinese hamster ovary cells. *Biotechnol Bioeng*, 62, 632–40.

Graham, F. L., Smiley, J., Russell, W. C. & Nairn, R. (1977) Characteristics of a human cell line transformed by DNA from human adenovirus type 5. *J Gen Virol*, 36, 59–74.

Hamilton, S. R., Bobrowicz, P., Bobrowicz, B., Davidson, R. C., Li, H., Mitchell, T., Nett, J. H., Rausch, S., Stadheim, T. A., Wischnewski, H., Wildt, S. & Gerngross, T. U. (2003) Production of complex human glycoproteins in yeast. *Science*, 301, 1244–6.

Hanania, E. G., Fieck, A., Stevens, J., Bodzin, L. J., Palsson, B. O. & Koller, M. R. (2005) Automated in situ measurement of cell-specific antibody secretion and laser-mediated purification for rapid cloning of highly-secreting producers. *Biotechnol Bioeng*, 91, 872–6.

Holmes, P. & Al-Rubeai, M. (1999) Improved cell line development by a high throughput affinity capture surface display technique to select for high secretors. *J Immunol Methods*, 230, 141–7.

Ishaque, A. & Al-Rubeai, M. (2002) Role of vitamins in determining apoptosis and extent of suppression by bcl-2 during hybridoma cell culture. *Apoptosis*, 7, 231–9.

Jayapal, K. P., Lian, W., Glod, F., Sherman, D. H. & Hu, W. S. (2007) Comparative genomic hybridizations reveal absence of large Streptomyces coelicolor genomic islands in Streptomyces lividans. *BMC Genomics*, 8, 229.

Jones, D., Kroos, N., Anema, R., Van Montfort, B., Vooys, A., Van Der Kraats, S., Van Der Helm, E., Smits, S., Schouten, J., Brouwer, K., Lagerwerf, F., Van Berkel, P., Opstelten, D. J., Logtenberg, T. & Bout, A. (2003) High-level expression of recombinant IgG in the human cell line per.c6. *Biotechnol Prog*, 19, 163–8.

Jostock, T. (2009) Reducing cycle time for cell line development. *Cell Line Development and Engineering Conference*. Berlin, Informa.

Jostock, T., Blinn, G., Renne, C., Kallen, K. J., Rose-John, S. & Mullberg, J. (1999) Immunoadhesins of interleukin-6 and the IL-6/soluble IL-6R fusion protein hyper-IL-6. *J Immunol Methods*, 223, 171–83.

Jostock, T., Dragic, Z., Fang, J., Jooss, K., Wilms, B. & Knopf, H. P. (2010a) Combination of the 2A/furin technology with an animal component free cell line development platform process. *Appl Microbiol Biotechnol*, 87, 1517–24.

Jostock, T., Knopf, H.-P., Wilms, B., Drori, S. & Assaraf, Y. G. A. (2010b) Evaluation and implementation of new expression technologies for enhanced CHO cell line development. *Antibody Development and Production Conference.* Carlsbad, CA, IBCLifeScience.

Jostock, T., Mullberg, J., Ozbek, S., Atreya, R., Blinn, G., Voltz, N., Fischer, M., Neurath, M. F. & Rose-John, S. (2001) Soluble gp130 is the natural inhibitor of soluble interleukin-6 receptor transsignaling responses. *Eur J Biochem*, 268, 160–7.

Jostock, T., Sewalt, R., Drewello, D., Lang, S., Dennler, S., Ortlepp, M., Wichter, J., Nommay, A., Dragic, Z., Van Der Vaart, H., Wilms, B. & Knopf, H. P. (2008) Evaluation of the STAR Technology. *European Biotechnology News*, 7, 33–6.

Jostock, T., Vanhove, M., Brepoels, E., Van Gool, R., Daukandt, M., Wehnert, A., Van Hegelsom, R., Dransfield, D., Sexton, D., Devlin, M., Ley, A., Hoogenboom, H. & Mullberg, J. (2004) Rapid generation of functional human IgG antibodies derived from Fab-on-phage display libraries. *J Immunol Methods*, 289, 65–80.

Kaloff, C. R. & Haas, I. G. (1995) Coordination of immunoglobulin chain folding and immunoglobulin chain assembly is essential for the formation of functional IgG. *Immunity*, 2, 629–37.

Kalwy, S., Rance, J. & Young, R. (2006) Toward more efficient protein expression: keep the message simple. *Mol Biotechnol*, 34, 151–6.

Kantardjieff, A., Nissom, P. M., Chuah, S. H., Yusufi, F., Jacob, N. M., Mulukutla, B. C., Yap, M. & Hu, W. S. (2009) Developing genomic platforms for Chinese hamster ovary cells. *Biotechnol Adv*, 27, 1028–35.

Kaufman, R. J., Davies, M. V., Wasley, L. C. & Michnick, D. (1991) Improved vectors for stable expression of foreign genes in mammalian cells by use of the untranslated leader sequence from EMC virus. *Nucleic Acids Res*, 19, 4485–90.

Kennard, M. L., Goosney, D. L., Monteith, D., Roe, S., Fischer, D. & Mott, J. (2009a) Auditioning of CHO host cell lines using the artificial chromosome expression (ACE) technology. *Biotechnol Bioeng*, 104, 526–39.

Kennard, M. L., Goosney, D. L., Monteith, D., Zhang, L., Moffat, M., Fischer, D. & Mott, J. (2009b) The generation of stable, high MAb expressing CHO cell lines based on the artificial chromosome expression (ACE) technology. *Biotechnol Bioeng*, 104, 540–53.

Kenney, J. S., Gray, F., Ancel, M. H. & Dunne, J. F. (1995) Production of monoclonal antibodies using a secretion capture report web. *Biotechnology (NY)*, 13, 787–90.

Kim, N. S. & Lee, G. M. (2000) Overexpression of bcl-2 inhibits sodium butyrate-induced apoptosis in Chinese hamster ovary cells resulting in enhanced humanized antibody production. *Biotechnol Bioeng*, 71, 184–93.

Kim, N. S. & Lee, G. M. (2002a) Inhibition of sodium butyrate-induced apoptosis in recombinant Chinese hamster ovary cells by constitutively expressing antisense RNA of caspase-3. *Biotechnol Bioeng*, 78, 217–28.

Kim, N. S. & Lee, G. M. (2002b) Response of recombinant Chinese hamster ovary cells to hyperosmotic pressure: effect of Bcl-2 overexpression. *J Biotechnol*, 95, 237–48.

Knopf, H. P. (2008) Keynote: speed to patient/market as key for success. *BioPharm SCALE-UP*. Geneva, Switzerland.

Kwaks, T. H., Barnett, P., Hemrika, W., Siersma, T., Sewalt, R. G., Satijn, D. P., Brons, J. F., Van Blokland, R., Kwakman, P., Kruckeberg, A. L., Kelder, A. & Otte, A. P. (2003) Identification of anti-repressor elements that confer high and stable protein production in mammalian cells. *Nat Biotechnol*, 21, 553–8.

Lee, G. (2008) Cell line development using a Per.C6 host cell line. *Cell Line Development and Engineering Conference*. IBC's 4th Annual Cell Line Development and Engineering, June 23–24, San Diego, USA.

Li, J., Menzel, C., Meier, D., Zhang, C., Dubel, S. & Jostock, T. (2007a) A comparative study of different vector designs for the mammalian expression of recombinant IgG antibodies. *J Immunol Methods*, 318, 113–24.

Li, J., Zhang, C., Jostock, T. & Dubel, S. (2007b) Analysis of IgG heavy chain to light chain ratio with mutant Encephalomyocarditis virus internal ribosome entry site. *Protein Eng Des Sel*, 20, 491–6.

Lindenbaum, M., Perkins, E., Csonka, E., Fleming, E., Garcia, L., Greene, A., Gung, L., Hadlaczky, G., Lee, E., Leung, J., Macdonald, N., Maxwell, A., Mills, K., Monteith, D., Perez, C. F., Shellard, J., Stewart, S., Stodola, T., Vandenborre, D., Vanderbyl, S. & Ledebur, H. C., Jr. (2004) A mammalian artificial chromosome engineering system (ACE System) applicable to biopharmaceutical protein production, transgenesis and gene-based cell therapy. *Nucleic Acids Res*, 32, e172.

Liu, X., Constantinescu, S. N., Sun, Y., Bogan, J. S., Hirsch, D., Weinberg, R. A. & Lodish, H. F. (2000) Generation of mammalian cells stably expressing multiple genes at predetermined levels. *Anal Biochem*, 280, 20–8.

Manz, R., Assenmacher, M., Pfluger, E., Miltenyi, S. & Radbruch, A. (1995) Analysis and sorting of live cells according to secreted molecules, relocated to a cell-surface affinity matrix. *Proc Natl Acad Sci USA*, 92, 1921–5.

Mastrangelo, A. J., Hardwick, J. M., Zou, S. & Betenbaugh, M. J. (2000) Part II. Overexpression of bcl-2 family members enhances survival of mammalian cells in response to various culture insults. *Biotechnol Bioeng*, 67, 555–64.

Mazur, X., Eppenberger, H. M., Bailey, J. E. & Fussenegger, M. (1999) A novel autoregulated proliferation-controlled production process using recombinant CHO cells. *Biotechnol Bioeng*, 65, 144–50.

Meents, H., Enenkel, B., Eppenberger, H. M., Werner, R. G. & Fussenegger, M. (2002) Impact of coexpression and coamplification of sICAM and antiapoptosis determinants bcl-2/bcl-x(L) on productivity, cell survival, and mitochondria number in CHO-DG44 grown in suspension and serum-free media. *Biotechnol Bioeng*, 80, 706–16.

Meissner, P., Pick, H., Kulangara, A., Chatellard, P., Friedrich, K. & Wurm, F. M. (2001) Transient gene expression: recombinant protein production with suspension-adapted HEK293-EBNA cells. *Biotechnol Bioeng*, 75, 197–203.

Montano, R. F. & Morrison, S. L. (2002) Influence of the isotype of the light chain on the properties of IgG. *J Immunol*, 168, 224–31.

Mutskov, V. & Felsenfeld, G. (2004) Silencing of transgene transcription precedes methylation of promoter DNA and histone H3 lysine 9. *Embo J*, 23, 138–49.

Nissom, P. M., Sanny, A., Kok, Y. J., Hiang, Y. T., Chuah, S. H., Shing, T. K., Lee, Y. Y., Wong, K. T., Hu, W. S., Sim, M. Y. & Philp, R. (2006) Transcriptome and proteome profiling to understanding the biology of high productivity CHO cells. *Mol Biotechnol*, 34, 125–40.

Olivier, S., Jacoby, M., Brillon, C., Bouletreau, S., Mollet, T., Nerriere, O., Angel, A., Danet, S., Souttou, B., Guehenneux, F., Gauthier, L., Berthome, M., Vie, H., Beltraminelli, N. & Mehtali, M. (2010) EB66 cell line, a duck embryonic stem cell-derived substrate for the industrial production of therapeutic monoclonal antibodies with enhanced ADCC activity. *MAbs*, 2, 405–15.

Otte, A. P., Kwaks, T. H., Van Blokland, R. J., Sewalt, R. G., Verhees, J., Klaren, V. N., Siersma, T. K., Korse, H. W., Teunissen, N. C., Botschuijver, S., Van Mer, C. & Man, S. Y. (2007) Various expression-augmenting DNA elements benefit from STAR-Select, a novel high stringency selection system for protein expression. *Biotechnol Prog*, 23, 801–7.

Oumard, A., Qiao, J., Jostock, T., Li, J. & Bode, J. (2006) Recommended method for chromosome exploitation: RMCE-based cassette-exchange systems in animal cell biotechnology. *Cytotechnology*, 50, 93–108.

O'gorman, S., Fox, D. T. & Wahl, G. M. (1991) Recombinase-mediated gene activation and site-specific integration in mammalian cells. *Science*, 251, 1351–5.

Pichler, J., Galosy, S., Mott, J. & Borth, N. (2011) Selection of CHO host cell subclones with increased specific antibody production rates by repeated cycles of transient transfection and cell sorting. *Biotechnol Bioeng*, 108(2), 386–94.

Porter, A. J., Dickson, A. J. & Racher, A. J. (2010a) Strategies for selecting recombinant CHO cell lines for cGMP manufacturing: realizing the potential in bioreactors. *Biotechnol Prog*, 26, 1446–54.

Porter, A. J., Racher, A. J., Preziosi, R. & Dickson, A. J. (2010b) Strategies for selecting recombinant CHO cell lines for cGMP manufacturing: improving the efficiency of cell line generation. *Biotechnol Prog*, 26, 1455–64.

Porteus, M. H. & Carroll, D. (2005) Gene targeting using zinc finger nucleases. *Nat Biotechnol*, 23, 967–73.

Powell, K. T. & Weaver, J. C. (1990) Gel microdroplets and flow cytometry: rapid determination of antibody secretion by individual cells within a cell population. *Biotechnology (NY)*, 8, 333–7.

Puck, T. T. (1958) Genetics of somatic mammalian cells. *J Exp Med*, 108, 945–55.

Rose, T., Winkler, K., Brundke, E., Jordan, I. & Sandig, V. (2005) Alternative Strategies and new cell lines for high-level production of biopharmaceuticals. IN Knäblein, J., (Ed.) *Modern Biopharmaceuticals.* Wiley-VCH, New York, NY.

Santiago, Y., Chan, E., Liu, P. Q., Orlando, S., Zhang, L., Urnov, F. D., Holmes, M. C., Guschin, D., Waite, A., Miller, J. C., Rebar, E. J., Gregory, P. D., Klug, A. & Collingwood, T. N. (2008) Targeted gene knockout in mammalian cells by using engineered zinc-finger nucleases. *Proc Natl Acad Sci USA*, 105, 5809–14.

Schiedner, G., Hertel, S., Bialek, C., Kewes, H., Waschutza, G. & Volpers, C. (2008) Efficient and reproducible generation of high-expressing, stable human cell lines without need for antibiotic selection. *BMC Biotechnol*, 8, 13.

Schlatter, S., Stansfield, S. H., Dinnis, D. M., Racher, A. J., Birch, J. R. & James, D. C. (2005) On the optimal ratio of heavy to light chain genes for efficient recombinant antibody production by CHO cells. *Biotechnol Prog*, 21, 122–33.

Seibler, J., Schubeler, D., Fiering, S., Groudine, M. & Bode, J. (1998) DNA cassette exchange in ES cells mediated by Flp recombinase: an efficient strategy for repeated modification of tagged loci by marker-free constructs. *Biochemistry*, 37, 6229–34.

Shaw, G., Morse, S., Ararat, M. & Graham, F. L. (2002) Preferential transformation of human neuronal cells by human adenoviruses and the origin of HEK 293 cells. *Faseb J*, 16, 869–71.

Simmons, L. C., Reilly, D., Klimowski, L., Raju, T. S., Meng, G., Sims, P., Hong, K., Shields, R. L., Damico, L. A., Rancatore, P. & Yansura, D. G. (2002) Expression of full-length immunoglobulins in Escherichia coli: rapid and efficient production of aglycosylated antibodies. *J Immunol Methods*, 263, 133–47.

Simpson, N. H., Singh, R. P., Perani, A., Goldenzon, C. & Al-Rubeai, M. (1998) In hybridoma cultures, deprivation of any single amino acid leads to apoptotic death, which is suppressed by the expression of the bcl-2 gene. *Biotechnol Bioeng*, 59, 90–8.

Sleiman, R. J., Gray, P. P., Mccall, M. N., Codamo, J. & Sunstrom, N. A. (2008) Accelerated cell line development using two-color fluorescence activated cell sorting to select highly expressing antibody-producing clones. *Biotechnol Bioeng*, 99, 578–87.

Suen, K. F., Turner, M. S., Gao, F., Liu, B., Althage, A., Slavin, A., Ou, W., Zuo, E., Eckart, M., Ogawa, T., Yamada, M., Tuntland, T., Harris, J. L. & Trauger, J. W. (2010) Transient expression of an IL-23R extracellular domain Fc fusion protein in CHO vs. HEK cells results in improved plasma exposure. *Protein Expr Purif*, 71, 96–102.

Umana, P., Jean-Mairet, J., Moudry, R., Amstutz, H. & Bailey, J. E. (1999) Engineered glycoforms of an antineuroblastoma IgG1 with optimized antibody-dependent cellular cytotoxic activity. *Nat Biotechnol*, 17, 176–80.

Urlaub, G. & Chasin, L. A. (1980) Isolation of Chinese hamster cell mutants deficient in dihydrofolate reductase activity. *Proc Natl Acad Sci USA*, 77, 4216–20.

Urlaub, G., Kas, E., Carothers, A. M. & Chasin, L. A. (1983) Deletion of the diploid dihydrofolate reductase locus from cultured mammalian cells. *Cell*, 33, 405–12.

Van Blokland, H. J., Kwaks, T. H., Sewalt, R. G., Verhees, J. A., Klaren, V. N., Siersma, T. K., Korse, J. W., Teunissen, N. C., Botschuijver, S., Van Mer, C., Man, S. Y. & Otte, A. P. (2007) A novel, high stringency selection system allows screening of few clones for high protein expression. *J Biotechnol*, 128, 237–45.

Van Craenenbroeck, K., Vanhoenacker, P. & Haegeman, G. (2000) Episomal vectors for gene expression in mammalian cells. *Eur J Biochem*, 267, 5665–78.

Van De Goor, J. (2004) Improvement of industrial cell culture processes by caspase-9 dominant negative and other apoptotic inhibitors. IN Al-Rubeai, M., Fussenegger, M., (Eds.) *Cell Engineering.* Kluwer Academic Publishers, Dodrecht.

Wlaschin, K. F., Seth, G. & Hu, W. S. (2006) Toward genomic cell culture engineering. *Cytotechnology*, 50, 121–40.

Wurm, F. M. (2004) Production of recombinant protein therapeutics in cultivated mammalian cells. *Nat Biotechnol*, 22, 1393–8.

Yamane-Ohnuki, N., Kinoshita, S., Inoue-Urakubo, M., Kusunoki, M., Iida, S., Nakano, R., Wakitani, M., Niwa, R., Sakurada, M., Uchida, K., Shitara, K. & Satoh, M. (2004) Establishment of FUT8 knockout Chinese hamster ovary cells: an ideal host cell line for producing completely defucosylated antibodies with enhanced antibody-dependent cellular cytotoxicity. *Biotechnol Bioeng*, 87, 614–22.

Yamane-Ohnuki, N., Yamano, K. & Satoh, M. (2008) Biallelic gene knockouts in Chinese hamster ovary cells. *Methods Mol Biol*, 435, 1–16.

Yee, J. C., De Leon Gatti, M., Philp, R. J., Yap, M. & Hu, W. S. (2008) Genomic and proteomic exploration of CHO and hybridoma cells under sodium butyrate treatment. *Biotechnol Bioeng*, 99, 1186–204.

Chapter 2
Bioreactor Systems for Producing Antibody from Mammalian Cells

Darrin Kuystermans and Mohamed Al-Rubeai

Abstract As the market demand for high quality monoclonal antibodies (mAbs) increases, the production of these biomolecules must meet these demands. In order to produce these complex biomolecules, researchers continue to evolve the design configurations of bioreactors to achieve the goals of easier scale-up, tighter control of the process parameters, and/or reductions in operational costs. This chapter reviews the current reactor technologies for producing mAbs using either suspension and anchorage-dependent cell lines, including any advantages and disadvantage of each.

2.1 Introduction

To optimise quality and quantity in the production of monoclonal antibodies (mAbs) from mammalian cells, researchers have developed many new methods. Considering that mAbs represent approximately 30% of total biopharmaceutical production (Walsh, 2003; 2006) and that high clinical dose requirements over long periods of time have pushed demand, the optimisation and scale-up of mAb production technologies have become quite important.

The majority of mAbs are produced in mammalian cells, which can produce complex folded proteins with correct posttranslational modifications. But the cost of producing hundreds to thousands of kilograms of mAbs per year make it worthwhile to improve the economic efficiency and scalability of this process (Rodrigues et al., 2010); thus, we must design cell culture cultivation systems with these goals in mind.

When mAbs came into prominence in the 1970s, they gave rise to a variety of bioreactor systems and cultivation techniques. Most of these systems aimed to overcome the low cell densities and increase product titres of the antibodies being secreted from the cells either in suspension or anchored to a solid support. Over the years, advancements in cell cultivation and engineering have led to a hundred-fold increase in volumetric productivity (Jain and Kumar, 2008). This chapter will discuss technologies and recent developments in bioreactor systems deployed to

D. Kuystermans and M. Al-Rubeai (✉)
School of Chemical and Bioprocess Engineering, and Conway Institute of Biomolecular and Biomedical Research, University College Dublin, Dublin, Ireland
e-mail: m.al-rubeai@ucd.ie

M. Al-Rubeai (ed.), *Antibody Expression and Production*, Cell Engineering 7,
DOI 10.1007/978-94-007-1257-7_2, © Springer Science+Business Media B.V. 2011

produce mAbs using either anchorage-dependent or suspension cells. We will cover the types of culture systems and their modes of operation, including the disposable systems utilised in biomanufacturing.

2.2 Anchorage-Dependent Systems

We can classify anchorage-dependent systems as either laboratory- or industrial-scale systems. The laboratory-scale systems use plastic wares capable of providing a suitable surface area to sustain mammalian cell growth, such as T-flasks, Petri dishes, and multiwell plates; industrial-scale systems tend to use larger surface areas for growth – to compensate for the limitation caused by cell concentration, which is directly proportional to the surface area – employing such equipment as roller bottles, multilayered stacked plate systems, like the Cell Factory (Thermo Scientific, Waltham MA, USA) and CellCube (Corning NY, USA). These systems can be disposable; also, roller bottle operations have used automated systems – i.e., the RollerCell (Cellon, S.A. Luxembourg) and Automation Partnership (TAP) CellMate (Hertfordshire, UK) – to reduce the risk of contamination and minimise labour intensive tasks (Christopher et al., 2004). However, these systems do not allow the control of pH and dissolved oxygen (DO) levels, and such systems as the CellCube have demonstrated heterogeneity of fluid flow, which can result in non-uniform cell growth patterns and shear stress (Aunins et al., 2003).

Although industrial-scale anchorage-dependent cultures produce higher concentrations and better productivity than traditional laboratory systems (Rodrigues et al., 2010), these systems introduce high costs in consumables, equipment, and labour, while allowing poor control of the culture parameters and posing a high risk of contamination (Ozturk and Hu, 2005). Microcarrier, membrane, and fiber systems were developed to overcome the problem of culturing cells lines that were difficult to adapt to suspension culture. Microcarriers, regarded as a major advance in culturing anchorage-dependent cell lines (Velden-de Groot, 1995), allow the cultivation of cells to occur in stirred tank bioreactors, airlifts, and fluidized bed bioreactors. The membrane and fiber systems rely on immobilising cells in a bioreactor run in perfusion or continuous mode.

2.2.1 Immobilised Systems

Van Wezel (1967) first described using microscopic particles with anion-exchange resin as a microcarrier to grow surface-dependent cells in a pseudo-suspension culture. This method gently agitated 0.2 mm particles in a homogenous environment. Modern microcarriers have advanced since the 1960s and come in many varieties, depending on application and cell line.

The development of microscopic particles that can act as a surface for cell growth gave a major boost to large-scale anchorage-dependent cell culture. These microcarriers can either be microporous or macroporous for the purposes of cell attachment. They are composed of nontoxic plastics, glass, silicone, or natural polymers and their derivatives, including fibrin, collagen, glutin, chitin, cellulose, or dextran (Bluml, 2007; Nam et al., 2007; Ozturk and Hu, 2005; Warnock and Al-Rubeai, 2005). Microcarriers should be small enough to maximize cell growth surface area and should be autoclavable for purposes of sterilisation. The microporous carriers have sufficiently small pore sizes not to allow cells to enter and colonise the interior; these carriers allow cells to grow on their external surfaces. Macroporous carriers have pores large enough to let cells colonize the interiors of the pores and to continue proliferating due to the increased surface area, allowing better overall maximum cell density, and consequently productivity, than microporous carriers. These carriers decrease the shear sensitivity of cells due to the extra protection provided by the pores' interiors (Ozturk and Hu, 2005; Butler, 2003).

Spinner flasks and stirred tank reactors (STRs) are most commonly used to maintain microcarrier cultures; one can achieve scale-up simply by adding new microcarriers and carefully controlling the microcarrier environment to stimulate bead to bead transfer (Wang and Ouyang, 1999; Ozturk and Hu, 2005; Butler, 2003). By increasing the agitation rate, mitotic cells may detach from colonised beads and reattach to fresh ones; this also allows the manipulation of pH and medium composition. As for medium composition, we know that increasing phosphates and decreasing Ca^{2+}, Mg^{2+}, and Mn^{2+} concentrations can cause the detachment of cells in some microcarrier systems. Exposure to a mixture of trypsin and ethylenediaminetetraacetic acid (EDTA) (Butler, 2003; Ozturk and Hu, 2005) or a cocktail of collagenase, Ca^{2+}, and Mg^{2+} (Bluml, 2007).

Microcarriers systems can also operate in fluidized bed configurations (Fig. 2.1), which use microporous carriers with a specific gravity (of greater than 1.6) to suspend cells in the upward flow (Rodrigues et al., 2010; Ozturk and Hu, 2005; Butler, 2003). These systems have the common design feature that an oxygenated upward flow of cell culture medium is kept circulating by an internal loop (although external loops had been used in the past). Fluidized bed reactors usually incorporate an internal circulation loop to avoid the formation of an oxygen gradient along the axis of the column (Biselli et al., 1995). Although the growth of anchorage-dependent cell lines is favourable in fluidized bed configurations, working at very large manufacturing scales can still present a technical challenge and remains very limited (Ozturk and Hu, 2005).

The alternative to the fluidized bioreactor is a fixed or packed bed bioreactor. These reactors tend to be used in perfusion mode and have a packed bed of microcarriers for cell growth. The packed bed bioreactors allow freshly oxygenated medium to pass through the bed to return to a medium reservoir where antibody product can be harvested either continuously or in batches (Yang et al., 2004). This reservoir can be internal or external to the packed bed compartment of the bioreactor. With the internal compartment, oxygenation of the medium usually happens in the internal central core of the bioreactor, moving upwards prior to circulating in a radial fashion through the packed bed before becoming reoxygenated at the bottom of the reactor and returning to the central core. Another mode of operation arranges the medium

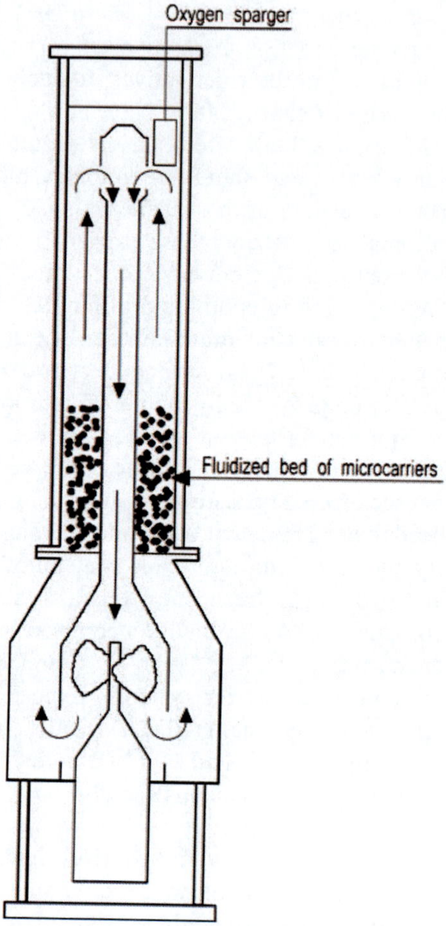

Fig. 2.1 Schematic of a microcarrier based fluidized bioreactor

flow to run parallel to its longitudinal axis within in internal packed bed system (Rodrigues et al., 2010). With an external reservoir, the oxygenation happens in the reservoir before moving onto the packed bed bioreactor.

Although packed bed bioreactors have improved over the years with the introduction of new carrier materials having higher internal porosities – such as non-woven polyester matrices (Gumusderelioglu et al., 2001), ceramic matrices (Grampp et al., 1996), glass fibers (Chiou et al., 1991) and non-woven fibers made from polydegradable polymers (Kim et al., 1998) – the maximum reported working volume achieved using packed bed bioreactors has only reached 30 L (Meuwly et al., 2007). Another system for the immobilisation of cells used to produce mAbs, the hollow fiber bioreactor, provides a high growth-surface-to-volume ratio and allows high cell densities. Cells are immobilized on the external surface of hollow fibres, while the oxygen and nutrients pass through the internal microcapillary fiber bundles. The fibers are usually composed of ultrafiltration or microfiltration membranes; this membrane,

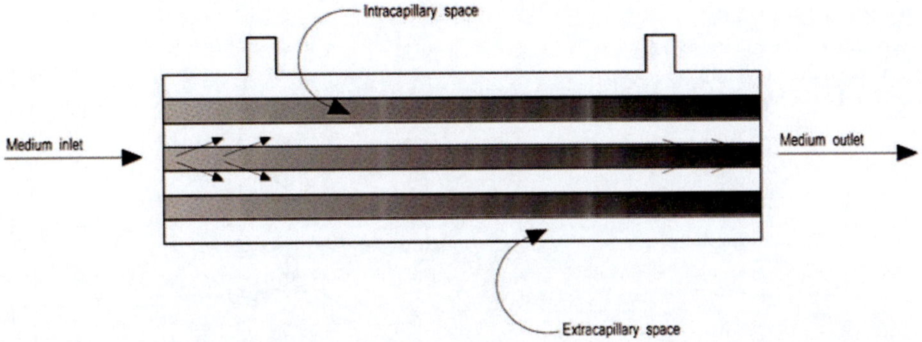

Fig. 2.2 Simplified representation of hollow fiber cartridge with only three fibers shown. This illustration shows the extracapillary space where the cells are seeded to receive nutrients that transverse the microcapillary fibers from the intracapillary space and release waste products that transverse the microcapillary fibers into the intracapillary space. Fresh media is continuously infused into the microcapillary fibers to introduce nutrients and carry waste products away from the cells

which separates the cells and main medium flow, provides a low shear environment (Gramer et al., 2003; Jackson et al., 1996). The medium is pumped from an oxygenated and pH-controlled reservoir to allow delivery to the hollow fiber unit, to flow through the microcapillary fiber bundles (Fig. 2.2). The pumped medium then exits the hollow fiber unit, removing waste products before returning to the reservoir for recirculation. The cell compartment of the hollow fiber unit retains the secreted mAbs; it can be removed together with dead cells and cell debris for purification. This system has the one major drawback that the units are susceptible to membrane fouling, making it difficult to operate in a consistent manner. Hence, cell viability and product quality can vary due to process instabilities (Dowd et al., 1999).

2.3 Suspension Systems

Suspension systems are very common in large-scale bioreactor cultures using suspension cells or microcarrier beads with anchorage-dependent cells. The stirred tank bioreactor (STR) is the most popular of the suspension systems; it is simpler to operate and easier to scale up than other systems (Warnock and Al-Rubeai, 2006). Suspension cell systems do not depend on surface area, making it possible to reach an increased cell density. Depending on the mode of operation and type of culture, STRs can culture to maximal cell densities of 2×10^6 to 15×10^7 cells mL^{-1} (Warnock and Al-Rubeai, 2006).

Stainless steel STRs, known as the main workhorses of the biopharmaceutical manufacturing industry, have been used at the largest scales. STRs provide efficient gas transfer through the use of typical marine or pitched blade impellers for axial flow, such as a tri-blade segment impeller for low shear and high aeration efficiency (Fig. 2.3), combined with the use of a gas sparger to release air, oxygen, and carbon

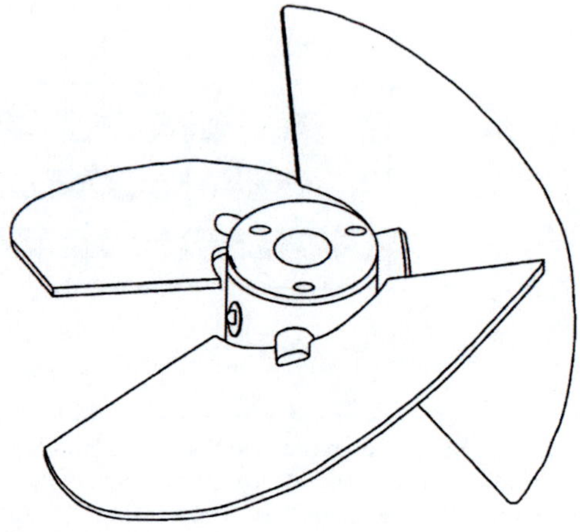

Fig. 2.3 Example of a segmented tri-blade impeller for mammalian cell culture in stirred tank reactors

dioxide into the medium. Compared to microbial cultures, mammalian cultures in bioreactors have much lower oxygen transfer capacities; mammalian cells demand 10–50 times less oxygen because they obtain lower cell mass (Zhou et al., 2009) than occur in microbial fermentation. Oxygen supply is critical. Although agitation can disperse the bubbles within the reactor, if conditions are not correct, gradients can form within the liquid bulk. Large-scale STRs usually can have multiple impeller configurations, such as upward-pumping axial-flow impellers or a combination of different impellers, to aid in oxygen mass transfer (Gogate et al., 2000; Nienow et al., 1994) since, in order to get good oxygen supplies, the mixing regimen must ensure that no dead zones exist and must prevent development of localised areas of high or low DO concentrations.

Apart from DO, pH, temperature, and nutrient gradients have to be avoided, which can cause non-ideal physiological conditions within a bioreactor (Fig. 2.4). One major concern, cell apoptosis, can be triggered by a lack of nutrients, non-ideal pH conditions, physical shear stress, and the absence or excess of DO (Al-Rubeai and Singh, 1998; Al-Rubeai, 1998; Arden and Betenbaugh, 2004; Ishaque and Al-Rubeai, 2002; 1998; Simpson et al., 1998; Singh, 1994; Singh et al., 1997). To reduce dead zones, agitation can be increased, but at a rate that does not generate eddies with a microscale of turbulence greater than the cell diameter (Nienow et al., 1994; Oh et al., 1992). Although sparging can provide for essential oxygenation of the media, direct sparging can damage the cells, due to the high energy that is liberated as gas bubbles burst when reaching the liquid surface. The gas entrainment and bursting, seen as the cause of most damage to mammalian cells from sparging (Al-Rubeai et al., 1993), can be partially alleviated by adding polymers, such as Pluronic F-68 and polyvinyl alchohol, to the medium, which reduces the attachment of cells to the bubble surface as they rise.

Three types of spargers are commonly used in laboratory-scale STR configurations: sintered stainless steel microspargers, open pipe spargers, and ring spargers with multiple holes. The microsparger has the advantage that it can provide more gas-medium interfaces and thus higher O_2 mass transfer coefficients (k_La's) due

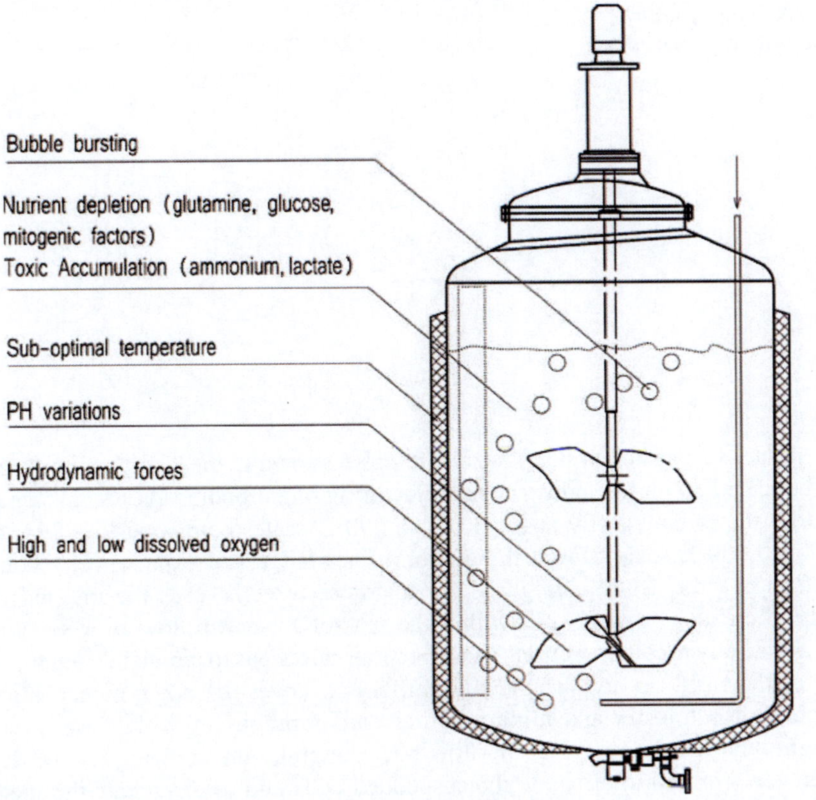

Fig. 2.4 Factors which influence cell survival in the bioreactor

to the 10–100 μM pore size, which produces swarms of small bubbles, allowing a larger interfacial area at a given flow rate. The advantage is that microsparging of pure oxygen need only occur at low flow rates, lower than 0.01 vvm (volume · volume^{-1} · min^{-1}), reducing foam formation and cell damage; it has the disadvantage that carbon dioxide can accumulate, decreasing the medium's pH levels, which can ultimately affect cell growth, productivity, and glycosylation antibodies. Using a microsparger with air can often cause difficulties with stable foam formation. When scaling up to larger STRs, open pipe spargers and ring spargers are preferred, since bubble coalescence takes place with microspargers, offsetting the advantage of smaller bubble generation (Zhou et al., 2009). Recent studies have shown that another factor which should be considered when implementing an STR is the gas velocity used, especially with the use of chemically defined or protein free media. Zhu et al. (2008) found that an exit velocity at the sparger site of greater than 30 ms^{-1} caused cell damage to NSO cells, reducing cell viability and hence antibody production. This becomes pertinent when scaling up and using high density cultures, where open pipe systems are commonly used and oxygen demand can be high.

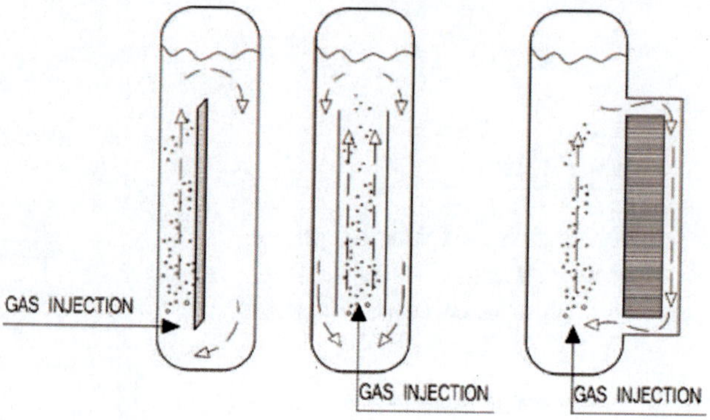

Fig. 2.5 Typical airlift reactors for mammalian cell cultures

Another suspension system, the airlift – also known as the airlift reactor (ALR) – can be scaled up for the industrial manufacturing of antibodies. ALRs have operated at large scales but mostly at the 1,000–5,000 L scale (compared to STRs, at the 1,000–25,000 L scale). From the authors' knowledge and experience, a common airlift bioreactor, when configured for mammalian cells, uses the internal single baffle to create a vertical split within the reactor for an up-flow riser section and down-flow down-comer section. These sections allow gas to circulate from the lower part of the riser, up and around the baffle, and down the down-comer, allowing coupled mass transfer and mixing. Other configurations of ALRs have a central draught tube which acts as the up-flow riser column, thus allowing for the density difference required to circulate the suspended cells and to oxygenate the medium. In these systems, the air flows through the riser, and the degassed liquid flows down the down-comer from the top of the tube. The other possibility, an external loop reactor, allows the riser and down-comer flow to act in two separate cylindrical sections of the reactor. Figure 2.5 illustrates some common ALR configurations. The typical aspect ratio (height-to-diameter ratio) is 5:1 or 6:1 for ALRs and 3:1 to 1:1 for STRs in large-scale mammalian cell culture systems (Varley and Birch, 1999; Kretzmer, 2002; Petrossian and Cortessis, 1990). Compared to STRs, ALRs have less CO_2 accumulation to toxic levels due to ballast gas. In addition, they have the advantage of needing no mechanical seals and moving parts, thus requiring less maintenance (Merchuk, 2003; 1990; Varley and Birch, 1999) and reducing the chances of mechanical failures.

2.4 Modes of Bioreactor Operation

A bioreactor can run in several modes of operation: batch, fed-batch, continuous, and perfusion (Fig. 2.6). The dominant mode of operation at industrial scale for STRs and ALRs is the fed-batch mode of operation. Industrial-scale STRs also run in perfusion mode for several processes, such as Remicade from J&J's Centocor and Advate from Baxter (Dowd et al., 2007) – prime examples of FDA-approved

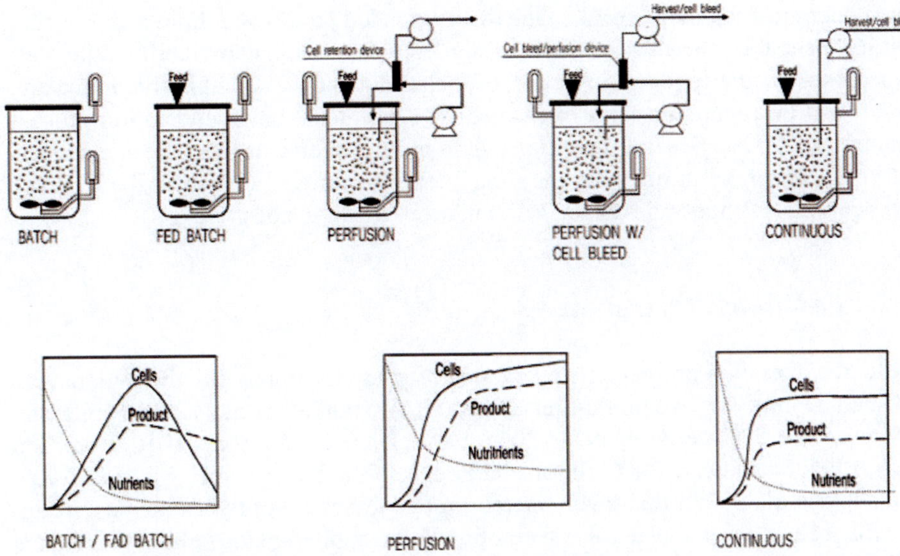

Fig. 2.6 Different modes of operation for bioreactors, (**a**) illustrating a typical STR in batch, fed-batch, perfusion (with and without cell bleed), and continuous operation. Corresponding growth curves (**b**) expected in each mode of operation

perfusion processes. The current trend for industrial-scale production runs in either fed-batch or perfusion mode (Chu and Robinson, 2001; Birch and Racher, 2006; Dowd et al., 2007).

2.4.1 Batch Operation

In batch operation of mammalian cell cultures, all nutrients are added at the start of the production cycle; at the end of the production cycle, all of the contents are harvested at one time. Here, the only continuous addition is the gas supply, which occurs by diffusion through the medium interface. The typical batch cultivation using standard medium can reach the rather low cell densities of approximately 1–4 $\times 10^6$ cells mL^{-1} before harvest. Either nutrient limitation or inhibition of growth by waste products limit maximum cell density in a given medium. As the simplest mode of operation, batch culture also has had the most studies done; investigations have considered how culture parameters affect culture kinetics, metabolism, media optimisation, and scale-up (Ozturk and Hu, 2005). Employing a feeding strategy to provide cells with nutrients while in culture can help increase cell density and product titres (Altamirano et al., 2004; Bibila and Robinson, 1995; Bibila et al., 1994; Robinson et al., 1994). A common feeding strategy, known as repeated batch operation, runs a culture in batch mode for an extended time, with intermittent harvests followed by fresh media feeds (Zhou et al., 2009; Jain and Kumar, 2008). Repeated batch operation can reduce costs. Due to an extended period of culture and intermittent harvests, the process needs to be designed to keep high cell viability at harvest in order to be able to leave a fraction of the media and cells within the bioreactor, which will be topped up with fresh media to repeat the batch cultivation process (Zhou et al., 2009). Several repetitions yield multiple harvests from the same bioreactor. It is essential to optimise the time needed to initiate a batch in the repeated batch process (Fenge and Lullau, 2005) in order to get the best yields.

2.4.2 Fed-Batch Operation

Fed-batch operation has become a key area of study for improving the mammalian cell culture process (Altamirano et al., 2004; Amanullah et al., 2010; Bibila and Robinson, 1995; Chee Furng et al., 2005; Farid, 2006; Huang et al., 2010; Robinson et al., 1994; Sauer et al., 2000; Sitton and Srienc, 2008; Spens and Haggstrom, 2007; Stansfield et al., 2007). Fed-batch operations can use either of two strategies: a continuous feed or a scheduled bolus feed of nutrients applied to the culture to enhance and prolong the exponential phase of cell growth and antibody production (Butler, 2005). Overall, a typical fed-batch mode of operation typically yields a higher antibody product concentration, due to the extended time during which the culture can sustain its exponential phase. The actual feeding of nutrients should be carefully controlled to avoid apoptosis (Al-Rubeai and Singh, 1998; Al-Rubeai, 1998; Cotter and Al-Rubeai, 1995; Singh and Al-Rubeai, 1998; Simpson et al., 1998) and to avoid inhibiting growth by reducing the formation of such waste metabolites (Yang and Butler, 2000; Chen et al., 2001; Tsao et al., 2005) as lactate and ammonia, while making sure not to deprive the culture of any essential nutrients (Altamirano et al., 2000; Jan et al., 1997; Martinelle et al., 1996). Some common nutrient feeds can contain glucose, glutamine, hydrolysate, and amino acids. In addition, concentrated basal media may also be part of a feed regimen. Fed-batch optimisation can depend on feed rate, feeding duration, the feed constituents, and culture conditions. Feed determination is best done by first analysing the spent medium for amino acids, lactate production, glucose consumption, integrated viable cells, and viability (Xie and Wang, 1997, 1996, 1994), while further parameters can be investigated as we learn more about culture requirements and kinetics. Studies have shown that vast improvements in antibody productivity can occur by switching to fed-batch mode (Bibila and Robinson, 1995). The major advantage of fed-batch operation is that it can vastly improve outcomes over batch operation, while being simpler to set up than perfusion or continuous type operations.

2.4.3 Perfusion Operation

When running in perfusion mode, a bioreactor is generally configured to withdraw spent medium and to introduce fresh medium at the same rate, while using a retention system to retain cells. Retaining the cells allows the concentration of the cell population; hence, the nutrient renewal and waste metabolite removal yield large cell populations. This mode of operation has been extensively studied in several system designs, including using suspension cultures with spin filters (Yabannavar et al., 1994; Emery et al., 1995), ultrasonic resonators (Trampler et al., 1994; Bierau et al., 1998), gravity cell settlers (Batt et al., 1990; 1990; Wen et al., 2000) centrifuge-based perfusion (Van Wie et al., 1991), and membrane- (Marx, 1998; Falkenberg, 1998) or hollow-fiber-based designs (Valdes et al., 2001). These systems can be classified as either filtration-based perfusion systems or sedimentation-based systems. Table 2.1 summarises the various systems commonly used, along with representative references.

Perfusion processes can viably achieve high cell densities of between 10^7 and 10^8 cells mL^{-1}. Depending on the type of configuration, perfusion can reduce cell viability at high densities due to oxygen mass transfer deficiencies that can occur and, if the perfusion system is susceptible to clogging from dead cells, a rapid buildup of toxins may occur (Al-Rubeai et al., 1992; Emery et al., 1995; Johnson et al., 1996; Mercille et al., 2000; 1994; Trampler et al., 1994). A successful perfusion culture can reduce the volumetric operating capacity by approximately 96% compared to batch operations with cell cultivation processes having similar product titres and quality requirements (Greenfield et al., 1991) using current manufacturing practices and under good conditions. The continuous cell lines used in perfusion mode will also influence how the perfusion operation can be maintained. Thus, some perfusion cultures have used a cell bleed to reduce the dead cell accumulation and increase cell viability at high cell densities, thereby extending the length of the bioreactor operation (Hiller et al., 1993). The cell bleed can occur as a continuous stream of the cell-containing medium, usually activated a few days into the culture period

Table 2.1 Common cell retention devices for perfusion bioreactors

Device	Highest approximate potential perfusion Rule	References
Filtration		
Spin filter	1,000 L day^{-1}	Emery et al., (1995), Yabannavar et al., (1994), Iding et al., (2000)
Cross-flow filter	50 L day^{-1}	Hiller et al., (1993), Mercille et al., (1994), Zhange et al., (1993)
Alternating tangential flow (ATF) filter	1,200 L day^{-1}	Yuk et al., (2004), Zouwenga et al., (2010)
Yortax flow fitration	250 L day^{-1}	Mercille et al., (1994), Voisard et al., (2002)
Sedimentation		
Gravity cell settler	1,000 L day^{-1}	Batt et al., (1990), Wen et al., (2000)
Ultrasonic/acoustic filter	1,000 L day^{-1}	Bierau et al., (1998), Trampler et al., (1994), Gorenflo et al., (2003)
Centrifuge	3,700 L day^{-1}	Van Wle et al., (1991), Johnson et al., (1996), Cruz et al., (2002)

to maintain cell viability (Dalm et al., 2004; Banik and Heath, 1996; Castilho and Medronho, 2002; Castilho et al., 2002). The use of perfusion has its advantages and disadvantages. Perfusion has the advantage of similar or higher productivity rates than fed-batch operation; also, the perfusion antibody product can be continuously harvested from the spent medium – useful for products sensitive to degradation from staying within the bioreactor for long time periods. Using long-term perfusion processes to cultivate mAbs with continuous mammalian cells has two disadvantages: over this time period, instabilities can maintain a steady state and processes can vary.

The most common filtration-based perfusion processes available for large-scale production of mAbs and other therapeutic proteins use either spin filters or alternating tangential filtration (ATF) (Voisard et al., 2003). Of the sedimentation perfusion systems, the best realisation of large-scale potential comes from either an ultrasonic/acoustic resonator (although sometimes referred to as ultrasonic/acoustic filters, they are based on the principals of sedimentation velocities) or centrifugal separators (Voisard et al., 2003; Su and Flickinger, 2009). Of these two types of perfusion systems, the filtration-based methods currently get used much more frequently at the larger scales of between 500 and 1,000 L.

Spin filter perfusion can rely on cell retention taking place either internally or externally of the bioreactor. The design comprises a rotating cylindrical filter screen with a porous screen usually made of stainless steel, although other materials have been used, with 10–120 μm porosity (Deo et al., 1996; Avgerinos et al., 1990; Iding et al., 2000). The cylindrical screen can be rotated either by the bioreactor agitator shaft (for internal spin filters) or by a separate drive mechanism (for both internal and external spin filters), which increases the complexity of the design and construction. The presence of a spin filter reduces the hydrodynamic damage that occurs in a STR bioreactor, consequently increasing cell viability and process productivity over processes lacking a spin filter (Jan et al., 1993). Using a spin filter to retain the cells in the bioreactor yet allowing spent medium to travel through the screen requires the action of several forces on both the cell particles and liquid medium: the gravitational force, the centrifugal force from the rotating filter, the axial force created by the impeller rotation, the radial force (drag) generated by the perfusion flux, and the hydrodynamic lift force created by the rotation to counteract the radial force (Castilho and Medronho, 2002; Su and Flickinger, 2009). Fouling or clogging of the spin filter can take place due to cell attachment, which depends on the cell type, screen material, hydrodynamic conditions, and cell density (Castilho and Medronho, 2002). The spin filter should be designed and operated to minimise the fouling that can occur over an extended period of operation. Another factor to consider when operating spin filters is the addition of a draft tube surrounding the internal spin filter making its function similar to an external spin filter since the forces exerted on the filter are similar while in operation (Su and Flickinger, 2009). The addition of a draft tube or the use of an external spin filter produces axial and rotational shear and Taylor vortices in the gap between the spin filter and external wall or draft tube, which in turn helps to reduce fouling. Taylor vortices occur when the angular velocity of the spin filter (the inner rotating cylinder) is increased above a certain threshold, this forces the regular couette flow to become unstable and a secondary steady state characterized by axisymmetric toroidal vortices (Taylor vortices) to materialize (Su and Flickinger, 2009). Taylor vortex flow also gets utilised in an external vortex flow filtration setup where a cylinder of filter membrane or a porous screen surrounds an impermeable rotating cylinder to allow the filtrate to pass through the membrane/screen and to let the retentate concentrated cell suspension return to the bioreactor (Mercille et al., 1994; Su and Flickinger, 2009; Castilho and Medronho, 2002).

The spin filter has been shown to operate comparably to an ultrasonic filter. Both perfusion systems can reach high cell densities; a Bcl-2 hybridoma cell line reached maximum cell densities of 1.58×10^7 cells mL^{-1} using the spin filter, compared to 1.21×10^7 cells mL^{-1} using the ultrasonic filter (Bierau et al., 1998). The largest spin filters for "off the-shelf" use are suitable for a 50 L bioreactor (Warnock and Al-Rubeai, 2006), but spin filters have also been scaled up to 500 L STRs successfully (Deo et al., 1996). Spin filters at larger scales have mainly used stainless steel screen meshes; they possess greater durability and reusability than other materials which can reduce fouling, such as hydrophobic polymers, which don't have the high surface charge that metals can have.

The other popular method of filtration-based perfusion uses crossflow filters (Zhang et al., 1993; Mercille et al., 1994) which can be based on either membranes or hollow fibers. A common type in operation today for mAb production is the alternating tangential filtration (ATF) system based on the crossflow filter design with some important modifications (Fig. 2.7). This system relies on an external membrane or hollow fiber; the system directs product flow tangentially along the surface of a membrane/fiber to allow the recovery of the spent medium and product and the return of a concentrated cell retentate to the bioreactor. Refine Technology (Edison, NJ, USA) developed this system to reduce the occurrence of fouling on tangential flow filtration modules. Using a diaphragm pump, which provides a back flush during each pump cycle under low shear conditions, reduces fouling. This reversal of culture flow across the filter module in turn reverses the pressure gradient across the filter wall to reduce the fouling and clogging problems associated with filtration-based perfusion systems. Yuk et al. (2004) used the ATF with volumetric perfusion rates of 3.5 culture volumes per day, a filter area of 0.46 m^2, 1 mm lumen, and a 0.2 μm pore rating to give a shear rate estimated at about 750 s^{-1} – a value low enough not to create shear damage to mammalian cells. The ATF perfusion system can also be used for microcarrier culture (Ozturk and Hu, 2005) by replacing the hollow fiber with a screen module.

Recently, a form of perfusion carried out using the ATF, referred to as concentrated fed-batch (CFB), has gained some attention with high cell densities of 1.75×10^8 cells mL^{-1} reached for a PER.C6 clone (high cell densities for CHO and NS0 cell lines were also claimed) and mAb product titres of 11.3 gL^{-1} on day 12 of the culture. This constituted a several-fold improvement over the fed-batch process; the two demonstrated processes used the same time scale, clone, STR, and basal medium (Zouwenga et al., 2010). The principle of this method is to run the ATF system with a MWCO membrane, which won't allow the mAbs to perfuse out of the system by exiting the bioreactor with the perfusate stream. This ensures that

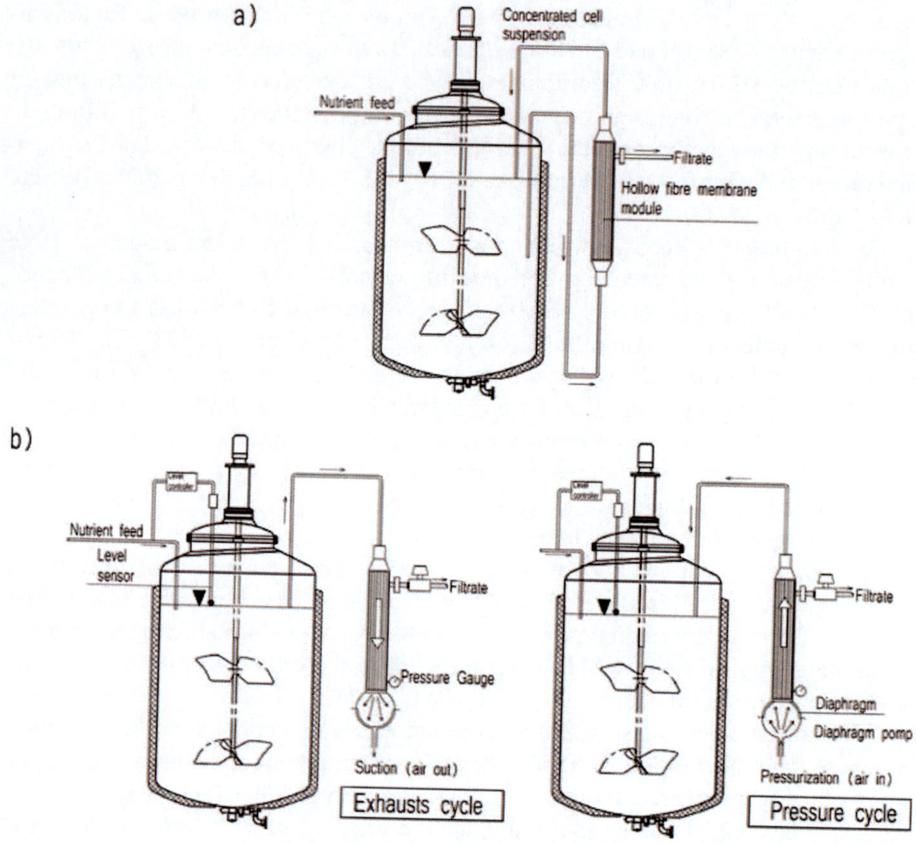

Fig. 2.7 A cross-flow filter configuration (**a**) where the flow of cells and medium undergo a unidirectional recirculation through an external cross-flow filter and (**b**) an alternating tangential filtration (ATF) system in comparison where a diaphragm pump is used to help reduce fouling and provide bidirectional flow

the mAb and cells stay in the bioreactor until harvest, while the media components, which are considered waste, get discarded via the filtrate output and the nutrient components get replenished.

An ultrasonic/acoustic filter is a cell retention device that uses the principle of sedimentation via ultrasonic field forces that promote cell aggregation. Cells suspended in medium that are exposed to an acoustic standing wave experience a time-averaged force that drives them to either the pressure nodes or anti-nodes of the wave. The node is a place where the fluid does not move and cells can be trapped causing cell aggregation. Thus due to the formation of resonant standing waves, cells are trapped in the pressure plane nodes of the wave while spent medium keeps flowing allowing filtration to take place. The standing waves form when an ultrasonic wave is reflected in the direction opposite to its propagation (Trampler et al., 1994; Su and Flickinger, 2009). This system has the advantages of using no mechanical moving parts and being less susceptible to fouling (Dalm et al., 2005). The system may require back flushing of cells in the acoustic chamber of the ultrasonic filter at high cell densities, when a tendency to get entrained cells exists, which may reduce the separation efficiency over time (Gorenflo et al., 2003). The pressure difference associated with the standing waves and pumping action may also increase shear stress to the cells (Gorenflo et al., 2003; 2002; Dalm et al., 2005) which may lead to decreased cell viability, decreased cell-specific productivity, and/or lower product quality (Merten, 2000; Warnock and Al-Rubeai, 2006; Senger and Karim, 2003; Shi et al., 1992; Petersen et al., 1988). Scaling this system up has been limited to 1,000 L per day of perfusion; ultrasonic/acoustic energy has the drawback that more is needed as the scale increases because it takes more power to sustain perfusion efficiency. The temperature increase within the medium, in turn, causes liquid convection, reducing the separation efficiency. Overcoming this requires a cooling system; thus, for example, the BioSEP 200 model from AppliSens has one.

Operating the ultrasonic/acoustic filter involves pumping the cell suspension from the bioreactor to the ultrasonic module below the separation chamber at a circulation rate twice that of the perfusion rate, as suggested by Gorenflo et al. (2002). Due to the possibility of having shear-sensitive cell lines in the system, the recirculation stream may be detrimental; thus, a back-flush of the settler using the perfusate can be initiated (Merten, 2000). This back-flush has the disadvantage that the cells may wash out at high perfusion rates, since the recirculation rate and perfusion rate are linked as a single parameter in this case. Another option for a back-flush involves using air, which has shown increased efficiency of up to 91% with 12–20 air back-flushes per hour (Gorenflo et al., 2003). This shows promise as a perfusion technology at the 1,000 L scale but still needs to be implemented for a sufficient time at a commercial scale to prove itself.

Of the scalable sedimentation systems, another one which shows promise is the centrifugal force perfusion bioreactor. Johnson et al. (1996) used such a system at the small scale to compare its rate of mAb production and cell growth to spinner flasks and continuous centrifugation coupled to a bioreactor. This system they obtained, called the Centritech centrifuge (Pneumatic Scale, Akron, OH, USA), had 98% viable cell retention at an average perfusion rate of 0.5 culture volumes per day. This perfusion rate was further increased to 2 volumes per day by Cruz et al. (2002). The results indicate that using a centrifuge to operate in perfusion mode could work, but scale-up studies will be needed to realise the true potential. Currently, Pneumatic Scale is selling the Centritech disposable centrifugal system on the commercial market as being capable of working in conjunction with bioreactors to a scale of 2,000 L.

2.4.4 Continuous Operation

Operating a bioreactor in continuous mode allows the harvest of mAbs as the culture progresses – like perfusion culture. Continuous culture differs because it can result in a chemostat culture where the cell growth is in a steady state (unlike perfusion) and can be limited by a single nutrient (glucose, oxygen, or glutamine, for example). This is accomplished by continuously adding fresh media to the bioreactor while removing cell-containing media at the same rate to keep the volume in the bioreactor constant. Carrying out continuous culture requires taking into account the dilution rate (feed rate to the vessel divided by the volume of vessel); the dilution rate should be close to the growth rate to avoid wash out. Many studies have used continuous cells, such as CHO, NSO, and hybridoma cells, in chemostat cultures (Miller et al., 1988; Barnes et al., 2001; Europa et al., 2000; Simpson et al., 1999; Tey and Al-Rubeai, 2005). However, due to the risk of unstable expression of mAbs after long periods of culture (Heath et al., 1990; Ozturk and Palsson, 1990; Barnes et al., 2001) continuous operation has not been very suitable for commercial production.

2.5 Disposable Bioreactors

Disposable technologies can provide many benefits, from the initial investment to the equipment required and the necessary facility design to use it – not to mention the reduced need to validate the process (Pierce and Shabram, 2004). The last few years have also seen an increase in the scale at which these disposal technologies can work, with systems capable of producing up to 2,000 L currently available and the possibility of 3,000 L systems on the horizon. The availability of disposable STRs has made it easier to transfer processes from stainless steel STRs to disposable STR bag systems.

The earliest disposable systems for mAb production, hollow fiber bioreactors, can operate at high cell densities reaching between 10^7 and 10^9 cells mL^{-1} but have a limit in terms of scale-up due to problems of keeping heterogeneity within the bioreactor (Jain and Kumar, 2008). The use of the CellCube for anchorage-dependent cells provides another example of a disposable system used for protein production. Again, it has the disadvantage that the system's can be scaled up to a limited degree (with the largest reported available surface area being 340,000 cm^2) and that parts of the system, such as the oxygenator and probes, require resterilisation between runs, making it difficult to validate (Warnock and Al-Rubeai, 2006; Merten, 2000; Kotani et al., 1994; Wikstrom et al., 2004; Jain and Kumar, 2008).

The fully disposable system, the Cell Factory, comprises a number of stacked chambers connected by a common vent and ports; it can be used for mAb production and provides a large flat growth surface area of up to 25,280 cm^2 for a 40-chamber unit. Each chamber represents 632 cm^2 at a working volume of 200 mL. The filling and emptying of medium or cell suspension can be done by an automatic cell factory manipulator (ACFM), an electronically and pneumatically controlled unit; the ability of each ACFM to handle up to 101,120 cm^2 reduces labour costs (Thermo Fisher Scientific, Waltham MA, USA). The Cell Factory has another advantage; it requires only one filling and emptying operation. Still, microcarrier-based systems offer far greater surface areas than systems with similar footprint sizes; thus, choosing the most suitable type of system really depends on the adherent cell line, optimum growth conditions, and economics.

Disposable technology for the production of mAbs also exits for static cultures in the form of a membrane bioreactor known as the CELLine flask (Trebak et al., 1999). These systems use a microporous membrane to create two compartments in a culture flask. A 10,000 MWCO dialysis membrane separates the top compartment (for media) from the bottom compartment (for cells), the bottom surface of which has a gas permeable silicon membrane for gas exchange. This design allows the top media compartment to provide fresh nutrients for the cells, while the waste products can traverse the dialysis membrane into the media compartment. Meanwhile, the cells get fresh oxygen and release carbon dioxide into the incubator through the silicon membrane. Studies have shown that regular cell densities of 10^7–10^8 cells mL^{-1} are reached with mAb titres of 0.7–2.5 mg mL^{-1} within a period of 2 months with 1–2 L of media (Trebak et al., 1999). A disadvantage of the system is that it has a maximum of 15 mL volume cell cultivation chamber, limiting the surface area of the chamber of the biggest flask currently on the market; thus, scale-up will need to be achieved in other technologies.

The disposable technologies available for suspension cells achieved a real breakthrough in development when bag systems started to be used. The WAVE bioreactor (GE Healthcare), developed in 1996 and available since 1998 (Brecht, 2010), provides an example of such a system; this system has the largest bag (CellBag) size, at 1,000 L, for a working volume of 500 L of mammalian cell culture. The system comprises a presterilised bag and a rocking platform with integrated heating for the bag to rest on when filled with medium and cells. This bioreactor uses a single fixed axle to create a back and forth rocking motion, generating waves at the liquid-air interface (see Fig. 2.8). The rocking motion of the CellBag also ensures good mixing without shear damage (Singh, 1999; Weber et al., 2002). The manufacturer has integrated pH and oxygen sensors into the bag design. Once filled to a maximum of 50% with culture medium, the rest of the CellBag is inflated with the process gas mixture (Genzel et al., 2006; Weber et al., 2002), controlled by a gas mixing system (Brecht, 2010). The system can also operate in a CO_2 incubator if needed (Singh, 1999).

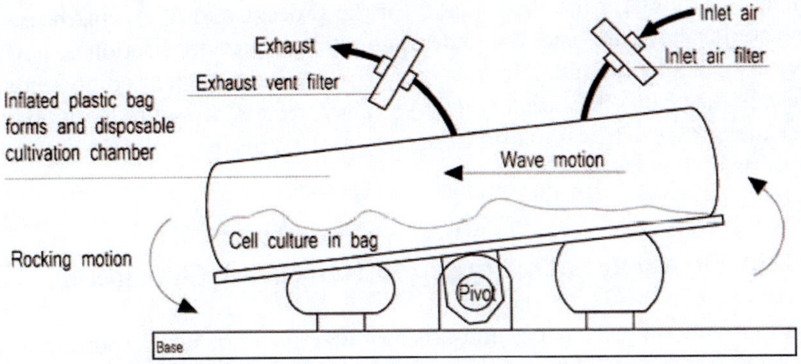

Fig. 2.8 A WAVE® bioreactor

The WAVE bioreactor can also be run in perfusion mode, using a load cell and a specially developed filtration system within the CellBag, where the wave motion helps prevent filter clogging. From the results of perfusion hybridoma cultures in a 1 L CellBag (Tang et al., 2007) the mAb volumetric productivity was 33.1 mg L^{-1} per day compared to batch culture of 20.3 mg L^{-1} per day. This perfusion mode of operation has been scaled up to 500 L for the WAVE system (Pierce and Shabram, 2004).

The largest-scale disposable systems have been in the form of STRs with culture volumes in the range of 1,000 L (Hyclone, Thermo Fisher, UT, USA) to 2,000 L (Xcellerex, Marlborough, MA, USA). These two similar systems use a disposable bag for the culture. Both house the bag in a stainless steel support vessel; both have bags that integrate all gas and liquid transfer tubes, gas filters, sparger, and ports for probes and sampling, and both include an agitation and temperature control system. The technologies differ in the agitation and temperature control technology utilised. The Hyclone system, marketed as the SUB (Single-Use Bioreactor), has an electrical heating jacket and an integrated pitch blade impeller at an angle, which is stabilised using a metal rod; in this system, the disposable bag acts as a barrier between the rod and cell culture space, which prevents direct contact. The XDR disposable bioreactor from Xcellerex, in comparison, uses a magnetically coupled agitator; multiple zones exist for water-jacketed heating, depending on the working volume in the disposable XDR component. An advantage of the Xcellerex system is its 5:1 turndown ratio (operational from 400 to 2,000 L), allowing the incorporation of scale-up without using a seed reactor at the 400 L scale for a 2,000 L system. At the moment, biopharmaceutical companies are adapting these systems, due to their similarities to STRs, which makes it simpler to transfer to larger (20,000 L) stainless steel STRs. In fact, Centocor is known to have included the SUB in a perfusion process design (Brecht, 2010) Other similar systems have been entering the market from stainless steel bioreactor manufacturers – such as Sartorius-Stedim, with the BIOSTAT CultiBag STR Plus line – as demand increases for disposables in biomanufacturing.

Although disposables are popular, concern must be raised about the possibilities that extractables and leachables from the materials used to manufacture disposable bag systems may affect the final quality of the process and of the mAb products. These materials have the potential to degrade after gamma sterilisation or long-term storage; also, at specific temperatures, they can release chemical compounds from the plastics that can affect quality (Jenke, 2007). Thus, it becomes important to conduct the appropriate validation to determine whether this will pose a problem before using at the GMP scale.

2.6 Main Parameters Optimised in Bioreactor Operations

For mAb production bioreactors and other similar systems, it is important to measure several parameters that can impact the quality and quantity of the mAbs. The primary use of a bioreactor is to allow the control of growth and production conditions using temperature, DO, and pH probes, while other parameters that control the homogeneity of the environment, such as the agitation and sparge rate in an STR, may be linked to these three probe readings. We can manipulate productivity and cell growth using pH, temperature and DO (Reuveny and Lazar, 1989; Varley and Birch, 1999). Understanding the impact of pH, temperature, and DO on the bioreactor culture can greatly enhance the process of optimising the bioreactor culture.

The introduction of probes and analytical systems with new measurement capabilities has opened the possibility of additional monitoring and control options for a bioreactor culture, beyond the standard oxygen, pH, and temperature controls. For example, Carvell and Dowd (2006) demonstrated the use of an RF impedance probe to determine viable biomass and to control the medium feed rate in a perfusion bioreactor (Carvell and Dowd, 2006). The use of an integrated analyser, such as the BioProfile FLEX (Nova Biomedical, Waltham, MA, USA), also provides information on the gas/electrolyte content, nutrient/metabolite concentration, IgG concentration, cell size, and density/viability, which may be used in feedback loops to control culture feeding and other growth conditions (Derfus et al., 2010). Although these parameters can facilitate the control of the cell culture process, optimising them in small-scale studies before scaling up makes the process more economical.

The typical mAb production process using mammalian cells will set the temperature to 36.5–37°C, unless the temperature is used to induce cell cycle arrest in order to extend production time and/or increase cell productivity, depending on the cell line used and on whether production is growth independent (Al-Fageeh et al., 2006). We need to determine how any shift in temperature affects the quality of the mAbs, as any change in culture conditions can impact the growth, productivity, and possibly quality of the mAb. On the basis of the Arrhenius equation, changes in temperature affect enzyme activity within the cell; the work of Satoshi Oguchi et al. (2006) shows that decreasing the temperature to 31°C can increase mRNA stability.

Introducing oxygen into the medium is a key requirement to enable mammalian cells to produce energy efficiently from the organic carbon sources available within the medium (Hanson et al., 2007). This demands the optimisation of DO content for efficient energy metabolism of the mammalian cells. DO can negatively affect cell metabolism, especially glycosylation of mAbs, if not in an optimum range (Kunkel et al., 1998). A typical range used for bioreactor culture is 40–60% DO, but this depends on the cell line and product quality required under the given conditions. A study has shown that continuous mammalian cell lines can grow with double the oxygen levels of air saturation (Oller et al., 1989).

The culture's pH can significantly affect cell growth and viable productivity. Intracellular pH tends to be sensitive to pH changes when the extracellular pH is below 6.8 or above 7.8 (Fellenz and Gerweck, 1988; Oguchi et al., 2006). A change in intracellular pH can affect enzymatic activity and protein productivity, especially when coupled with temperature shifts in the culture (Fellenz and Gerweck, 1988). Lowering the pH is known to decrease specific glucose consumption while reducing the lactate production rate (Sauer et al., 2000). While a reduction in glucose consumption can impact cell growth, it also can reduce the amount of glucose needed by the culture. Sauer et al. found that, for certain CHO cell lines, lowering the pH set-point in the bioreactor from 7.2 to 7.0 increased mAb productivity and integral viable cell density, resulting in a 2.4-fold average increase in final antibody concentration. It also decreased specific glucose consumption and reduced the ratio of lactate produced to glucose consumed for all cell lines. Many studies have observed similar relationships using other types of mammalian cells than CHO cells, such as murine hybridoma cell lines (Ozturk and Palsson, 1991a), Sp2/0-derived mouse hybridomas (Miller et al., 1988) and HL60 cultures (McDowell and Papoutsakis, 1998).

Most cell culture media have an osmolality of 260–320 mOsm kg^{-1} (Freshney, 2005). Research has shown that changes in osmolality affect the cell volume and also significantly affect cell productivity (Ozturk and Palsson, 1991b). Osmolality changes may cause some nutrients in the medium formulation to become overabundant or limiting since it affects osmotic pressure which helps regulate the flow of metabolites (in the media) in and out of the cell. Thus a change in osmolality may result in a metabolic shift. This makes it important to understand how osmolality affects particular cell lines in the bioreactor, especially with fed-batch cultures, where concentrated feeds and by-product accumulations may increase the osmolality (Bibila and Robinson, 1995; Robinson et al., 1994). Also, the addition of base to the culture may increase osmolality. High osmolality generally decreases the growth rate, but it can have the positive effect of increasing specific productivity for the required protein (Min Lee et al., 2009; Oh et al., 1996). Over the years, those seeking to optimise bioreactor cultures have looked at using osmoprotective agents, such as glycine betaine and L-proline, to protect against the decreased growth rate caused by high osmolality (Oyaas et al., 1994a, b; Oyaas et al., 1995; Ryu et al., 2000; Kim et al., 2000; Schmelzer and Miller, 2002). The use of L-proline is also a regular constituent of media, which thus should pose no regulatory problems. We can also adapt cell lines to a higher osmolality by adding potassium or sodium chloride to the medium to allow fed-batch cultivations where osmolality can reach over 500 mOsm kg^{-1} and to increase specific productivity without impacting growth (Min Lee et al., 2009). By over-expressing bcl-2 it has been shown that hybridoma cells can be adapted to 400 mOsm medium from 300 mOsm medium while increasing antibody productivity by 100% compared to the control which could not be adapted (Perani et al., 1998).

Optimisation of bioreactor conditions can be done on the micro-bioreactor scale with the ability to control DO, pH, and temperature. For instance, the ambr™ system (The Automation Partnership, UK) comprises 48 micro-bioreactors (arranged in four culture stations containing sets of 12). Each station has a miniature internal impeller. Also, each allows individual closed-loop control of DO and pH, along with the ability to independently control each culture station for temperature and agitation, while the addition of feed and base can be automated. Another example system, the SimCell Micro Bioreactor system (BioProcessors Corp., Woburn, MA, USA) which uses gas-permeable chambers based on a microfluidic design to contain six sub-millilitre bioreactors in one array, with each bioreactor housing non-invasive optical sensors for pH and DO. A high-capacity robotic system manipulates the arrays to automate inoculation, sampling, incubation, feeding, and culture process monitoring and control. Amanullah et al. (2010), compared the SimCell system with shake flasks, as well as bench-top and 100 L pilot-scale bioreactors, for fed-batch CHO cultures; the series of 19 SimCell arrays exhibited good reproducibility with under 10% CV, a similar variability to well-controlled bioreactor runs. Micro-bioreactor systems allow the optimisation of bioreactor cultures, making possible the use of such statistical tools as Design of Experiments to optimise bioreactor conditions and to study large-scale changes by means of scaled-down models.

New probes and automated analytical systems combined with micro-bioreactor process information have brought with them the possibility of further automating process control in order to increase mAb quality. This avenue of technology, labelled process analytical technology (PAT), has only reached the early stages of application to mAb production with bioprocess bioreactor runs at a manufacturing scale (Rathore and Winkle, 2009; Read et al., 2010a, b). Applying these new technologies to current bioreactor operations may facilitate further optimisations in the manufacturing of high quality mAbs.

2.7 Conclusion

The production of mAbs from continuous cell lines can be accomplished using several bioreactor configurations, each of which was designed to satisfy differing operational, product quality, economic, and production scale requirements. The demand for improved bioreactor designs over the years has also fuelled these design variations. Typical bioreactor types include hollow fiber reactors, stirred tank reactors, airlift reactors, membrane reactors, fluidized bed and fixed bed reactors, and wave reactors, not to mention the range of small-scale systems available. These systems allow further operational possibilities, depending on both the mode of operation employed – such as batch, fed-batch, perfusion, and continuous culture – and on culture suitability. No one bioreactor configuration and operation mode can fit all cell lines, due to product stability and cell line diversity, although the STR could be considered the most common type of cell culture bioreactor, with its ability to apply many different operational modes to either suspension or adherent cell lines.

References

Al-Fageeh M. B., Marchant R. J., Carden M. J., Smales C. M. 2006. The cold-shock response in cultured mammalian cells: harnessing the response for the improvement of recombinant protein production. Biotechnol Bioeng 93:829–835.

Al-Rubeai M. 1998. Apoptosis and cell culture technology. Adv Biochem Eng Biotechnol 59: 225–249.

Al-Rubeai M., Emery A. N., Chalder S., Goldman M. H. 1993. A flow cytometric study of hydrodynamic damage to mammalian cells. J Biotechnol 31:161–177.

Al-Rubeai M., Emery A. N., Chalder S., Jan D. C. 1992. Specific monoclonal antibody productivity and the cell cycle-comparisons of batch, continuous and perfusion cultures. Cytotechnology 9:85–97.

Al-Rubeai M., Singh R. P. 1998. Apoptosis in cell culture. Curr Opin Biotechnol 9:152–156.

Altamirano C., Paredes C., Cairo J. J., Godia F. 2000. Improvement of CHO cell culture medium formulation: simultaneous substitution of glucose and glutamine. Biotechnol Prog 16:69–75.

Altamirano C., Paredes C., Illanes A., Cairo J. J., Godia F. 2004. Strategies for fed-batch cultivation of t-PA producing CHO cells: substitution of glucose and glutamine and rational design of culture medium. J Biotechnol 110:171–179.

Amanullah A., Otero J. M., Mikola M., Hsu A., Zhang J., Aunins J., Schreyer H. B., Hope J. A., Russo A. P. 2010. Novel micro-bioreactor high throughput technology for cell culture process development: reproducibility and scalability assessment of fed-batch CHO cultures. Biotechnol Bioeng 106:57–67.

Arden N., Betenbaugh M. J. 2004. Life and death in mammalian cell culture: strategies for apoptosis inhibition. Trends Biotechnol 22:174–180.

Aunins J. G., Bader B., Caola A., Griffiths J., Katz M., Licari P., Ram K., Ranucci C. S., Zhou W. 2003. Fluid mechanics, cell distribution, and environment in CellCube bioreactors. Biotechnol Prog 19:2–8.

Avgerinos G. C., Drapeau D., Socolow J. S., Mao J. I., Hsiao K., Broeze R. J. 1990. Spin filter perfusion system for high density cell culture: production of recombinant urinary type plasminogen activator in CHO cells. Biotechnology (NY) 8:54–58.

Banik G. G., Heath C. A. 1996. High-density hybridoma perfusion culture. Limitation vs inhibition. Appl Biochem Biotechnol 61:211–229.

Barnes L. M., Bentley C. M., Dickson A. J. 2001. Characterization of the stability of recombinant protein production in the GS-NS0 expression system. Biotechnol Bioeng 73:261–270.

Batt B. C., Davis R. H., Kompala D. S. 1990. Inclined sedimentation for selective retention of viable hybridomas in a continuous suspension bioreactor. Biotechnol Prog 6:458–464.

Bibila T. A., Ranucci C. S., Glazomitsky K., Buckland B. C., Aunins J. G. 1994. Monoclonal antibody process development using medium concentrates. Biotechnol Prog 10:87–96.

Bibila T. A., Robinson D. K. 1995. In pursuit of the optimal fed-batch process for monoclonal antibody production. Biotechnol Prog 11:1–13.

Bierau H., Perani A., Al-Rubeai M., Emery A. N. 1998. A comparison of intensive cell culture bioreactors operating with hybridomas modified for inhibited apoptotic response. J Biotechnol 62:195–207.

Birch J. R., Racher A. J. 2006. Antibody production. Adv Drug Deliv Rev 58:671–685.

Biselli M., Born C., Scr Oder B., Wandrey C. 1995. Production of monoclonal antibodies in a pilot scale fluidized bed bioreactor. Beuvery E, Griffiths J, Zeijlemaker W, editors. Animal Cell Technology: developments towards the 21st century. Meeting of the European Society for Animal Cell Technology and the Japanese Association for Animal Cell Technology, pp 683–686. Dordrecht: Kluwer Academic Publishers.

Bluml G. 2007. Microcarrier Cell Culture Technology, In: Portner R., editor. *Animal Cell Biotechnology*. Humana Press, Totowa, NJ. pp 149–178.

Brecht R. 2010. Disposable Bioreactors: Maturation into Pharmaceutical Glycoprotein Manufacturing, In: Eibl R., Eibl D., editors. *Disposable Bioreactors*. Springer, Berlin/ Heidelberg. pp 1–31.

Butler M. 2003. Modes of Culture for High Cell Densities, In: Butler M., editor. *Animal Cell Culture and Technology*, 2nd Edition. Bios Scientific, Oxford, Routledge. pp 175–194.

Butler M. 2005. Animal cell cultures: recent achievements and perspectives in the production of biopharmaceuticals. Appl Microbiol Biotechnol 68:283–291.

Carvell J. P., Dowd J. E. 2006. On-line measurements and control of viable cell density in cell culture manufacturing processes using radio-frequency impedance. Cytotechnology 50:35–48.

Castilho L. R., Anspach F. B., Deckwer W. D. 2002. An integrated process for mammalian cell perfusion cultivation and product purification using a dynamic filter. Biotechnol Prog 18: 776–781.

Castilho L. R., Medronho R. A. 2002. Cell retention devices for suspended-cell perfusion cultures. Adv Biochem Eng Biotechnol 74:129–169.

Chee Furng W. D., Tin Kam W. K., Tang G. L., Kiat H. C., Gek Sim Y. M. 2005. Impact of dynamic online fed-batch strategies on metabolism, productivity and N-glycosylation quality in CHO cell cultures. Biotechnol Bioeng 89:164–177.

Chen K., Liu Q., Xie L., Sharp P. A., Wang D. I. 2001. Engineering of a mammalian cell line for reduction of lactate formation and high monoclonal antibody production. Biotechnol Bioeng 72:55–61.

Chiou T. W., Murakami S., Wang D. I., Wu W. T. 1991. A fiber-bed bioreactor for anchorage-dependent animal cell cultures: part I. Bioreactor design and operations. Biotechnol Bioeng 37:755–761.

Christopher J. B., David C., Lauren B., Sukhanya J., Andrew B., Angela C. 2004. Adjunct automation to the cellmate cell culture robot. JALA 9(4):209–217. 8-1-2004 [Abstract].

Chu L., Robinson D. K. 2001. Industrial choices for protein production by large-scale cell culture. Curr Opin Biotechnol 12:180–187.

Cotter T. G., Al-Rubeai M. 1995. Cell death (apoptosis) in cell culture systems. Trends Biotechnol 13:150–155.

Cruz H. J., Conradt H. S., Dunker R., Peixoto C. M., Cunha A. E., Thomaz M., Burger C., Dias E. M., Clemente J., Moreira J. L., Rieke E., Carrondo M. J. 2002. Process development of a recombinant antibody/interleukin-2 fusion protein expressed in protein-free medium by BHK cells. J Biotechnol 96:169–183.

Dalm M. C., Cuijten S. M., van Grunsven W. M., Tramper J., Martens D. E. 2004. Effect of feed and bleed rate on hybridoma cells in an acoustic perfusion bioreactor: part I. Cell density, viability, and cell-cycle distribution. Biotechnol Bioeng 88:547–557.

Dalm M. C., Jansen M., Keijzer T. M., van Grunsven W. M., Oudshoorn A., Tramper J., Martens D. E. 2005. Stable hybridoma cultivation in a pilot-scale acoustic perfusion system: long-term process performance and effect of recirculation rate. Biotechnol Bioeng 91:894–900.

Deo Y. M., Mahadevan M. D., Fuchs R. 1996. Practical considerations in operation and scale-up of spin-filter based bioreactors for monoclonal antibody production. Biotechnol Prog 12:57–64.

Derfus G. E., Abramzon D., Tung M., Chang D., Kiss R., Amanullah A. 2010. Cell culture monitoring via an auto-sampler and an integrated multi-functional off-line analyzer. Biotechnol Prog 26:284–292.

Dowd J., Cosentino G., Chao S.-B. 2007. New Technologies in Biophamrceutical Upstream Processing, In: Roshni L. D., Jeno M. S., editors. *Advanced Technologies in Biopharmaceutical Processing*. Blackwell Publishing Professional, Ames, IA. pp 97–126.

Dowd J. E., Weber I., Rodriguez B., Piret J. M., Kwok K. E. 1999. Predictive control of hollow-fiber bioreactors for the production of monoclonal antibodies. Biotechnol Bioeng 63: 484–492.

Emery A. N., Jan D. C., Al-Rubeai M. 1995. Oxygenation of intensive cell-culture system. Appl Microbiol Biotechnol 43:1028–1033.

Europa A. F., Gambhir A., Fu P. C., Hu W. S. 2000. Multiple steady states with distinct cellular metabolism in continuous culture of mammalian cells. Biotechnol Bioeng 67:25–34.

Falkenberg F. W. 1998. Production of monoclonal antibodies in the miniPERM bioreactor: comparison with other hybridoma culture methods. Res Immunol 149:560–570.

Farid S. S. 2006. Established bioprocesses for producing antibodies as a basis for future planning. Adv Biochem Eng Biotechnol 101:1–42.

Fellenz M. P., Gerweck L. E. 1988. Influence of extracellular pH on intracellular pH and cell energy status: relationship to hyperthermic sensitivity. Radiat Res 116:305–312.

Fenge C., Lullau E. 2005. Cell Culture Bioreactors, In: Ozturk S. S., Hu W. S., editors. *Cell Culture Technology for Pharmaceutical and Cell Based Therapies*. CRC Press Taylor and Francis Group, Boca Raton, FL. pp 155–224.

Freshney R. 2005. *Culture of Animal Cells: A Manual of Basic Technique*. Wiley-Liss, New York, NY. pp 118–119.

Genzel Y., Olmer R. M., Schafer B., Reichl U. 2006. Wave microcarrier cultivation of MDCK cells for influenza virus production in serum containing and serum-free media. Vaccine 24: 6074–6087.

Gogate P. R., Beenackers A. A. C. M., Pandit A. B. 2000. Multiple-impeller systems with a special emphasis on bioreactors: a critical review. Biochem Eng J 6:109–144.

Gorenflo V. M., Angepat S., Bowen B. D., Piret J. M. 2003. Optimization of an acoustic cell filter with a novel air-backflush system. Biotechnol Prog 19:30–36.

Gorenflo V. M., Smith L., Dedinsky B., Persson B., Piret J. M. 2002. Scale-up and optimization of an acoustic filter for 200 L/day perfusion of a CHO cell culture. Biotechnol Bioeng 80:438–444.

Gramer M. J., Maas J., Lieberman M. M. 2003. Use of hollow fiber systems for rapid and direct scale up of antibody production from hybridoma cell lines cultured in CL-1000 flasks using BD Cell MAb medium. Cytotechnology 42:155–162.

Grampp G. E., Applegate M. A., Stephanopoulos G. 1996. Cyclic operation of ceramic-matrix animal cell bioreactors for controlled secretion of an endocrine hormone. A comparison of single-pass and recycle modes of operation. Biotechnol Prog 12:837–846.

Greenfield P. F., Guillaume J. -M., Randerson D. H., Smith C. G. 1991. Experience in scale-up of homogeneous perfusion culture for hybridomas. Bioprocess Biosyst Eng 6:213–219.

Gumusderelioglu M., Aslankaraoglu E., Gurhan S. I. 2001. Rabies virus production in non-woven polyester fabric(NWPF) packed-bed reactors. Biotechnol Appl Biochem 33:167–172.

Hanson M. A., Ge X., Kostov Y., Brorson K. A., Moreira A. R., Rao G. 2007. Comparisons of optical pH and dissolved oxygen sensors with traditional electrochemical probes during mammalian cell culture. Biotechnol Bioeng 97:833–841.

Heath C., Dilwith R., Belfort G. 1990. Methods for increasing monoclonal antibody production in suspension and entrapped cell cultures: biochemical and flow cytometric analysis as a function of medium serum content. J Biotechnol 15:71–89.

Hiller G. W., Clark D. S., Blanch H. W. 1993. Cell retention-chemostat studies of hybridoma cells-analysis of hybridoma growth and metabolism in continuous suspension culture in serum-free medium. Biotechnol Bioeng 42:185–195.

Huang Y. M., Hu W., Rustandi E., Chang K., Yusuf-Makagiansar H., Ryll T. 2010. Maximizing productivity of CHO cell-based fed-batch culture using chemically defined media conditions and typical manufacturing equipment. Biotechnol Prog 26:1400–1410.

Iding K., Lutkemeyer D., Fraune E., Gerlach K., Lehmann J. 2000. Influence of alterations in culture condition and changes in perfusion parameters on the retention performance of a 20 μm spinfilter during a perfusion cultivation of a recombinant CHO cell line in pilot scale. Cytotechnology 34:141–150.

Ishaque A., Al-Rubeai M. 1998. Use of intracellular pH and annexin-V flow cytometric assays to monitor apoptosis and its suppression by bcl-2 over-expression in hybridoma cell culture. J Immunol Methods 221:43–57.

Ishaque A., Al-Rubeai M. 2002. Role of vitamins in determining apoptosis and extent of suppression by bcl-2 during hybridoma cell culture. Apoptosis 7:231–239.

Jackson L. R., Trudel L. J., Fox J. G., Lipman N. S. 1996. Evaluation of hollow fiber bioreactors as an alternative to murine ascites production for small scale monoclonal antibody production. J Immunol Methods 189:217–231.

Jain E., Kumar A. 2008. Upstream processes in antibody production: evaluation of critical parameters. Biotechnol Adv 26:46–72.

Jan D. C. H., Emery A. N., Al-Rubeai M. 1993. Use of a spin-filter can reduce disruption of hybridoma cells in a bioreactor. Biotechnol Tech 7:351–356.

Jan D. C. H., Petch D. A., Huzel N., Butler M. 1997. The effect of dissolved oxygen on the metabolic profile of a murine hybridoma grown in serum-free medium in continuous culture. Biotechnol Bioeng 54:153–164.

Jenke D. 2007. Evaluation of the chemical compatibility of plastic contact materials and pharmaceutical products – safety considerations related to extractables and leachables. J Pharm Sci 96:2566–2581.

Johnson M., Lanthier S., Massie B., Lefebvre G., Kamen A. A. 1996. Use of the Centritech Lab centrifuge for perfusion culture of hybridoma cells in protein-free medium. Biotechnol Prog 12:855–864.

Kim B. S., Putnam A. J., Kulik T. J., Mooney D. J. 1998. Optimizing seeding and culture methods to engineer smooth muscle tissue on biodegradable polymer matrices. Biotechnol Bioeng 57:46–54.

Kim T. K., Ryu J. S., Chung J. Y., Kim M. S., Lee G. M. 2000. Osmoprotective effect of glycine betaine on thrombopoietin production in hyperosmotic Chinese hamster ovary cell culture: clonal variations. Biotechnol Prog 16:775–781.

Kotani H., Newton P. B. III, Zhang S., Chiang Y. L., Otto E., Weaver L., Blaese R. M., Anderson W. F., McGarrity G. J. 1994. Improved methods of retroviral vector transduction and production for gene therapy. Hum Gene Ther 5:19–28.

Kretzmer G. 2002. Industrial processes with animal cells. Appl Microbiol Biotechnol 59:135–142.

Kunkel J. P., Jan D. C., Jamieson J. C., Butler M. 1998. Dissolved oxygen concentration in serum-free continuous culture affects N-linked glycosylation of a monoclonal antibody. J Biotechnol 62:55–71.

Martinelle K., Westlund A., Haggstrom L. 1996. Ammonium ion transporta cause of cell death. Cytotechnology 22:251–254.

Marx U. 1998. Membrane-based cell culture technologies: a scientifically and economically satisfactory alternative to malignant ascites production for monoclonal antibodies. Res Immunol 149:557–559.

McDowell C. L., Papoutsakis E. T. 1998. Increased agitation intensity increases CD13 receptor surface content and mRNA levels, and alters the metabolism of HL60 cells cultured in stirred tank bioreactors. Biotechnol Bioeng 60:239–250.

Merchuk J. C. 1990. Why use air-lift bioreactors? Trends Biotechnol 8:66–71.

Merchuk J. C. 2003. Airlift bioreactors: review of recent advances. Can J Chem Eng 81:324–337.

Mercille S., Johnson M., Lanthier S., Kamen A. A., Massie B. 2000. Understanding factors that limit the productivity of suspension-based perfusion cultures operated at high medium renewal rates. Biotechnol Bioeng 67:435–450.

Mercille S., Johnson M., Lemieux R., Massie B. 1994. Filtration-based perfusion of hybridoma cultures in protein-free medium: reduction of membrane fouling by medium supplementation with DNase I. Biotechnol Bioeng 43:833–846.

Merten O. W. 2000. Constructive improvement of the ultrasonic separation device ADI 1015. Cytotechnology 34:175–179.

Meuwly F., Ruffieux P. A., Kadouri A., von Stockar U. 2007. Packed-bed bioreactors for mammalian cell culture: bioprocess and biomedical applications. Biotechnol Adv 25: 45–56.

Miller W. M., Blanch H. W., Wilke C. R. 1988. A kinetic analysis of hybridoma growth and metabolism in batch and continuous suspension culture: effect of nutrient concentration, dilution rate, and pH. Biotechnol Bioeng 32:947–965.

Min Lee G., Koo J., Flickinger M. C. 2009. *Osmolarity Effects, Chinese Hamster Ovary Cell Culture*. Wiley, New York, NY.

Nam J. H., Ermonval M., Sharfstein S. T. 2007. Cell attachment to microcarriers affects growth, metabolic activity, and culture productivity in bioreactor culture. Biotechnol Progress 23: 652–660.

Nienow A. W., Hunt G., Buckland B. C. 1994. A fluid dynamic study of the retrofitting of large agitated bioreactors: turbulent flow. Biotechnol Bioeng 44:1177–1185.

Oguchi S., Saito H., Tsukahara M., Tsumura H. 2006. pH Condition in temperature shift cultivation enhances cell longevity and specific hMab productivity in CHO culture. Cytotechnology 52:199–207.

Oh S., Chua F. K. F., Al-Rubeai M. 1996. Flow Cytometric Studies of Osmotically Stressed and Sodium Butyrate-Treated Hybridoma Cells, In: Al-Rubeai M., Emery A. N., editors. *Flow Cytometry Applications in Cell Culture*. Marcel Dekker, Inc., New York, NY. pp 101–119.

Oh S. K., Nienow A. W., Al-Rubeai M., Emery A. N. 1992. Further studies of the culture of mouse hybridomas in an agitated bioreactor with and without continuous sparging. J Biotechnol 22:245–270.

Oller A. R., Buser C. W., Tyo M. A., Thilly W. G. 1989. Growth of mammalian cells at high oxygen concentrations. J Cell Sci 94(Pt 1):43–49.

Oyaas K., Ellingsen T. E., Dyrset N., Levine D. W. 1994a. Utilization of osmoprotective compounds by hybridoma cells exposed to hyperosmotic stress. Biotechnol Bioeng 43:77–89.

Oyaas K., Ellingsen T. E., Dyrset N., Levine D. W. 1994b. Hyperosmotic hybridoma cell cultures: increased monoclonal antibody production with addition of glycine betaine. Biotechnol Bioeng 44:991–998.

Oyaas K., Ellingsen T. E., Dyrset N., Levine D. W. 1995. Transport of osmoprotective compounds in hybridoma cells exposed to hyperosmotic stress. Cytotechnology 17:143–151.

Ozturk S. S., Hu W. S. 2005. *Cell Culture Technology for Pharmaceutical and Cell Based Therapies*. CRC press Taylor and Francis Group, New York, NY. pp 1–776.

Ozturk S. S., Palsson B. O. 1990. Loss of antibody productivity during long-term cultivation of a hybridoma cell line in low serum and serum-free media. Hybridoma 9:167–175.

Ozturk S. S., Palsson B. O. 1991a. Growth, metabolic, and antibody production kinetics of hybridoma cell culture: 2. Effects of serum concentration, dissolved oxygen concentration, and medium pH in a batch reactor. Biotechnol Prog 7:481–494.

Ozturk S. S., Palsson B. O. 1991b. Effect of medium osmolarity on hybridoma growth, metabolism, and antibody production. Biotechnol Bioeng 37:989–993.

Perani A., Singh R. P., Chauhan R., Al-Rubeai M. 1998. Variable functions of bcl-2 in mediating bioreactor stress-induced apoptosis in hybridoma cells. Cytotechnology 28:177–188.

Petersen J. F., Mcintire L. V., Papoutsakis E. T. 1988. Shear Sensitivity of Cultured Hybridoma Cells (Crl-8018) Depends on Mode of Growth, Culture Age and Metabolite Concentration. J Biotechnol 7:229–246.

Petrossian A., Cortessis G. P. 1990. Large-scale production of monoclonal antibodies in defined serum-free media in airlift bioreactors. Biotechniques 8:414–422.

Pierce L., Shabram P. 2004. Scalability of a disposable bioreactor from 25 L–500 L run in a perfusion mode with a CHO-based cell line: a tech review. Bioprocess J 3, 51–56.

Rathore A. S., Winkle H. 2009. Quality by design for biopharmaceuticals. Nat Biotechnol 27: 26–34.

Read E. K., Park J. T., Shah R. B., Riley B. S., Brorson K. A., Rathore A. S. 2010b. Process analytical technology (PAT) for biopharmaceutical products: part I. concepts and applications. Biotechnol Bioeng 105:276–284.

Read E. K., Shah R. B., Riley B. S., Park J. T., Brorson K. A., Rathore A. S. 2010a. Process analytical technology (PAT) for biopharmaceutical products: part II. Concepts and applications. Biotechnol Bioeng 105:285–295.

Reuveny S., Lazar A. 1989. Equipment and procedures for production of monoclonal antibodies in culture. Adv Biotechnol Processes 11:45–80.

Robinson D. K., Seamans T. C., Gould S. L., DiStefano D. J., Chan C. P., Lee D. K., Bibila T., Glazomitsky K., Munshi S., Daugherty B. 1994. Optimization of a fed-batch process for production of a recombinant antibody. Ann N Y Acad Sci 745:285–296.

Rodrigues M. E., Costa A. R., Henriques M., Azeredo J., Oliveira R. 2010. Technological progresses in monoclonal antibody production systems. Biotechnol Prog 26:332–351.

Ryu J. S., Kim T. K., Chung J. Y., Lee G. M. 2000. Osmoprotective effect of glycine betaine on foreign protein production in hyperosmotic recombinant chinese hamster ovary cell cultures differs among cell lines. Biotechnol Bioeng 70:167–175.

Sauer P. W., Burky J. E., Wesson M. C., Sternard H. D., Qu L. 2000. A high-yielding, generic fed-batch cell culture process for production of recombinant antibodies. Biotechnol Bioeng 67:585–597.

Schmelzer A. E., Miller W. M. 2002. Effects of osmoprotectant compounds on NCAM polysialylation under hyperosmotic stress and elevated pCO2. Biotechnol Bioeng 77:359–368.

Senger R. S., Karim M. N. 2003. Effect of shear stress on intrinsic CHO culture state and glycosylation of recombinant tissue-type plasminogen activator protein. Biotechnol Progress 19:1199–1209.

Shi Y., Ryu D. D., Park S. H. 1992. Performance of mammalian cell culture bioreactor with a new impeller design. Biotechnol Bioeng 40:260–270.

Simpson N. H., Singh R. P., Emery A. N., Al-Rubeai M. 1999. Bcl-2 over-expression reduces growth rate and prolongs G1 phase in continuous chemostat cultures of hybridoma cells. Biotechnol Bioeng 64:174–186.

Simpson N. H., Singh R. P., Perani A., Goldenzon C., Al-Rubeai M. 1998. In hybridoma cultures, deprivation of any single amino acid leads to apoptotic death, which is suppressed by the expression of the bcl-2 gene. Biotechnol Bioeng 59:90–98.

Singh M. A. 1994. Cell death in bioreactors: a role for apoptosis. Biotechnol Bioeng 44:720–726.

Singh V. 1999. Disposable bioreactor for cell culture using wave-induced agitation. Cytotechnology 30:149–158.

Singh R. P., Al-Rubeai M. 1998. Apoptosis and bioprocess technology. Adv Biochem Eng Biotechnol 62:167–184.

Singh R. P., Finka G., Emery A. N., AlRubeai M. 1997. Apoptosis and its control in cell culture systems. Cytotechnology 23:87–93.

Sitton G., Srienc F. 2008. Mammalian cell culture scale-up and fed-batch control using automated flow cytometry. J Biotechnol 135:174–180.

Spens E., Haggstrom L. 2007. Defined protein and animal component-free NS0 fed-batch culture. Biotechnol Bioeng 98:1183–1194.

Stansfield S. H., Allen E. E., Dinnis D. M., Racher A. J., Birch J. R., James D. C. 2007. Dynamic analysis of GS-NS0 cells producing a recombinant monoclonal antibody during fed-batch culture. Biotechnol Bioeng 97:410–424.

Su W. W., Flickinger M. C. 2009. *Bioreactors, Perfusion*. Wiley, New York, NY.

Tang Y. J., Ohashi R., Hamel J. F. 2007. Perfusion culture of hybridoma cells for hyperproduction of IgG(2a) monoclonal antibody in a wave bioreactor-perfusion culture system. Biotechnol Prog 23:255–264.

Tey B. T., Al-Rubeai M. 2005. Effect of Bcl-2 overexpression on cell cycle and antibody productivity in chemostat cultures of myeloma NS0 cells. J Biosci Bioeng 100:303–310.

Trampler F., Sonderhoff S. A., Pui P. W., Kilburn D. G., Piret J. M. 1994. Acoustic cell filter for high density perfusion culture of hybridoma cells. Biotechnology (N Y) 12:281–284.

Trebak M., Chong J. M., Herlyn D., Speicher D. W. 1999. Efficient laboratory-scale production of monoclonal antibodies using membrane-based high-density cell culture technology. J Immunol Methods 230:59–70.

Tsao Y. S., Cardoso A. G., Condon R. G., Voloch M., Lio P., Lagos J. C., Kearns B. G., Liu Z. 2005. Monitoring Chinese hamster ovary cell culture by the analysis of glucose and lactate metabolism. J Biotechnol 118:316–327.

Valdes R., Ibarra N., Gonzalez M., Alvarez T., Garcia J., Llambias R., Perez C. A., Quintero O., Fischer R. 2001. CB.Hep-1 hybridoma growth and antibody production using protein-free medium in a hollow fiber bioreactor. Cytotechnology 35:145–154.

van Wezel A. L. 1967. Growth of cell-strains and primary cells on micro-carriers in homogeneous culture. Nature 216:64–65.

Van Wie B. J., Brouns T. M., Elliot M. L., Davis W. C. 1991. A novel continuous centrifugal bioreactor for high-density cultivation of mammalian and microbial cells. Biotechnol Bioeng 38:1190–1202.

Varley J., Birch J. 1999. Reactor design for large scale suspension animal cell culture. Cytotechnology 29:177–205.

Velden-de Groot C. A. M. 1995. Microcarrier technology, present status and perspective. Cytotechnology 18:51–56.

Voisard D., Meuwly F., Ruffieux P. A., Baer G., Kadouri A. 2003. Potential of cell retention techniques for large-scale high-density perfusion culture of suspended mammalian cells. Biotechnol Bioeng 82:751–765.

Walsh G. 2003. Biopharmaceutical benchmarks – 2003. Nat Biotechnol 21:865–870.

Walsh G. 2006. Biopharmaceutical benchmarks 2006. Nat Biotechnol 24:769–776.

Wang Y., Ouyang F. 1999. Bead-to-bead transfer of Vero cells in microcarrier culture. Cytotechnology 31:221–224.

Warnock J. N., Al-Rubeai M. 2005. Production of Biologics from Animal Cell Cultures, In: Nedovic V., Willeart R., editors. *Applications of Cell Immobilisation Biotechnology*. Series Focus on Biotechnology, vol. 8B. Springer, Dordrecht. Chapter 23, pp 423–438. ISBN: 1-4020-3229-3.

Warnock J. N., Al-Rubeai M. 2006. Bioreactor systems for the production of biopharmaceuticals from animal cells. Biotechnol Appl Biochem 45:1–12.

Weber W., Weber E., Geisse S., Memmert K. 2002. Optimisation of protein expression and establishment of the Wave Bioreactor for Baculovirus/insect cell culture. Cytotechnology 38:77–85.

Wen Z. Y., Teng X. W., Chen F. 2000. A novel perfusion system for animal cell cultures by two step sequential sedimentation. J Biotechnol 79:1–11.

Wikstrom K., Blomberg P., Islam K. B. 2004. Clinical grade vector production: analysis of yield, stability, and storage of gmp-produced retroviral vectors for gene therapy. Biotechnol Prog 20:1198–1203.

Xie L., Wang D. I. 1994. Fed-batch cultivation of animal cells using different medium design concepts and feeding strategies. Biotechnol Bioeng 43:1175–1189.

Xie L., Wang D. I. 1996. High cell density and high monoclonal antibody production through medium design and rational control in a bioreactor. Biotechnol Bioeng 51:725–729.

Xie L., Wang D. I. 1997. Integrated approaches to the design of media and feeding strategies for fed-batch cultures of animal cells. Trends Biotechnol 15:109–113.

Yabannavar V. M., Singh V., Connelly N. V. 1994. Scaleup of spinfilter perfusion bioreactor for mammalian cell retention. Biotechnol Bioeng 43:159–164.

Yang M., Butler M. 2000. Effects of ammonia on CHO cell growth, erythropoietin production, and glycosylation. Biotechnol Bioeng 68:370–380.

Yang S. T., Luo J., Chen C. 2004. *A Fibrous-Bed Bioreactor for Continuous Production of Monoclonal Antibody by HybridomaBiomanufacturing*. Springer, Berlin/Heidelberg. pp 61–96.

Yuk I. H., Olsen M. M., Geyer S., Forestell S. P. 2004. Perfusion cultures of human tumor cells: a scalable production platform for oncolytic adenoviral vectors. Biotechnol Bioeng 86:637–642.

Zhang S., Handa-Corrigan A., Spier R. E. 1993. A comparison of oxygenation methods fro high-density perfusion culture of animal cells. Biotechnol Bioeng 41:685–692.

Zhou W., Seth G., Guardia M. J., Hu W. S., Flickinger M. C. 2009. *Mammalian Cell Bioreactors*. Wiley, New York, NY.

Zhu Y., Cuenca J. V., Zhou W., Varma A. 2008. NSO cell damage by high gas velocity sparging in protein-free and cholesterol-free cultures. Biotechnol Bioeng 101:751–760.

Zouwenga R., D'Avino A., Improving Z. G. 2010. Productivity in bioreactors. Genet Eng Biotechnol News 30(7). 1-4-2010.

Chapter 3
Production of Antibody in Insect Cells

Hideki Yamaji

Abstract Insect cells have proven to be an excellent platform for the production of recombinant antibodies. The baculovirus – insect cell system directs transient expression of recombinant antibodies in batch culture upon infection of insect cells with a recombinant baculovirus, while stably transformed insect cells allow constitutive or inducible production. Both systems provide rapid, simple ways of producing considerable amounts of recombinant antibody molecules with biological activities, in particular antibody fragments, such as scFv and Fab fragments, and their derivatives. Therefore, insect cell expression systems would be highly valuable for high-throughput antibody production. In addition, the display of heterologous proteins, including antibody fragments, on the baculovirus surface might be a useful tool for the generation and production of monoclonal antibodies with high affinity and specificity.

3.1 Introduction

In recent years, insect cells have been used extensively as a platform for the production of a wide variety of biologically active recombinant proteins. Insect cells are easy to culture, as compared with mammalian cells. They can be maintained at 25–28°C without CO_2 supplementation in the atmosphere. Insect cells have traditionally been cultured in basal media supplemented with around 10% vertebrate serum, most usually fetal bovine serum (FBS), instead of insect hemolymph. However, insect cells are easily adaptable to serum-free medium and can be grown to high densities in suspension culture. The doubling times of insect cells are typically 18–24 h, but the growth rate is slower at lower temperatures. Insect cells are capable of producing heterologous proteins through post-translational processing and modifications of higher eukaryotes. The baculovirus – insect cell system is the most commonly used insect cell-based expression system. Upon infection with a recombinant baculovirus carrying the foreign gene of interest, insect cells in culture often express large quantities of the foreign protein, which retains its functional activity, during the very late stage of infection (Luckow, 1995). Recombinant

H. Yamaji (✉)
Department of Chemical Science and Engineering, Graduate School of Engineering, Kobe University, Kobe 657-8501, Japan
e-mail: yamaji@kobe-u.ac.jp

M. Al-Rubeai (ed.), *Antibody Expression and Production*, Cell Engineering 7,
DOI 10.1007/978-94-007-1257-7_3, © Springer Science+Business Media B.V. 2011

baculoviruses have also been used for the production of various recombinant proteins in insect larvae or pupae (Kato et al., 2010).

In the baculovirus – insect cell system, recombinant protein is transiently expressed and commonly produced in batch culture, because continuous production is virtually impossible due to the lytic nature of the viral infection process. By contrast, recombinant protein can be synthesized continuously by integrating a constitutively expressed gene into the genome of a host cell line upon transfection with plasmid vectors. Stably transformed insect cell lines have been employed as an attractive system for the continuous production of recombinant proteins (Douris et al., 2006; McCarroll and King, 1997; Pfeifer, 1998). The stably transformed insect cell system is especially useful for the production of secreted complex proteins, since the protein processing machinery and secretory pathway of the host insect cell are not damaged by baculovirus infection.

A display of heterologous peptides or proteins on the surface of virus particles is invaluable for the selection of genes that encode desired properties from combinatorial libraries based on protein interactions. In viral surface display systems, viruses are engineered to display foreign peptides or proteins on their surface by fusing the DNA encoding the peptides or proteins with a gene encoding a viral coat protein. The resulting recombinant viruses become vehicles that carry the DNA encoding expressed peptides or proteins, and replicate to high titers after infection of host cells. Using selection methods based on interactions between expressed peptides/proteins and target molecules, virus particles displaying peptides or proteins that bind specifically to the target are isolated from a large library of different expression clones, and specific virus clones can then be amplified in host cells. A phage display system, using filamentous bacteriophages that infect *Escherichia coli*, has proven to be a powerful tool for the selection of polypeptides with novel functions (Smith, 1985; Smith and Petrenko, 1997). Phage display, however, has limitations to the successful display of eukaryotic proteins that require complex folding and extensive post-translational processing and modifications. A variety of strategies have recently been developed for the display of heterologous proteins on the surface of baculovirus particles (Boublik et al., 1995; Grabherr et al., 2001; Mäkelä and Oker-Blom, 2006). Insect cells have been used as host cells for the generation and amplification of recombinant baculoviruses displaying foreign proteins on their surface. Baculovirus display would be valuable in functional genomics and proteomics, and in drug discovery, because it allows the presentation of complex proteins through post-translational processing and modifications of higher eukaryotes on the viral surface.

The technologies mentioned above, using insect cells, are available for the generation and production of antibodies (Fig. 3.1). Like phage display, baculovirus display could be used to generate monoclonal antibodies that recognize a specific antigen by screening a library of baculoviruses expressing antibody molecules for the baculoviruses that bind to the target antigen. Meanwhile, baculovirus surface display has been employed for the generation of monoclonal antibodies by immunizing mice with baculoviruses displaying protein immunogens (Lindley et al., 2000; Mäkelä and Oker-Blom, 2006). Furthermore, the baculovirus – insect cell

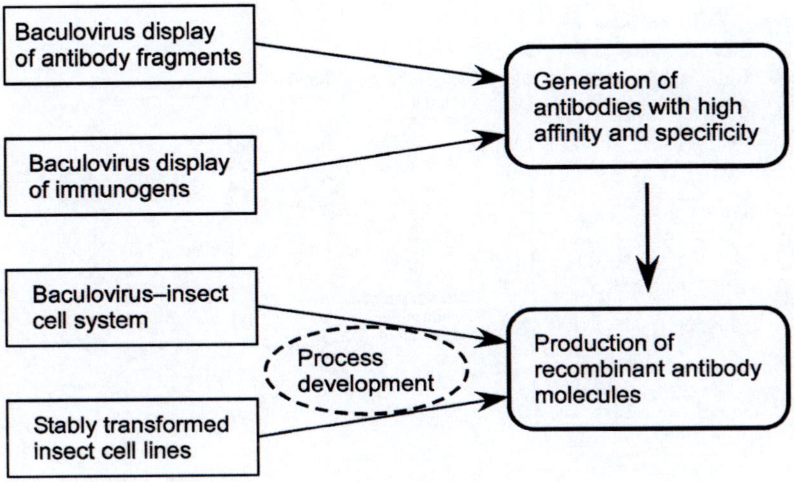

Fig. 3.1 Generation and production of antibodies using insect cell-based technologies

system and the stably transformed insect cell system have been widely used for the efficient expression of antibody genes. This chapter gives an overview of these insect cell-based technologies for the generation and production of recombinant antibody molecules (Fig. 3.1).

3.2 Antibody Production in the Baculovirus – Insect Cell System

The baculovirus – insect cell system is the most widely established and commonly used insect cell-based expression system, as it can produce large amounts of the foreign protein of interest. Baculoviruses are a family of rod-shaped viruses with a circular double-stranded DNA genome, which exclusively infects invertebrates (mainly insects). The baculoviruses commonly used in the baculovirus – insect cell system are in the genus *Nucleopolyhedrovirus*, such as *Autographa californica* nucleopolyhedrovirus (AcNPV) and *Bombyx mori* nucleopolyhedrovirus (BmNPV), which are pathogenic for lepidopteran insects such as butterflies and moths. Three insect cell lines derived from lepidopteran insects are most commonly used with AcNPV in the baculovirus – insect cell system. The Sf21 cell line and its clonal isolate Sf9 are derived from the pupal ovarian tissue of the fall army worm, *Spodoptera frugiperda*. The BTI-TN-5B1-4 (High Five) cell line originated from the ovarian cells of the cabbage looper, *Trichoplusia ni*.

Upon infection of a susceptible insect with a nucleopolyhedrovirus, progeny virus particles are released by budding off from the cell membrane of infected cells during the early phase of infection, and then these particles spread the infection throughout the insect. During the very late stage of infection, infected cells synthesize immense quantities of a viral protein called polyhedrin (Fig. 3.2).

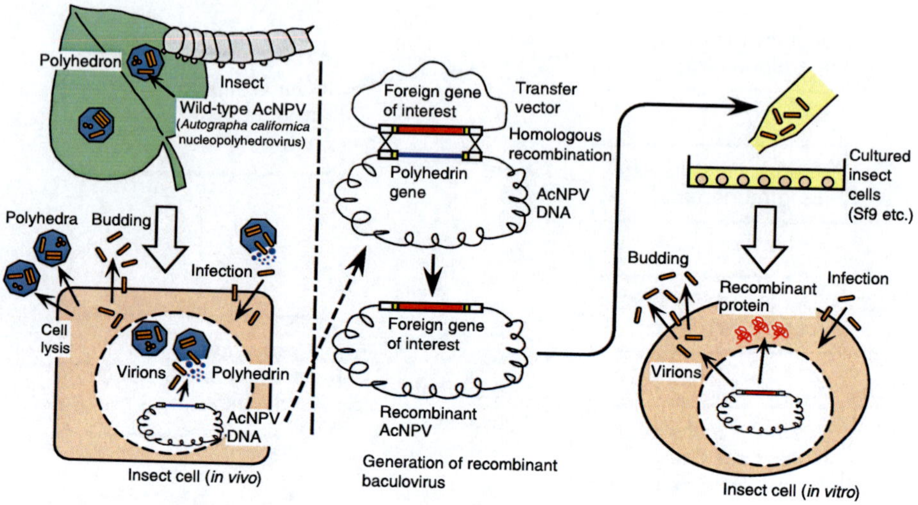

Fig. 3.2 Recombinant protein production in the baculovirus – insect cell system

Polyhedrin molecules assemble into large particles called occlusion bodies or polyhedra, in which multiple progeny virions are embedded. These particles are released into the environment for horizontal transmission of the virus after cell lysis and the death of a host insect. Polyhedrin is important in the life cycle of a wild-type baculovirus because it protects the embedded virus particles from inactivation by environmental factors until ingestion by a susceptible insect. The promoter of the nucleopolyhedrovirus polyhedrin gene is remarkably strong. However, the polyhedrin gene is not essential for infection or for replication of the virus. In the baculovirus – insect cell system, therefore, a recombinant nucleopolyhedrovirus is constructed, in which the polyhedrin gene is usually replaced with the foreign gene of interest. Subsequent infection of cultured lepidopteran insect cells with the recombinant baculovirus often leads to the expression of extremely large quantities of the foreign protein, instead of polyhedrin, under the control of the very strong polyhedrin promoter (Fig. 3.2). The strong and "very late" promoter of the nucleopolyhedrovirus p10 gene has also been used in the baculovirus – insect cell system.

Baculovirus-infected insect cells perform most of the post-translational processing and modifications of higher eukaryotes, including phosphorylation, glycosylation, correct signal peptide cleavage, proteolytic processing, and fatty acid acylation (Luckow, 1995). The highly restricted host range of the baculovirus makes this expression system safe. A wide variety of recombinant proteins, including both secreted and membrane-bound proteins, have been successfully produced in the baculovirus – insect cell system (Luckow, 1995; Massotte, 2003). With respect to glycosylation, insect cells can transfer oligosaccharide side chains to the same sites in recombinant proteins as those in native mammalian proteins. However, insect cells mostly produce simpler *N*-glycans with terminal mannose residues, paucimannose *N*-glycans, while mammalian cells produce more complex *N*-glycans containing terminal sialic acids. This difference in the *N*-glycosylation between insect and mammalian cells may limit the application of recombinant glycoproteins produced in insect cells to therapeutics for human use, although recent genetic and metabolic engineering of the insect *N*-glycosylation pathway has been shown to successfully produce complex, terminally sialylated *N*-glycans (Harrison and Jarvis, 2006; Kost et al., 2005; Tomiya et al., 2004).

Recombinant baculoviruses have traditionally been constructed based on in vivo inefficient homologous recombination between a transfer vector carrying the foreign gene of interest and the nucleopolyhedrovirus DNA genome, both of which are co-transfected into cultured insect cells (Fig. 3.2, middle panel). The resulting recombinant virus progeny are isolated from wild-type parental baculoviruses by time-consuming and subjective plaque assay with microscopic observation. This traditional procedure for the generation of a recombinant baculovirus has been improved to simplify the overall process, and rapid and efficient methods to generate recombinant baculovirus have been developed (Jarvis, 2009; Kost et al., 2005; Luckow, 1995). These include an approach using site-specific transposon-mediated insertion of the gene of interest into a baculovirus genome (bacmid) in *E. coli* (Luckow et al., 1993). This system is commercially available as the Bac-to-Bac baculovirus expression system (Invitrogen, Carlsbad, CA, USA). A faster and easier method for direct transfer of the gene of interest into the baculovirus genome in vitro, based on bacteriophage lambda site-specific recombination, has also been developed and is commercially available as the BaculoDirect baculovirus expression system (Invitrogen).

A large number of intact immunoglobulins in a biologically active form have been expressed at high yields using the baculovirus – insect cell system (Edelman et al., 1997; Hasemann and Capra, 1990; Liang et al., 1997, 2001; Nesbit et al., 1992; Poul et al., 1995; Song et al., 2010; Tan and Lam, 1999; zu Putlitz et al., 1990) (Table 3.1). Functional Fab fragments (Abrams et al., 1994; Furuta et al., 2010) and scFv fragments (Brocks et al., 1997; Kretzschmar et al., 1996; Laroche et al., 1991; Lemeulle et al., 1998; Yoshida et al., 1999) have also been expressed successfully in the baculovirus – insect cell system. Fab fragments consist of two chains comprised of the variable and the first constant domains of an immunoglobulin, VH + CH1 (Fd fragment) and VL + CL (light chain), which need to assemble with a disulphide bond (Fig. 3.3). On the other hand, scFv fragments are contiguous polypeptides that consist of the variable heavy and light chain domains (VH, VL) of an immunoglobulin, linked through a flexible peptide linker such as $(Gly_4Ser)_3$. Whereas scFv fragments can be efficiently produced in *E. coli*, expressed proteins often form insoluble aggregates called inclusion bodies in the cytoplasm, which requires proper refolding of the inactive proteins. In contrast, insect cells effectively secrete scFv fragments with antigen-binding activity into the extracellular fluid. Comparable or higher yields of scFv fragments were achieved in baculovirus-infected insect cells as compared with *E. coli* cells (Brocks et al., 1997; Lemeulle et al., 1998). Furthermore, the baculovirus – insect cell system has been used to express scFv fusion proteins (Bei et al., 1995; Brocks et al., 1997; Peipp et al., 2004) and bispecific scFv antibodies (Bruenke et al., 2004; Yoshida et al., 2003a).

Table 3.1 Production of recombinant antibodies in the baculovirus – insect cell system

Product	Cell line	Expression level	Note	References
Whole antibodies				
Mouse IgG1	Sf9	5 μg mL^{-1}	Two polyhedrin promoters, heavy and light chain signal peptides	Hasemann and Capra (1990)
Mouse IgG2a	*S. frugiperda*	25–30 μg mL^{-1}	Two polyhedrin promoters	zu Putlitz et al. (1990)
Mouse IgG2a	Sf9	25–30 μg mL^{-1}	Two polyhedrin promoters, heavy and light chain signal peptides	Nesbit et al. (1992)
Mouse/human chimeric IgG	Sf9	15–20 μg/(10^{6} cells)	Polyhedrin and p10 promoters, mouse VH signal peptide	Poul et al. (1995)
Human IgG1	Sf9	10 μg mL^{-1}	Polyhedrin and p10 promoters, mouse heavy and light chain signal peptides, spinner culture	Edelman et al. (1997)
Human IgG1	Sf9	9 μg mL^{-1}	Polyhedrin and p10 promoters, human heavy and light chain signal peptides	Liang et al. (1997)
Mouse/human chimeric IgG	Sf21 High Five	40 μg mL^{-1} (Sf21) >70 μg mL^{-1} (High Five)	Polyhedrin and p10 promoters, honeybee melittin and *B. mori* larval serum protein signal peptides	Tan and Lam (1999)
Human IgG1	High Five	6–18 μg mL^{-1}	Baculovirus expression cassette vectors for cloning of heavy and light chain genes of Fab or scFv, polyhedrin and p10 promoters, human heavy and light chain signal peptides	Liang et al. (2001)
Mouse IgG	Sf9	5–10 μg mL^{-1}	Bac-to-Bac baculovirus expression system, polyhedrin and p10 promoters, mouse heavy and light chain signal peptides, suspension culture	Song et al. (2010)
Fab fragments				
Mouse fab	Sf9	10–20 μg mL^{-1}	Co-infection with two recombinant baculoviruses, polyhedrin promoter, honeybee melittin signal peptides, spinner flask culture	Abrams et al. (1994)
Mouse fab	High Five	600 μg mL^{-1}	Bac-to-Bac baculovirus expression system, polyhedrin and p10 promoters, AcNPV gp64 signal peptide, shake-flask culture	Furuta et al. (2010)

Table 3.1 (continued)

Product	Cell line	Expression level	Note	References
scFv fragments and related proteins				
Mouse scFv	Sf9	15 μg mL^{-1}	Polyhedrin promoter, mouse light chain signal peptide	Laroche et al. (1991)
Human scFv	Sf9	32 μg mL^{-1} (purified scFv)	Polyhedrin promoter, AcNPV gp64 signal peptide, bioreactor culture	Kretzschmar et al. (1996)
Mouse scFv	Sf9	10 μg mL^{-1} (purified scFv)	p10 promoter, mouse heavy chain signal peptide, spinner culture	Lemeulle et al. (1998)
Mouse scFv	High Five	12 μg/(10^6 cells) [10–15% of total cellular protein]	Polyhedrin promoter, honeybee melittin signal peptide, scFv was not secreted into medium	Yoshida et al. (1999)
Mouse scFv joined to human Fc γ1 (hinge-CH2-CH3) and its fusion with hIL-2	Sf9	9 μg mL^{-1} (scFv-Fc) 3 μg mL^{-1} (scFv-Fc-IL-2)	Polyhedrin promoter	Bei et al. (1995)
Mouse scFv and its fusion with human IgG1 Fc (hinge-CH2-CH3)	Sf 158	0.2 μg mL^{-1} (scFv) 0.6 μg mL^{-1} (scFv-Fc)	Polyhedrin promoter, signal peptide from hIL-6 receptor	Brocks et al. (1997)
Mouse scFv fused with green fluorescent protein	Sf9	0.5–1 μg mL^{-1}	Bac-to-Bac baculovirus expression system, polyhedrin promoter, honeybee melittin signal peptide	Peipp et al. (2004)
Mouse scFv and bispecific scFv	High Five	–	Polyhedrin promoter, honeybee melittin signal peptide	Yoshida et al. (2003a)
Mouse bispecific scFv	Sf21	0.4–0.5 μg mL^{-1} (purified bispecific scFv)	Bac-to-Bac baculovirus expression system, polyhedrin promoter, honeybee melittin signal peptide	Bruenke et al. (2004)

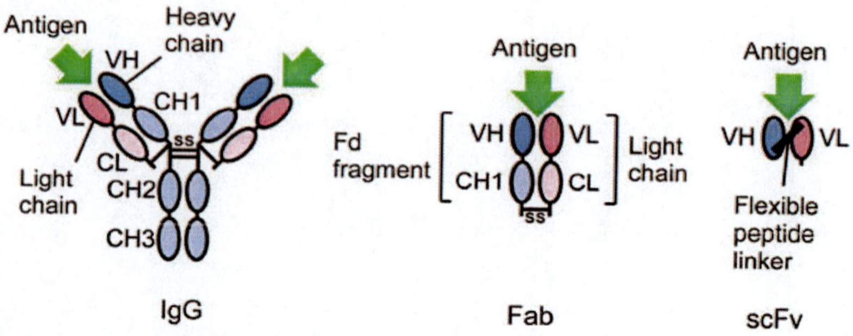

Fig. 3.3 Schematic representation of an antibody molecule (human IgG1) and its fragments. Fab fragments can be made by papain digestion of an IgG molecule. Single-chain Fv (scFv) fragments are a recombinant molecule consisting of the VH and VL domains of an IgG molecule, linked through a flexible peptide linker

To express an intact IgG and an Fab fragment, the genes that encode both the heavy chain and the light chain must be expressed in the same cell. Although expression of both the heavy- and light-chain genes can be achieved by double infection of insect cells with two recombinant baculoviruses, one carrying the heavy chain gene and the other the light chain gene (Abrams et al., 1994; Hasemann and Capra, 1990; Nesbit et al., 1992), this strategy requires the generation of two recombinant baculoviruses and careful adjustment of the infection conditions. Therefore, recombinant baculoviruses that contain both the heavy- and light-chain cDNA have been generated and used for the expression of intact IgG molecules (Liang et al., 1997, 2001; Song et al., 2010; Tan and Lam, 1999; zu Putlitz et al., 1990) and an Fab fragment (Furuta et al., 2010).

Prior to the generation of a recombinant baculovirus, factors, such as the promoter and the signal peptide, should be considered, because they affect the expression level of the recombinant protein. Various dual-expression vectors (Luckow, 1995), which are designed to accept and express two foreign genes, are available for the generation of a single recombinant baculovirus that contains both the heavy- and light-chain genes. When a dual-expression vector that has two different promoters is employed, the expression level may depend on the combination of promoter and gene. For the production of an Fab fragment, Furuta et al. (2010) generated recombinant baculoviruses using the Bac-to-Bac baculovirus expression system (Invitrogen) with the donor plasmid pFastBac Dual, which contains two multiple cloning sites to allow simultaneous expression of two heterologous genes in a single recombinant baculovirus. In the pFastBac Dual, one gene is expressed under the control of the AcNPV polyhedrin promoter, while the other is under the control of the AcNPV p10 promoter. Both the polyhedrin and p10 promoters are strong and "very late" promoters, but the promoter (polyhedrin or p10) and gene (heavy chain or light chain) combination may affect the expression level of the Fab fragment. High Five cells infected with a recombinant baculovirus, in which the light- and heavy-chain genes were located downstream of the p10 and polyhedrin promoters, respectively, produced a higher Fab fragment yield than those with a baculovirus in which the heavy- and light-chain genes were downstream of the p10 and polyhedrin promoters, respectively (Furuta et al., 2010).

Signal peptides that target secreted proteins into the endoplasmic reticulum and through the secretory pathway may influence the production of antibody molecules by baculovirus-infected insect cells. DNA sequences encoding authentic signal peptides of the heavy chain and the light chain have been used (Hasemann and Capra, 1990; Liang et al., 1997, 2001; Nesbit et al., 1992; Song et al., 2010), because the signal peptides are properly processed in the baculovirus-infected insect cells (Hasemann and Capra, 1990). Signal peptides of insect proteins like honeybee melittin (Abrams et al., 1994; Tan and Lam, 1999) and *B. mori* larval serum protein (Tan and Lam, 1999) have also been employed for the production of an IgG and an Fab fragment. Furuta et al. (2010) compared the effects of three different signal peptides of honeybee melittin, *Drosophila* BiP, and AcNPV gp64 on the secretory production of a mouse Fab fragment by baculovirus-infected High Five cells. The use of the DNA sequence encoding baculovirus gp64 signal peptide upstream of the heavy- and light-chain genes resulted in a higher yield of the secreted Fab fragment than the sequences of insect-derived BiP and melittin signal peptides, but there were no notable differences in the secretion efficiency of the different signal peptides. A signal peptide from honeybee melittin has been successfully used to express an scFv fusion protein (Peipp et al., 2004) and bispecific scFv antibodies (Bruenke et al., 2004; Yoshida et al., 2003a) as secreted proteins, while scFv fragments linked to this signal peptide were not secreted into a culture medium (Reavy et al., 2000; Yoshida et al., 1999). A signal peptide from AcNPV gp64 has been employed for the secretory expression of an scFv fragment (Kretzschmar et al., 1996). In addition, a monovalent scFv and a bivalent scFv fusion protein with a human IgG1 Fc region were expressed and secreted after correctly processing a signal peptide from human IL-6 receptor (Brocks et al., 1997).

Once a recombinant baculovirus has been generated, the recombinant protein can be obtained upon infection of cultured insect cells. Batch culture is generally employed for recombinant protein production by the baculovirus – insect cell system, because of the lytic nature of the viral infection process. Production of antibody molecules, as well as other recombinant proteins, is affected by a number of factors, which include host insect cells, culture medium, culture method, dissolved oxygen concentration, the multiplicity of infection (MOI), and cell density at the time of infection (Taticek et al., 1995). High Five cells have been shown to be better for the expression of secreted recombinant proteins than Sf9 or Sf21 cells (Davis et al., 1992; Taticek et al., 2001; Wickham and Nemerow, 1993), but exceptions have also been reported (Chai et al., 1996; Rhiel et al., 1997; Sugiura and Amann, 1996). Baculovirus-infected High Five cells in serum-free medium Express Five (Invitrogen) produced a higher Fab fragment yield than Sf9 cells in TNM-FH supplemented with 10% FBS, in both static and shake-flask cultures (Furuta et al., 2010). When infection is performed at high cell densities, recombinant protein production in batch culture is often limited by nutrient depletion in the culture medium (Radford et al., 1997; Taticek and Shuler, 1997; Yamaji et al., 1999). Hence, it is important to choose an appropriate combination of MOI and cell density at the time

Table 3.2 Production of recombinant antibodies in insect larvae and pupae

Product	Insect	Expression level	Note	References
Mouse IgG2a	*B. mori* larvae	800 μg mL^{-1}-hemolymph	Infection with recombinant BmNPV	Reis et al. (1992)
Human IgG1	*B. mori* larvae	46 μg mL^{-1}-hemolymph (36 μg/larvae)	Injection of recombinant cysteine protease- and chitinase-deficient BmNPV bacmid	Park et al. (2009)
Human IgG1	*B. mori* larvae and pupae	43 μg mL^{-1}-hemolymph (30 μg/larvae) 78 μg/pupae 239 μg mL^{-1}-hemolymph (co-expression of calreticulin)	Injection of recombinant cysteine protease- and chitinase-deficient BmNPV bacmid, co-expression of human molecular chaperone calreticulin	Dojima et al. (2010)
Mouse Fab	*T. ni* larvae	1.1 mg g^{-1}-larvae	Oral infection with recombinant AcNPV, automated insect rearing system	O'Connell et al. (2007)
Human scFv and green fluorescent protein-scFv fusion	*B. mori* larvae	188 μg mL^{-1}-hemolymph [132 μg/larva](scFv) 19 μg mL^{-1}-hemolymph (GFP-scFv)	Injection of recombinant cysteine protease- and chitinase-deficient BmNPV bacmid	Ishikiriyama et al. (2009)
Mouse scFv	*B. mori* larvae	650 μg mL^{-1}-hemolymph (0.5 mg/larvae) [purified scFv]	Injection of recombinant BmNPV bacmid	Sakamoto et al. (2010)
Mouse IgG1	Transgenic *B. mori*	2.4 mg g^{-1}-cocoon (purified IgG) [1.1% in cocoons]	Secretion into sericin layer of silk fiber of cocoons	Iizuka et al. (2009)

of infection, so that the recombinant protein production is completed before nutrient depletion in the culture medium occurs (Power et al., 1994; Wong et al., 1996; Yamaji et al., 1999; Yang et al., 1996).

The expression of a functional IgG may be accompanied by the accumulation of insoluble immunoglobulin aggregates in baculovirus-infected insect cells (Hasemann and Capra, 1990), indicating a limitation in the processing and secretory pathway of the insect cells. In this case, co-expression of an enzyme protein disulfide isomerase (PDI), an endoplasmic reticulum molecular chaperone BiP, or a cytosolic chaperon hsp70 may improve the solubility and secretion of recombinant IgG in the baculovirus – insect cell system, as previously reported (Ailor and Betenbaugh, 1998; Hsu and Betenbaugh, 1997; Hsu et al., 1996).

Because a recombinant baculovirus can also infect insect larvae or pupae, they have been used for recombinant protein production as "protein factories" (Kato et al., 2010; Maeda et al., 1985; Medin et al., 1990). Upon infection of silkworm larvae using a needle moistened with a recombinant BmNPV, a functional IgG has been expressed at a concentration of 800 $\mu g\ mL^{-1}$ in the hemolymph (Reis et al., 1992) (Table 3.2). A functionally active Fab fragment has been produced, with a yield of 1.1 mg g^{-1}-larvae, upon infection of *T. ni* larvae after the ingestion of a diet containing a recombinant AcNPV (O'Connell et al., 2007). More recently, an scFv fragment and an IgG have been produced rapidly in silkworm larvae and pupae injected with recombinant BmNPV bacmid (Dojima et al., 2010; Ishikiriyama et al., 2009; Park et al., 2009; Sakamoto et al., 2010). A significantly high level expression of recombinant antibodies can be achieved in insect larva and pupae. Whereas an automated insect rearing system and a method for the oral infection of larvae with recombinant baculovirus have been developed (O'Connell et al., 2007), efficient methods for the injection of bacmids into insects and for the recovery and purification of products from insects should be developed for the large-scale production of recombinant antibodies.

3.3 Antibody Production in Recombinant Insect Cells

Continuous protein production provides good lot-to-lot reproducibility and facilitates large-scale production in bioreactors. In the baculovirus – insect cell system, however, continuous protein production is virtually impossible because of the lytic nature of the viral infection process. The lysis of host insect cells following baculovirus infection, and the resultant release of intracellular proteins, may also result in protein degradation by proteases and the need for complicated downstream processing and purification of products. Proteolysis control in the baculovirus – insect cell system has therefore been investigated, including the addition of protease inhibitors and the use of baculoviruses that lack protease genes (Gotoh et al., 2001a, b; Ikonomou et al., 2003).

Stably transformed insect cell lines have emerged as attractive alternative platforms for the continuous production of complex recombinant proteins (Douris et al., 2006; McCarroll and King, 1997; Pfeifer, 1998). In the stably transformed insect cell

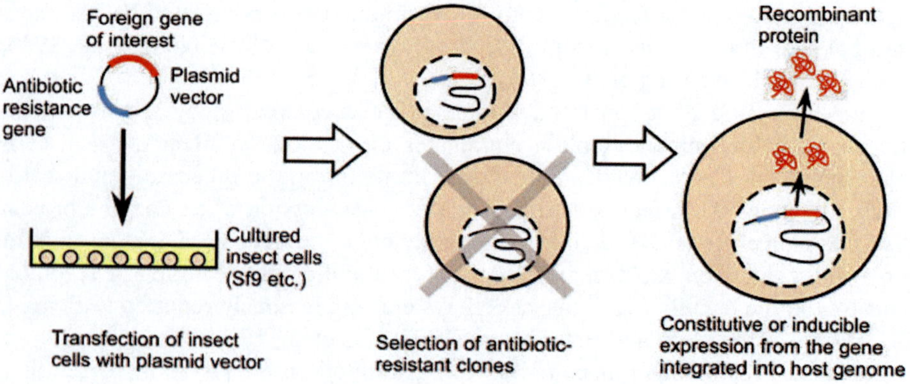

Fig. 3.4 Recombinant protein production by stably transformed insect cell lines

system, host insect cells are transfected with a plasmid vector into which the foreign gene of interest is cloned under the control of an appropriate promoter (Fig. 3.4). If the introduced vector integrates into the chromosomal DNA of the host cell, the foreign protein can be synthesized either constitutively or upon induction. In order to identify a small fraction of the stably transformed cells, antibiotic resistance genes are used as selectable markers and are co-transfected together with the heterologous gene of interest. This system is especially useful for the production of secreted complex proteins, as the protein synthesis and processing machinery of the host insect cell is not compromised by baculoviral infection.

In the stably transformed insect cell system, cell lines derived from dipteran insects, such as fruit flies and mosquitoes, have often been employed as host cells. In particular, the *Drosophila melanogaster* Schneider 2 (S2) cell line has been most commonly used for inducible expression with the *Drosophila* metallothionein promoter. An antibiotic resistance gene, such as a hygromycin, blasticidin, or neomycin resistance gene, has been used as a selectable marker and co-transfected with the heterologous gene of interest downstream of the metallothionein promoter. Hundreds of copies of the expression plasmid stably integrate into the genome of *Drosophila* S2 cells in a single transfection-selection event (Johansen et al., 1989). The metallothionein promoter is tightly regulated at high copy numbers, and is capable of directing high levels of expression when induced by copper sulfate (Kirkpatrick and Shatzman, 1999). After co-transfection of the heavy- and light-chain cDNA with an antibiotic resistance gene, human and humanized monoclonal antibodies with antigen-binding activity have been efficiently expressed and secreted under the control of the *Drosophila* metallothionein promoter in stably transformed *Drosophila* S2 cells (Johansson et al., 2007; Kirkpatrick et al., 1995) (Table 3.3). Secretory production of functional scFv fragments has also been achieved in recombinant S2 cells (Mahiouz et al., 1998; Reavy et al., 2000). In addition, constitutive secretory expression of a biologically active scFv fragment using the *Drosophila* actin 5C promoter in S2 cells has been demonstrated (Gupta et al., 2001).

Table 3.3 Production of recombinant antibodies in stably transformed insect cells

Product	Cell line	Expression level	Note	References
Humanized IgG1	S2	≥1 μg mL^{-1}	Co-transfection of heavy and light chain cDNA with a hygromycin resistance gene, *Drosophila* metallothionein promoter	Kirkpatrick et al. (1995)
Human IgG1	S2	5–35 μg mL^{-1}	Use of plasmid carrying both the heavy and light chain cDNA, *Drosophila* metallothionein promoter, *Drosophila* BiP signal peptide, shake-flask culture	Johansson et al. (2007)
Mouse scFv	S2	0.2–0.4 μg mL^{-1}	*Drosophila* metallothionein promoter, hHLA A2 signal peptide	Mahiouz et al. (1998)
Mouse scFv and its fusion with human κ chain	S2	25 μg mL^{-1} (scFv) 20 μg mL^{-1} (scFv-κ)	*Drosophila* metallothionein promoter, honeybee melittin signal peptide, T-flask culture	Reavy et al. (2000)
Mouse scFv	S2	200 μg mL^{-1} (purified scFv)	*Drosophila* actin 5C promoter for constitutive expression, κ chain signal peptide	Gupta et al. (2001)
Human IgG1	High Five	0.06 μg mL^{-1}	Co-transfection of heavy and light chain genes with a neomycin resistance gene, AcNPV hr5 enhancer and AcNPV immediate early promoter IE1, human heavy and light chain signal peptides	Guttieri et al. (2000)
Mouse IgG1 and its fusion with human TNFα or EGF	Sf9	0.5–1.0 μg mL^{-1}	Co-transfection of heavy and light chain genes with a Zeocin resistance gene, OpNPV immediate early 2 promoter OpIE2, mouse heavy and light chain signal peptides	Li et al. (2001)
Mouse fab	High Five	300 μg mL^{-1}	Use of plasmid containing BmNPV IE-1 transactivator, BmNPV HR3 enhancer, and *B. mori* actin promoter with either a blasticidin or a neomycin resistance gene, *Drosophila* BiP signal peptide, shake-flask culture	Yamaji et al. (2008)

Stable transformation of lepidopteran insect cells, such as Sf9 and High Five cells, has been investigated as well. In this case, the choice of a promoter to drive the heterologous gene expression is important, as the use of weak promoters results in low recombinant protein yields. Reportedly, the protein yields in stably transformed lepidopteran cells have often been considerably lower than those obtained in the baculovirus – insect cell system (Ivanova et al., 2007; Jarvis et al., 1990). Guttieri et al. (2000) separately cloned the heavy- and light-chain genes of a monoclonal antibody downstream of the AcNPV hr5 enhancer and the AcNPV immediate early promoter IE1 in two plasmid vectors. High Five cells were then co-transfected with the constructs and a selection plasmid carrying a neomycin resistance gene. After selection, stably transformed cells continuously secreted biologically active IgG, although the yield of IgG in the culture supernatant was considerably lower (0.06 $\mu g\ mL^{-1}$) than that obtained in the baculovirus – insect cell system (9 $\mu g\ mL^{-1}$) (Table 3.3). Li et al. (2001) cloned immunoglobulin heavy- and light-chain genes downstream of the *Orgyia pseudotsugata* NPV immediate early 2 promoter OpIE2 in two plasmid vectors containing a Zeocin resistance gene. Stably transformed Sf9 cells secreted functional antibodies at yields of approximately 1 $\mu g\ mL^{-1}$.

Recently, the expression vector pIE1/153A has been developed for the continuous high-level expression of secreted proteins by transformed lepidopteran insect cells (Farrell et al., 1998, 1999; Keith et al., 1999). pIE1/153A utilizes the *B. mori* cytoplasmic actin promoter, from which foreign gene expression is stimulated with the BmNPV IE-1 transactivator and the BmNPV HR3 enhancer. Use of the IE-1 transactivator and the HR3 enhancer has resulted in a more than 1,000-fold increase in the stimulation of foreign gene expression through the actin promoter (Lu et al., 1997). Lepidopteran insect cells are first co-transfected with pIE1/153A containing the heterologous gene and a plasmid vector carrying a selectable marker, after which antibiotic selection and isolation of highly productive clones is carried out. Stable cell lines expressing secreted proteins, including tissue plasminogen activator and secreted alkaline phosphatase, have shown higher expression levels than the baculovirus – insect cell system (Farrell et al., 1999; Jardin et al., 2007; Kato et al., 2004).

When the expression vector mentioned above, utilizing the BmNPV IE-1 transactivator, the BmNPV HR3 enhancer, and the *B. mori* actin promoter, is employed for the expression of an intact IgG or an Fab fragment (Fig. 3.3), host insect cells must be co-transfected with a total of three plasmid vectors: two sets of expression vectors that separately contain the genes encoding the heavy chain and light chain of the antibody molecule, and a plasmid carrying a selectable marker. Co-transfection with three plasmid vectors inevitably results in low efficiency in obtaining stably transformed cells producing the antibody molecule. On the basis of the expression vector utilizing the IE-1 transactivator, the HR3 enhancer, and the actin promoter, Yamaji et al. (2008) constructed two plasmid vectors, pIHAbla and pIHAneo, which contain either a blasticidin or a neomycin resistance gene, for use as a selectable marker (Fig. 3.5). After co-transfection with these plasmid vectors, into which the heavy- and light-chain genes of an Fab fragment were separately inserted, High Five cells secreting a high concentration of the Fab fragment were efficiently generated

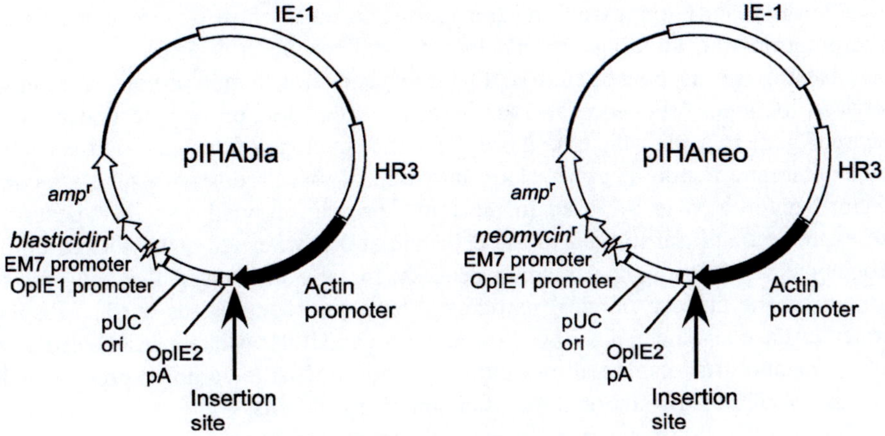

Fig. 3.5 Expression vectors for stable transformation of lepidopteran insect cells

by incubation in the presence of blasticidin and G418. High yields of more than 300 μg mL^{-1} of the Fab fragment were produced in a shake-flask culture of the recombinant insect cells.

Transgenic silkworms that synthesize a functional IgG into the sericin layer of silk fibers in their cocoons have recently been generated (Iizuka et al., 2009) (Table 3.2). The IgG produced was extractable from the cocoons with a buffer containing 3 M urea.

3.4 Baculovirus Display for Antibody Generation

Phage display has been successfully used to select antibodies that recognize specific antigens from diverse libraries (Bradbury and Marks, 2004; McCafferty et al., 1990). In antibody phage display, antibody fragments, such as scFv fragments and Fab fragments, are expressed on the surface of filamentous bacteriophages by fusion of the genes encoding the antibody fragments to one of the phage coat proteins. Specific antibody-expressing phages can then be amplified in *E. coli* after multiple rounds of affinity selection for the antigen. When displayed on the phage surface, Fab fragments have a tendency to be more functional than the corresponding scFv fragments; some scFv fragments show a lower affinity than the corresponding Fab fragments (Bird and Walker, 1991; Bradbury and Marks, 2004). However, Fab fragments are often produced at significantly lower yields in *E. coli*, because they are twice the size of scFv and require the assembly of two protein chains.

Recently, baculoviruses have been successfully used for the display of heterologous proteins on the surface of viral particles by fusing the protein to the major baculoviral envelope glycoprotein gp64 (Boublik et al., 1995; Grabherr et al., 2001; Mäkelä and Oker-Blom, 2006). gp64, also known as gp67, is essential for viral entry into insect cells via baculovirus-cell fusion. During the infection cycle, gp64 is abundantly expressed and transported to the surface of infected cells for incorporation into budding virions. In a typical baculovirus display, a recombinant baculovirus is constructed so that the gene of a target protein is inserted between the signal sequence and the mature protein domain of a second copy of the gp64 gene to anchor the fusion protein in the envelope of the baculovirus via the transmembrane region of gp64. Upon infection of insect cells with such a recombinant baculovirus, gp64 fused to the target protein, as well as wild-type gp64, are expressed and transported to the cell membrane, where they are picked up by progeny viruses during the budding process, thereby displaying the gp64-fusion protein on the surface of viral particles. Alternative baculovirus surface display strategies have also been developed (Grabherr et al., 2001; Mäkelä and Oker-Blom, 2006). Baculovirus display allows the presentation of large complex proteins with the eukaryotic posttranslational modification of insect cells.

Murine scFv fragments specific to the hapten 2-phenyloxazolone and human scFv against carcinoembryonic antigen were successfully displayed in a functional form on the AcNPV surface by fusion to the N-terminus of gp64 (Mottershead et al., 2000; Ojala et al., 2001). These results suggest that baculovirus particles displaying scFv fragments could be useful in targeting to a specific antigen for gene or drug delivery, although there would be little advantage to the use of baculoviruses displaying scFv fragments for the selection of specific antibodies, because scFv phage display has proven effective.

Generation of a recombinant baculovirus displaying antibody Fab fragments on its surface has been investigated (Yamaji and Fukuda, 2007). A recombinant baculovirus was designed so that either the heavy chain (Fd fragment) or the light chain of a mouse Fab fragment (Fig. 3.3) was expressed as a fusion protein to the N-terminus of AcNPV gp64, while at the same time the other chain of the Fab fragment was expressed as a secretion protein. For expression as a gp64-fusion protein, the cDNA fragment encoding the heavy- or light-chain gene of the Fab fragment was inserted between the signal sequence and the mature protein domain of a second copy of the AcNPV gp64 gene. On the other hand, the *Drosophila* BiP signal sequence was employed upstream of the light- or heavy-chain gene for secretory expression. The resulting DNA fragments were PCR-cloned into a donor plasmid, pFastBac Dual (Invitrogen), which contained two multiple cloning sites. A recombinant baculovirus was then generated using the Bac-to-Bac baculovirus expression system (Invitrogen). When the culture supernatant, after infection of Sf9 cells with the recombinant baculovirus, was analyzed by enzyme-linked immunosorbent assay (ELISA) using antigen-coated 96-well plates and either an anti-AcNPV gp64 antibody or an anti-mouse IgG, a relatively strong signal showing antigen-binding activity was obtained in each case. In western blot analysis of the culture supernatant under reducing conditions, a specific protein band was detected at an electrophoretic mobility coinciding with the molecular weight of the gp64-fusion protein. The Fab fragments were successfully detected on the surface of Sf9 cells infected with the recombinant baculovirus by flow cytometry using fluorescein isothiocyanate (FITC)-labeled anti-mouse IgG (Fig. 3.6). These results suggest that antibody Fab

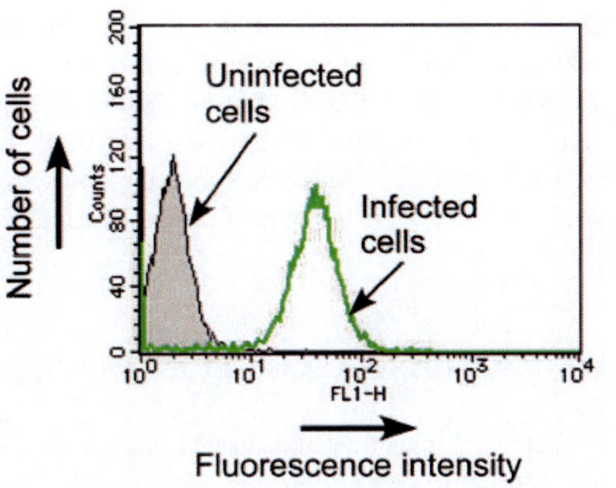

Fig. 3.6 Flow cytometric analysis of Sf9 cells infected with a recombinant baculovirus displaying mouse Fab fragments on its surface. Recombinant baculovirus carrying the genes of the Fd fragment and the light chain-gp64 fusion was used. Infected cells 24 h post-infection and uninfected cells were treated with fluorescein isothiocyanate (FITC)-labeled anti-mouse IgG before flow cytometry

fragments can be displayed on the surface of baculovirus particles in a functional form, and that baculoviruses displaying target Fab fragments could be selected using a fluorescence-activated cell sorter (FACS) with a fluorescence-labeled antigen. The use of baculovirus display technologies for the generation and screening of cDNA expression libraries has been successfully demonstrated (Crawford et al., 2004, 2006; Ernst et al., 1998), although the methods have not yet been applied to antibody selection. While further developments are needed, baculovirus display of antibody molecules, including Fab fragments, may provide an opportunity for the selection of functional antibodies with high affinity and specificity.

Baculoviruses displaying gp64-fusion proteins have been successfully used as effective immunogens for the generation of monoclonal antibodies (Lindley et al., 2000; Mäkelä and Oker-Blom, 2006). Lindley et al. (2000) constructed recombinant baculoviruses, which displayed N-terminal domains of human nuclear receptors LXRβ and FXR on the surface, by insertion of the corresponding gene fragment between the signal sequence and the mature protein domain of the AcNPV gp64 gene. After recombinant baculoviruses were pelleted by ultra-centrifugation and filter sterilized, mice were immunized with the whole baculovirus displaying respective gp64/nuclear receptor fusion proteins to produce hybridoma cells. Monoclonal antibodies that recognized LXRβ or FXR were successfully generated with a low background of monoclonal antibodies against viral proteins, except for gp64. This method allows for the rapid production of immunogens without rigorous purification once the DNA sequence is available. Since fusion proteins are displayed on the viral surface in a native form, due to the similarities in protein processing and post-translational modification between insect and mammalian systems, this method may enhance the generation of antibodies capable of recognizing antigens under non-denaturing conditions (Lindley et al., 2000). Similarly, baculoviruses displaying gp64-fusion proteins have been successfully used to generate monoclonal antibodies against human peroxisome proliferator-activated receptors (Tanaka et al., 2002). Baculovirus-displaying proteins, including *Plasmodium berghei* circumsporozoite protein (Yoshida et al., 2003b), foot-and-mouth disease virus antigen (Tami et al., 2004), bovine herpesvirus-1 glycoprotein D (Peralta et al., 2007), avian influenza virus hemagglutinin (Yang et al., 2007), and the E2 envelope glycoprotein of classical swine fever virus (Xu and Liu, 2008), have also proven to effectively elicit protective immune responses in mice, indicating that baculovirus display could be useful for the development of potential vaccine candidates.

3.5 Conclusions and Future Developments

A variety of insect cell-based technologies are currently available for the generation and production of antibodies (Fig. 3.1). Baculoviruses displaying foreign proteins on their surface have proven useful as effective immunogens for the immunization of mice to obtain hybridoma cells that produce monoclonal antibodies. The display of antibody Fab fragments on the baculovirus surface might also offer a novel approach for the selection of monoclonal antibodies with high affinity and specificity from diverse libraries. Insect cell-based expression systems, both the well developed baculovirus – insect cell system and the relatively new stably transformed cell system, provide a rapid, simple way to produce considerable amounts of recombinant antibody molecules that retain their functional activities (Tables 3.1 and 3.3). Optimization of the generation of recombinant baculovirus and stably transformed insect cells, and further development of culture media and cell lines, will lead to higher yields of recombinant antibodies in these systems. High-throughput antibody production in insect cell systems would have particular value for the discovery and development of efficacious therapeutic antibody leads, and the development of antibody microarrays in proteomics and clinical diagnostics. Advances have been made in the development of bioprocesses for the large-scale, high-density culture of insect cells (Chan et al., 1998; Chico and Jäger, 2000; Ikonomou et al., 2003; Jardin et al., 2007; Marteijn et al., 2003; Taticek et al., 1995; Yamaji et al., 2006), which also result in increased yields of recombinant antibodies. With respect to biomedical applications of insect cell expression systems, the baculovirus – insect cell system is employed for the industrial-scale production of a human papillomavirus vaccine, Cervarix, which has been approved for use against cervical cancer (Schiller et al., 2008). In the case of recombinant antibodies, however, the paucimannose type *N*-glycosylation typically observed in insect cells may limit the use of whole antibodies produced in insect cells as therapeutics for human use, while the engineering of the insect *N*-glycosylation pathway has been shown to allow the production of mammalianized *N*-glycoproteins (Harrison and Jarvis, 2006; Kost et al., 2005; Tomiya et al., 2004). For the present, therefore, insect cells can serve as a platform for the efficient production of antibody fragments, such as scFv and Fab fragments, and related proteins, but not intact IgG, for use as therapeutic applications. Future developments in insect cell and baculovirus engineering will provide further opportunities for enhancement of the utility and applications of recombinant antibodies produced in insect cell expression systems.

References

Abrams C, Deng Y-J, Steiner B, O'Toole T, Shattil SJ (1994) Determinants of specificity of a baculovirus-expressed antibody Fab fragment that binds selectively to the activated form of integrin $\alpha_{IIb}\beta_3$. J Biol Chem 269:18781–18788

Ailor E, Betenbaugh MJ (1998) Overexpression of a cytosolic chaperone to improve solubility and secretion of a recombinant IgG protein in insect cells. Biotechnol Bioeng 58:196–203

Bei R, Schlom J, Kashmiri SVS (1995) Baculovirus expression of a functional single-chain immunoglobulin and its IL-2 fusion protein. J Immunol Methods 186:245–255

Bird RE, Walker BW (1991) Single chain antibody variable regions. Trends Biotechnol 9:132–137

Boublik Y, Di Bonito P, Jones IM (1995) Eukaryotic virus display: engineering the major surface glycoprotein of the *Autographa californica* nuclear polyhedrosis virus (AcNPV) for the presentation of foreign proteins on the virus surface. Bio/Technology 13:1079–1084

Bradbury ARM, Marks JD (2004) Antibodies from phage antibody libraries. J Immunol Methods 290:29–49

Brocks B, Rode H-J, Klein M, Gerlach E, Dübel S, Little M, Pfizenmaier K, Moosmayer D (1997) A TNF receptor antagonistic scFv, which is not secreted in mammalian cells, is expressed as a soluble mono- and bivalent scFv derivative in insect cells. Immunotechnology 3:173–194

Bruenke J, Fischer B, Barbin K, Schreiter K, Wachter Y, Mahr K, Titgemeyer F, Niederweis M, Peipp M, Zunino SJ, Repp R, Valerius T, Fey GH (2004) A recombinant bispecific single-chain Fv antibody against HLA class II and FcγRIII (CD16) triggers effective lysis of lymphoma cells. Br J Haematol 125:167–17

Chai H, Al-Rubeai M, Chua KL, Oh SKW, Yap MGS (1996) Insect cell line dependent gene expression of recombinant human tumor necrosis factor–β. Enzyme Microb Technol 18: 126–132

Chan LCL, Greenfield PF, Reid S (1998) Optimising fed-batch production of recombinant proteins using the baculovirus expression vector system. Biotechnol Bioeng 59:178–188

Chico E, Jäger V (2000) Perfusion culture of baculovirus-infected BTI-Tn-5B1-4 insect cells: a method to restore cell-specific β-trace glycoprotein productivity at high cell density. Biotechnol Bioeng 70:574–586

Crawford F, Huseby E, White J, Marrack P, Kappler JW (2004) Mimotopes for alloreactive and conventional T cells in a peptide – MHC display library. PLoS Biol 2:e90

Crawford F, Jordan KR, Stadinski B, Wang Y, Huseby E, Marrack P, Slansky JE, Kappler JW (2006) Use of baculovirus MHC/peptide display libraries to characterize T-cell receptor ligands. Immunol Rev 210:156–170

Davis TR, Trotter KM, Granados RR, Wood HA (1992) Baculovirus expression of alkaline phosphatase as a reporter gene for evaluation of production, glycosylation and secretion. Bio/Technology 10:1148–1150

Dojima T, Nishina T, Kato T, Uno T, Yagi H, Kato K, Ueda H, Park EY (2010) Improved secretion of molecular chaperone-assisted human IgG in silkworm, and no alterations in their *N*-linked glycan structures. Biotechnol Prog 26:232–238

Douris V, Swevers L, Labropoulou V, Andronopoulou E, Georgoussi Z, Iatrou K (2006) Stably transformed insect cell lines: tools for expression of secreted and membrane-anchored proteins and high-throughput screening platforms for drug and insecticide discovery. Adv Virus Res 68:113–156

Edelman L, Margaritte C, Chaabihi H, Monchâtre E, Blanchard D, Cardona A, Morin F, Dumas G, Petres S, Kaczorek M (1997) Obtaining a functional recombinant anti-rhesus (D) antibody using the baculovirus – insect cell expression system. Immunology 91:13–19

Ernst W, Grabherr R, Wegner D, Borth N, Grassauer A, Katinger H (1998) Baculovirus surface display: construction and screening of a eukaryotic epitope library. Nucleic Acids Res 26: 1718–1723

Farrell PJ, Behie LA, Iatrou K (1999) Transformed lepidopteran insect cells: new sources of recombinant human tissue plasminogen activator. Biotechnol Bioeng 64:426–433

Farrell PJ, Lu M, Prevost J, Brown C, Behie L, Iatrou K (1998) High-level expression of secreted glycoproteins in transformed lepidopteran insect cells using a novel expression vector. Biotechnol Bioeng 60:656–663

Furuta T, Ogawa T, Katsuda T, Fujii I, Yamaji H (2010) Efficient production of an antibody Fab fragment using the baculovirus – insect cell system. J Biosci Bioeng 110:577–581

Gotoh T, Miyazaki Y, Kikuchi K-I, Bentley WE (2001a) Investigation of sequential behavior of carboxyl protease and cysteine protease activities in virus-infected Sf-9 insect cell culture by inhibition assay. Appl Microbiol Biotechnol 56:742–749

Gotoh T, Miyazaki Y, Sato W, Kikuchi K-I, Bentley WE (2001b) Proteolytic activity and recombinant protein production in virus-infected Sf-9 insect cell cultures supplemented with carboxyl and cysteine protease inhibitors. J Biosci Bioeng 92:248–255

Grabherr R, Ernst W, Oker-Blom C, Jones I (2001) Developments in the use of baculoviruses for the surface display of complex eukaryotic proteins. Trends Biotechnol 19:231–236

Gupta S, Eastman J, Silski C, Ferkol T, Davis PB (2001) Single chain Fv: a ligand in receptor-mediated gene delivery. Gene Ther 8:586–592

Guttieri MC, Bookwalter C, Schmaljohn C (2000) Expression of a human, neutralizing monoclonal antibody specific to Puumala virus G2-protein in stably-transformed insect cells. J Immunol Methods 246:97–108

Harrison RL, Jarvis DL (2006) Protein *N*-glycosylation in the baculovirus – insect cell expression system and engineering of insect cells to produce "mammalianized" recombinant glycoproteins. Adv Virus Res 68:159–191

Hasemann CA, Capra JD (1990) High-level production of a functional immunoglobulin heterodimer in a baculovirus expression system. Proc Natl Acad Sci USA 87:3942–3946

Hsu T-A, Betenbaugh MJ (1997) Coexpression of molecular chaperone BiP improves immunoglobulin solubility and IgG secretion from *Trichoplusia ni* insect cells. Biotechnol Prog 13:96–104

Hsu T-A, Watson S, Eiden JJ, Betenbaugh MJ (1996) Rescue of immunoglobulins from insolubility is facilitated by PDI in the baculovirus expression system. Protein Expr Purif 7:281–288

Iizuka M, Ogawa S, Takeuchi A, Nakakita S, Kubo Y, Miyawaki Y, Hirabayashi J, Tomita M (2009) Production of a recombinant mouse monoclonal antibody in transgenic silkworm cocoons. FEBS J 276:5806–5820

Ikonomou L, Schneider Y-J, Agathos SN (2003) Insect cell culture for industrial production of recombinant proteins. Appl Microbiol Biotechnol 62:1–20

Ishikiriyama M, Nishina T, Kato T, Ueda H, Park EY (2009) Human single-chain antibody expression in the hemolymph and fat body of silkworm larvae and pupae using BmNPV bacmids. J Biosci Bioeng 107:67–72

Ivanova MM, Mattingly KA, Klinge CM (2007) Estrogen receptor beta yield from baculovirus lytic infection is higher than from stably transformed Sf21 cells. Appl Microbiol Biotechnol 74:1256–1263

Jardin BA, Montes J, Lanthier S, Tran R, Elias C (2007) High cell density fed batch and perfusion processes for stable non-viral expression of secreted alkaline phosphatase (SEAP) using insect cells: comparison to a batch Sf-9-BEV system. Biotechnol Bioeng 97:332–345

Jarvis DL (2009) Baculovirus – insect cell expression systems. Methods Enzymol 463:191–222

Jarvis DL, Fleming J-AGW, Kovacs GR, Summers MD, Guarino LA (1990) Use of early baculovirus promoters for continuous expression and efficient processing of foreign gene products in stably transformed lepidopteran cells. Bio/Technology 8:950–955

Johansen H, van der Straten A, Sweet R, Otto E, Maroni G, Rosenberg M (1989) Regulated expression at high copy number allows production of a growth-inhibitory oncogene product in *Drosophila* Schneider cells. Genes Dev 3:882–889

Johansson DX, Drakenberg K, Hopmann KH, Schmidt A, Yari F, Hinkula J, Persson MAA (2007) Efficient expression of recombinant human monoclonal antibodies in *Drosophila* S2 cells. J Immunol Methods 318:37–46

Kato T, Kajikawa M, Maenaka K, Park EY (2010) Silkworm expression system as a platform technology in life science. Appl Microbiol Biotechnol 85:459–470

Kato T, Murata T, Usui T, Park EY (2004) Comparative analysis of GFP$_{UV}$-β1,3-*N*-acetylglucosaminyltransferase 2 production in two insect-cell-based expression systems. Protein Expr Purif 35:54–61

Keith MB, Farrell PJ, Iatrou K, Behie LA (1999) Screening of transformed insect cell lines for recombinant protein production. Biotechnol Prog 15:1046–1052

Kirkpatrick RB, Ganguly S, Angelichio M, Griego S, Shatzman A, Silverman C, Rosenberg M (1995) Heavy chain dimers as well as complete antibodies are efficiently formed and secreted from *Drosophila* via a BiP-mediated pathway. J Biol Chem 270:19800–19805

Kirkpatrick RB, Shatzman A (1999) *Drosophila* S2 system for heterologous gene expression. In: Fernandez JM, Hoeffler JP (eds), Gene expression systems: using nature for the art of expression. Academic Press, San Diego, pp 289–330

Kost TA, Condreay JP, Jarvis DL (2005) Baculovirus as versatile vectors for protein expression in insect and mammalian cells. Nat Biotechnol 23:567–575

Kretzschmar T, Aoustin L, Zingel O, Marangi M, Vonach B, Towbin H, Geiser M (1996) High-level expression in insect cells and purification of secreted monomeric single-chain Fv antibodies. J Immunol Methods 195:93–101

Laroche Y, Demaeyer M, Stassen J-M, Gansemans Y, Demarsin E, Matthyssens G, Collen D, Holvoet P (1991) Characterization of a recombinant single-chain molecule comprising the variable domains of a monoclonal antibody specific for human fibrin fragment D-dimer. J Biol Chem 266:16343–16349

Lemeulle C, Chardès T, Montavon C, Chaabihi H, Mani J-C, Pugnière M, Cerutti M, Devauchelle G, Pau B, Biard-Piechaczyk M (1998) Anti-digoxin scFv fragments expressed in bacteria and in insect cells have different antigen binding properties. FEBS Lett 423:159–166

Li E, Brown SL, Dolman CS, Brown GB, Nemerow GR (2001) Production of functional antibodies generated in a nonlytic insect cell expression system. Protein Expr Purif 21:121–128

Liang M, Dübel S, Li D, Queitsch I, Li W, Bautz EKF (2001) Baculovirus expression cassette vectors for rapid production of complete human IgG from phage display selected antibody fragments. J Immunol Methods 247:119–130

Liang M, Guttieri M, Lundkvist Å, Schmaljohn C (1997) Baculovirus expression of a human G2-specific, neutralizing IgG monoclonal antibody to Puumala virus. Virology 235:252–260

Lindley KM, Su J-L, Hodges PK, Wisely GB, Bledsoe RK, Condreay JP, Winegar DA, Hutchins JT, Kost TA (2000) Production of monoclonal antibodies using recombinant baculovirus displaying gp64–fusion proteins. J Immunol Methods 234:123–135

Lu M, Farrell PJ, Johnson R, Iatrou K (1997) A baculovirus (*Bombyx mori* nuclear polyhedrosis virus) repeat element functions as a powerful constitutive enhancer in transfected insect cells. J Biol Chem 272:30724–30728

Luckow VA (1995) Protein production and processing from baculovirus expression vectors. In: Shuler ML, Wood HA, Granados RR, Hammer DA (eds), Baculovirus expression systems and biopesticides. Wiley-Liss, New York, pp 51–90

Luckow VA, Lee SC, Barry GF, Olins PO (1993) Efficient generation of infectious recombinant baculoviruses by site-specific transposon-mediated insertion of foreign genes into a baculovirus genome propagated in *Esherichia coli*. J Virol 67:4566–4579

Maeda S, Kawai T, Obinata M, Fujiwara H, Horiuchi T, Saeki Y, Sato Y, Furusawa M (1985) Production of human α-interferon in silkworm using a baculovirus vector. Nature 315:592–594

Mahiouz DL, Aichinger G, Haskard DO, George AJT (1998) Expression of recombinant anti-E-selectin single-chain Fv antibody fragments in stably transfected insect cell lines. J Immunol Methods 212:149–160

Marteijn RCL, Jurrius O, Dhont J, de Gooijer CD, Tramper J, Martens DE (2003) Optimization of a feed medium for fed-batch culture of insect cells using a genetic algorithm. Biotechnol Bioeng 81:269–278

Massotte D (2003) G protein-coupled receptor overexpression with the baculovirus – insect cell system: a tool for structural and functional studies. Biochim Biophys Acta 1610:77–89

Mäkelä AR, Oker-Blom C (2006) Baculovirus display: a functional technology for gene delivery and eukaryotic library development. Adv Virus Res 68:91–112

McCafferty J, Griffiths AD, Winter G, Chiswell DJ (1990) Phage antibodies: filamentous phage displaying antibody variable domains. Nature 348:552–554

McCarroll L, King LA (1997) Stable insect cell cultures for recombinant protein production. Curr Opin Biotechnol 8:590–594

Medin JA, Hunt L, Gathy K, Evans RK, Coleman MS (1990) Efficient, low-cost protein factories: expression of human adenosine deaminase in baculovirus-infected insect larvae. Proc Natl Acad Sci USA 87:2760–2764

Mottershead DG, Alfthan K, Ojala K, Takkinen K, Oker-Blom C (2000) Baculoviral display of functional scFv and synthetic IgG-binding domains. Biochem Biophys Res Commun 275: 84–90

Nesbit M, Fu ZF, McDonald-Smith J, Steplewski Z, Curtis PJ (1992) Production of a functional monoclonal antibody recognizing human colorectal carcinoma cells from a baculovirus expression system. J Immunol Methods 151:201–208

Ojala K, Mottershead DG, Suokko A, Oker-Blom C (2001) Specific binding of baculoviruses displaying gp64 fusion proteins to mammalian cells. Biochem Biophys Res Commun 284: 777–784

O'Connell KP, Kovaleva E, Campbell JH, Anderson PE, Brown SG, Davis DC, Valdes JJ, Welch RW, Bentley WE, van Beek NA (2007) Production of a recombinant antibody fragment in whole insect larvae. Mol Biotechnol 36:44–51

Park EY, Ishikiriyama M, Nishina T, Kato T, Yagi H, Kato K, Ueda H (2009) Human IgG1 expression in silkworm larval hemolymph using BmNPV bacmids and its *N*-linked glycan structure. J Biotechnol 139:108–114

Peipp M, Saul D, Barbin K, Bruenke J, Zunino SJ, Niederweis M, Fey GH (2004) Efficient eukaryotic expression of fluorescent scFv fusion proteins directed against CD antigens for FACS applications. J Immunol Methods 285:265–280

Peralta A, Molinari P, Conte-Grand D, Calamante G, Taboga O (2007) A chimeric baculovirus displaying bovine herpesvirus-1 (BHV-1) glycoprotein D on its surface and their immunological properties. Appl Microbiol Biotechnol 75:407–414

Pfeifer TA (1998) Expression of heterologous proteins in stable insect cell culture. Curr Opin Biotechnol 9:518–521

Poul M-A, Cerutti M, Chaabihi H, Devauchelle G, Kaczorek M, Lefranc M-P (1995) Design of cassette baculovirus vectors for the production of therapeutic antibodies in insect cells. Immunotechnology 1:189–196

Power JF, Reid S, Radford KM, Greenfield PF, Nielsen LK (1994) Modeling and optimization of the baculovirus expression vector system in batch suspension culture. Biotechnol Bioeng 44:710–719

Radford KM, Reid S, Greenfield PF (1997) Substrate limitation in the baculovirus expression vector system. Biotechnol Bioeng 56:32–44

Reavy B, Ziegler A, Diplexcito J, Macintosh SM, Torrance L, Mayo M (2000) Expression of functional recombinant antibody molecules in insect cell expression systems. Protein Expr Purif 18:221–228

Reis U, Blum B, von Specht B-U, Domdey H, Collins J (1992) Antibody production in silkworm cells and silkworm larvae infected with a dual recombinant *Bombyx mori* nuclear polyhedrosis virus. Bio/Technology 10:910–912

Rhiel M, Mitchell-Logean CM, Murhammer DW (1997) Comparison of *Trichoplusia ni* BTI-Tn-5B1-4 (High Five) and *Spodoptera frugiperda* Sf-9 insect cell line metabolism in suspension cultures. Biotechnol Bioeng 55:909–920

Sakamoto S, Pongkitwitoon B, Nakamura S, Maenaka K, Tanaka H, Morimoto S (2010) Efficient silkworm expression of single-chain variable fragment antibody against ginsenoside Re using *Bombyx mori* nucleopolyhedrovirus bacmid DNA system and its application in enzyme-linked immunosorbent assay for quality control of total ginsenosides. J Biochem 148:335–340

Schiller JT, Castellsagué X, Villa LL, Hildesheim A (2008) An update of prophylactic human papillomavirus L1 virus-like particle vaccine clinical trial results. Vaccine 26S:K53–K61

Smith GP (1985) Filamentous fusion phage: novel expression vectors that display cloned antigens on the virion surface. Science 228:1315–1317

Smith GP, Petrenko VA (1997) Phage Display. Chem Rev 97:391–410

Song M, Park D-Y, Kim Y, Lee K-J, Lu Z, Ko K, Choo YK, Han YS, Ahn M-H, Oh D-B, Ko K (2010) Characterization of *N*-glycan structures and biofunction of anti-colorectal cancer monoclonal antibody CO17-1A produced in baculovirus-insect cell expression system. J Biosci Bioeng 110:135–140

Sugiura T, Amann E (1996) Properties of two insect cell lines useful for the baculovirus expression system in serum-free culture. Biotechnol Bioeng 51:494–499

Tami C, Peralta A, Barbieri R, Berinstein A, Carrillo E, Taboga O (2004) Immunological properties of FMDV-gP64 fusion proteins expressed on SF9 cell and baculovirus surfaces. Vaccine 23:840–845

Tan W, Lam PHY (1999) Expression and purification of a secreted functional mouse/human chimaeric antibody against bacterial endotoxin in baculovirus-infected insect cells. Biotechnol Appl Biochem 30:59–64

Tanaka T, Takeno T, Watanabe Y, Uchiyama Y, Murakami T, Yamashita H, Suzuki A, Aoi R, Iwanari H, Jiang S-Y, Naito M, Tachibana K, Doi T, Shulman AI, Mangelsdorf DJ, Reiter R, Auwerx J, Hamakubo T, Kodama T (2002) The generation of monoclonal antibodies against human peroxisome proliferator-activated receptors (PPARs). J Atheroscler Thromb 9:233–242

Taticek RA, Choi C, Phan S-E, Palomares LA, Shuler ML (2001) Comparison of growth and recombinant protein expression in two different insect cell lines in attached and suspension culture. Biotechnol Prog 17:676–684

Taticek RA, Hammer DA, Shuler ML (1995) Overview of issues in bioreactor design and scale-up. In: Shuler ML, Wood HA, Granados RR, Hammer DA (eds), Baculovirus expression systems and biopesticides. Wiley-Liss, New York, pp 131–174

Taticek RA, Shuler ML (1997) Effect of elevated oxygen and glutamine levels on foreign protein production at high cell densities using the insect cell – baculovirus expression system. Biotechnol Bioeng 54:142–152

Tomiya N, Narang S, Lee YC, Betenbaugh MJ (2004) Comparing N-glycan processing in mammalian cell lines to native and engineered lepidopteran insect cell lines. Glycoconj J 21:343–360

Wickham TJ, Nemerow GR (1993) Optimization of growth methods and recombinant protein production in BTI-Tn-5B1-4 insect cells using the baculovirus expression system. Biotechnol Prog 9:25–30

Wong KTK, Peter CH, Greenfield PF, Reid S, Nielsen LK (1996) Low multiplicity infection of insect cells with a recombinant baculovirus: the cell yield concept. Biotechnol Bioeng 49: 659–666

Xu X-G, Liu H-J (2008) Baculovirus surface display of E2 envelope glycoprotein of classical swine fever virus and immunogenicity of the displayed proteins in a mouse model. Vaccine 26:5455–5460

Yamaji H, Fukuda H (2007) Display of functional antibody Fab fragment on the baculovirus surface. J Biotechnol 131:S54

Yamaji H, Manabe T, Kitaura A, Izumono E, Fukuda H (2006) Efficient production of recombinant protein in immobilized insect cell culture using serum-free basal media after baculovirus infection. Biochem Eng J 28:67–72

Yamaji H, Manabe T, Watakabe K, Muraoka M, Fujii I, Fukuda H (2008) Production of functional antibody Fab fragment by recombinant insect cells. Biochem Eng J 41:203–209

Yamaji H, Tagai S, Fukuda H (1999) Optimal production of recombinant protein by the baculovirus – insect cell system in shake-flask culture with medium replacement. J Biosci Bioeng 87:636–641

Yang D-G, Chung Y-C, Lai Y-K, Lai C-W, Liu H-J, Hu Y-C (2007) Avian influenza virus hemagglutinin display on baculovirus envelope: cytoplasmic domain affects virus properties and vaccine potential. Mol Ther 15:989–996

Yang J-D, Gecik P, Collins A, Czarnecki S, Hsu H-H, Lasdun A, Sundaram R, Muthukumar G, Siberklang M (1996) Rational scale-up of a baculovirus – insect cell batch process based on medium nutritional depth. Biotechnol Bioeng 52:696–706

Yoshida S, Kobayashi T, Matsuoka H, Seki C, Gosnell WL, Chang SP, Ishii A (2003a) T-cell activation and cytokine production via a bispecific single-chain antibody fragment targeted to blood-stage malaria parasites. Blood 101:2300–2306

Yoshida S, Kondoh D, Arai E, Matsuoka H, Seki C, Tanaka T, Okada M, Ishii A (2003b) Baculovirus virions displaying *Plasmodium berghei* circumsporozoite protein protect mice against malaria sporozoite infection. Virology 316:161–170

Yoshida S, Matsuoka H, Luo E, Iwai K, Arai M, Sinden RE, Ishii A (1999) A single-chain antibody fragment specific for the *Plasmodium berghei* ookinete protein Pbs21 confers transmission blockade in the mosquito midgut. Mol Biochem Parasitol 104:195–204

zu Putlitz J, Kubasek WL, Duchêne M, Marget M, von Specht B-U, Domdey H (1990) Antibody production in baculovirus-infected insect cells. Bio/Technology 8:651–654

Chapter 4
Production of Monoclonal Antibodies in Glycoengineered *Pichia pastoris*

Dongxing Zha, Thomas Linden, and Thomas Potgieter

Abstract Although improvements in antibody expression by mammalian cells are nearing maturation, efforts to improve antibody efficacy through glycoengineering are rapidly expanding. For example, the production of full length monoclonal antibodies with uniform human N-linked glycans by glycoengineered yeast has been used to optimize antibody effector function. The glycoengineered yeast expression platform not only enables elucidation of structure function relationships but also offers a robust and economically viable alternative to mammalian cell expression. This chapter provides an overview of glycobiology, engineering of *P. pastoris* to secrete recombinant proteins with uniform human N-linked glycans as well as bioprocess considerations in the production of full length monoclonal antibodies by a yeast based expression system.

4.1 Introduction

Growth of the diagnostic and therapeutic antibody market has fueled major improvements in mammalian cell expression levels (Beck et al., 2008, 2010, Sheridan, 2010, Walsh, 2010). Yields on the order of 5 g/L are now common and production bottlenecks for some high volume products has shifted to downstream processing (Walsh, 2010, Browne and Al-Rubeai, 2007, Durocher and Butler, 2009, Liu and Downey, 2009). Although efforts to improve upstream productivity may be nearing maturation, approaches to enhance antibody efficacy through glycoengineering is rapidly expanding (Walsh, 2010).

Glycosylation is one of the most common post-translational protein modifications found in natural and recombinant proteins and could impact folding, stability, solubility, bioavailability, trafficking, immunogenicity and functional activity (Li and d'Anjou, 2009). Antibodies are N-glycosylated in the CH2 domain of the Fc fragment and about 30% of circulating human IgGs are also glycosylated in the

D. Zha (✉)
GlycoFi Inc. (A wholly-owned subsidiary of Merck & Co Inc.), Lebanon, NH 03766, USA
e-mail: Dongxing_Zha@merck.com

T. Linden and T. Potgieter
Bioprocess Development, Merck & Co, Rahway, NJ 07065, USA

M. Al-Rubeai (ed.), *Antibody Expression and Production*, Cell Engineering 7,
DOI 10.1007/978-94-007-1257-7_4, © Springer Science+Business Media B.V. 2011

Fab region (Wright and Morrison, 1997, Jeffries, 2007, Raju, 2008). Fc glycans of human IgGs are mainly core fucosylated, complex, biantennary-type glycans with heterogeneity because of the presence or absence of terminal galactosylation and sialylation. It is know that Fc glycosylation of antibodies impacts effector function, half life, immunogenicity and antigen binding (Raju, 2008, Jeffries, 2007, Nimmerjahn and Ravetch, 2008, Wright and Morrison, 1997, Jeffries, 2005, Malhotra et al., 1995).

Due to the importance of glycosylation for antibody efficacy, most recombinant antibodies are produced by mammalian cell culture. Chinese hamster ovary cells (CHO) are the most common expression system for the production of recombinant monoclonal antibodies, although other expression systems such as murine hybridomas (NS/0 and SP2/0), baby hamster kidney cells (BHK) and human cells (HEK293 and PER.C6) are also used (Jones et al., 2003, Petricciani and Sheets, 2008, Durocher and Butler, 2009, Hossler et al., 2009, Walsh, 2010). A notable exception is the αIL-6 monoclonal antibody from Alder Biopharmaceuticals produced by wild type *Pichia pastoris*. ALD518 is a full length aglycosylated, humanized antibody that has advanced to phase II in clinical development (Brief, 2010).

The glycosylation profile of recombinant proteins is not only dependent on the expression system but also on cell culture process conditions. Hossler et al. (2009) reviewed the factors that enable optimal and consistent protein glycosylation in mammalian cell culture (Hossler et al., 2009). Factors such as the cell line, the clone, the manufacturing mode (batch, fed-batch, repeated fed-batch or semi-continuous perfusion), process operating conditions (such as pH, dissolved oxygen level, dissolved CO_2) and media composition influence the composition of glycoforms.

Glycoengineering efforts have attempted to exert control over glycosylation profiles, independent of expression host or culture conditions, in order to exploit carbohydrate-related structure function relationships (Beck et al., 2010). Earlier approaches involved downstream processing to enrich or modify oligosaccharides. For example glucocerebrosidase, a replacement therapy used to treat Gaucher's disease, a rare lysosomal storage disorder, is currently either extracted from placental tissue or produced recombinantly by CHO cells. These products are then treated with an exoglycosidase enzyme as part of the downstream process to expose terminal mannose residues, facilitating direct product uptake by macrophages via cell surface mannose receptors (Barton et al., 1991).

Protein serum half-life and efficacy have been improved by the incorporation of additional glycosylation sites into the protein backbone. For example, with Aranesp®, Amgen added additional N-linked glycosylation sites to erythropoietin to improve its pharmacokinetic profile. Similarly, hyperglycosylated interferon-α displayed an increase in plasma half-life (Ceaglio et al., 2008), whereas hyperglycosylated FSH, enhanced ovulation and embryo maturation in female mice (Trousdale et al., 2009).

Another glycoengineering approach entails the chemical conjugation of presynthesized oligosaccharides to the protein backbone. For example the conjugation of an oligosaccharide with terminal mannose-6-phosphate improves cellular uptake of lysosomal α-glucosidase (Zhu et al., 2008, 2005).

The most direct approach to glycoengineering involves the direct engineering of the glycosylation pathways of producer cell types. For example a CHO cell line (BioWa, Princeton NJ) has been developed which is capable of producing completely afucosylated antibodies, displaying improved ADCC (antibody dependent cell-mediated cytotoxicity) in animal models (Natsume et al., 2009). Similarly, improved ADCC has been demonstrated for antibodies with bisecting afucosylated glycans, produced in genetically engineered CHO cells (Umana et al., 1999).

Recent advances have also been reported in the glycoengineering of plant, bacterial and yeast cells. Glycoprotein expression in plant based systems typically results in oligosaccharides containing xylose and fucose moieties that are immunogenic in humans. Greenovation Biotech (Heilbronn, Germany) has developed a glycoengineered moss (*Physcomitrella patens*) lacking core xylose and fucose transferase activity (Huether et al., 2005). Biolex Therapeutic (Pittsboro, NC USA) has developed an alternative system based on engineered duckweed (*Lemna minor*) in which the endogenous fucosyl and xylosyl transferase activities are inhibited by means of RNA interference (Cox et al., 2006).

Engineering of the N-linked glycosylation system of *Campylobacter jejuni* into *Escherichia coli* made it possible to produce glycoproteins in a prokaryotic expression host (Feldman et al., 2005, Ihssen et al., 2010).

Glycoprotein expression in yeast results in the attachment of mannose-enriched sugar chains without terminal sialic acid which are readily cleared from circulation and can be immunogenic. Recently GlycoFi (a wholly owned subsidiary of Merck & Co.) reported on their glycoengineering efforts in *Pichia pastoris* that enables production of uniformly sialylated glycoproteins. Li and co-workers reported the use of glycoengineered strains of *P. pastoris* to express a functional full length antibody with uniform human N-linked glycans to optimize antibody effector function (Li et al., 2006).

This chapter focuses on the use of glycoengineered *P. pastoris* to produce full length monoclonal antibodies with uniform human N-linked glycans. It includes an overview of glycosylation, followed by a summary of the glycoengineering approach taken to yield a library of host strains each capable of secreting recombinant proteins with a uniform N-linked glycosylation profile. Expression of full length monoclonal antibodies by glycoengineered yeast in a platform cultivation process is realized through the integration of an expression vector into the chromosome of the engineered host strain, followed by high throughput screening and selection. The cultivation process is then further optimized through a multivariate design of experiments approach combined with a fundamental understanding of expression kinetics. Considerations in the development of a purification platform are discussed with emphasis on the impact of high cell density on primary recovery. Finally, physicochemical characteristics of the yeast-produced monoclonal antibody are benchmarked against a CHO-produced antibody with the same amino acid sequence.

4.2 An Overview of Eukaryotic Glycosylation

In glycoproteins, most oligosaccharides are attached via an N-glycosidic bond to asparagine residues or via an O-glycosidic bond to serine or threonine residues. N-linked glycosylation in both fungi and mammals involves attachment of a specific oligosaccharide to the asparagine residue in the consensus sequence Asn-X-Ser/Thr (X represent any amino acid except proline). The assembly of a lipid-linked oligosaccharide ($Glc_3Man_9GlcNAc_2$) followed by the transfer to the nascent protein and the removal of three glucose and one mannose residues to yield a $Man_8GlcNAc_2$ structure is conserved between eukaryotes. The biosynthetic glycosylation pathways diverge between yeast and mammals once the glycoprotein leaves the endoplasmic reticulum (ER) and is shuttled through the Golgi. Yeast and other fungi typically produce high-mannose-type N-glycans by adding up to 100 mannose sugars including beta-linked mannoses and mannosylphosphates whereas the formation of mammalian glycans generally involve removal of mannose followed by the sequential addition of N-acetyl-glucosamine, galactose, fucose and sialic acid.

While the pathway for N-linked glycosylation has been the subject of much analysis, the process and function of O-linked glycosylation in higher eukaryotes is not as well understood. In contrast to N-linked glycosylation, a distinct consensus sequence for O-glycosylation has not been found. In general however this linkage is found in clusters of threonine and serine residues with a β-turn near proline and at a distance from charged amino acids (Spiro, 2002). Furthermore, in addition to the abundant O-GalNAc or mucin type glycans, several other O-linked glycoforms have been reported such as O-linked fucose, glucose, N-acetylglucosamine, xylose, galactose, arabinose and mannose (Endo, 2004). In fungal systems O-mannosylation is common and typically consists of linear chains of α1,2 linked mannose and beta linked mannose that can be modified with mannosyl phosphate (Willer et al., 2003).

Recombinant monoclonal antibodies are typically sparsely O-glycosylated and therefore glycoengineering efforts to improve antibody efficacy have focused mostly on N-linked glycosylation.

4.3 Glycoengineering of *P. pastoris* to Produce Therapeutic Proteins with Uniform Human N-Linked Glycans

4.3.1 Glycoengineering to Modulate N-Linked Oligosaccharide Structures

The history of glycoengineering of yeast cells to enable production of human N-linked glycans have been described by Wildt & Gerngross (Wildt and Gerngross, 2005). Hamilton et al. (2006) reported the production of human glycoproteins with fully sialylated N-glycans (Hamilton et al., 2006). This was achieved by deletion of four genes to eliminate yeast specific glycosylation and introduction of 14 heterologous genes (Hamilton et al., 2006).

Glycoengineering of the yeast pathway starts by knocking out genes responsible for the addition of mannose residues to the $Man_8GlcNAc_2$ structure that exits from the ER. OCH1, an α1,6-mannose transferase is one of the key enzymes that initiates hypermannosylation. Elimination of the OCH1 gene prevents hypermannosylation and creates the substrate required for further processing.

Glycan processing though the Golgi occurs in a sequential order mediated by a set of type II membrane proteins that are arranged so that the substrate for each modification is created by a preceding enzyme and that the product is modified by the next enzyme located along the secretory pathway. Type II membrane proteins consists of 4 main subdomains: cytoplasmic tail, membrane domain, stem region and catalytic domain. Choi et al. (2003) as well as Hamilton et al. (2003) described the use of combinatorial libraries of fusions of leader sequences (cytoplasmic tail, membrane domain and stem region) to catalytic domains to optimize localization and activity of glycosylation enzymes within the secretory pathway (Choi et al., 2003, Hamilton et al., 2003). The combinatorial approach allowed for the selection of host strains capable of producing glycoproteins with a specific human glycoform at high uniformity. For example, a host strain capable of producing a full length monoclonal antibody with more than 90% $Man_5GlcNAc_2$ as Fc glycan was achieved by the co-expression of mannosidases (Potgieter et al., 2009). Similarly the proteins with the hybrid glycan GlcNAcMan5GlcNAc2 can be secreted after optimized expression of GnT1 (Li et al., 2006). The first complex human glycan structure on glycoproteins secreted by yeast was $GlcNAc_2Man_3GlcNAc_2$ achieved by engineering Manosidase II and GnT2 into the secretory pathway (Hamilton et al., 2003). This created the basis for the addition of galactose.

The addition of galactose residues not only required optimization of galactose transferase but also required the conversion of UDP-glucose to UDP-galactose by a recombinant galactose epimerase as well as the transport of UDP-galactose to the Golgi (Bobrowicz et al., 2004). Similarly, the addition of sialic acid residues to glycoproteins secreted by glycoengineered yeast, required synthesis of CMP-sialic acid from UDP-GlcNAc, translocation to the Golgi and transfer onto terminal galactose. (Hamilton et al., 2006).

Glycoengineereing enabled the creation of a library of host strains each capable of secreting recombinant glycoproteins with a specific N-linked glycan at high uniformity. This library can be used to elucidate structure-function relationships for a specific glycoform. Li et al. (2006) illustrated this principle by optimizing the glycosylation profile of rituximab (αCD20) to maximize binding to the FcγRIIIa receptor to enhance ADCC activity (Li et al., 2006).

4.3.2 Glycoengineering to Reduce O-Linked Glycosylation

Antibodies secreted by mammalian cells generally contain no O-linked glycans. Yeast can O-mannosylate aberrant proteins in the ER to increase their solubility and prevent aggregation (Harty et al., 2001, Nakatsukasa et al., 2004). As a result, heterologous proteins which are not O-glycosylated in mammalian cells may be O-mannosylated in yeast (Harty et al., 2001, Nakatsukasa et al., 2004). In addition the structure of O-linked sugars in yeast differ from that of mammalian cells (Goto, 2007, Willer et al., 2003) and it is generally considered that yeast O-linked mannose structures might be immunogenic in humans. As a result, strategies to engineering O-linked glycosylation in yeast focused mostly on elimination of O-linked glycosylation. Kuroda et al. (2008) reported the chemical inhibition of O-mannosyltransferases in yeast (Kuroda et al., 2008). They not only demonstrated that a chemical mannosyltransferase inhibitor could partially reduce O-mannosylation on a yeast secreted antibody but also reported an increase in the amount of assembled antibody in addition to enhanced antigen binding affinity with decreasing O-mannosylation (Kuroda et al., 2008).

4.4 Expression of Full Length Monoclonal Antibodies in Glycoengineered *P. Pastoris*

Figure 4.1 shows a typical full length antibody expression vector. The vector contains three main components:

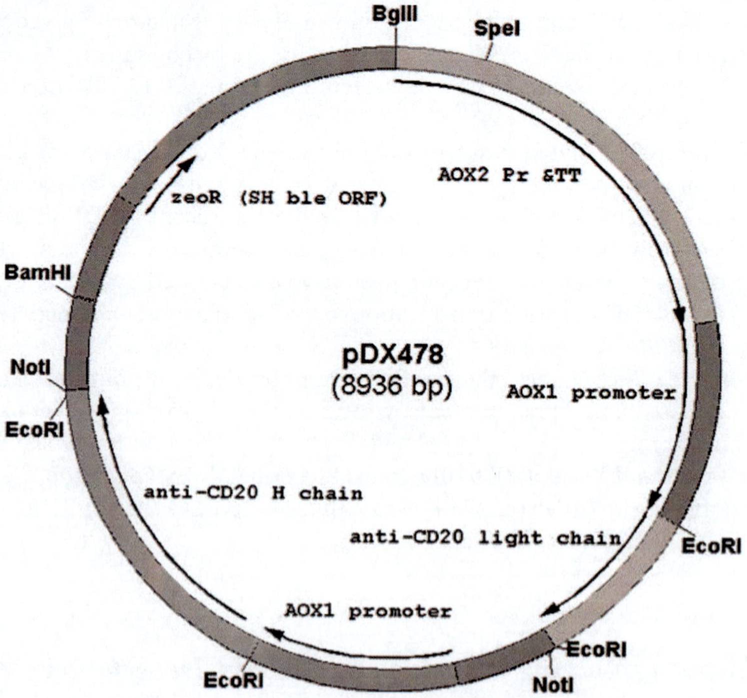

Fig. 4.1 Typical expression vector for the production of a full length monoclonal antibody in glycoengineered yeast. Electroporation after linearization with SpeI allows selection of transformants resistant to Zeocin with the heavy and light chain integrated into the host chromosome into the AOX2 locus, downstream of the AOX1 inducible promoter (from Li et al., 2006)

- Region encoding the heavy and light chain under control of the inducible AOX1 promoter
- A gene encoding for drug resistance to enable selection of transformants
- A fragment of a *P. pastoris* gene such as the AOX2 promoter and terminator to enable integration of the vector onto the chromosome.

The expression vector is integrated into the chromosome by recombination after linearization followed by electroporation. Transformats are isolated as single colonies from agar plates containing the compound for the corresponding resistance marker.

Several host strain modifications have been reported to improve the expression level of full length monoclonal antibodies in yeast. These improvements included (i) increased recombinant gene copy number, (ii) improvement of mRNA stability through optimization of 5′ and 3′ UTRs, (iii) codon optimization, (iv) screening of secretion signal peptides (Gasser et al., 2006), (v) co-expression of folding chaperones (Damasceno et al., 2007) such as immunoglobulin binding protein (BiP) and protein disulfide isomerase (PDI), (vi) protease gene deletions (Brankamp et al., 1995) and (vii) the use of fusion proteins (Reitinger et al., 2008). These improvements combined with optimization of the cultivation platform enabled the production of monoclonal antibodies at expression levels in excess of 1 g/L after only 5 years of development (Potgieter et al., 2009). At this rate of development volumetric antibody productivities from yeast based expression could match or exceed that of mammalian cells in the near future.

4.5 Development of a Cultivation Platform for Expression of Monoclonal Antibodies in Glycoengineered Yeast

Due to the large failure rates (60–80%) of new molecular entities during clinical development, pipeline optimization strategies rely on increased project throughput combined with minimal early development resource investment. Once proof of concept has been demonstrated, development resources are employed to develop the commercial production process and establish the necessary process understanding to ensure reliable manufacturing. Platform development enables learning from earlier programs to be applied to newer programs and improves process understanding during the early development phases in a resource efficient manner.

The glycoengineered yeast cultivation platform, summarized in Fig. 4.2, consists of a structured development path:

(a) strain screening and selection to identify robust, high producing cell lines;
(b) evaluation of these cell lines in a predefined cultivation process, followed by;
(c) generation and characterization of a research cell bank to establish proof of genetic stability;

(d) a systematic application of design of experiments to optimize process conditions (pH, temperature & dissolved oxygen, etc.);
(e) measurement of the antibody expression kinetics to optimize the methanol feeds rate, and;
(f) scale-up (Potgieter et al., 2010, Barnard et al., 2010, Potgieter et al., 2009).

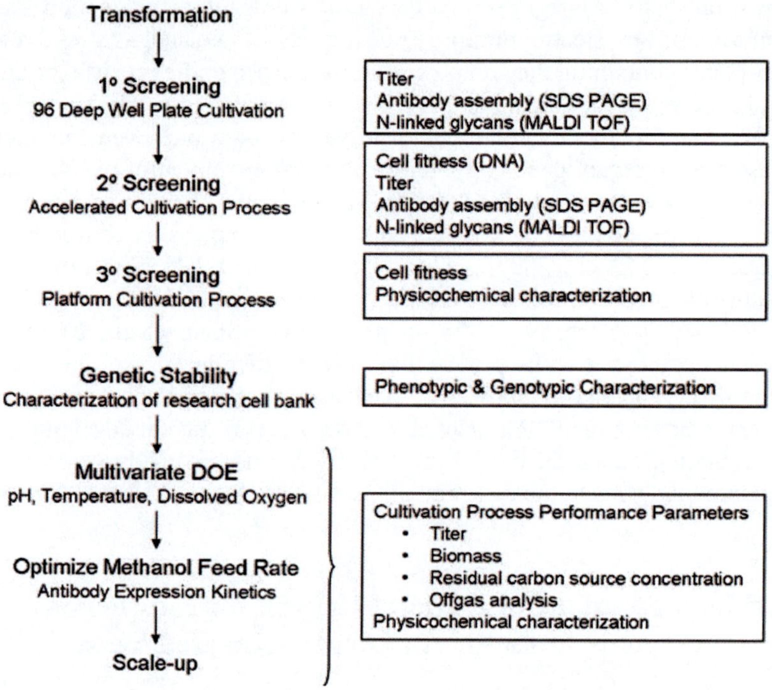

Fig. 4.2 The glycoengineered yeast cultivation platform consists of a structured development path that consists of strain screening and selection to identify robust, high producing cell lines, evaluation of these clones in a predefined cultivation process followed by generation and characterization of a research cell bank to demonstrate genetic stability, a systematic application of design of experiments to optimize process conditions (pH, temperature & dissolved oxygen), measurement of the antibody expression kinetics to optimize the methanol feeds rate and scale-up (Potgieter et al., 2010, Barnard et al., 2010, Potgieter et al., 2009)

4.5.1 Screening and Selection of a Robust Production Strain

Identification of robust high-producing cell lines is essential for reliable therapeutic protein production. High throughput screening methods for *S. cerevisiae* (Holz et al., 2002) and *P. pastoris* clones have been described (Boettner et al., 2002, Boettner and Lang, 2004, Weis et al., 2004). These methods rely on cultivation in well plate systems. However, these systems do not adequately predict performance of clones in bioreactors due to differences in biomass density, pH and dissolved oxygen control. Barnard et al. (2010) described a screening approach that combines deep-well plate and small-scale bioreactor screening methods to isolate glycoengineered *P. pastoris* clones that could be successfully scaled up to bioreactors (Barnard et al., 2010).

The primary screening step relied on cultivation in 96-deep-well plates followed by titer estimation by ELISA and a semi-automated small scale protein A purification to enable determination of N-linked glycan uniformity by mass spectrometry (MALDI-TOF) and antibody assembly by non-reducing SDS-PAGE (Barnard et al., 2010).

Some high yielding strains from the deep-well plate screen would not be developable in bioreactors. Therefore, an accelerated 3-day cultivation protocol was developed to better predict performance of the clones in the platform cultivation process. Clones were selected based on the extent of cell lysis, determined from the DNA concentration in supernatant as well as titer, N-linked glycan uniformity and antibody assembly from protein A purified samples (Barnard et al., 2010).

Figure 4.3 shows an example of screening and selection results for the development of mAb1. During the execution of this project, 13,000 clones were

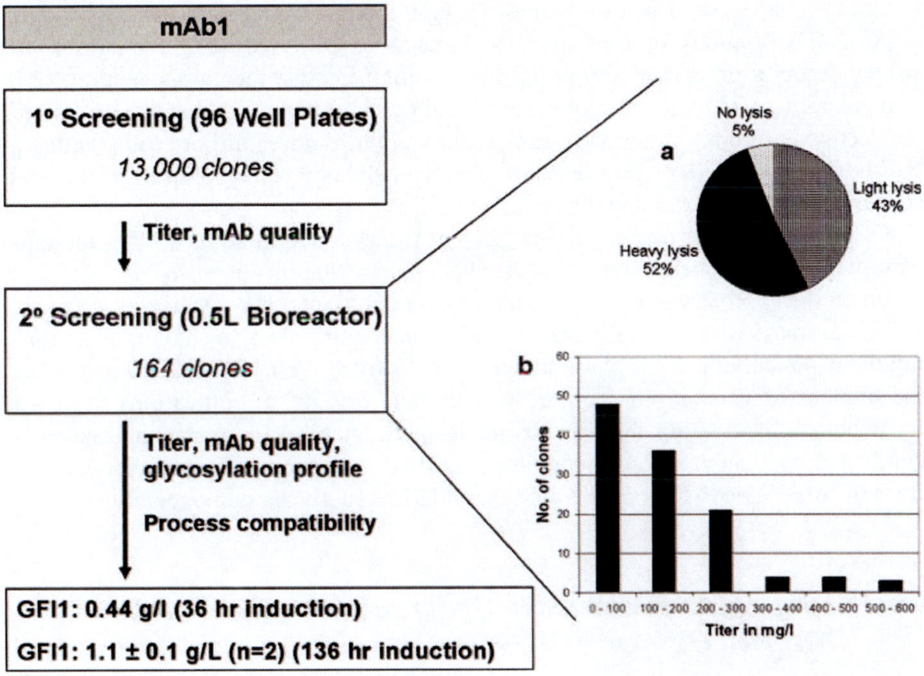

Fig. 4.3 Secondary screening results for mAb1. (**a**) 164 glycoengineered *P. pastoris* clones expressing mAb1 were tested in 0.5-L bioreactors (secondary screening). Of the 164 clones, only 5% showed no degree of cell lysis while 43% showed a light degree of cell lysis (acceptable). However, 52% showed excessive cell lysis (unacceptable). (**b**) The 164 clones were distributed among five final fermentation titer distribution categories and shown as a bar graph. All titers are the titers after protein A purification, normalized to the bioreactor working volume. Only four clones yielded a titer greater than 500 mg/L in the fermentation screening protocol. The titer of 48 clones was not measured as these fermentations were prematurely terminated due to excessive foaming associated with excessive cell lysis. With kind permission from Springer Science+Business Media, Barnard et al. (2010)

screened (Barnard et al., 2010). At the end of primary screening, 164 promising clones were identified for further evaluation in secondary screening and 52% of the expressing clones were discarded due to a significant degree of cell lysis. Most of the clones exhibited very low expression levels. Thus, most clones were unsuitable for antibody production because of low titer and/or low cell viability in cultivation. Clone YGLY4140 was identified as a promising candidate strain expressing mAb1 and cultivated in the platform cultivation process (Potgieter et al., 2009). In the platform cultivation process this clone produced an average titer of 1.26 ± 0.05 g/L (Potgieter et al., 2009).

4.5.2 Production and Characterization of a Research Cell Bank

Once a set of suitable strains are identified, research cell banks (RCB) are created and characterized for purity, viability and authenticity as well as genetic stability as previously described (Potgieter et al., 2009).

Purity is assessed based on methods adapted from Plantz et al. (2003) (Plantz et al., 2003). Viability of the frozen stocks is assessed by counting the number of colony forming units after serial dilution on non-selective media plates after 72 h of incubation at 24. Authenticity is confirmed phenotypically by characterization of the secreted antibody in terms of molecular weight, binding affinity to its antigen, N-linked glycosylation profile and genotypically by PCR amplification and sequencing of the recombinant genes.

Genetic stability refers to the integrity of the production strain during the cultivation process (Schenerman et al., 1999). Genetic stability is tested at 3-L scale based on the approach reported by Schenerman and co-workers (Schenerman et al., 1999). Biomass is sequentially transferred from shaking flask to shaking flask during the exponential growth phase to age the cells by at least 60 generations (double the number of generations required to complete a 2,000 L cultivation) from the RCB before inoculating the bioreactors. Genetic stability is then phenotypically confirmed by comparison of volumetric antibody productivity (after recovery by rProtein A relative to the broth volume) and N-linked glycan homogeneity.

4.5.3 Design of Experiments to Optimize pH, Temperature and Dissolved Oxygen

The cultivation process is further optimized through a multivariate assessment of the impact of pH, temperature and the level of dissolved oxygen on productivity, N-linked glycan homogeneity and antibody assembly. Experiments are typically conduced on 3 L scale and the resultant optimal condition scaled to 30 L.

Figure 4.4 shows an example contour plot of titer as a function of pH and temperature at a dissolved oxygen level of 20% relative to saturation at atmospheric

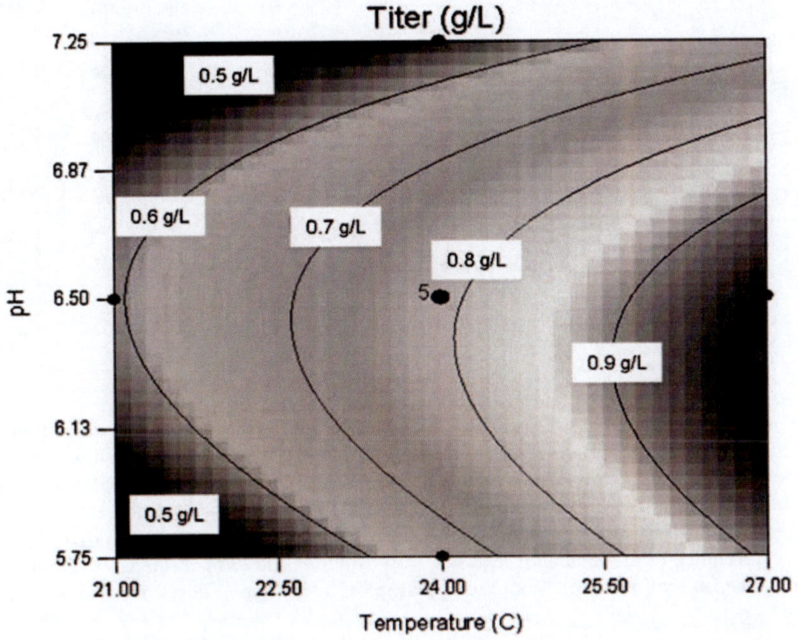

Fig. 4.4 Contour plot generated from a central composite design of experiments showing the impact of pH and temperature on expression level for mAb1. Expression at a pH of 6.2 and a temperature of 27°C shows a 20% improvement over center point conditions of pH 6.5 at a temperature of 24°C. The contour plot indicates that extension of the temperature range above 28°C may yield further improvements in productivity

conditions for mAb1. The optimal condition predicts a 20% improvement over the platform midpoint. This improvement was realized in a subsequent cultivation on 30 L scale (data not shown).

4.5.4 Optimizing Antibody Expression Kinetics by Manipulation of the Methanol Feed Rate

The next development stage involves optimization of the methanol feed rate. The relationship between the observed specific growth rate (μ) of a cellular population and the specific methanol uptake rate (ν_C), the specific oxygen uptake rate (q_O) and the specific productivity (q_P) can be experimentally determined from a series of fed-batch cultivations conducted with different exponential methanol feed rates as described by Potgieter et al. (2010). Figure 4.5 shows an example of a fed-batch cultivation at 3-L scale with a targeted growth rate of 0.01 h^{-1} during induction.

The relationship between specific growth rate and specific productivity is non-linear and not predictable a-priori (Cunha et al., 2004; Kobayashi et al., 2000;

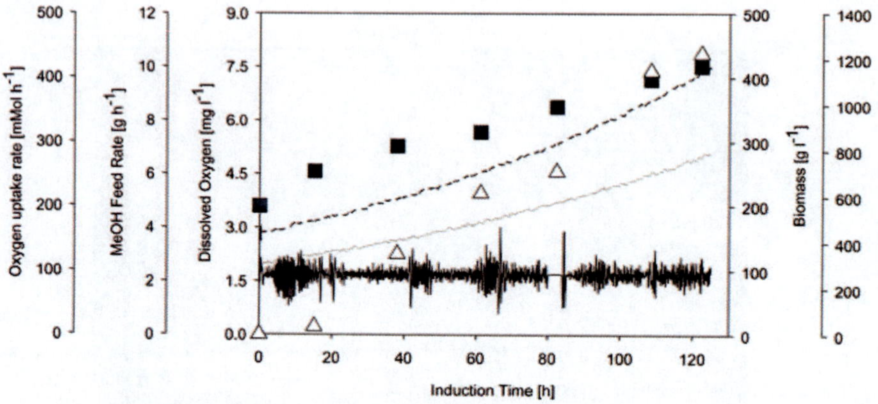

Fig. 4.5 Cultivation done on 3-L scale of YGLY4140 using the bioreactor cultivation protocol at a targeted growth rate of 0.01–1 h. Biomass concentration as measured by wet cell weight in [g/L] as a function of the cultivation time is indicated by the symbol: ■, Antibody titer after capture by rProtein relative to the broth volume in [mg/L] as a function of cultivation time is indicated by the symbol: Δ, The solid line indicates the level of dissolved oxygen in [mg/L]. The *dotted line* (.......) indicates the methanol feed rate in [g/h] while the *dashed line* (——) indicates the oxygen uptake rate in [mmol/h]. With kind permission from John Wiley & Sons, Potgieter et al. (2010)

Schenk et al., 2008; Zhang et al., 2000b). For example, for mAb1, the specific productivity has been modeled as a third order polynomial function in specific growth rate (Equation 1).

$$q_p = 2934\mu^3 - 238\mu^2 + 4.31\mu + 0.019 \tag{1}$$

where q_p is the specific productivity in [mg g WCW^{-1} h^{-1}] and μ is the specific growth rate in [h^{-1}] (Potgieter et al., 2010).

The specific methanol uptake rate and the specific oxygen uptake rates can be modeled as linear functions of the specific growth rate (Equations 2 and 3) (Potgieter et al., 2010).

$$\upsilon_M = \frac{\mu}{Y_{X/M}} + m_{S,M} \tag{2}$$

where υ_M is the specific methanol uptake rate in [g g WCW^{-1} h^{-1}], $Y_{X/M}$ is the yield of biomass from methanol in [gWCW g MeOH^{-1}] and $m_{S,M}$ is the methanol maintenance coefficient in [g g WCW^{-1} h^{-1}] and

$$q_O = \frac{\mu}{Y_{X/O}} + m_{S,O} \tag{3}$$

where q_O is the specific oxygen uptake rate in [mmol g WCW^{-1} h^{-1}], $Y_{X/O}$ is the yield of biomass from oxygen in [gWCW mmol^{-1}] and $m_{S,O}$ is the oxygen maintenance coefficient in [mmol g WCW^{-1} h^{-1}].

For a mAb1 producing strain, the yield of biomass from methanol was found to be 1.36 ± 0.05 g WCW g $MeOH^{-1}$ with a maintenance coefficient of 0.004 ± 0.001 g MeOH g WCW^{-1} h^{-1}. The yield of biomass from oxygen determined from the inverse of the slope of the plot of specific oxygen uptake rate against the specific growth rate was 0.052 ± 0.004 g WCW mmol ${O_2}^{-1}$ with a maintenance coefficient of 0.13 ± 0.02 mmol O_2 g WCW^{-1} h^{-1}.

These correlations can be used in a simple mass balance based model to predict the cultivation performance of carbon-limited cultivations under oxygen-transfer-limited conditions (Potgieter et al., 2010).

4.5.5 Scale-Up of the Cultivation Process

Scale-up of microbial cultivations are typically based on maintaining equivalent gas-liquid mass transfer rates and in particular oxygen transfer rates (OTR) or variables that influence oxygen transfer rate such as power per unit volume or the oxygen mass transfer coefficient (kla). For high cell density cultures, heat transfer can also become limiting upon scale-up (Hensing et al., 1995, Jenzch et al., 2004, Jungo et al., 2007, Schilling et al., 2001). Heat evolution and oxygen consumption rates are closely related during aerobic growth independent of growth rate but depending slightly on the carbon source and the organism involved. Therefore it can be assumed that between 430 and 520 kJ of heat is generated per mole of oxygen consumed (Bailey and Ollis, 1986, Cooney et al., 1968, Jungo et al., 2007). This implies that the heat transfer rate of a bioreactor can be constrained by limiting the oxygen uptake rate. As a result, Potgieter et al., 2009 proposed a scale-up approach for gly coengineered yeast cultivations that was based on maintaining an equivalent level of dissolved oxygen as well as limiting the maximum oxygen uptake rate by limiting the methanol feed rate. Figure 4.6 shows that the biomass profile and the antibody titer have been maintained across the range of scales (0.5–40 L) using this approach.

4.6 Development of a Purification Platform for the Recovery of Monoclonal Antibodies Produced by Glycoengineered Yeast

Downstream processing platforms for purification of monoclonal antibodies from mammalian cell culture have been established to leverage similarities in the biochemical properties and chromatographic behavior of these molecules (Shukla, 2007, Kelley, 2007, Kozlowski and Swann, 2006, Low, 2007, Kelley, 2009). These platforms typically involve cell removal and purification through sequential chromatographic and membrane filtration steps to consistently reduce product and non-product related impurities.

Unlike upstream processing, the use of a downstream platform does not imply the implementation of a templated process for every antibody since even subtle differences in antibody properties can impact their purification behavior (Shukla, 2007).

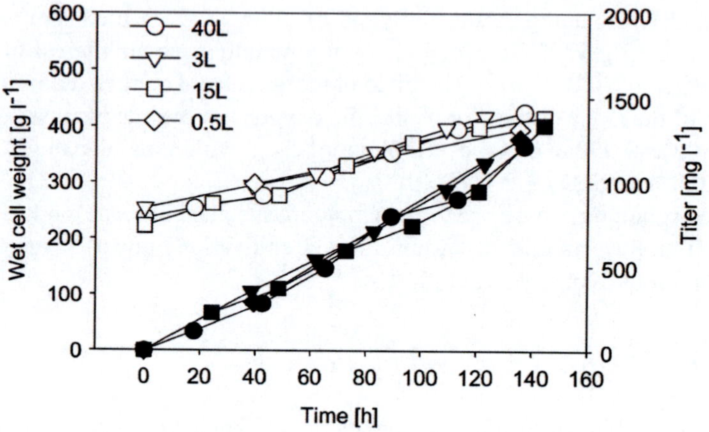

Fig. 4.6 Comparison of biomass concentration in g/L wet cell weight (*open symbols*) and antibody titer in mg/L (*filled symbols*) relative to the broth volume as a function of induction time for cultivations of YGLY4140 done on 0.5, 3.0, 15 and 40-L scale. Reprinted from Potgieter et al. (2009), with permission from Elsevier

Purification platforms strive towards the creation of a common philosophy and alignment over the types of unit operations to include in the downstream process. A schematic diagram for a typical downstream process for purification of antibodies from mammalian cell culture, including other filtration and viral-clearance-oriented process steps, is depicted in Fig. 4.7.

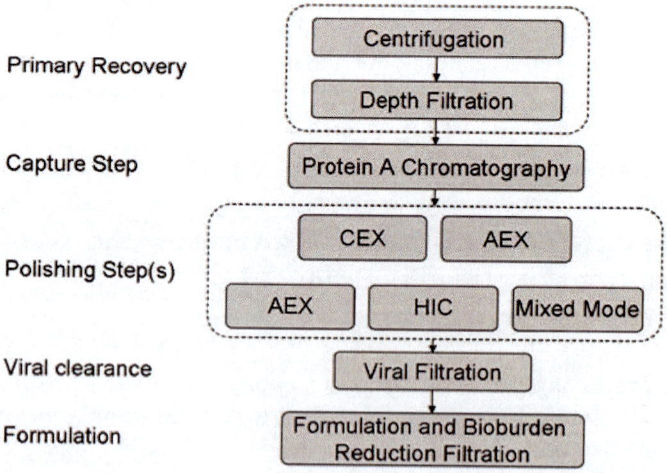

Fig. 4.7 Monoclonal antibody downstream process from mammalian cell culture. The process depicted in this figure starts with cell separation by centrifugation followed by depth filtration, followed by capture with Protein A and includes two subsequent polishing chromatographic steps for impurity removal. (Platform processes utilizing only a single polishing step have also been reported). The process also includes two dedicated orthogonal steps for viral clearance: low pH viral inactivation after Protein A chromatography and viral filtration. The final process step is ultrafiltration/diafiltration (UF/DF) to formulate and concentrate the product

A typical harvest procedure for mammalian cell culture utilizes centrifugation followed by depth and membrane filters to remove cells and cell debris (Shukla and Kandula, 2008). The use of depth filters enable removal of host cell proteins and DNA in addition to particulates (Yigzaw, 2006). Post cell separation, most downstream platforms for antibody purification rely on the use of Protein A affinity chromatography to further remove impurities and yield a >90% improvement in purity in a single step. The polishing steps in the downstream process aim to reduce process and product-related impurities, particularly host cell proteins, DNA and high-molecular-weight aggregates to acceptable levels in the drug substance (Fahrner et al., 1999). Several chromatographic separation technologies have been employed including anion exchange, size exclusion, hydroxyapatite, immobilized metal affinity chromatography (IMAC) (Gagnon, 1995) and cation-exchange chromatography (CEX) in either a bind and elute or in flow-though mode (Fahrner et al., 2001). Fahrner et al. (2001) reported the use of cation-exchange chromatography (CEX) and anion-exchange chromatography (AEX) operated in the flow-through mode (Fahrner et al., 2001). The CEX step cleared host cell proteins, aggregates and leached Protein A while the AEX flowthrough step removed DNA and achieved further reduction in host cell protein impurities. This sequence of steps has been applied as purification scheme for a number of monoclonal antibodies.

Viral clearance is assured through two orthogonal steps; low pH viral inactivation after protein A and viral filtration. In the final step, the drug substance is formulated and filtered to reduce bioburden.

Downstream processing of antibodies from glycoenineered *P. pastoris* follows a similar platform approach with a few notable exceptions (Fig. 4.8):

(a) The feedstock originating from the *P. pastoris* cultivation processes is characterized by high cell content, (>40% wet cell weight). This increases the complexity of the primary recovery step.
(b) A second feedstock characteristic is the high salt concentration resulting in conductivities between 30 and 50 mS cm^{-1} requiring dilution or buffer exchange prior to most chromatographic separation steps.
(c) Viral clearance steps can be omitted for *P. pastoris* cultures, resulting in significant cost savings over mammalian produced antibodies.

Centrifugation and filtration are typically applied for cell separation of high cell density cultures. Jahic et al. (2006) reviewed the different continuous centrifugal separators that can be employed for cell removal (Jahic et al., 2006). Thoemmes et al. (2001) investigated the filterability of a *P. pastoris* suspension in small-scale batch cake filtration and characterized it as problematic (Thoemmes et al., 2001). However, the authors have established preliminary proof of concept for cell removal with crossflow microfiltration (0.45 μm) in a diafiltration mode (non-published results). A flux of 30–50 L m^{-2} h^{-1} and a target protein recovery of >90% were achieved in a process applying five wash volumes of buffer. Similar results were reported by Wang et al. (2006). They compared microfiltration to disk-stack centrifugation (two passes) combined with depth filtration and disk-stack

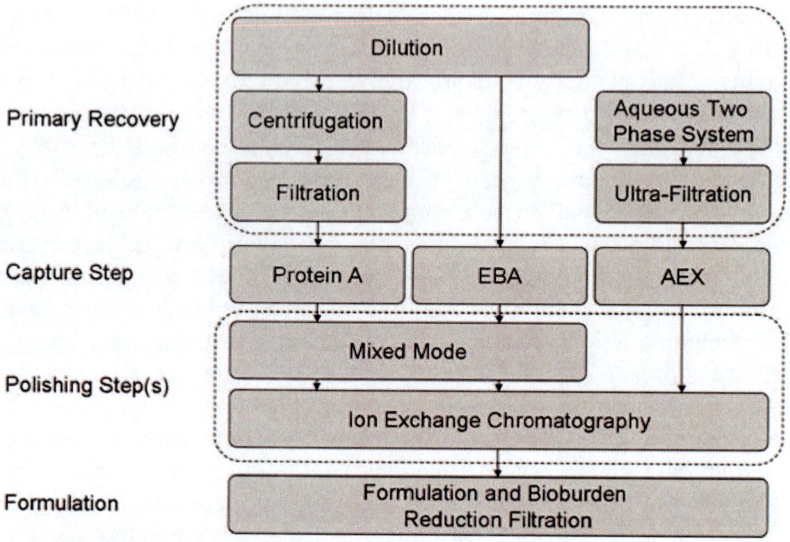

Fig. 4.8 Downstream process options for recovery of monoclonal antibody from high cell density cultures of glycoengineered yeast

centrifugation (one pass) combined with filter-aid enhanced depth filtration (Wang et al., 2006).

The high cell density associated with *P. pastoris* makes the use of integrated approaches which combines solid-liquid separation with initial fractionation more attractive. These integrated approaches include expanded bed adsorption (EBA) and aqueous two-phase partitioning systems (ATPS).

ATPSs can be used to extract monoclonal antibodies from suspension with up to 50% wet weight without compromising capacity or resolution. Furthermore these systems are scaleable and allows for continuous operation (Veide et al., 1984, Kula, 1990, Rosa et al., 2007, Azevedo et al., 2009, Low, 2007). Adoption of this technology has been limited due to a lack of understanding of the mechanism of solute partitioning combined with the lack of experience in terms of installation, validation and operation (Kelley, 2007, Rito-Palomares, 2004, Srinivas et al., 2002, Azevedo et al., 2009, Rosa et al., 2007).

The recombinant antibody can also be captured directly from the high density glycoengineered yeast culture by EBA (Anspach et al., 1999, Hjorth, 1997). EBA typically requires about five times dilution of the culture to reduce the conductivity, resulting in a reduced productivity (Shepard et al., 2000, Trinh et al., 2000, Shepard et al., 2001, Jahic et al., 2006).

Thoemmes et al. (2001) compared ATPS to EBA for recovery of a recombinant protein from *P. pastoris* culture broth (Thoemmes et al., 2001). They concluded that although ATPS required an additional ion exchange step to achieve a similar

Table 4.1 Sample data from the evaluation of different recovery options for the recovery of a monoclonal antibody from high cell density cultures of glycoengineered yeast

Primary recovery option	Purity improvement (%)	Yield (%)	Cycle time (h)
ATPS	>40	96	24
EBA	>70	94	11
Centrifugation & filtration	90	77	50
Centrifugation	–	85	18
Depth filtration	–	97	2
Sterile filtration	<5	98	2
Protein A	90	95	28

ATPS – Aqueous two phase system with PEG/Citrate/NaCl; EBA – Expanded bed adsorption with cation exchange resin from Upfront Chromatography™, Centrifugation & Filtration Process.

purity than EBA, it was more suitable for high cell density culture broths from fed-batch cultivations while EBA was more suited to continuous cultures at lower cell densities.

Linden et al. (2010) presented a comparison of (i) centrifugation followed by filtration and protein A to (ii) EBA technology from Upfront Technologies™ and to (iii) ATPS using a PEG/CitrateNaCl solvent system to purify a monoclonal antibody produced by glycoengineered yeast. The results are summarized in Table 4.1. From these results it is clear that EBA and ATPS offers advantages in both recovery and processing time over the conventional centrifugation and filtration approach (Linden et al., 2010).

4.7 Physicochemical Characterization & Analytical Comparability to An Approved CHO-Produced Therapeutic IgG1

Antibodies produced by glycoengineered yeast is physicochemically comparable to their CHO counterpart as assessed by reduced and non-reduced SDS PAGE, size exclusion chromatograms and antigen binding affinity (Potgieter et al., 2009) (Figs. 4.9 and 4.10). However antibodies derived from CHO cells typically exhibit a heterogeneous mixture of N-linked glycoforms while glycosylation profile of the glycoengineered yeast antibody is more homogeneous. For example analysis of an approved CHO produced IgG1 revealed the following glycosylation profile: 5% G0 ($GlcNAc_2Man_3GlcNAc_2$), 39% fucosylated G0, 10% G1 (GalGlcNAc2Man3GlcNAc2), 40% fucosylated G1 ($GalGlcNAc_2Man_3GlcNAc_2F$) and 6% fucosylated G2 ($Gal_2GlcNAc_2Man_3GlcNAc_2F$). The N-linked glycans from glycoengineered yeast producing the same antibody sequence was highly uniform containing more than 90% $Man_5GlcNAc_2$ (Potgieter et al., 2009).

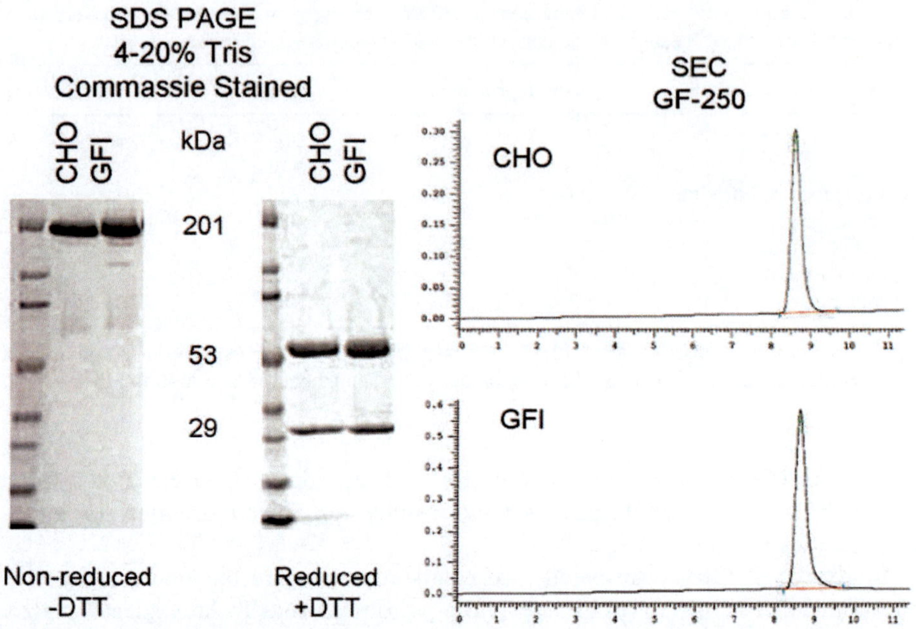

Fig. 4.9 Coomassie blue stained SDS-PAGE under reduced and non-reduced conditions of IgG1 produced by YGLY4140 (GFI) compared to marketed therapeutic IgG1 produced by CHO cells (CHO). The left lane of each gel is a molecular weight standard fromBio-Rad (Catalog #161-0374, Hercules, CA). The non-reduced condition shows intact antibody while the reduced condition shows the heavy and the light chains. Antibody assembly is compared by the size exclusion chromatograms (SEC) from a GF-250 column. Reprinted from Potgieter et al. (2009), with permission from Elsevier

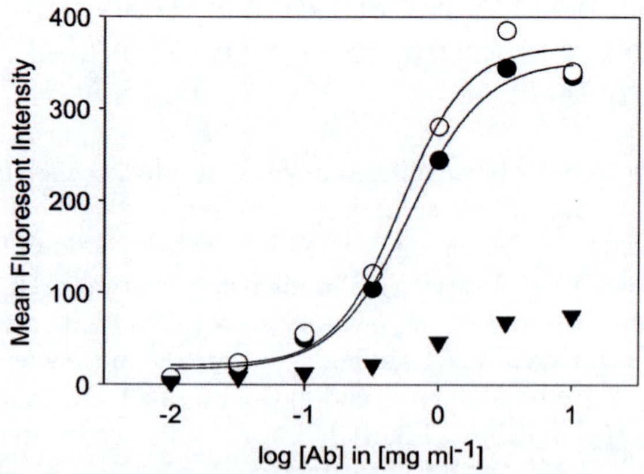

Fig. 4.10 Comparison of antigen binding affinity of glycoengineered yeast and commercially available CHO cell produced IgG1 by mean fluorescence intensity of antibody binding to target antigen expressed on mammalian cell surface. The *filled circles* (●) refers to the IgG1 produced by CHO cells, the *open circles* (○) refers to the glycoengineered yeast strain produced IgG1 while the *filled triangles* (▲) refers to the IgG1 isotype negative control. Reprinted from Potgieter et al. (2009), with permission from Elsevier

4.8 Conclusions

The production of full length monoclonal antibodies with uniform human N-linked glycans by glycoengineered yeast has not only enabled the elucidation of structure function relationships but also offers a robust and economically viable alternative to mammalian cell expression.

References

Anspach, F. B. C., R. Hartmann, et al. (1999). "Expanded-bed chromatography in primary protein purification." J Chromatogr **865**:129–144.

Azevedo, A. M., P. A. J. Rosa, et al. (2009). "Chromatography-free recovery of biopharmaceuticals through aqueous two-phase processing." Trends Biotechnol **27**:240.

Bailey, J., D. Ollis (1986). Biochemical Engineering Fundamentals, New York, NY, McGraw-Hill.

Barnard, G. C., A. R. Kull, et al. (2010). "High-throughput screening and selection of yeast cell lines expressing monoclonal antibodies." J Ind Microbiol Biotechnol **37**:961–971.

Barton, N. W., R. O. Brady, et al. (1991). "Replacement therapy for inherited enzyme defficiency – macrophage targetted glucocerebrosidase for Gaucher's disease." N Eng J Med **324**(May 23, 1991):1464.

Beck, A., O. Cochet, et al. (2010). "GlycoFi's technology to control the glycosylation of recombinant therapeutic proteins." Expert Opin Drug Discov **5**(1):95–111.

Beck, A., T. Wurch, et al. (2008). "Therapeutic antibodies and derivatives: from the bench to the clinic." Curr Pharm Biotechnol **9**(6):421–422.

Bobrowicz, P., R. C. Davidson, et al. (2004). "Engineering of an artificial glycosylation pathway blocked in core oligosaccharide assembly in the yeast Pichia pastoris: production of complex humanized glycoproteins with terminal galactose." Glycobiology **14**(9):757–766.

Boettner, M., C. Lang (2004). "High-throughput expression in microplate format in *Pichia pastoris*." Methods Mol Biol **267**:277–286.

Boettner, M., B. Prinz, et al. (2002). "Highthroughput screening for expression of heterologous proteins in the yeast *Pichia pastoris*." J Biotechnol **99**:51–62.

Brankamp, R. G., K. Sreekrishna, et al. (1995). "Expression of a synthetic gene encoding the anticoagulant-antimetastatic protein ghilanten by the methylotropic yeast Pichia pastoris." Protein Expr Purif **6**(6):813–820.

Brief, B. (2010). "Deal watch: BMS acquires rights for IL-6 inhibitor." Nat Rev Drug Discov **9**:10.

Browne, S. M., M. Al-Rubeai (2007). "Selection methods for high producing cell lines." Trends Biotechnol **25**:425–432.

Ceaglio, N., et al. (2008). "Novel long-lasting interferon alpha derivatives designed by glycoengineering." Biochimie **90**:437–449.

Choi, B. K., P. Bobrowicz, et al. (2003). "Use of combinatorial genetic libraries to humanize N-linked glycosylation in the yeast Pichia pastoris." Proc Natl Acad Sci USA **100**(9):5022–5027.

Cooney, C. L., D. I. C. Wang, et al. (1968). "Measurement of heat evolution and correlation with oxygen consumption during microbial growth." Biotechnol Bioeng **11**:269–281.

Cox, K. M., J. D. Sterling, et al. (2006). "Glycan optimization of a human monoclonal antibody in the aquatic plant Lemna minor." Nat Biotechnol **24**(12):1591–1597.

Cunha, A. E., J. J. Clemente, R. Gomes, F. Pinto, M. Thomaz, S. Miranda, R. Pinto, D. Moosmayer, P. Donner, M. J. Carrondo (2004). "Methanol induction optimization for SCFV antibody fragment production in *Pichia pastoris*." Biotechnol Bioeng **86**(4):458–467.

Damasceno, L. M., K. A. Anderson, et al. (2007). "Cooverexpression of chaperones for enhanced secretion of a single-chain antibody fragment in Pichia pastoris." Appl Microbiol Biotechnol **74**(2):381–389.

Durocher, Y., M. Butler (2009). "Expression systems for therapeutic glycoprotein production." Curr Opin Biotechnol **20**:700–707.

Endo, T. (2004). "Structure, function and pathology of O-mannosyl glycans." Glycoconj J **21**:3–7.

Fahrner, R. L., H. Iyer, et al. (1999). "The optimal flow rate and column length for maximum production rate of protein A affinity chromatography." Bioproc Eng **21**(4):287–292.

Fahrner, R. L., H. L. Knudsen, et al. (2001). "Industrial purification of pharmaceutical antibodies: development, operation, and validation of chromatography processes." Biotechnol Gen Eng Rev **18**:301–327.

Feldman, M. F., M. Wacker, et al. (2005). "Engineered N-linked protein glycosylation with diverse O antigen liposaccharide structures in *Eschercichia coli*." Proc Natl Acad Sci USA **102**(8):3016–3021.

Gagnon, P. (1995). Purification Tools for Monoclonal Antibodies. Validated Biosystems, Vol. 33. Tucson, AZ, Omstead, Biopharm. Int.

Gasser, B., M. Maurer, et al. (2006). "Engineering of Pichia pastoris for improved production of antibody fragments." Biotechnol Bioeng **92**(2):353–361.

Goto, M. (2007). "Protein O-glycosylation in fungi: diverse structures and multiple functions." Biosci Biotechnol Biochem **71**:1415–1427.

Hamilton, S. R., P. Bobrowicz, et al. (2003). "Production of complex human glycoproteins in yeast." Science **301**(5637):1244–1246.

Hamilton, S. R., R. C. Davidson, et al. (2006). "Humanization of yeast to produce complex terminally sialylated glycoproteins." Science **313**(5792):1441–1443.

Harty, C., S. Strahl, et al. (2001). "O-mannosylation protects mutant alpha-factor precursor from endoplasmic reticulum-associated degradation." Mol Biol Cell **12**:1093–1101.

Hensing, M. C. M., R. J. Rouwenhorst, et al. (1995). "Physiological and technological aspects of large scale heterologous protein production with yeast." Anton Leeuw Int J G **67**(3):261–279.

Hjorth, R. (1997). "Expanded-bed adsorption in industrial bioprocessing: recent developments." Trends Biotechnol **15**:230–235.

Holz, C., O. Hesse, et al. (2002). "A microscale process for high-throughput expression of cDNAs in the yeast *Saccharomyces cerevisiae*." Protein Expr Purif **25**:372–378.

Hossler, P., S. F. Khattak, et al. (2009). "Optimal and consistent protein glycosylation in mammalian cell culture." Glycobiology **19**(9):936–949.

Huether, C. M., O. Lienhart, et al. (2005). "Glyco-engineering of moss lacking plant-specific sugar residues." Plant Biology **7**(3):292–299.

Ihssen, J., M. Kowarik, et al. (2010). "Production of glycoprotein vaccines in *Escherichia coli*." Microb Cell Fact **9**:61.

Jahic, M., J. Knoblechner, et al. (2006). "Interfacing Pichia pastoris cultivation with expanded bed adsorption." Biotechnol Bioeng **93**:1040–1049.

Jeffries, R. (2005). "Glycosylation of recombinant antibody therapeutics." Biotechnol Prog **21**: 11–16.

Jeffries, R. (2007). "Antibody therapeutics: isotype and glycoform selection." Expert Opin Biol Ther **7**:1401–1413.

Jenzch, M., M. Lange, et al. (2004). "Bioreactor retrofitting to avoid aeration with oxygen in Pichia pastoris cultivation processes for recombinant protein production." Chem Eng Res Des **82**(A9):1144–1152.

Jones, D., N. Kroos, et al. (2003). "High level expression of recombinant IgG in human cell line PER.C6." Biotechnol Prog **19**:163–168.

Jungo, C., J. Schenk, et al. (2007). "A quantitative analysis of the benefits of mixed feeds of sorbitol and methanol for the production of recombinant avidin with Pichia pastoris." J Biotechnol **131**(1):57–66.

Kelley, B. (2007). "Very large-scale monoclonal antibody purification – the case for conventional unit operations." Biotechnol Prog **23**:995–1008.

Kelley, B. (2009). "Industrialization of mAb production technology – the bioprocessing industry at a crossroads." MAbs **1**:1–10.

Kobayashi, K., S. Kuwae, T. Ohya, T. Ohda, M. Ohyama, K. Tomomitsu (2000). "High level secretion of recombinant human serum albumin by fedbatch fermentation of the methylotrophic yeast, *Pichia pastoris*, based on optimal methanol feeding strategy." J Biosci Bioeng **90**(3):280–288.

Kozlowski, S., P. Swann (2006). "Current and future issues in the manufacturing and development of monoclonal antibodies." Adv Drug Deliv Rev **58**:707–722.

Kula, M. R. (1990). "Trends and future prospects of aqueous two-phase extraction." Bioseparation **1**:181–189.

Kuroda, K., K. Kobayashi, et al. (2008). "Efficient antibody production upon suppression of O mannosylation in the yeast *Ogataea minuta*." Appl Environ Microbiol **74**(2):446–453.

Li, H., N. Sethuraman, et al. (2006). "Optimization of humanized IgGs in glycoengineered *Pichia pastoris*." Nat Biotechnol **24**(2):210.

Li, H., M. d'Anjou (2009). "Pharmacological significance of glycosylation in therapeutic proteins." Curr Opin Biotechnol **20**:678–684.

Linden, T., D. Roush, et al. (2010). Optimization and Scale-up of Downstream Processing of Biopharmaceutical Proteins Produced bv Glycoengineered Pichia Pastoris. Recovery of Biological Products XIV, Lake Tahoe, California, USA.

Liu, C., W. Downey (2009). "Contract manufacturing demands remain strong." GEN **29**:53–59.

Low, D., R. O'Learly, et al. (2007). "Future of antibody purification." J Chromatogr B **848**(1):48–63.

Malhotra, R., M. R. Wormald, et al. (1995). "Glycosylation changes of IgG associated with rheumatoid arthritis can activate complement via the mannose binding protein." Nat Med **1**:237–243.

Nakatsukasa, K., S. Okada, et al. (2004). "Roles of O-mannosylation of aberrant proteins in reduction of the load for endoplasmic reticulum chaperones in yeast." J Biol Chem **279**:49762–49772.

Natsume, A., R. Niwa, et al. (2009). "Improving effector functions of antibodies for cancer treatment: enhancing ADCC and CDC." Drug Des Devel Ther **3**:7–16.

Nimmerjahn, F., J. V. Ravetch (2008). "Fc gamma receptors as regulators of immune responses." Nat Rev Immunol **8**:34–47.

Petricciani, J., R. Sheets (2008). "An overview of animal cell substrates for biological products." Biologicals **36**:359–362.

Plantz, B. A., J. Andersen, et al. (2003). "Detection of non-host viable contaminants in *Pichia pastoris* cultures and fermentation broths." J Ind Microbiol Biotechnol **30**(11):643–650.

Potgieter, T. I., M. Cukan, et al. (2009). "Production of monoclonal antibodies by glycoengineered Pichia pastoris." J Biotechnol **139**(4):318–325.

Potgieter, T. I., S. D. Kersey, et al. (2010). "Antibody Expression Kinetics in Glycoengineered *Pichia Pastoris*." Biotechnol Bioeng **106**(6):918.

Raju, T. S. (2008). "Terminal Sugar of Fc glycans influence antibody effector functions of IgGs." Curr Opin Immunol **20**:471–478.

Reitinger, S., T. Boroviak, et al. (2008). "High-yield recombinant expression of the extremophile enzyme, bee hyaluronidase in Pichia pastoris." Protein Expr Purif **57**(2):226–233.

Rito-Palomares, M. (2004). "Practical application of aqueous two-phase partition to process development for the recovery of biological products." J Chromatogr B **807**(1):3–11.

Rosa, P. A. J., A. M. Azevedo, et al. (2007). "Application of central composite design to the optimisation of aqueous two-phase extraction of human antibodies." J Chromatogr A **1141**(2):50–60.

Schenerman, M. A., J. N. Hope, et al. (1999). "Comparability testing of a humanized monoclonal antibody (synagis) to support cell line stability process validation and scale-up for manufacturing." Biologicals **27**:203–215.

Schenk, J., K. Balazs, C. Jungo, J. Urfer, C. Wegmann, A. Zocchi, I. W. Marison, U. von Stockar (2008). "Influence of specific growth rate on specific productivity and glycosylation of a recombinant avidin produced by a *Pichia pastoris* mut$^+$ strain." Biotechnol Bioeng **99**(2):368–377.

Schilling, B. M., J. C. Goodrick, et al. (2001). "Scale-up of a high cell density continuous culture with Pichia pastoris X33 for the constitutive expression of rh-Chitinase." Biotechnol Prog **17**:629–633.

Shepard, S. R. B., A. Gregory, et al. (2001). "Routine manufacture of recombinant proteins using expanded bed adsorption chromatography." Bioseparation **10**:51–56.

Shepard, S. R., B. Robert, et al. (2000). "Large scale purification of recombinant human angiostatin." Protein Expression Purif **20**:216–227.

Sheridan, C. (2010). "Fresh from the biologic pipeline – 2009." Nat Biotechnol **28**(4): 307–310.

Shukla, A. (2007). "Downstream processing of monoclonal antibodies -application of platform approaches." J Chromatogr B **848**:28–39.

Shukla, A., J. Kandula (2008). "Harvest and recovery of monoclonal antibodies from large-scale mammalian cell culture." Biopharm Int **7**:34–45.

Spiro, R. G. (2002). "Protein glycosylation: nature, distribution, enzymatic formation and disease implications of glycopeptide bonds." Glycobiology **12**(4):43R–56R.

Srinivas, N. D., A. V. Narayan, et al. (2002). "Mass transfer in a spray column during two-phase extraction of horseradish peroxidase." Process Biochem **38**(3):387–391.

Thoemmes, J., M. Halfar, et al. (2001). "Human chymotrypsinogen B production from Pichia pastoris by integrated development of fermentation and downstream processing. Part 2." Biotechnol Prog **17**:503–512.

Trinh, L. N., B. Santosh, et al. (2000). "Recovery of mouse endostatin produced by Pichia pastoris using expanded bed adsorption." Bioseparation **9**:223–230.

Trousdale, R. K., et al. (2009). "Efficacy of native and hyperglycosylated follicle-stimulating hormone analogs for promoting fertility in female mice." Fertil Steril **91**:265–270.

Umana, P., J. Jean-Mairet, et al. (1999). "Engineered glycoforms of an antineuroblastoma IgG1 with optimized antibody dependent cellular cytotoxicity activity." Nat Biotechnol **17**(2): 176–180.

Veide, A., L. Torgny, et al. (1984). "Continuous extraction of â-D-galactosidase from Escherichia coli in an aqueous two-phase system: effects of biomass concentration on partitioning and mass transfer." Enzyme Microb Technol **6**:325–330.

Walsh, G. (2010). "Biopharmaceutical benchmarks 2010." Nat Biotechnol **28**(9):917–924.

Wang, A., R. Lewus, et al. (2006). "Comparison of different options for harvest of a therapeutic protein product from high cell density yeast fermentation broth." Biotechnol Bioeng **94**: 91–104.

Weis, R., R. Luiten, et al. (2004). "Reliable high-throughput screening with *Pichia pastoris* by limiting yeast cell death phenomena." FEMS Yeast Res **5**:179–189.

Wildt, S., T. U. Gerngross (2005). "The humanization of N-glycosylation pathways in yeast." Nat Rev Microbiol **3**(2):119–128.

Willer, T., M. C. Valero, et al. (2003). "O-mannosyl glycans: from yeast to novel associations with human disease." Curr Opin Struct Biol **13**:621–630.

Wright, A., S. L. Morrison (1997). "Effect of glycosylation on antibody function: implications for genetic engineering." Trends Biotechnol **15**:26–32.

Yigzaw, Y. (2006). "Exploitation of the adsorptive properties of depth filters for host cell protein removal during monoclonal antibody purification." Biotechnol Prog **22**:288–296.

Zhang, W. H., M. A. Bevins, B. A. Plantz, L. A. Smith, M. M. Meagher (2000b). "Modeling *Pichia pastoris* growth on methanol and optimizing the production of a recombinant protein, the heavy-chain fragment C of botulinum neurotoxin, serotype A." Biotechnol Bioeng **70**(1):1–8.

Zhu, Y., et al. (2005). "Carbohydrate-remodelled acid alpha-glucosidase with higher affinity for the cation-independent mannose 6 phosphate receptor demonstrates improved delivery of muscles of Pompe mice." Biochem J **389**:619–628.

Zhu, Y., et al. (2008). "Glycoengineered acid alpha-glucosidase with improved efficacy at correcting the metabolic aberrations and motor function deficits in a mouse model of Pompe disease." Mol Ther **17**:954–963.

Chapter 5
Production of Antibodies in *Hansenula polymorpha*

Georg Melmer, Stephan Hellwig, Grit Hehmann, and Ulrike Dahlems

Abstract The increasing aging of the population and the steady progress in understanding molecular processes underlying diseases leads to an ever increasing demand for new drugs. A paradigm shift from small molecules to recombinant proteins results in an escalating demand in regard to diversity as well as amount of such proteins. Recombinant production of protein based pharmaceuticals is still a challenge, although there is a fast progress in this area with many products on the market, yet. The wide variety of potential product candidates including antibodies leads to a steady improvement of existing production platforms and a search for new ones. Antibodies represent a large portion with increasing impact since there are many strategies to use these molecules in treatment of cancer and chronic inflammatory diseases. Besides the well established bacterial and mammalian systems, yeasts are gaining increasing interest as production organisms, since they combine the advantages of both, bacteria and mammal cells. High productivity, use of defined chemical media, cheap production and robustness combined with natural protein modification, devoid of endotoxins and pathogens are characteristic features of yeast systems. While *Saccharomyces cerevisiae* is used for a long time for the production of insulin and *Hansenula polymorpha* for the production of Hepatitis B vaccine, these are only limited success stories. However, we recently succeeded with the expression of antibodies in *H. polymorpha*, which might give the expression system a new boost of applications.

5.1 Recombinant Production of Pharmaceuticals

In the last 3 decades a wide range of recombinant proteins, especially pharmaceuticals, have been produced based on heterologous gene expression in bacterial organisms, mammalian cells and several yeasts and fungi (Gellissen et al. 2005; Melmer 2005; Yin et al. 2007). Production processes had to be developed that employ platforms which meet both, the demand for efficient mass production and criteria of safety as well as authenticity of the produced compounds.

G. Melmer (✉), S. Hellwig, G. Hehmann, and U. Dahlems
PharmedArtis GmbH, Forckenbeckstr. 6, D-52074 Aachen, Germany
e-mail: georg.melmer@pharmedartis.de

M. Al-Rubeai (ed.), *Antibody Expression and Production*, Cell Engineering 7,
DOI 10.1007/978-94-007-1257-7_5, © Springer Science+Business Media B.V. 2011

5.1.1 The Different Expression Systems for Recombinant Production of Pharmaceuticals – Comparison of E. Coli, Yeasts and Mammalian Cells

Prokaryotic expression systems are widely used and lots of recombinant pharmaceuticals have been launched since the first product, insulin (marketed now in different form such as Humulin) was released in 1982 by Genentech (Tof 1994) (News Release Genentech 1978). *E. coli* is by far the most often employed host due to its well known genetics, physiology, productivity and simplicity. The whole genome is sequenced and cultivation in simple, inexpensive media is fast and easy. The cells grow rapidly to high densities and heterologous proteins are expressed in high levels and can account for up to 30% of their total soluble protein.

However, there are serious limitations in using prokaryotic cells for the production of eukaryotic proteins. Many essential proteins undergo a variety of posttranslational modifications like formation of disulfide bridges, glycosylation and phosphorylation, which are often indispensable for proper function. *E. coli* and other prokaryotic hosts have no capacity to glycosylate proteins in a eukaryotic manner. Furthermore, proteins expressed in large amounts in the cell tend to form insoluble aggregates, called inclusion bodies.

To gain access to the active protein, time consuming solubilisation and renaturation procedures have to be performed which in addition often cause a remarkable loss of yield as well as functionality. Lysis of the cells furthermore releases proteases leading to degraded products and the challenge of further purification steps. In addition, endotoxins and pyrogens (LPS) are almost inherent contaminations of the *E. coli* system and for any pharmaceutical application must be removed from the final product. The recombinant protein can be directed into the periplasmic space by using an appropriate leader sequence. In contrast to cytoplasmic expression this enables the isolation of purer and properly folded product, since the periplasmic space contains only few proteins and represents an oxidative environment allowing formation of disulfide bonds.

Mammalian cell cultures, like CHO (Chinese hamster ovary), BHK (baby hamster kidney), cell lines derived from mouse myeloma (NSO) or human embryo kidney (HEK) have gained regulatory approval for recombinant protein production (Chu and Robinson 2001; Wurm 2004) and offer the highest degree of fidelity when mammalian proteins have to be expressed. Therefore they are regarded as the system of choice when absolute authenticity is essential for clinical efficacy. Nevertheless there are some draw-backs, too.

There are differences in glycosylation pattern in rodents compared to humans, and even the use of human cell lines does not guarantee the right glycoslation profile since the transfection event to produce a stable cell line may result in altered glycosylation patterns. It is also well described that the growth media used as well as the proliferation status of the cells can have an influence on the extent as well as the pattern of glycosylation. Mammalian expression techniques are accident-sensitive, time consuming and difficult to perform on a large scale. The media used are highly complex, and expensive. In addition low product yields, compared to microbial systems, are typical. Another major drawback of mammalian cells is the well-documented contamination with oncogenic and/or viral DNA leading to a serious issue of overall safety. Especially this last issue makes many regulatory precautions necessary. This not only prolongs development times, but also is a major cost contribution to the final marketed drug.

With regard to efficient production, safety and authenticity of the produced compounds yeasts offer considerable advantages over alternative microbial and mammalian cell systems. Yeasts are low-cost screening and production systems for authentically processed and modified proteins. They combine several advantageous properties of bacterial and mammalian systems. Yeasts are eukaryotic organisms and have a similar cellular structure and biochemical composition to other eukaryotes. This facilitates post-translational modification of protein products such as disulfide bond formation, endoproteolytic cleavage, glycosylation and multimeric assembly required to produce authentic and bioactive mammalian proteins. Yet, yeasts retain the advantages of unicellular microorganisms with respect to rapid growth and ease of genetic manipulation. They are easy to cultivate and robust compared to mammalian cells. Another important factor is their ability to secrete heterologous proteins into the extracellular environment in basically the same way as higher animal and plant cells. This is preferred mainly because recovery and purification is much easier compared to cytoplasmic localization of the recombinant protein. In addition only proteins that leave the cell via the endoplasmatic reticulum and the Golgi experience the right environment (enzymes and oxidative conditions) for glycosylation and the formation of disulfide bridges. Moreover the methionin normally positioned at the N-terminus of intracellular produced proteins is missing in secreted ones like in the authentic naturally secreted protein.

Yeasts of the genus *Saccharomyces* and *Kluyveromyces* are free of endotoxins and classified as GRAS (Generally Regarded As Safe) microorganisms which is an important factor for the production of pharmaceutical products. Many biomedical products such as hormones, blood proteins and recombinant vaccines have been produced from yeasts and are in current therapeutic uses. The first genetically engineered vaccine licensed by the American Food and Drug Administration (FDA) for administration to humans, Hepatitis B Surface Antigen (HBsAg) was produced in *Saccharomyces cerevisiae* (Valenzuela et al. 1982). Therefore yeasts are a well accepted system for the production of pharmaceutical proteins.

5.2 The Production Platform *Hansenula polymorpha*

Over the years the methylotrophic yeast *Hansenula polymorpha* became a recognized producer of biopharmaceuticals and other recombinant proteins (Barnes et al. 2001; Janowicz et al. 1991; Mayer et al. 1999; Müller et al. 2002; Weydemann et al. 1995). *H. polymorpha* (*Pichia angusta*) belongs to a limited number of yeast species that are able to utilize methanol as a sole energy and carbon source.

5.2.1 *Hansenula polymorpha* Host Strains

Two out of three basic strains with unclear relationships, different features, and independent origins are biotechnologically applied: strain CBS4732 (CCY38-22-2; ATCC34438, NRRL-Y-5445) and DL-1 (NRRL-Y-7560; ATCC26012) and auxotrophic derivatives thereof (Kang and Gellissen 2005). The entire genome of strain CBS4732 has been sequenced (Ramezani-Rad et al. 2003) and a derivative of this strain is a well established host for the production of successfully marketed recombinant proteins like hepatitis B surface antigen, hirudin, phytase, saratin and interferon alpha 2a (Barnes et al. 2001; Janowicz et al. 1991; Mayer et al. 1999; Müller et al. 2002; Weydemann et al. 1995) (Table 5.1).

5.2.2 Promoters for Recombinant Gene Expression

Heterologous gene expression in *Hansenula polymorpha* is linked to strong and regulable promoters derived from genes of the methanol utilization pathway (Hartner and Glieder 2006), most commonly elements derived from the alcohol oxidase and the formiate dehydrogenase gene are used.

Figure 5.1 show the conversion of methanol, which starts in the peroxisome.

The enzyme components of this pathway and their control have been reviewed extensively in the recent past (Hartner and Glieder 2006; Yurimoto et al. 2002). The genes of this pathway are described to be tightly regulated; they are highly repressed in the presence of non-limiting concentrations of glucose and strongly induced if methanol is used as a carbon source (Hartner and Glieder 2006). Methylotrophic growth is furthermore accompanied by a massive proliferation of peroxisomes in which several methanol-metabolizing enzymes are compartmentalized (Klei and Veenhuis 2002; Yurimoto et al. 2002).

However, it soon became evident that activation of methanol pathway promoters did not depend on the presence of methanol in *H. polymorpha* in contrast to the situation in the other methylotrophic yeasts (Gellissen 2000). For all other methylotrophic yeast species an inductive activation of such promoters has been stated that is strictly dependent on the presence of methanol (Klei and Veenhuis 2002) which is critical with respect to safety issues. As a consequence several *H. polymorpha*-based industrial fermentation processes have been defined that lean on glucose or glycerol supplementation in suitable concentrations to a culture broth without any methanol additions (Gellissen 2000).

5.2.3 Selection System

Promoters for inducible or constitutive expression are available (Amuel et al. 2000; Janowicz et al. 1988). The gene of interest is cloned between the promoter and the MOX-terminator using a multi cloning site and standard techniques (Fig. 5.2).

Table 5.1 Selection of *H. polymorpha* host strains

Strain	Genotype	Phenotype	Source
Parental strain			
DL-1	Wild-type (NRRL-Y-7560, ATCC26012)		Levine and Cooney (1973)
Auxotrophic strains			
DL-1-L	*leu2*	Leu⁻	Sohn et al. (1996)
uDL10	*leu2 ura3*	Leu⁻ Ura⁻	KRIBB
DL-LdU	*leu2 Δura3::lacZ*	Leu⁻ Ura⁻	KRIBB
DL1Δ-A	*leu2 Δade2*	Leu⁻ Ade⁻	CRC
DL1Δ-L	*Δade2 Δleu2::ADE2*	Leu⁻	CRC
DL1Δ-U	*leu2 Δade2 Δura3::ADE2*	Leu⁻ Ura⁻	CRC
Protease-deficient strains			
uDLB11	*leu2 ura3 Δpep4::lacZ*	Leu⁻ Ura⁻ Pep4⁻	KRIBB
uDLB12	*leu2 ura3 Δprc1::lacZ*	Leu⁻ Ura⁻ Prc1⁻	KRIBB
uDLB13	*leu2 ura3 Δkex1::lacZ*	Leu⁻ Ura⁻ Kex1⁻	KRIBB
uDLB14	*leu2 ura3 Δpep4::lacZ Δprc1::lacZ*	Leu⁻ Ura⁻ Pep4⁻ Prc1⁻	KRIBB
uDLB15	*leu2 ura3 Δpep4::lacZ Δkex1::lacZ*	Leu⁻ Ura⁻ Pep4⁻ Kex1⁻	KRIBB
uDLB16	*leu2 ura3 Δprc1::lacZ Δkex1::lacZ*	Leu⁻ Ura⁻ Prc1⁻ Kex1⁻	KRIBB
uDLB17	*leu2 ura3 Δpep4::lacZ Δprc1::lacZ Δkex1::lacZ*	Leu⁻ Ura⁻ Pep4⁻ Prc1⁻ Kex1⁻	KRIBB
Parental strain			
CBS4732	Wild-type (CCY38-22-2, ATCC34438, NRRL-Y-5445)		Morais and Maia (1959)
Auxotrophic strains			
LR9	*ura3-1 (odc1)*	Ura⁻	Roggenkamp et al. (1986)
RB11	*ura3-1*	Ura⁻	Weydemann et al. (1995)
RB12	*ura3 leu1-1*⁻	Ura⁻ Leu⁻	Rhein Biotech, unpublished
RB13	*ura3 leu1-1*[a] *mox*	Ura⁻ Leu⁻ Mox⁻	Rhein Biotech, unpublished
RB14	*ura3 mox*	Ura⁻ Mox⁻	Rhein Biotech, unpublished
RB15	*leu1-1*[a] *mox*	Leu⁻ Mox⁻	Rhein Biotech, unpublished
RB17	*haro7*	Tyr⁻	Krappmann et al. (2000)
RC296	*ade*	Ade⁻	Rhein Biotech, unpublished
A16	*leu2 trp3 mox*	Leu⁻ Trp⁻ Mox⁻	Veale et al. (1992)
1B	*ade2-88 leu2-2*	Ade⁻ Leu⁻	Bogdanova et al. (1998)
1-HP065	*ade2-88 ura2-1 met4-220*	Ade⁻ Leu⁻ Met⁻	Mannazzu et al. (1997)
14C	*leu2-2 cat1-14*	Leu⁻ Cat⁻	Lahtchev (2002)
5C-HP156	*leu2-88*	Ade⁻	Lahtchev (2002)
8 V	*leu2*	Leu⁻	Agaphonov et al. (1995)

[a]*leu1-1* and *leu2* correspond to the same gene.
From: Kang and Gellissen (2005, pp. 111–142).

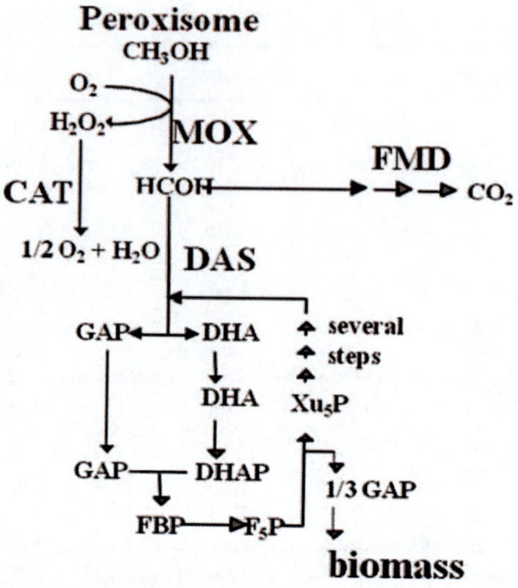

Fig. 5.1 Methanol utilization pathway in *Hansenula polymorpha*. Methanol is oxidized by methanol oxidase (MOX) to generate formaldehyde and hydrogen peroxide which is decomposed by catalase (CAT) to water and oxygen. The formaldehyde is oxidized to carbon dioxide in the cytoplasm by two subsequent dehydrogenase reactions, the latter one a formate dehydrogenase (FMD). For biomass generation the formaldehyde reacts with xylulose-5-phosphate (Xu_5P) by the action of dihydroxyacetone synthase (DAS) to form glyceraldehyde-3-phosphate (GAP) and dihydroxyacetone (DHA). DHA is phosphorylated to dihydroxyacetone-phosphate which, in reaction with GAP, generate fructose 1,6-bisphosphate (FBP). In further steps of the pentose phosphate cycle, fructose-5-phosphate and xylulose-5-phosphate are finally generated

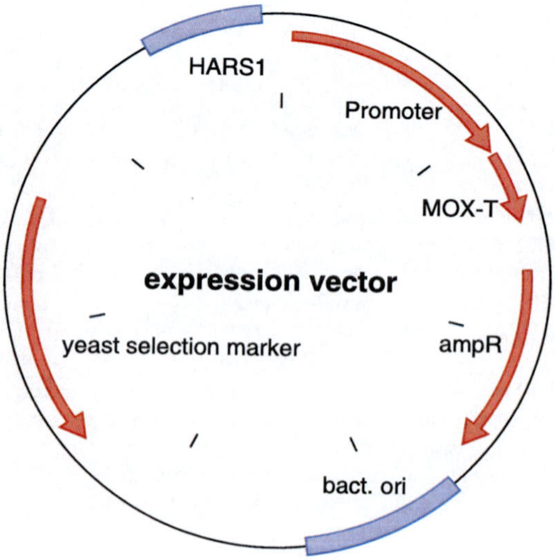

Fig. 5.2 General design of an *H. polymorpha* integration/expression vector. The elements characterizing a *H. polymorpha* expression vector are: promoter, inducible (e.g. MOX, FMD) or constitutive (TPS1); MOX-terminator (MOX-T); *Hansenula* autonomous replication site (HARS1); *E. coli* autonomous replication site; Marker gene for yeast (e.g. URA3, LEU2); Marker gene for *E. coli* (ampR)

Selection in *E. coli* is facilitated by the gene for ß-lactamase. The orotididine-5′-phosphate (ODC1) gene (URA3) or the ß-isopropyl malate dehydrogenase (Leu2) gene from *S. cerevisiae* are frequently used as selection markers in yeast. However, nowadays vectors are available without any bacterial sequence, but still allowing selection in such cases by using pyrF⁻ *E. coli* strains, which can be complemented by the yeast derived ODC1 gene. This offers the important opportunity to generate stable recombinant *Hansenula* strains without any DNA sequences coding for resistance to antibiotics in their genome. This is an indispensable prerequisite for the production of food additives and might also add safety to the production of biopharmaceuticals and the health care system. Food additives, orally taken, are suspected to transfer resistance to antibiotics to the microbial flora of individuals.

A hexose oxidase derived from a red alga (*Chindrus crispus*) was successfully expressed using the adapted vector and the product launched recently (Cook and Thygesen 2003; Smith and Olempska-Beer 2004).

5.2.4 Transformation/Integration

Strain generation is a simple but time consuming process since a selection of individual transformants has to be cultivated under selective pressure for at least 80 generations (passaging) before the cells can be stabilized. Stabilization in addition needs 20–30 generations of growth in complete medium. This input of time and work pays off since the resulting multi-copy strains are absolutely stable without further application of any selective pressure (Gatzke et al. 1995) (Fig. 5.3).

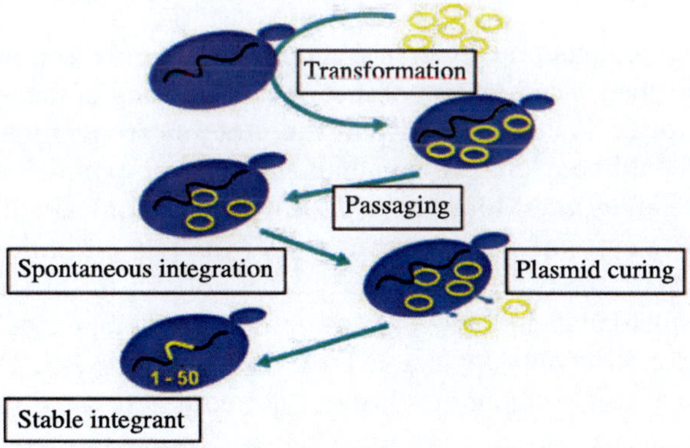

Fig. 5.3 Preparation of stable recombinant *H. polymorpha* strains. *Hansenula* cells are transformed with intact plasmid. The transformed cells are kept under constant selection pressure by continuous cultivation in minimal medium with regular inoculation of fresh medium (passaging). During this step the plasmid is amplified and at a certain time (which is different for all individual clones) integrated into the genome. Shift to complete medium induces the loss of the non-integrated plasmid copies (plasmid curing), whereas the integrated ones become an absolutely stable part of the chromosome. The process can theoretically be repeated as often as additional markers for transformation are available

The integration is often observed to occur into homologous regions of the genome (FMD-promoter in case of HBsAg, Rhein Biotech unpublished results). The plasmid copies are clustered and orientated in a head-to-tail arrangement. By introduction of a portion of the rDNA into the plasmid and linearization within this region prior to transformation the integration is directed into the rDNA of the cell (Klabunde et al. 2002).

5.2.5 Increase of Productivity

Hansenula polymorpha can be transformed with intact or linearized plasmids. The transformed DNA is amplified during passaging and integrates at a certain time into the genome. Upon depletion of selective pressure the non-integrated DNA copies are no longer replicated and finally degraded. Consequently, a set of stable transformants with different copy numbers is generated. Up to 60 copies can be integrated in one step. Theoretically this process can be repeated as often as additional selection markers are available. This offers the possibility to increase the copy number of a certain heterologous gene and by this the gene dosage dependant strength of expression.

5.2.6 Co-Expression – Use of Helper Genes

Even more valuable is the fact that a second or a third gene can be introduced. Due to the variety of the resulting transformants an optimal ratio of expression of these genes can be accomplished. This offers the chance to build up a multi-stage process in a single cell and further offers the opportunity for optimal co-expression of helper proteins.

In this way the production of secreted and correctly processed interferon-alpha could be accomplished. The problem of incorrect processing of the leader sequence which was due to the low accessibility of the protease recognition side by sterical interference could be overcome by simultaneous over-expression of the KEX2 protease from *S. cerevisiae* (Müller et al. 2002). The N-terminal amino acid of the mature protein is a cytosine and is part of one of the two disulfide bridges of the molecule. This makes the N-terminus poorly accessible for the processing protease (Fig. 5.4) and resulted in an inhomogeneity of product, several bands of slightly different length where observed. Upon co-expression of the *S. cerevisiae* derived Kex2 protease in an appropriate amount exclusively correctly processed interferon could be produced.

A further example is the conversion of glycolate oxidase to glyoxylate, an enzymatic reaction which produces hydrogen peroxide as a side product, which is toxic for the cell. This problem could be solved by co-expression of a catalase from *S. cerevisiae* in a suitable amount. A robust production process could be established in which the generated hydrogen peroxide is continuously degraded and intoxication of the cell avoided (Gellissen et al. 1996). The twice transformed cells were used as whole-cell biocatalysts for the described conversion of glycolic acid. Even though

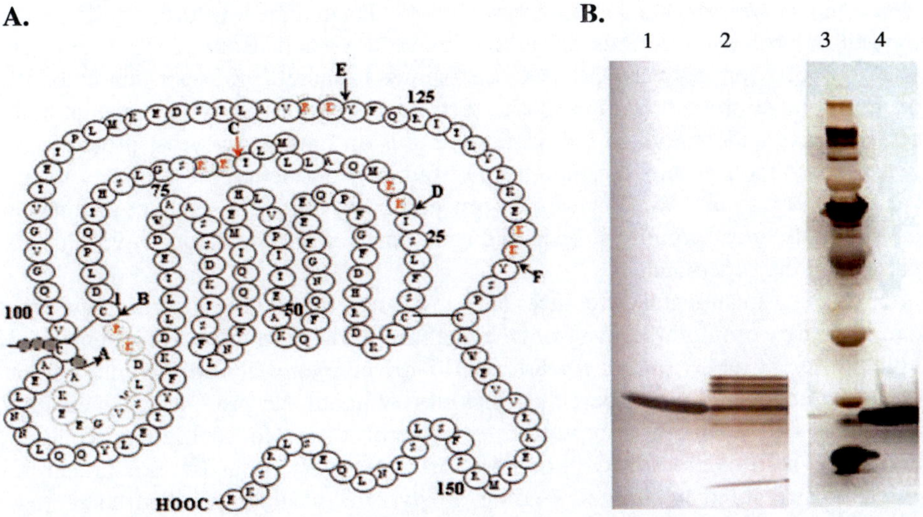

Fig. 5.4 Protein co-expression effecting product processing. **(a)** The interferon alpha molecule, **(b)** silver stained polyacrylamide gels, *1*: Interferon standard, *2*: Recombinant interferon produced without co-expression of Kex2, *3*: Mol. Weight marker, *4*: Recombinant interferon produced with co-expression of Kex2

the cells have to be permeabilized for use they can be stored at –80°C without loss of activity and can be re-used three to four times in the production process. This additionally demonstrates the exceptional robustness of the system.

Further applications allow for the production of mixed structures in a natural composition from one cell in a single process. The production of mixed HBsAg particles carrying L- and S-type surface antigens of the hepatits B virus on its surface is an impressive demonstration of this feature of the *Hansenula polymorpha* expression system (Janowicz et al. 1991). The co-expression of genes in one cell additionally offers the chance to solve expression problems which are caused by limitations of the secretion machinery of a cell, when the expression level overstrains the capacity of the secretory pathway, which leads to aggregation of the proteins in the ER. Those limitations can for example be overcome by overexpression of the protein disulfide isomerase (PDI) in the production cell. This enzyme is located in the ER and ensures the proper folding of proteins by detection of incorrectly folded protein, solving of disulfide bridges and properly refolding the protein. This could be of outstanding importance for proteins with many disulfide bridges like e.g. BSA or HSA.

5.2.7 Production by Fermentation

Hansenula polymorpha is, similar to its close relative *Pichia pastoris*, a very robust organism that can be grown in a wide pH and temperature range on defined media. *Hansenula* tolerates temperatures up to 50°C and pH values as low as 2.5.

Hansenula polymorpha and *Pichia pastoris* both also tolerate low levels of dissolved oxygen or prolonged periods of carbon source depletion. Both yeasts feature an effective secretion pathway and have been shown to secrete gram-per-liter amounts of recombinant protein. Because both yeasts do not excessively break even at high cell densities and prolonged cultiviation periods and not many yeast proteins are secreted into the medium, the supernatant from these fermentations provides a convenient starting point for downstream processing in which the product of interest is usually the most prominent band and sometimes can account for 75–95% of all proteins in the supernatant.

However, fundamental differences exist in the way that expression of foreign proteins driven by promotors taken from the methanol utilization pathways is controlled in the different methylotrophic yeasts. *AOX1*-driven expression in *Pichia* does not occur beyond what would be called leakiness without the presence of methanol as inducer and seems to be repressed by glycerol at low concentrations (Hellwig et al. 2001). In contrast, MOX- or FMD-driven expression in Hansenula can be efficiently triggered by limited feeding of glycerol, a strategy called "derepression". Thus, *Hansenula polymorpha* fermentation strategies can completely avoid methanol-fed-batch phases and technical implications that come with the use of a toxic and flammable liquid in production facilities. It should also be noted that the physiological implications of growth on methanol as the only carbon and energy source are dramatic. With no building blocks other than methanol, the yeast cell has to channel every atom used in the catabolic and anabolic pathways through the peroxisomal pathway. Consequently, the process of adaptation to methanol utilization takes 3–6 h in which the process must be closely monitored.

5.2.7.1 Media and Fermentation Strategies

The defined medium Syn 6 (Gellissen 2000) is the most widely used synthetic medium for Hansenula polymorpha fermentations. This medium contains all basal salts to support growth at high cell densities. The original fermentation strategy is to grow the cells using glycerol as a carbon source in a batch phase, to boost the cell mass to a desired density in a fed-batch phase, and finally to induce expression of the recombinant gene driven by the MOX or the FMD promoter (Hodgkins et al. 1993; Jenzelewski 2002; Kang et al. 2001).

Using the initial concentration of 0.12 mol L^{-1} of phosphate in the basal salts formulation as a benchmark, Syn6 is approximately one third of the concentration of the synthetic defined medium recommended for *Pichia* and approximately twice as concentrated as salt-reduced media that have been used for *Pichia* (Hellwig et al. 2001) – however, all of these media have been used to obtain high cell densities in the range of 70–100 g DW/L. Although from a regulatory point of view, addition of complex compounds such as Yeast extract or peptones is not desirable, it has been reported for at least one gene of interest that product quality was compromised in defined media compared to complex media (Degelmann et al. 2002). For *Pichia*, the most abundantly used promoter is AOX1, while for *Hansenula polymorpha*, MOX and FMD promoters have been used more or less equally. The TPS1 promoter which is derived from the trehalose phosphate synthase 1 gene, coding for a heat shock protein, draws more and more attention, because it offers the possibility of strong and constitutive expression (Amuel et al. 2000).

In all cases, production by fermentation usually relies on a batch phase using glycerol as the carbon and energy source followed by a glycerol fed-batch phase in which biomass is ramped up to high-cell densities and the induction phase, in which either glycerol is fed at limiting rates or methanol or both are fed. The actual biomass level that is targeted prior to induction or derepression is largely depending on the limitations of the bioreactor used in providing adequate mixing, oxygen transfer and cooling. Generally, the system should be able to deliver an OTR of higher than 50 mmol L^{-1} h^{-1} and be able to sufficiently agitate a fermentation broth containing greater than 25% of solids. The recommendation that dissolved oxygen levels should be kept higher than 20% should be verified for each production strain, since the methylotrophs will not produce large amounts of ethanol through fermentative pathways and it has been shown several times that productivity was actually higher in oxygen-limited cultures. Very recently, a study on cultivation of *Pichia* in different oxygen conditions was presented that suggests an influence of hypoxic conditions on the membrane composition and thus the secretion mechanism that could stimulate productivity.

With respect to fermentation and control strategies, two simple and efficient modes of controlling the glycerol or methanol feed rate in *Hansenula* fermentations should be mentioned. A non-limiting glycerol feed, the rate of which is self-adjusting to the culture growth, can be realized by mixing ammonia and glycerol in a fixed stoichiometric ratio (Stockmann et al. 2003) and using this solution to control the pH. A limiting feed of glycerol or methanol can be obtained by using the feed as the actuating variable in a dO_2-control circuit: At high cell densities, the dO_2 spike upon carbon source depletion is fast enough to use it as the control variable for the feed pump. Using careful parameterization of the control circuit, the pump will feed glycerol or methanol in a way that a limiting concentration is not exceeded on the one hand side and carbon source depletion does not occur long enough to cause physiological changes (Fig. 5.5).

Derepression of MOX- and FMD-driven expression in *Hansenula polymorpha* without the addition of methanol can not only be achieved by limiting glycerol fed-batch strategies, but also by limiting glucose concentrations, In the context of methylotrophic yeasts this is an outstanding property of *Hansenula*. In limiting glucose fed-batch fermentations, no glucose must be measurable in the supernatant. Using this strategy, identical productivities can be reached as with limited feed of glycerol, thus offering an additional economical benefit over the use of glycerol as the sole carbon source (Mayer et al. 1999).

5.2.8 Secretion Capacity

Secretion and processing of proteins in yeasts is similar to that in higher eukaryotic cells, giving yeasts a major advantage over bacteria as potential expression systems

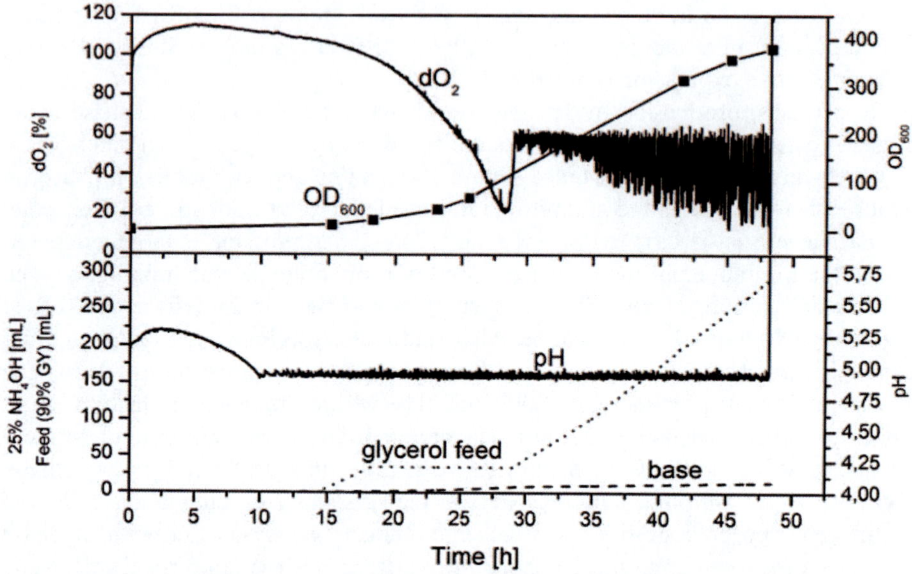

Fig. 5.5 Scheme for fermentation using the derepression mode in *Hansenula polymorpha*

for secretory proteins. Moreover, *Hansenula polymorpha* secrets only very low levels of endogenous proteins. However, as protein secretion is a complex process performing post translational modifications such as glycosylation and proteolytic cleavage, several problems may be unexpectedly encountered during the development of a production system involving heterologous protein secretion. To improve the quantity and quality of the secretory recombinant proteins produced from yeasts, various strategies can be applied including:

- Changing gene dosage
- Optimizing the expression cassette
- Molecular manipulation of host strains
- Adjusting medium and growth conditions.

The robustness and power of the system can impressively be illustrated by the production of phytase which is used as an additive in animal feed. Phytase is a phosphatase that degrades phytic acid, a storage compound for phosphate in plant cells. Non-ruminant animals that are fed with cereals are not able to degrade phytase and excrete it into the environment which leads to pollution problems.

To solve this problem a consensus phytase designed by combining several fungal phytase sequences and using also a fungal consensus signal sequence for secretion was used for expression in *Hansenula polymorpha* (Mayer et al. 1999). The resulting recombinant *Hansenula* cell secreted high amounts of phytase into the supernatant. A powerful fermentation process was established based on the use of

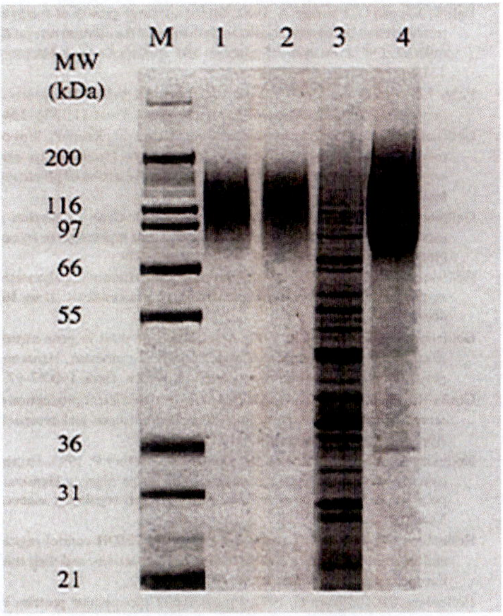

Fig. 5.6 Product characterisation by SDS-PAGE, *M*: molecular weight marker, lane *1*: purified product (phytase), lane *2*: fermentation sample, supernatant, lane *3*: fermentation sample, soluble fraction of cells, lane *4*: fermentation samples, supernatant: lane overload to visualize low levels of other secreted proteins

glucose as a sole carbon source resulting in 13.5 g L^{-1} of phytase directed into the medium. Since *Hansenula polymorpha* secretes very few endogenous proteins consequently the purity of the recombinant protein was already 95% before any purification step was applied. This is shown in Fig. 5.6.

The process could easily be scaled up to fermentation volumes of 5,000 L without any loss of productivity.

5.2.9 Glycosylation in Hansenula polymorpha

About 70% of all therapeutic proteins are glycoproteins and require the attachment of sugar structures (i.e. glycosylation) to attain full therapeutic activity. Current manufacturing methods based on mammalian cell culture do momentarily not completely allow for the control of glycosylation and usually produce a mixture of different glycoforms. Some of these variants are more active than others and some of which have no activity at all. Overall such differences and variations cause major problems in the reproducibility of the production of biopharmaceuticals. This results in a substantial increase in the cost of products mainly due to permanent analysis of thc glycosylation pattern in quality control.

Therefore glycosylation has extensively been studied in the different microbial organisms that are used for recombinant protein production. Detailed analysis of glycosylation has been performed for *S. cerevisiae*, *Hansenula polymorpha* and *Pichia pastoris* (Kim et al. 2006). Like all yeasts they are capable of N- and O-glycosylation. In contrast to the complex glyco-pattern in mammalian proteins N-glycosylation in yeast is restricted to the “high mannose type”. Whereas in

S. cerevisiae a tendency to hyperglycosylation, an addition of 50–150 mannose residues to the glycosylation core, is often observed this is rarely reported from methylotrophic yeasts. The outer mannose chain normally ranges in length between 8 and 14 molecules (Kang and Gellissen 2005). In *S. cerevisiae* the terminal mannose residue is linked by an α1,3 bond which is considered to be allergenic. In contrary, a non-allergenic α1,2 bond is present in methylotrophs such as *Hansenula polymorpha* or *Pichia pastoris*. There is extensive work going on to humanize the pattern of N-glycosylation in recombinant proteins produced by eukaryotic microbial organisms. This could be done by engineering the genes for glycosylation by adding human type enzymes and/or blocking yeast enzymes. This strategy is successfully applied by GlycoFi Inc., a biotech company, now part of Merck &Co. Inc. (Wildt and Gerngross 2005). Another approach is to restrict the addition of sugar moieties to the core glycosylation to make the protein acceptable for application in humans (Oh et al. 2008).

5.3 Production of Antibodies and Antibody Fragments

5.3.1 Need of Antibodies

Recombinant antibodies and their fragments currently represent over 30% of all biological proteins undergoing clinical trials for diagnosis and therapy (Akhtar and Maghfoor 2002; Keating et al. 2002). These reagents dominate the cancer-targeting field. Important advances have been made in the design, selection and production of recombinant antibodies. The natural immune repertoire and somatic cell affinity maturation has been superseded by large antibody display libraries and rapid molecular evolution strategies. These novel libraries and selection methods have enabled the rapid isolation of high-affinity cancer targeting and antiviral antibodies, the latter capable of redirecting viruses for gene therapy applications. In alternative strategies for cancer diagnosis and therapy, recombinant antibody fragments have been fused to radioisotopes, drugs, toxins, enzymes and biosensor surfaces. Multi-specific antibodies have been effective for cytotoxic T-cell recruitment and antibody-fusion proteins have delivered enhanced immunotherapeutic and vaccination strategies.

The described properties of the *Hansenula polymorpha* expression system favour this organism among the microbes currently used for production of pharmaceuticals for the production of antibodies or derivatives thereof.

5.3.2 Production of HCV Antibodies

Hepatitis C, a liver disease caused by the lymhpotropic and hepatotropic hepatitis C virus (HCV) was first described in 1975 as the non-A non-B hepatitis virus. Meanwhile, HCV infection became a major worldwide public health problem.

According to the newest WHO estimation, about 170 million people are infected with HCV worldwide (3% of the world population). The HCV-dependent mortality rate is projected to show a twofold to threefold increase within the next 2 decades (4 times higher than in case of HIV e.g.), as hepatitis C virus-infected patients develop cirrhosis (20–30%) and/or hepatocellular carcinoma (5–10%), additionally to chronic infection.

Basic therapeutic strategies include monotherapy with interferon alfa (cytokine) or its combination therapy with ribavirin (nucleoside analogon). In order to detect the disease and avoid its spreading various tests have been established for diagnosis and monitoring of hepatitis C infection: (a) in the indirect serotyping method, antibodies against the virus are detected by enzyme immunoassays (EIA) and recombinant immunoblot assays (RIBA); (b) as a model of a direct method the polymerase chain reaction (RT-PCR) assay permits direct quantification of the virus RNA in the blood and permits determination of HCV genotypes by sequence analysis, reverse hybridization to genotype-specific oligonucleotide probes and restriction length polymorphism analysis.

There is a very small chance of clearing the virus spontaneously in chronic HCV carriers (0.5–0.74% per year). However, the majority of patients with chronic hepatitis C will not clear it without treatment. Approximately 50% of these do not respond to therapy. Current treatment is a combination of pegylated interferon-alpha-2a or pegylated interferon-alpha-2b and the antiviral drug ribavirin, a nucleoside analog. So, there is an ongoing need for development of diagnostic and therapeutic agents, especially because attempts to develop an efficient vaccine still failed until today. Antiviral antibodies capable of redirecting viruses for gene therapy applications are an alternative treatment option.

However, the virus is very actively mutating and several different genotypes are reported, which also show different geographic distributions. Since the therapeutic outcome depends very much on the prevailing genotype of the HCV infection we aimed to generate antibodies specific for the different genotypes and able to differentiate the genotypes (manuscript in preparation).

The Hepatitis C virus is the only known member of the *hepacivirus* genus in the family *Flaviviridae*. Six major genotypes of the hepatitis C virus and thirty subtypes are described. It is a small (50 nm in size), enveloped, single-stranded, positive sense RNA virus. One single poly-protein is made from the RNA molecule which is processed into nine different proteins by viral and cellular peptidases. The two envelope proteins E1 and E2 are glycoproteins which are embedded in the viral membrane (core) and function as trans-membrane proteins that enable the binding of the virus to the receptor proteins of the target cell and the membrane fusion. Core and envelope proteins are potent antigens for antibody related immunological diagnosis of HCV. Antibodies or fragments thereof directed against exposed and accessible parts of the virus, like E2, can be also used to reduce viral load in the body.

In order to use such antibodies for the production of a diagnostic chip being marketed in developing countries such as India or Indonesia a high efficient production system is necessary.

From the several dozens of antibodies generated we choose in a pilot study three different deduced single chain antibodies (scFv), which were cloned in five different yeast species. Using a special cloning strategy (patent application in preparation) we transferred the coding sequences in the proprietary vector system pCoMed11, a yeast vector harboring all elements described in Fig. 5.2 and part of rDNA as a target for homologous integration.

We transfected the vectors into *Hansenula polymorpha, Pichia pastoris, Kluyveromyces lactis, Arxula adeninivorans and Saccharomyces cerevisiae* and stable strains were generated using the established protocol of passaging and stabilization as described above. The LeuUra selection marker was used (manuscript in preparation).

So far the best productivity was seen in *Hansenula polymorpha* and therefore we recommend using it for the production of these specific antibodies (manuscript in preparation).

5.3.3 Production of Antibody Fusion Proteins

Monoclonal antibodies (mAbs) are widely used as therapeutic agents in cancer and severe inflammatory diseases. Several mAbs for various indications are nowadays approved for clinical trials (Deonarain 2008) and some already entered the market. These include rituximab (Akhtar and Maghfoor 2002), and cetuximab, (Goldberg 2005; Vincenzi et al. 2010) members of the four top ten cancer therapy drugs, which are blockbuster products with sales of over one billion US$. Because of their high specificity and their strong binding characteristics antibodies can be employed as a tool for targeting cells. This opens the chance of circumventing the problem of systemic toxification as it is seen with conventional treatment strategies like chemotherapy or irradiation. One class of this kind of fusion proteins are immunotoxins, chimeric proteins with a cell-selective ligand chemically linked or genetically fused to a toxin moiety initiating cell death by the inhibition of protein synthesis or induction of apoptosis by the modification of signal transduction pathways. In general tumor-specific ligands, predominantly cytokines, monoclonal antibodies or derivatives thereof, like single chain variable fragments (scFv), directed against cancer associated cell surface antigens are used to achieve a tumor cell-specific binding, triggering internalisation (Pastan et al. 2007).

The first immunotoxin consisting of a scFv directed against the IL-2 receptor which is overexpressed on certain tumor cells fused to Pseudomonas Exotoxin (ETA) was generated in 1989 (Chaudhary et al. 1989). By now a series of immunotoxins based on very potent protein toxins of bacterial or plant origin like ETA and diphtheria toxin (DT) or saporin and ricin have been used for the design of highly effective immunotoxins (Barth et al. 2000; Ma et al. 1997; Yip et al. 2007; Zhou et al. 2010). Some of these have already been evaluated in clinical trials (Pastan et al. 2006, 2007). In 1999 the FDA approved the first recombinant immunotoxin composed of DT and interleukin-2 (ONTAK) for the specific treatment of patient with cutaneous T-cell lymphoma (Foss 2000).

The major drawback of immunotoxins of non-human origin however is their immunogenic potential limiting a repeated application and therefore significantly reduce their therapeutic applications so far.

To solve the problem of immunogenicity the next generation of immunotoxins was developed based on cytotoxic human enzymes like proteases and RNases ensuring a dramatic decrease in immunogenic potential (Mathew and Verma 2009). The toxic moiety is in general fused to a ligand of human origin or a humanized scFv to generate fully human or humanized immunotoxins.

Functional expression of human immunotoxins have been shown to be successful in bacteria (Hetzel et al. 2008; Liu et al. 2003), yeasts (Dalken et al. 2006; Gurkan and Ellar 2003; Liu et al. 2005; Woo et al. 2002; 2004), mammalian cells like HEK293T, CHO-K1 and derivatives thereof (Liu et al. 2000; Stahnke et al. 2008; Woo et al. 2004), as well as insect cells (Choo et al. 2002). Functional expression of immunotoxins in the yeast *Pichia pastoris* includes a diphteria toxin based fusion protein for the treatment of T cell leukemia, autoimmune diseases and tolerance induction (Woo et al. 2002) and a Granzyme B-based fusion protein for the specific elimination of EGFR/Her2 positive tumors (Dalken et al. 2006).

The functional expression of human enzymes that requires a free N-terminus to be fully active, like the proapoptotic serine protease granzyme B (Stahnke et al. 2008), that is initially expressed as an inactive zymogen containing an N-terminal propeptide, and processed by local enzyme activity in the cytotoxic granules of natural killer cells and cytotoxic T lymphocytes (Smyth et al. 1995), has been shown to be advantageous in expression systems that ensure a correct in vivo processing (see below). In prior studies in vitro activation of a granzyme B variant modified by the genetic fusion of an enterokinase recognition site to the N-terminus of the mature sequence has been shown to be successful (Stahnke et al. 2008). But the in vitro processing of granzyme B by enterokinase cleavage is cost intensive and time consuming. Successful expression and in vivo activation of a granzyme B variant with the sequence for the propeptide deleted has been reported for COS (Caputo et al. 1993; Smyth et al. 1995) and insect cells (Xia et al. 1998). The functional secretion of granzyme B and a granzyme B-based immunotoxin into the culture supernatant of *P. pastoris* by genetic fusion of the MFα factor prepro leader sequence and the sequence of mature granzyme B was achieved earlier (Giesubel et al. 2006; Pham et al. 1998; Sun et al. 1999). This offers the opportunity of an economic expression of immunotoxins based on human enzymes that depends on the activation of their inactive zymogens in yeasts in general.

We conclude that the secretory expression of activated recombinant immunotoxins into the cell culture supernatant of the yeast *H. polymorpha* would not only avoid labor-intensive cell lysis procedures, therefore not only keeping the risk of contamination with cellular proteins low facilitating purification, but reducing costs resulting from further essential downstream processes.

References

Agaphonov, M. O., M. Y. Beburov, M. D. Ter-Avanesyan and V. N. Smirnov 1995. "A disruption-replacement approach for the targeted integration of foreign genes in Hansenula polymorpha." Yeast 11(13):1241–1247.

Akhtar, S. and I. Maghfoor. 2002. "Rituximab plus CHOP for diffuse large-B-cell lymphoma." N Engl J Med 346(23):1830–1831; author reply 1830–1831.

Amuel, C., G. Gellissen, C. P. Hollenberg and M. Suckow 2000. "Analysis of heat shock promoters in Hansenula polymorpha: TPS1, a novel element for heterologous gene expression." Biotechnol Bioprocess Eng 5:247–252.

Barnes, C. S., B. Krafft, M. Frech, U. R. Hofmann, A. Papendieck, U. Dahlems, G. Gellissen and M. F. Hoylaerts. 2001. "Production and characterization of saratin, an inhibitor of von Willebrand factor-dependent platelet adhesion to collagen." Semin Thromb Hemost 27(4): 337–348.

Barth, S., M. Huhn, B. Matthey, S. Tawadros, R. Schnell, T. Schinkothe, V. Diehl and A. Engert. 2000. "Ki-4(scFv)-ETA', a new recombinant anti-CD30 immunotoxin with highly specific cytotoxic activity against disseminated Hodgkin tumors in SCID mice." Blood 95(12):3909–3914.

Bogdanova, A. I., O. S. Kustikova, M. O. Agaphonov and M. D. Ter-Avanesyan. 1998. "Sequences of Saccharomyces cerevisiae 2 microns DNA improving plasmid partitioning in Hansenula polymorpha." Yeast 14(1):1–9.

Caputo, A., R. S. Garner, U. Winkler, D. Hudig and R. C. Bleackley. 1993. "Activation of recombinant murine cytotoxic cell proteinase-1 requires deletion of an amino-terminal dipeptide." J Biol Chem 268(24):17672–17675.

Chaudhary, V. K., C. Queen, R. P. Junghans, T. A. Waldmann, D. J. FitzGerald and I. Pastan. 1989. "A recombinant immunotoxin consisting of two antibody variable domains fused to Pseudomonas exotoxin." Nature 339(6223):394–397.

Choo, A. B., R. D. Dunn, K. W. Broady and R. L. Raison. 2002. "Soluble expression of a functional recombinant cytolytic immunotoxin in insect cells." Protein Expr Purif 24(3):338–347.

Chu, L. and D. K. Robinson. 2001. "Industrial choices for protein production by large-scale cell culture." Curr Opin Biotechnol 12(2):180–187.

Cook, M. W. and H. V. Thygesen. 2003. "Safety evaluation of a hexose oxidase expressed in Hansenula polymorpha." Food Chem Toxicol 41(4):523–529.

Dalken, B., U. Giesubel, S. K. Knauer and W. S. Wels. 2006. "Targeted induction of apoptosis by chimeric granzyme B fusion proteins carrying antibody and growth factor domains for cell recognition." Cell Death Differ 13(4):576–585.

Degelmann, A., F. Muller, H. Sieber, V. Jenzelewski, M. Suckow, A. W. Strasser and G. Gellissen. 2002. "Strain and process development for the production of human cytokines in Hansenula polymorpha." FEMS Yeast Res 2(3):349–361.

Deonarain, M. P. 2008. "Recombinant antibodies for cancer therapy." Expert Opin Biol Ther 8(8):1123–1141.

Foss, F. M. 2000. "DAB(389)IL-2 (ONTAK): a novel fusion toxin therapy for lymphoma." Clin Lymphoma 1(2):110–116; discussion 117.

Gatzke, R., U. Weydemann, Z. A. Janowicz and C. P. Hollenberg. 1995. "Stable multicopy integration of vector sequences in Hansenula polymorpha." Appl Microbiol Biotechnol 43(5):844–849.

Gellissen, G. 2000. "Heterologous protein production in methylotrophic yeasts." Appl Microbiol Biotechnol 54(6):741–750.

Gellissen, G., G. Kunze, C. Gaillardin, J. M. Cregg, E. Berardi, M. Veenhuis and I. van der Klei. 2005. "New yeast expression platforms based on methylotrophic Hansenula polymorpha and Pichia pastoris and on dimorphic Arxula adeninivorans and Yarrowia lipolytica – a comparison." FEMS Yeast Res 5(11):1079–1096.

Gellissen, G., M. Piontek, U. Dahlems, V. Jenzelewski, J. E. Gavagan, R. DiCosimo, D. L. Anton and Z. A. Janowicz. 1996. "Recombinant Hansenula polymorpha as a biocatalyst: coexpression of the spinach glycolate oxidase (GO) and the S. cerevisiae catalase T (CTT1) gene." Appl Microbiol Biotechnol 46(1):46–54.

Genentech. 1978. "First successful laboratory production of human insulin announced" News Release. Genentech 1978-09-06.

Giesubel, U., B. Dalken, H. Mahmud and W. S. Wels. 2006. "Cell binding, internalization and cytotoxic activity of human granzyme B expressed in the yeast Pichia pastoris." Biochem J 394(Pt 3):563–573.

Goldberg, R. M. 2005. "Cetuximab." Nat Rev Drug Discov Suppl:S10–11.

Gurkan, C. and D. J. Ellar. 2003. "Expression in Pichia pastoris and purification of a membrane-acting immunotoxin based on a synthetic gene coding for the Bacillus thuringiensis Cyt2Aa1 toxin." Protein Expr Purif 29(1):103–116.

Hartner, F. S. and A. Glieder. 2006. "Regulation of methanol utilisation pathway genes in yeasts." Microb Cell Fact 5:39.

Hellwig, S., F. Emde, N. P. Raven, M. Henke, P. van Der Logt and R. Fischer. 2001. "Analysis of single-chain antibody production in Pichia pastoris using on-line methanol control in fed-batch and mixed-feed fermentations." Biotechnol Bioeng 74(4):344–352.

Hetzel, C., C. Bachran, R. Fischer, H. Fuchs, S. Barth and M. Stocker. 2008. "Small cleavable adapters enhance the specific cytotoxicity of a humanized immunotoxin directed against CD64-positive cells." J Immunother 31(4):370–376.

Hodgkins, M., D. Mead, D. J. Ballance, A. Goodey and P. Sudbery. 1993. "Expression of the glucose oxidase gene from Aspergillus niger in Hansenula polymorpha and its use as a reporter gene to isolate regulatory mutations." Yeast 9(6):625–635.

Janowicz, Z. A., K. Melber, A. Merckelbach, E. Jacobs, N. Harford, M. Comberbach and C. P. Hollenberg. 1991. "Simultaneous expression of the S and L surface antigens of hepatitis B, and formation of mixed particles in the methylotrophic yeast, Hansenula polymorpha." Yeast 7(5):431–443.

Janowicz, Z. A., A. Merckelbach, M. Eckart, U. Weydemann, R. Roggenkamp and C. P. Hollenberg. 1988. "Expression system based on the methylotrophic yeast Hansenula polymorpha." Yeast 4:431–443.

Jenzelewski, V. 2002. "Fermentation and primary product recovery." In: Hansenula Polymorpha – Biology and Application, ed. Gellissen, G. Weinheim: Wiley-VCH.

Kang, H. A. H. and G. Gellissen. 2005. "Hansenula polymorpha." In: Production of Recombinant Proteins – Novel Microbial and Eukaryotic Expression Systems, ed. Gellissen, G. Weinheim: Wiley-VCH.

Kang, H. A., W. Kang, W. K. Hong, M. W. Kim, J. Y. Kim, J. H. Sohn, E. S. Choi, K. B. Choe and S. K. Rhee. 2001. "Development of expression systems for the production of recombinant human serum albumin using the MOX promoter in Hansenula polymorpha DL-1." Biotechnol Bioeng 76(2):175–185.

Keating, M. J., I. Flinn, V. Jain, J. L. Binet, P. Hillmen, J. Byrd, M. Albitar, L. Brettman, P. Santabarbara, B. Wacker and K. R. Rai. 2002. "Therapeutic role of alemtuzumab (Campath-1H) in patients who have failed fludarabine: results of a large international study." Blood 99(10):3554–3561.

Kim, M. W., E. J. Kim, J. Y. Kim, J. S. Park, D. B. Oh, Y. Shimma, Y. Chiba, Y. Jigami, S. K. Rhee and H. A. Kang. 2006. "Functional characterization of the Hansenula polymorpha HOC1, OCH1, and OCR1 genes as members of the yeast OCH1 mannosyltransferase family involved in protein glycosylation." J Biol Chem 281(10):6261–6272.

Klabunde, J., A. Diesel, D. Waschk, G. Gellissen, C. P. Hollenberg and M. Suckow. 2002. "Single-step co-integration of multiple expressible heterologous genes into the ribosomal DNA of the methylotrophic yeast Hansenula polymorpha." Appl Microbiol Biotechnol 58(6):797–805.

Klei, I. Jv. ander and M. Veenhuis 2002. "Hansenula polymorpha – a versatile model in peroxisome research." In: Hansenula Polymorpha – Biology and Applications, ed. Gellissen, G. Weinheim: Wiley-VCH.

Krappmann, S., R. Pries, G. Gellissen, M. Hiller and G. H. Braus. 2000. "HARO7 encodes chorismate mutase of the methylotrophic yeast Hansenula polymorpha and is derepressed upon methanol utilization." J Bacteriol 182(15):4188–4197.

Lahtchev, K. 2002. "Basic genetics of Hansenula polymorpha." In: Hansenula Polymorpha – Biology and Application, ed. Gellissen, G. Weinheim: Wiley-VCH.

Levine, D. W. and C. L. Cooney. 1973. "Isolation and characterization of a thermotolerant methanol-utilizing yeast." Appl Microbiol 26(6):982–990.

Liu, Y., L. H. Cheung, W. N. Hittelman and M. G. Rosenblum. 2003. "Targeted delivery of human pro-apoptotic enzymes to tumor cells: In vitro studies describing a novel class of recombinant highly cytotoxic agents." Mol Cancer Ther 2(12):1341–1350.

Liu, Y. Y., I. Gordienko, A. Mathias, S. Ma, J. Thompson, J. H. Woo and D. M. Neville, Jr. 2000. "Expression of an anti-CD3 single-chain immunotoxin with a truncated diphtheria toxin in a mutant CHO cell line." Protein Expr Purif 19(2):304–311.

Liu, Y. Y., J. H. Woo and D. M. Neville, Jr. 2005. "Overexpression of an anti-CD3 immunotoxin increases expression and secretion of molecular chaperone BiP/Kar2p by Pichia pastoris." Appl Environ Microbiol 71(9):5332–5340.

Ma, S., H. Hu, J. Thompson, S. Stavrou, J. Scharff and D. M. Neville, Jr. 1997. "Genetic construction and characterization of an anti-monkey CD3 single-chain immunotoxin with a truncated diphtheria toxin." Bioconjug Chem 8(5):695–701.

Mannazzu, I., E. Guerra, R. Strabbioli, A. Masia, G. B. Maestrale, M. A. Zoroddu and F. Fatichenti. 1997. "Vanadium affects vacuolation and phosphate metabolism in Hansenula polymorpha." FEMS Microbiol Lett 147(1):23–28.

Mathew, M. and R. S. Verma. 2009. "Humanized immunotoxins: a new generation of immunotoxins for targeted cancer therapy." Cancer Sci 100(8):1359–1365.

Mayer, A. F., K. Hellmuth, H. Schlieker, R. Lopez-Ulibarri, S. Oertel, U. Dahlems, A. W. Strasser and A. P. van Loon. 1999. "An expression system matures: a highly efficient and cost-effective process for phytase production by recombinant strains of Hansenula polymorpha." Biotechnol Bioeng 63(3):373–381.

Melmer, G. 2005. "Biopharmaceuticals and the industrial environment." In: Production of Recombinant Proteins – Novel Microbial and Eukaryotic Expression Systems, ed. Gellissen, G. Weinheim: Wiley-VCH.

Morais, J. O. F. and M. H. D. Maia 1959. "Estudos de microorganismos encontrados em leitos dedespejos de caldas de destilarias dePernambuco. II. Una nova especie de *Hansenula*: *H. polymorpha* Anais de Escola Superior de Quimica de Universidade do Recife 1:15–20.

Müller, F., A. Tieke, D. Waschk, F. Müller, M. Seigelchifer, A. Pesce, V. Jenzelewski and G. Gellissen. 2002. "Production of IFNα-2a in Hansenula polymorpha." Process Biochem 38(1):15–25.

Oh, D. B., J. S. Park, M. W. Kim, S. A. Cheon, E. J. Kim, H. Y. Moon, O. Kwon, S. K. Rhee and H. A. Kang. 2008. "Glycoengineering of the methylotrophic yeast Hansenula polymorpha for the production of glycoproteins with trimannosyl core N-glycan by blocking core oligosaccharide assembly." Biotechnol J 3(5):659–668.

Pastan, I., R. Hassan, D. J. FitzGerald and R. J. Kreitman. 2007. "Immunotoxin treatment of cancer." Annu Rev Med 58:221–237.

Pastan, I., R. Hassan, D. J. Fitzgerald and R. J. Kreitman. 2006. "Immunotoxin therapy of cancer." Nat Rev Cancer 6(7):559–565.

Pham, C. T., D. A. Thomas, J. D. Mercer and T. J. Ley. 1998. "Production of fully active recombinant murine granzyme B in yeast." J Biol Chem 273(3):1629–1633.

Ramezani-Rad, M., C. P. Hollenberg, J. Lauber, H. Wedler, E. Griess, C. Wagner, K. Albermann, J. Hani, M. Piontek, U. Dahlems and G. Gellissen. 2003. "The Hansenula polymorpha (strain CBS4732) genome sequencing and analysis." FEMS Yeast Res 4(2):207–215.

Roggenkamp, R., H. Hansen, M. Eckart, Z. Janowicz and C. P. Hollenberg. 1986. "Transformation of the methylotrophic yeast Hansenula polymorpha by autonomous replication and integration vectors." Mol Gen Genet 202:302–308.

Smith, J. and Z. Olempska-Beer. 2004. "Hexose oxidase from Chondrus crispus expressed in Hansenula polymorpha." Chemical and technical assessment (FAO 2004).

Smyth, M. J., M. J. McGuire and K. Y. Thia. 1995. "Expression of recombinant human granzyme B. A processing and activation role for dipeptidyl peptidase I." J Immunol 154(12):6299–6305.

Sohn, J. H., E. S. Choi, C. H. Kim, M. O. Agaphonov, M. D. Ter-Avanesyan, J. S. Rhee and S. K. Rhee. 1996. "A novel autonomously replicating sequence (ARS) for multiple integration in the yeast Hansenula polymorpha DL-1." J Bacteriol 178(15):4420–4428.

Stahnke, B., T. Thepen, M. Stocker, R. Rosinke, E. Jost, R. Fischer, M. K. Tur and S. Barth. 2008. "Granzyme B-H22(scFv), a human immunotoxin targeting CD64 in acute myeloid leukemia of monocytic subtypes." Mol Cancer Ther 7(9):2924–2932.

Stockmann, C., U. Maier, T. Anderlei, C. Knocke, G. Gellissen and J. Buchs. 2003. "The oxygen transfer rate as key parameter for the characterization of Hansenula polymorpha screening cultures." J Ind Microbiol Biotechnol 30(10):613–622.

Sun, J., C. H. Bird, M. S. Buzza, K. E. McKee, J. C. Whisstock and P. I. Bird. 1999. "Expression and purification of recombinant human granzyme B from Pichia pastoris." Biochem Biophys Res Commun 261(2):251–255.

Tof, I. 1994. Recombinant DNA Technology in the Synthesis of Human Insulin. Little Tree Publishing, retrieved 19.9.2010.

Valenzuela, P., A. Medina, W. J. Rutter, G. Ammerer and B. D. Hall. 1982. "Synthesis and assembly of hepatitis B virus surface antigen particles in yeast." Nature 298(5872):347–350.

Veale, R. A., M. L. F. Guiseppin, H. M. J. Van Eijk, P. E. Sudbery and C. T. Verrips. 1992. "Development of a strain of *Hansenula polymorpha* for the efficient expression of Guar β-galactosidase." Yeast 8:361–372.

Vincenzi, B., A. Zoccoli, F. Pantano, O. Venditti and S. Galluzzo. 2010. "Cetuximab: from bench to bedside." Curr Cancer Drug Targets 10(1):80–95.

Weydemann, U., P. Keup, M. Piontek, A. W. Strasser, J. Schweden, G. Gellissen and Z. A. Janowicz. 1995. "High-level secretion of hirudin by Hansenula polymorpha – authentic processing of three different preprohirudins." Appl Microbiol Biotechnol 44(3–4):377–385.

Wildt, S. and T. U. Gerngross. 2005. "The humanization of N-glycosylation pathways in yeast." Nat Rev Microbiol 3(2):119–128.

Woo, J. H., Y. Y. Liu, A. Mathias, S. Stavrou, Z. Wang, J. Thompson and D. M. Neville, Jr. 2002. "Gene optimization is necessary to express a bivalent anti human anti T cell immunotoxin in Pichia pastoris." Protein Expr Purif 25(2):270–282.

Woo, J. H., Y. Y. Liu, S. Stavrou and D. M. Neville, Jr. 2004. "Increasing secretion of a bivalent anti-T-cell immunotoxin by Pichia pastoris." Appl Environ Microbiol 70(6):3370–3376.

Wurm, F. M. 2004. "Production of recombinant protein therapeutics in cultivated mammalian cells." Nat Biotechnol 22(11):1393–1398.

Xia, Z., C. M. Kam, C. Huang, J. C. Powers, R. J. Mandle, R. L. Stevens and J. Lieberman. 1998. "Expression and purification of enzymatically active recombinant granzyme B in a baculovirus system." Biochem Biophys Res Commun 243(2):384–389.

Yin, J., G. Li, X. Ren and G. Herrler. 2007. "Select what you need: a comparative evaluation of the advantages and limitations of frequently used expression systems for foreign genes." J Biotechnol 127(3):335–347.

Yip, W. L., A. Weyergang, K. Berg, H. H. Tonnesen and P. K. Selbo. 2007. "Targeted delivery and enhanced cytotoxicity of cetuximab-saporin by photochemical internalization in EGFR-positive cancer cells." Mol Pharm 4(2):241–251.

Yurimoto, H., Y. Sakai and N. Kato. 2002. "Methanol metabolism." In: Hansenula Polymorpha – Biology and Applications, ed. Gellissen, G. Weinheim: Wiley-VCH.

Zhou, X. X., F. Ji, J. L. Zhao, L. F. Cheng and C. F. Xu. 2010. "Anti-cancer activity of anti-p185HER-2 ricin A chain immunotoxin on gastric cancer cells." J Gastroenterol Hepatol 25(7):1266–1275.

Chapter 6
Production of Antibody by Transgenic Avians

Yoshinori Kawabe and Masamichi Kamihira

Abstract Transgenic avian bioreactors are being proposed as a powerful way of addressing the growing need for recombinant biopharmaceutical production. Avian systems as transgenic bioreactors have several advantages including high protein productivity in eggs, a relatively short period of sexual maturation, and similar protein glycosylation patterns to those of humans. Several examples of successful viral biopharmaceuticals are already being produced as human vaccines. In this chapter, we describe the generation of genetically manipulated (GM) avians producing pharmaceutical proteins including antibodies using retroviral vectors for gene transfer, the analysis of glycosylation patterns of recombinant antibodies produced in the serum and eggs of GM chickens, and the recovery of recombinant antibodies and Fc-fusion proteins mediated by yolk transport. The development of transgenic avian bioreactors promises to be an important procedure for production of therapeutic proteins.

6.1 Introduction

In recent years, many pharmaceutical protein products represented by therapeutic antibodies against cancer and immune diseases have appeared on the market (Shukla and Thömmes, 2010). Such proteins have been produced primarily by recombinant animal cell culture employing Chinese hamster ovary (CHO) cells and mouse myeloma line NS0 cells as the host cells because of their ability for complex post-translational modification such as proper protein folding and glycosylation (Wurm, 2004). However, the increased demand for recombinant therapeutic proteins requires the development of effective and inexpensive systems for mass production of pharmaceutical proteins, since the production cost using animal cell culture is extremely high (Dyck et al., 2003).

Transgenic livestock animals have been proposed as living bioreactors for the production platforms of biopharmaceuticals, which are produced mainly in the milk of mammals or in the eggs of avians (Houdebine, 2009). Transgenic technologies

Y. Kawabe and M. Kamihira (✉)
Department of Chemical Engineering, Faculty of Engineering, Kyushu University, Fukuoka 819-0395, Japan
e-mail: kamihira@chem-eng.kyushu-u.ac.jp

M. Al-Rubeai (ed.), *Antibody Expression and Production*, Cell Engineering 7,
DOI 10.1007/978-94-007-1257-7_6, © Springer Science+Business Media B.V. 2011

in mammals have been well developed using mouse models. Similarly, the generation of transgenic farm mammals has been established by DNA microinjection into embryo pronuclei and by using viral vectors or transposons for gene transfer (Kues and Niemann, 2004; Whitelaw, 2004; Wheeler, 2007). Using this technique, high concentrations (14 mg mL^{-1}-milk) of monoclonal antibody was produced in the milk of a transgenic goat (Pollock et al., 1999). Recently, anti-thrombin produced in the milk of transgenic goats was approved by the Food and Drug Administration (FDA) (Lavine, 2009). An increasing number of biopharmaceuticals may be produced using transgenic mammalian bioreactors in the future. However, the transgenic mammalian bioreactors have several drawbacks. These include the requirement of long-term sexual maturation, a relatively large area for breeding (Whitelaw, 2004), and the risk of bovine spongiform encephalitis (BSE).

On the other hand, avian species, including chicken and quail, have attracted attention as alternative transgenic bioreactors for the production of pharmaceutical proteins in their eggs because of short generation period, high production yield of egg proteins, ease of breeding and the absence of the prion problem (Ivarie, 2003; Sang, 2004; Lillico et al., 2005). Furthermore, it is known that oligosaccharide chain structures of glycoproteins produced by chickens show considerable similarities to those of humans. For example, the terminal sialic acid of oligosaccharide chains for both humans and chickens is composed of *N*-acetylneuraminic acid, whereas most other animals use *N*-glycolylneuraminic acid or a combination of the two (Raju et al., 2000). In addition, vaccines for infectious diseases such as influenza have been produced by embryonated egg culture using chicken specific pathogen free species. However, the development of transgenic technologies for avian species lags far behind that of mammals. Therefore, it is important to develop the technology required for the generation of transgenic chickens and for the expression systems for pharmaceutical protein production to establish efficient transgenic avian bioreactors.

Many research groups have attempted to generate genetically manipulated (GM) avian species by retroviral vector injection into the blastodermal stage of embryos (Mizuarai et al., 2001; Harvey et al., 2002; Mozdziak et al., 2003; Rapp et al., 2003; McGrew et al., 2004; Kamihira et al., 2005; Lillico et al., 2007; Kwon et al., 2010) or by microinjection into the single-cell stage of fertilized eggs (Love et al., 1994; Sherman et al., 1998). Chicken embryonic stem (ES) cells were also used for generation of GM chickens, although chicken ES cells did not contribute to germ line cells (Zhu et al., 2005). Cultured primordial germ cells may become a promising cell source for the generation of transgenic chickens because of germ line transmission of engineered cells (van de Lavoir et al., 2006). Recently, sperm-mediated gene delivery in chickens was reported to produce transgenic progeny (Harel-Markowitz et al., 2009). Using these techniques GM chickens producing pharmaceutical proteins have been generated (Table 6.1). As a pioneering work, α-2b human interferon was produced in the serum and egg white of GM chickens generated by using an avian leukosis virus (ALV)-based retroviral vector for gene transfer (Rapp et al., 2003), although the productivity and the germ line transmission efficiency were not high (2.7–4.5 $\mu g\ mL^{-1}$ and 0.06%, respectively).

Table 6.1 Pharmaceutical proteins produced by GM avian

Method	Target gene	Promoter	Expression level	Transmission efficiency	Remarks column	References
Virus						
ALV (retrovirus)	hIFN α-2b	CMV	G_2 egg white; 2.7–4.5 μg mL^{-1}	0.063% (1/1,597)	Chickens 65,000 particles/350 embryos	Rapp et al. (2003)
MSCV (retrovirus)	scFv-Fc	Chicken β-actin	G_1,G_2 egg white; 0.05–1.5 mg mL^{-1} G_1,G_2 egg yolk; 0.17–0.52 mg mL^{-1}	3.3% (6/181)	Chickens 0.6–1.8 × 10^9 IU mL^{-1}	Kamihira et al. (2005)
MSCV (retrovirus)	scFv-Fc	Chicken β-actin	G_0 egg white; 0.2–1.2 mg mL^{-1} G_0 egg yolk; 0.1–0.7 mg mL^{-1}	–[a]	Quails 0.3–1.8 × 10^9 IU mL^{-1}	Kawabe et al. (2006a)
EIAV (lentivirus)	scFv-Fc, hIFNβ1a	OVA	G_1,G_2 egg white; 15–50 μg mL^{-1} (scFv-Fc) 3.5–426 μg mL^{-1} (hIFNβ1a)	4.1% (19/463)	Chickens –[b]	Lillico et al. (2007)
MoMLV (retrovirus)	hPTH	RSV	G_0 blood; 6.3–8.2 ng mL^{-1}	–[a]	Chickens 1 × 10^9 cfu mL^{-1}	Lee et al. (2007)
MSCV (retrovirus)	hEpo	Chicken β-actin	G_0 egg white; 1,250–5,670 IU mL^{-1}	0% (0/282)	Chickens 0.03–5.0 × 10^{10} IU mL^{-1}	Kodama et al. (2008)
MSCV (retrovirus)	TNFR/Fc	Chicken β-actin	G_0 egg white; 0.2–1.4 μg mL^{-1} G_0 egg yolk; 0.7–25 μg mL^{-1}	–[a]	Chickens 1.5× 10^7–6.6× 10^9 IU mL^{-1}	Kyogoku et al. (2008)

Table 6.1 (continued)

Method	Target gene	Promoter	Expression level	Transmission efficiency	Remarks column	References
MoMLV (retrovirus)	hG-CSF	CMV	G_0 blood; 0.1–3.4 μg mL^{-1}	8% (8/101)	Chickens 1 × 10^9 cfu mL^{-1}	Kwon et al. (2008)
MSCV (retrovirus)	Monoclonal antibody	Chicken β-actin	G_0 egg white; 18.4–669 μg mL^{-1} G_0 egg yolk; 1.7–129 μg mL^{-1}	–[a]	Chickens 4–8 × 10^8 IU mL^{-1}	Kamihira et al. (2009)
MSCV (retrovirus)	hEpo/Fc	Chicken β-actin	G_0 egg white; 9–66 μg mL^{-1} G_0 egg yolk; 12–41 μg mL^{-1}	–[a]	Chickens 0.11–2.2 × 10^8 IU mL^{-1}	Penno et al. (2010)
HIV (lentivirus)	rhIL1RN	OVA	G_4 egg white; 40–234 ng mL^{-1}	1.5% (1/68)	Quails 1 × 10^8 TU mL^{-1}	Kwon et al. (2010)
MoMLV (retrovirus)	hEpo	TREtight	G_1 blood; 300 IU mL^{-1} egg white; 1.8 IU mL^{-1}	1.5% (6/412)	Chickens 1 × 10^9 cfu mL^{-1}	Koo et al. (2010)
Non-virus						
Genetically modified chicken ES cells	Monoclonal antibody	OVA	G_0 egg white; 0.2–148 μg mL^{-1}	0% (0/8,862)	Electroporation for chicken ES cells Injection to chicken stage X	Zhu et al. (2005)

Table 6.1 (continued)

Method	Target gene	Promoter	Expression level	Transmission efficiency	Remarks column	References
Lipofected sperm	hFSH	CMV	G_1 blood; <0.05–1.2 Units mL^{-1}	89.4% (17/19) for GFP gene 100% (8/8) for hFSH gene	Lipofection of restriction enzyme digested-plasmid into chicken sperm	Harel-Markowitz et al. (2009)

[a]Not determined.
[b]The used viral titer was not described.
EIAV, equine infectious anemia virus; HIV, human immunodeficiency virus; hIFN, human interferon; hPTH, human parathormone; hG-CSF, human granulocyte-colony stimulating factor; rhIL1RN, recombinant human interleukin 1 receptor antagonist; hFSH, human follicle-stimulating hormone; CMV, cytomegalovirus; OVA, ovalbumin; TRE, tetracycline responsive element.

We have developed a retroviral method for the generation of GM avian species. Moloney murine leukemia virus (MoMLV)-based retroviral vectors pseudotyped with vesicular stomatitis virus G protein (VSV-G) were used for gene transfer. VSV-G pseudotyped retroviral vectors exhibit broad host range specificity (Burns et al., 1993). This system can be applied to many types of cells (pantropic) including mammalian, avian, fish and insect cells, because VSV-G interacts with phospholipids

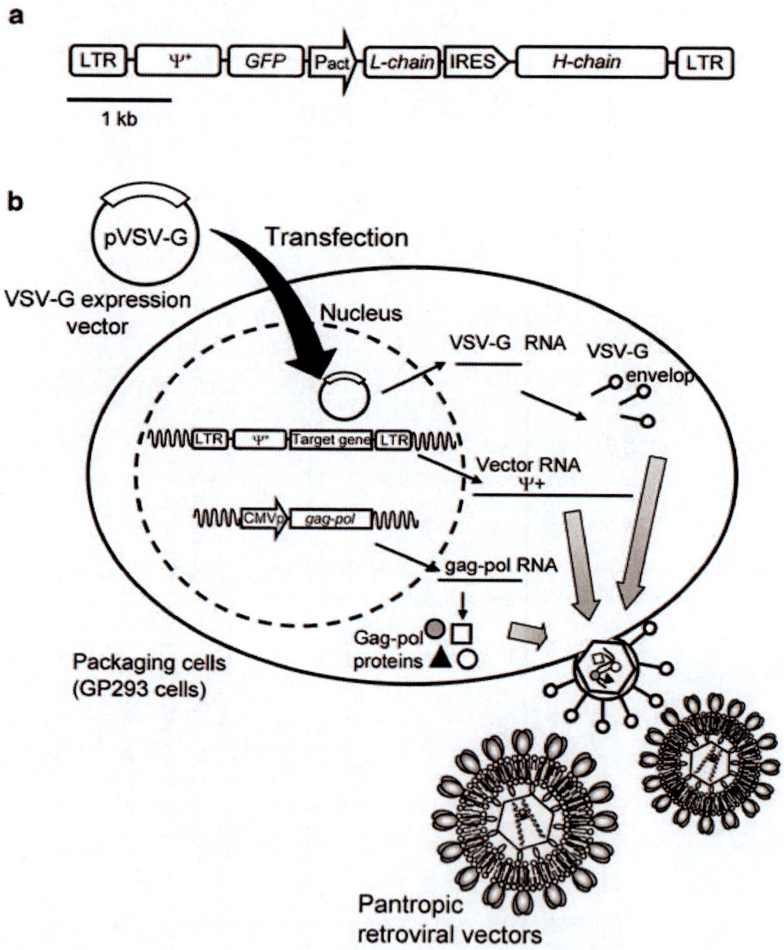

Fig. 6.1 Schematic drawing of the production of replication-defective pantropic retroviral vector pseudotyped with VSV-G. (**a**) Provirus structure of a retroviral vector for the production of chimeric monoclonal antibodies. LTR, long terminal repeat; Ψ^+, virus packaging signal sequence; GFP, green fluorescent protein gene; Pact, chicken β-actin promoter; L-chain, antibody light chain gene; IRES, internal ribosomal entry site sequence from EMCV; H-chain, antibody heavy chain gene. (**b**) Production of pantropic retroviral particles using virus packaging cells. Retroviral vector particles were produced by transfection of a VSV-G expression plasmid to GP293 packaging cells expressing viral genome and gag-pol

found in the cell membrane. In addition, VSV-G improves mechanical strength of retroviral vectors, and viral titers can be increased up to 1,000-fold by ultracentrifugation. Figure 6.1 shows a schematic diagram of a typical retroviral vector construct used in our research together with the production method of retroviral vectors. After the establishment of virus-producer cells (GP293 packaging cells), viral vector particles were produced by transient transfection of a VSV-G expression vector. The viral solution concentrated by ultracentrifugation was injected into avian embryos. The injected embryos were then allowed to hatch (Fig. 6.2). In the initial experiments concentrated retroviral solution was injected into the subgerminal cavity of blastodermal stage quail embryos. The viral sequence was detected in the tissues of all quails that hatched. Although the efficiency of germ line transmission was very high, transgene expression was very low (Mizuarai et al., 2001). To maximize transgene expression, retroviral injection at various developmental stages was attempted using a reporter vector. The maximal expression of the reporter gene was

Fig. 6.2 Retroviral injection into a chicken embryo and embryo culture. (**a**) The retrovirus-producer cells (GP293 packaging cells) are established and retroviral vector particles are produced as shown in Fig. 6.1. (**b**) After concentrating the retroviral solutions by ultracentrifugation, the retroviral solution is microinjected into developing chicken embryos. (**c**) The injected embryos are transferred to surrogate chicken eggshells and cultured in an incubator equipped with an automatic rocking device. (**d**) One day before the expected day of hatching, rocking is stopped and the chick hatches

observed when the retroviral vector was injected into the heart of embryos after 55 and 48 h incubation for chicken and quail, respectively (Kamihira et al., 2005; Kawabe et al., 2006a). Using this procedure, we generated GM chickens and quails producing a single chain antibody fragment fused with the Fc region of human IgG (scFv-Fc) (Kamihira et al., 2005; Kawabe et al., 2006a), chimeric monoclonal antibodies (Kamihira et al., 2009), human erythropoietin (hEpo) (Kodama et al., 2008), Fc-fusion of extracellular domain of tumor necrosis factor (TNF) receptor 2 (TNFR/Fc) (Kyogoku et al., 2008) and hEpo/Fc (Penno et al., 2010). In this chapter, we describe the generation of GM avian animals, the glycosylation pattern analysis of recombinant antibodies produced in the serum and eggs of GM chickens, and the recovery of recombinant antibodies and Fc-fusion proteins mediated by yolk transport.

6.2 Exogenous Gene Transduction to Avian Embryos Using Retroviral Vectors

For the generation of GM avian species using retroviral vectors, the blastodermal stage (stage X) embryos just after laying have been most commonly used because of the availability of many eggs compared with the use of single-cell stage fertilized eggs obtained by sacrificing hens. Since the blastodermal stage embryos develop to approximately 60,000 cells, retroviral methods for gene transfer are preferable to other methods because of the high integration efficiency of gene transduction into the host chromosome (Harvey et al., 2002; Mozdziak et al., 2003). By using VSV-G pseudotyped retroviral vectors, it became possible to apply mammalian retroviral vectors in avian species and retrieve high viral titer by ultracentrifugation. In the previous study, a MoMLV-based retroviral vector pseudotyped with VSV-G was injected into quail embryos at the blastodermal stage and the manipulated embryos were cultured up to and including hatching (Mizuarai et al., 2001). All quails (G_0) were positive for the retroviral DNA sequence by genomic PCR. The average germ-line transmission efficiency of G_0 quails mated with non-transgenic quails was more than 80%. However, the transgene expression under the control of an internal promoter (Rous sarcoma virus (RSV) promoter) in G_1 and G_2 transgenic quails was detectable but very weak. Since this gene suppression was observed even in G_0 quails and since the transgene expression was detected in primary chicken embryonic fibroblasts transduced using the retroviral vector, the retroviral gene suppression seemed to occur at an early stage of embryonic development. Thus, we examined the optimal timing of retroviral injection during embryonic development to maximize viral transduction and transgene expression (Kamihira et al., 2005; Kawabe et al., 2006a). A MoMLV-derived mouse stem cell virus (MSCV)-based retroviral vector encoding a bacterial β-galactosidase gene under the control of chicken β-actin promoter was injected at various developmental stages. When the retroviral injection was performed for blastodermal stage embryos, no evidence of β-galactosidase expression was observed in spite of the detection of retroviral sequence in the cells. Maximal reporter gene expression was observed when the retroviral solution was injected into the heart of embryos after 55 and 48 h of incubation for chicken and quail, respectively. This incubation time corresponds to around stage 15 in the staging categories of Hamburger and Hamilton (1951). By injecting the viral solution into the heart of an embryo, a viral vector can be

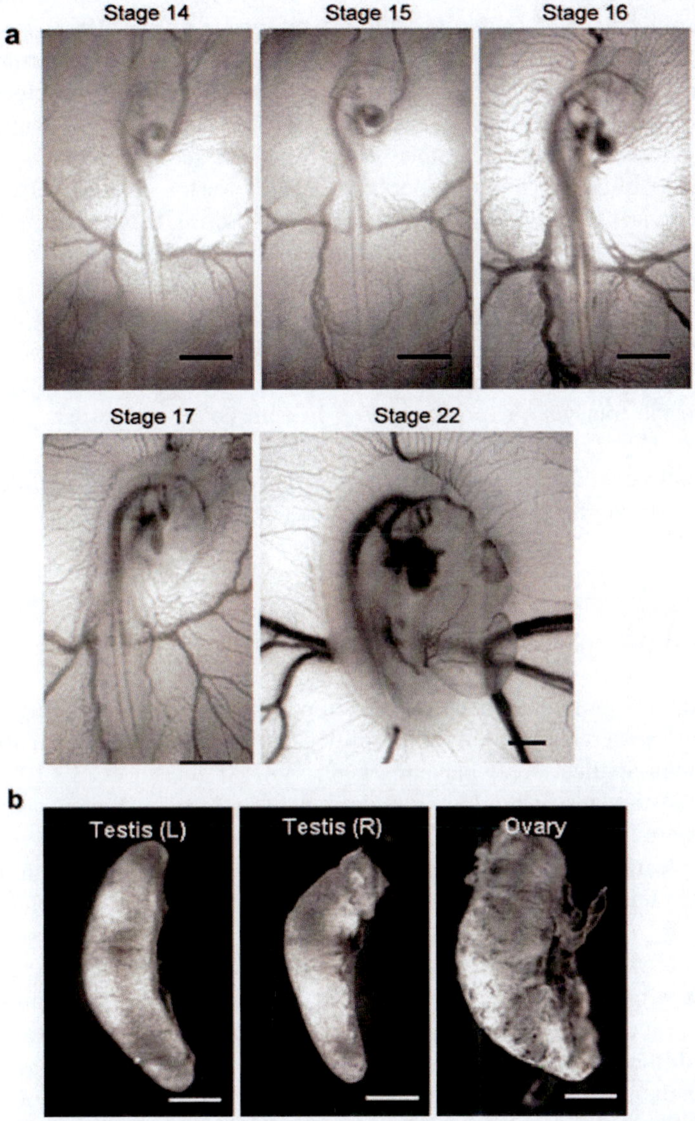

Fig. 6.3 Optimization of the timing for effective gene transduction into chicken gonads. (**a**) Photographs of developing chicken embryos staged according to Hamburger and Hamilton (1951). (**b**) In situ X-gal staining of gonads isolated from embryos just before hatching. Staining was carried out for the testis and ovary after isolation from embryos injected at stage 15. Scale bars: 1 mm (from the authors own work, Kawabe et al., 2008)

efficiently delivered to the whole embryonic body through blood circulation, and high-level expression can be achieved. To achieve germ line transmission of the transgene, the retroviral vector must be infected into primordial germ cells (PGCs). PGCs first appear in the upper cell layer in the center region of the blastoderm and then migrate to the germinal crescent at stage 4 (after 20 h for chicken). The cells circulate through the bloodstream from the germinal crescent to the gonadal anlage at around stage 15 (after 50–55 h for chicken) (Kuwana, 1993; Tajima et al., 1999). Thus, the timing of retroviral injection was also examined in terms of effective gene transduction into the chicken gonads (Fig. 6.3). To prevent retroviral gene suppression, later stage embryos are a better target for viral injection. However, efficient gene transduction may become difficult for later stage embryos composed of larger numbers of cells. The transduction efficiency for the gonadal cells was highest when the viral solution at a viral titer over 10^9 IU mL^{-1} was injected at stages 14–15 (Kawabe et al., 2008). We obtained the transgenic progeny at the frequency of approximately 3% when the equivalent titer of retroviral particles was injected at the heart of embryos at this stage (Kamihira et al., 2005). These results suggest that PGCs migrating into the gonads from the bloodstream are effective viral targets for generating transgenic avian animals using retroviral vectors. Conversely, the copy number and the transgene expression level in embryonic bodies and gonads were negligible at stage 22 (Kawabe et al., 2008). Because the blood volume at this stage is much greater than that during stages 14–17, the viral solution was diluted and the viral particles might be taken up by the abundant red blood cells.

6.3 Production of scFv-Fc Fusion Proteins

As an example of practical protein production, GM chickens producing anti-prion scFv-Fc were generated. scFv is a simplified antibody molecule potentially useful for various application (Clackson et al., 1991; Tsumoto et al., 1998). There are several advantages in producing recombinant proteins as fusion proteins with the Fc region of human IgGs such as easily purification using affinity procedures and longer clearance time in blood. Many Fc-fused proteins have been used therapeutically in humans (Steurer et al., 1995; Murray and Dahl, 1997; Enever et al., 2009).

A retroviral vector encoding an expression cassette for scFv-Fc under the control of the chicken β-actin promoter was injected into the heart of developing chicken embryos at around stage 15. A total of 51 chicken embryos in 4 experiments were injected with the retroviral solution (1.5–3.5 μL) with titers of 0.6–1.8 × 10^9 IU mL^{-1}. After the retroviral vector injection, the embryos were cultured to hatching. The average hatchability of the embryos was 63% (32/51). The birds that hatched exhibited no apparent abnormalities during breeding and female chickens produced eggs after sexual maturation. The viral DNA sequence and the expression of scFv-Fc were observed in the various tissues such as brain, heart, liver, muscle and gonads of all G_0 chickens by genomic PCR and Western blotting, respectively. Furthermore, the transgene sequence was detected in the sperm of G_0 roosters by genomic PCR, indicating that transgenic offspring can be obtained. scFv-Fc was expressed in the serum at the concentration of 0.2–3.6 mg mL^{-1} and stably produced throughout the measurement period. The expression level of GM chickens was dependent on the injected retroviral titer. After sexual maturation, the scFv-Fc protein was also produced in the eggs. The protein was detected in both egg white and yolk, and the concentration of scFv-Fc in the egg white was 3–8 mg mL^{-1}, corresponding to 0.2 g per egg.

Since the transgene was detected in the sperm of G_0 roosters, a single rooster was crossed with a non-transgenic hen to generate G_1 transgenic chickens. Six transgenic chickens were obtained in the analysis of 181 chicks. Five of them produced scFv-Fc at a concentration of 0.5–1.9 mg mL^{-1} in serum. To determine the copy number, intactness and chromosomal location of G_1 transgenic chickens, Southern blot and FISH analyses were performed. Among the six transgenic chickens, five had one copy and one had two copies of the transgene (Fig. 6.4). It was found that the transgene inserted in the chromosome was the same in three G_1 chickens, suggesting that these chickens were derived from the same genetic population of PGCs. Although the expression level of scFv-Fc varied among G_1 transgenic hens, one hen showed a higher concentration of 1.5 mg mL^{-1} in the egg white. This result may be related to the integration site of the transgene in the genome, the so-called position effect (Wilson et al., 1990). It has been reported that the position of chromosomal insertion influenced transgene expression in transgenic mice (Robertson et al., 1995). Therefore, a selection of high-producer chickens in terms of productivity in serum and eggs will be necessary for commercial production using transgenic

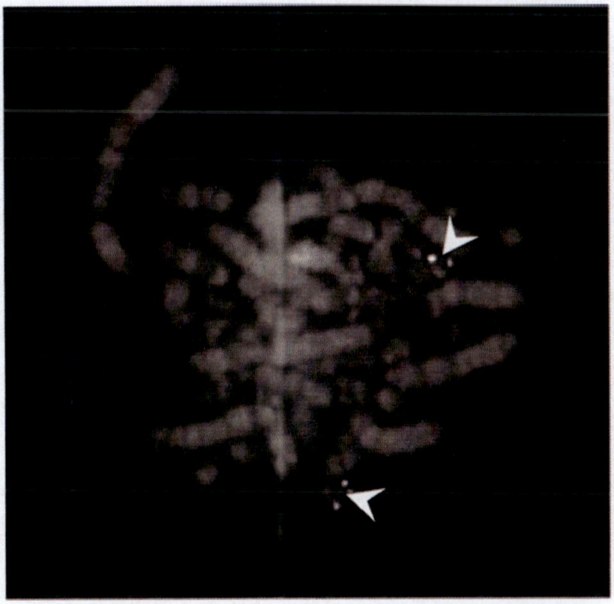

Fig. 6.4 FISH analysis for the determination of the chromosomal location of the scFv-Fc gene in G_1 transgenic chicken. The chromosome was stained using 4′,6-diamidino-2-phenylindole (DAPI) and the transgene was detected using a fluorescein isothiocyanate (FITC)-labeled probe. The *arrowheads* show the location of the transgene and show that two transgenes were integrated into the chromosome (from the authors own work, Kamihira et al., 2005)

progeny. The efficiency of germ-line transmission from G_1 to G_2 progeny mated with a non-transgenic chicken was consistent with the copy number of parental chickens according to Mendel's law. The mating between transgenic chickens could lead to the enhancement of the expression level in transgenic progeny for the accumulation of transgene copies based on the chromosomal location of the transgene. Thus, we could generate GM chickens producing scFv-Fc at an expression level of several mg mL^{-1} in egg white. Transgenic progeny producing the target protein were also obtained.

6.4 Production of Recombinant Monoclonal Antibodies

The establishment of the hybridoma technique by Köhler and Milstein (1975) and recombinant DNA technology revolutionized the field of therapeutic antibody production. Therapeutic antibodies can potentially be used for treating various diseases such as cancer, autoimmune disorders and viral infections. A large number of therapeutic antibodies are currently available in the market (Reichert and Valge-Archer, 2007). It is very important to develop effective mass production systems for therapeutic antibodies because the production cost of animal cell culture is high and administration of antibodies at a high dose is required. Transgenic avian bioreactors can be an effective production system for therapeutic antibodies. Thus, we attempted to establish the production of whole antibodies in GM chickens (Kamihira et al., 2009).

Since antibodies (immunoglobulin G; IgG) are hetero-tetramers comprised of two polypeptides (the heavy (H)- and light (L)-chains), the expression of corresponding genes at the same time is required to produce the antibody molecules. We constructed retroviral vectors in which the bicistronic expression of the genes for the H- and L-chains of anti-CD2 and anti-prion peptide (PrP) chimeric monoclonal antibodies were expressed using an internal ribosomal entry site (IRES) derived from the encephalomyocarditis virus (EMCV) (Fig. 6.1a). We previously reported that it was important to express H- and L-chains polypeptides in a balanced manner in recombinant CHO cells (Hotta et al., 2004). A lower efficiency from IRES-mediated translation was reported compared with that from 5′Cap-mediated translation (Mizuguchi et al., 2000). Since an excess amount of H-chain expression in cells affected cell growth (Köhler, 1980), the H-chain gene was positioned downstream of the IRES and the L-chain gene was positioned downstream of the actin promoter (i.e., Pact-L-IRES-H). The concentrated viral vectors at two concentrations (9×10^8 (anti-CD2) and 4×10^8 (anti-PrP) IU mL^{-1}) were injected into the heart of 46 developing chicken embryos at around stage 15. The hatchability of manipulated embryos was 27% (3/11) and 31% (11/35), respectively. Figure 6.5 shows a photograph of hatched GM chickens producing anti-CD2 chimeric monoclonal antibody. A retroviral DNA sequence was detected by genomic PCR in all hatched chickens. Southern blot analysis revealed that the intact transgene was incorporated into the genome without deletion or recombination in the retroviral sequence and integrated into the genome of various tissue cells such as brain, heart, liver, muscles, oviduct and gonads as a mosaic. Since the expression of chimeric

Fig. 6.5 Photographs of GM chickens producing anti-CD2 chimeric monoclonal antibody. The GM chickens grew to maturity without apparent abnormality and the chimeric monoclonal antibody was stably produced in the serum and eggs throughout the breeding period

antibodies was controlled under the constitutive chicken β-actin promoter, the production of antibodies could be observed in the whole body of the chicken. The chimeric anti-CD2 and anti-PrP antibodies were produced in the serum of GM chickens at a concentration of 8–30 and 64–350 $\mu g\ mL^{-1}$, respectively. At 5–6 months after hatching, the GM hens were sexually mature and produced eggs. Then, the concentration of chimeric antibodies in the eggs was measured. For both target proteins, GM hens produced the chimeric antibodies in both the egg white and yolk. More than 600 $\mu g\ mL^{-1}$ of anti-PrP antibody was produced in the egg white of some GM hens. Western blotting in reducing and non-reducing conditions showed the molecular weight and structure of chimeric antibodies to be correct in the serum, egg white and yolk of GM chickens. scFv-Fc produced in yolk of GM chickens was partially digested upstream of the hinge region between scFv and Fc-region (Kamihira et al., 2005). Although this digestion seemed to occur during transport from the blood, such digestion was not observed in whole antibodies and an intact form was maintained in the yolk. The antibodies produced by GM chickens exhibited antigen-binding activities, since the anti-CD2 and PrP antibodies recognized the CD2 produced by Jurkat cells and a human prion synthetic peptide, respectively.

6.5 Analysis of *N*-Linked Oligosaccharide Chain Structures in Recombinant Antibodies Produced by GM Chickens

A potential advantage of chickens as transgenic bioreactors is that the glycosylation patterns of antibodies are very similar to those of humans (Raju et al., 2000). For example, the terminal sialic acid of oligosaccharide chains added to human glycoproteins is composed of only *N*-acetylneuraminic acid. The chicken is the only livestock animal sharing the same structure of terminal sialic acid. Other animals mainly use *N*-glycolylneuraminic acid as the terminal sialic acid. The oligosaccharide chain structures influence the efficacy of pharmaceutical proteins. Therefore, it was important to identify the oligosaccharide chain structures of recombinant antibodies produced by GM chickens. We analyzed the *N*-linked oligosaccharide chain structures added to the Fc-region of recombinant chimeric antibody and scFv-Fc proteins produced in the serum, egg white and yolk of GM chickens and their progeny.

Table 6.2 shows the oligosaccharide structures of recombinant chimeric antibody and scFv-Fc fusion proteins produced in the egg white of chickens. The

Table 6.2 Oligosaccharide structures of chimeric antibody (anti-PrP) and scFv-Fc fusion proteins produced in egg white of chickens

	Relative quantity[b] (%)			
		scFv-Fc[c]		
Oligosaccharide structure[a]	Chimeric antibody	G_0	G_2	G_3
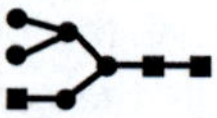	13.8	2.8	9.8	11.3
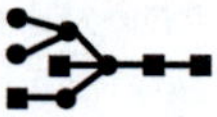	15.3	2.2	4.6	6.5
	22.5	21.8	36.2	38.9
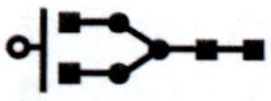 or 	12.8	4.5	3.5	5.9
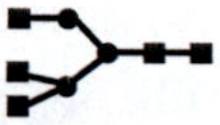	4.6	7.5	5.2	5.8
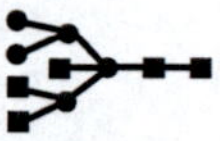	1.3	1.5	–[d]	–[d]
	14.6	20.1	24.0	18.9
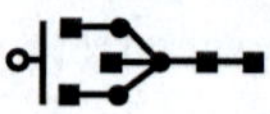 or 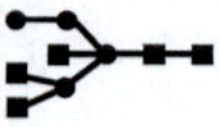	4.5	3.7	1.3	1.3
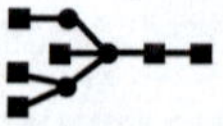	10.7	36.0	15.4	11.4

[a]Symbols: closed squares, *N*-acetyl glucosamine; closed circles, mannose; open circles, galactose.
[b]Total amounts of identified structures for each sample are represented as 100%.
[c]ScFv-Fc fusion proteins were obtained from the egg white produced by the vector-infected first generation of hen (G_0), and the third (G_2) and fourth (G_3) generations of transgenic progeny hens (Kamihira et al., 2005).
[d]Not detected.
From the authors own work, Kamihira et al. (2009)

Table 6.3 Oligosaccharide structures of chimeric antibody (anti-PrP) and scFv-Fc fusion proteins produced in serum and yolk of chickens

		Relative quantity[b] (%)		
		Yolk		Serum
	Oligosaccharide structure[a]	Chimeric antibody	scFv-Fc	scFv-Fc
Neutral	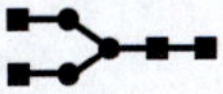	4.7	4.5	4.7
	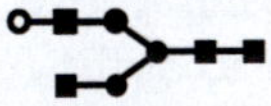	2.9	3.6	1.1
	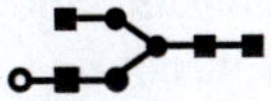	2.0	1.4	1.1
	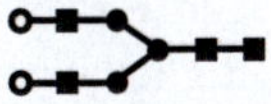	3.3	5.4	9.1
	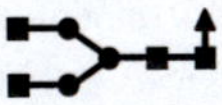	9.2	20.8	23.0
	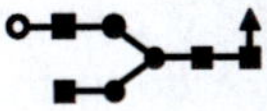 or 	8.9	10.0	7.2
	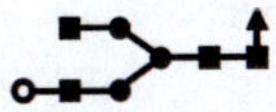	11.7	10.1	7.2
	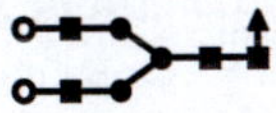	20.3	15.2	11.1
	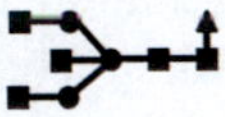	11.1	10.4	4.0
	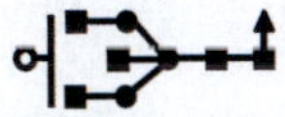	17.6	11.5	5.6
	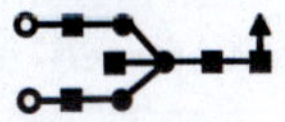	8.3	7.2	3.8
	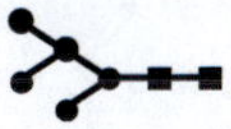	–[c]	–[c]	5.3
	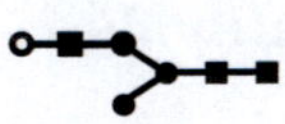	–[c]	–[c]	4.4
Monosialyl	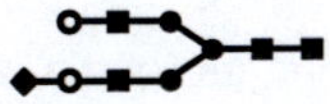	Not determined	Not determined	4.9
Disialyl	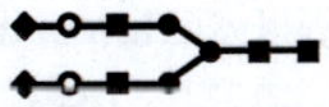	Not determined	Not determined	7.6

[a]Symbols: closed squares, *N*-acetyl glucosamine; closed circles, mannose; open circles, galactose; closed trangles, fucose; closed diamonds, sialic acid (*N*-acetyl neuraminic acid).
[b]Total amounts of identified structures for each sample are represented as 100%.
[c]Not detected.
From the authors own work, Kamihira et al. (2009).

antibodies produced in egg white of GM chickens possessed the oligosaccharide chain structures which were mostly asialo-oligosaccharides terminated with *N*-acetyl glucosamine (GlcNAc) and hybrid-type structures including a bisecting GlcNAc. A small proportion of oligosaccharides were terminated with galactose. These results were consistent with the previous study (Zhu et al., 2005). The oligosaccharide chain structures were also analyzed using scFv-Fc produced in the egg white of GM chickens and their offspring. The major oligosaccharide chain structures in scFv-Fc were almost the same as those of the chimeric antibodies and were retained over the generations. However, there were some differences in the contents of the individual oligosaccharide chains between the antibodies and scFv-Fc proteins, indicating that the structure of Fab or the fusion partner with the Fc-region may influence the synthesis of oligosaccharide chains of Fc-regions.

As shown in Table 6.3, oligosaccharide chain structures of the chimeric antibodies and scFv-Fc proteins produced in serum and egg yolk of chickens were analyzed. In the antibodies produced in the yolk, higher variation of oligosaccharide structures were observed compared with those produced in the egg white and their profiles were similar to human antibodies produced in the serum. A previous report showed that the oligosaccharide profiles of human IgGs are comprised of 16, 4 and 3 kinds of neutral, monosialyl and disialyl oligosaccharide chains, respectively (Takahashi et al., 1987). The eleven neutral oligosaccharide chains detected in the yolk of GM chicken were the same as those of human IgGs. Sialyl-oligosaccharides were also detected although their structure was not determined. The majority of the oligosaccharide chains were terminated with galactose. A high content of fucosylation was observed in this study, although a low rate of fucosylation in chicken IgGs was found (Raju et al., 2000). The profile of structures and contents of the oligosaccharide chain for scFv-Fc proteins showed high similarity between serum and yolk. This result indicates that recombinant antibodies or Fc-fusion proteins detected in yolk had been transported and accumulated from the serum, and that the structures of oligosaccharide chains were maintained during transport into the yolk.

6.6 Yolk Transport of Recombinant Antibodies and Fc-Fusion Proteins

Mammalian fetuses and neonates obtain maternal antibodies by transport through the placenta from the blood stream during fetal development and through the intestine from breast milk after birth. Similarly, the newly-hatched birds use the maternal antibody (IgY) accumulated in the egg yolk from the maternal blood stream as mediated by Fc receptors on the ovum (Loeken and Roth, 1983). Based on this mechanism, it was reported that human and mouse IgGs and fusion proteins with the Fc region of human IgG injected into the blood stream of birds could be transported to the yolk of eggs via IgY accumulation (Mohammed et al., 1998; Morrison et al., 2002; Kawabe et al., 2006b; Kitaguchi et al., 2008; Bae et al., 2010). In GM chickens and quails producing scFv-Fc proteins or recombinant monoclonal antibodies in the whole body, the recombinant antibodies were detected not only in the serum and egg white, but also in the yolk (Kamihira et al., 2005; Kawabe et al., 2006a; Kamihira et al., 2009). On the other hand, hEpo was produced in the serum and egg white of GM chickens, but production in the yolk was negligible (Kodama et al., 2008). These results indicate that the recombinant antibodies and the Fc-fusion proteins detected in the yolk of GM chickens seemed to be transported from the blood stream via the yolk transport mechanism.

The bioactivity of hEpo produced in the serum of GM chickens was higher than that of hEpo produced in the egg white. In fact, the oligosaccharide chain structures of serum-derived hEpo favored bioactivity (Kodama et al., 2008). When the *N*-linked oligosaccharide chains added to the Fc-region of antibodies produced by GM chickens were analyzed, the oligosaccharide chain structures, (including sialyloligosaccarides), produced in the yolk showed higher variation compared with those produced in the egg white. The structure and contents of oligosaccharides of antibodies produced in the serum was very similar to that in yolk (Kamihira et al., 2009). Thus, if hEpo was produced as Fc-fusion protein, the hEpo/Fc produced in the serum could be recovered from the yolk via maternal antibody transport mechanisms retaining the higher bioactivity (Fig. 6.6). For this purpose, we constructed a retroviral vector encoding an expression cassette of hEpo/Fc gene under the control of chicken β-actin promoter (Penno et al., 2010). Retroviral injection into 74 chicken embryos was performed in four experiments and the average hatchability of GM chickens was 28% (21/74). The hEpo/Fc was stably produced not only in the serum and egg white, but also in the egg yolk as expected. The expression levels

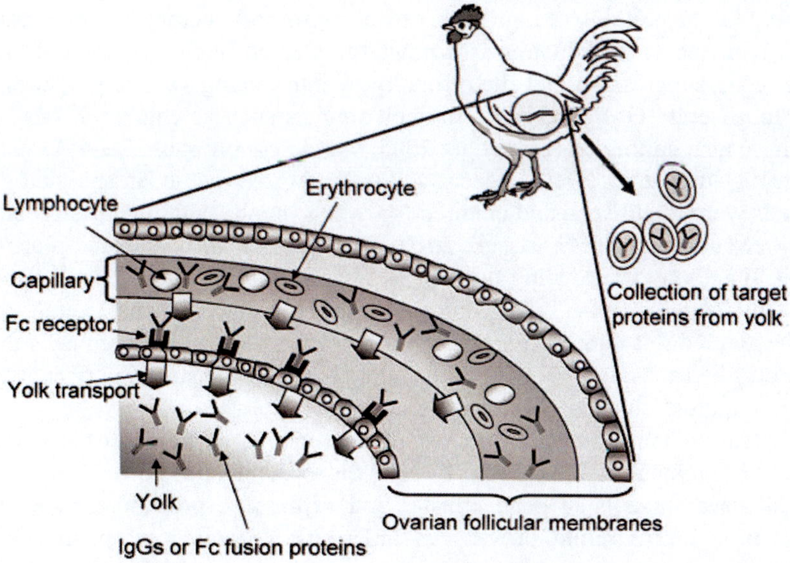

Fig. 6.6 Schematic drawing of transport of IgGs and Fc fusion proteins into chicken egg yolk. The target proteins produced in the serum of GM chickens were transferred and accumulated in the yolk

of hEpo/Fc in serum, egg white and yolk were in the range of 27–144, 9–66 and 12–41 $\mu g\ mL^{-1}$, respectively. The hEpo/Fc produced in the serum could be transported to and accumulated in the egg yolk. hEpo/Fc was expressed in the whole body of GM chickens without toxic effect. Recombinant hEpo production using transgenic mammals showed a physiological effect on the animal body (Massoud et al., 1996). In this regard, the production of human proteins such as cytokines and hormones in chickens may be more preferable to those produced in mammals. We measured in vitro bioactivity of hEpo/Fc partially purified from serum, egg white and yolk using Epo-dependent cells. The yolk-derived hEpo/Fc was almost equivalent to the serum-derived hEpo/Fc. Lectin blot analysis revealed that the terminal galactose was incorporated into the hEpo/Fc produced by GM chickens and the α2-6-linked sialic acid was detected in the serum- and yolk-derived hEpo/Fc, whereas these structures were negligible in the egg white-derived hEpo/Fc. Thus, hEpo/Fc produced in the serum of GM chickens was recovered from the egg yolk and retained some of the desired biological characteristics. The yolk transport of Fc fusion protein may represent an effective strategy for the production of therapeutic glycoproteins in transgenic chicken bioreactors.

6.7 Concluding Remarks

Transgenic avian bioreactors possess great potential for the production of recombinant pharmaceutical proteins. To date, many groups have developed procedures for the generation of transgenic chickens. We have achieved generation of GM chickens producing target proteins at high levels using retroviral vectors for gene transfer. Although retroviral transduction is a powerful method, there are some problems such as gene suppression and the difficulty of introducing very large genes. The efforts to generate GM chickens using cell sources such as chicken ES cells and PGCs have also continued (Zhu et al., 2005; van de Lavoir et al., 2006; Lavial and Pain, 2010; Song et al., 2010). Recently, it was reported that avian spermatogonial stem cells were established and cultured for over 2 months although transgenic offspring were not obtained (Yu et al., 2010). The technology of induced pluripotent stem (iPS) cells by transducing the reprogramming factors (Oct4, Sox2, Klf4 and c-Myc) into somatic cells (Takahashi and Yamanaka, 2006; Takahashi et al., 2007) could be applied to avian species to establish chicken pluripotent stem cells. In this case, it may be necessary to identify the factors for inducing stem cell phenotype by reprogramming of chicken somatic cells.

Transgenic avian bioreactors represent an attractive approach for the production of recombinant antibodies. Although the system still needs further technological developments in gene transfer and expression procedures, the unique features of chickens could provide an indispensable source of pharmaceutical proteins.

Acknowledgment This work was supported in part by Grant-in Aid for Scientific Research (no. 20360376) from the Japan Society for the Promotion of Science (JSPS).

References

Bae HD, Kobayashi M, Horio F, Murai A (2010) Identification of the amino acid residues involved in human IgG transport into egg yolks of Japanese quail (*Coturnix japonica*). Mol Immunol 47:1404–1410.

Burns JC, Friedmann T, Driever W, Burrascano M, Yee JK (1993) Vesicular stomatitis virus G glycoprotein pseudotyped retroviral vectors: concentration to very high titer and efficient gene transfer into mammalian and nonmammalian cells. Proc Natl Acad Sci USA 90:8033–8037.

Clackson T, Hoogenboom HR, Griffiths AD, Winter G (1991) Making antibody fragments using phage display libraries. Nature 352:624–628.

Dyck MK, Lacroix D, Pothier F, Sirard MA (2003) Making recombinant proteins in animals-different systems, different applications. Trends Biotechnol 21:394–399.

Enever C, Batuwangala T, Plummer C, Sepp A (2009) Next generation immunotherapeutics–honing the magic bullet. Curr Opin Biotechnol 20:405–411.

Hamburger V, Hamilton HL (1951) A series of normal stages in the development of the chick embryo. J Morphol 88:49–92.

Harel-Markowitz E, Gurevich M, Shore LS, Katz A, Stram Y, Shemesh M (2009) Use of sperm plasmid DNA lipofection combined with REMI (restriction enzyme-mediated insertion) for production of transgenic chickens expressing eGFP (enhanced green fluorescent protein) or human follicle-stimulating hormone. Biol Reprod 80:1046–1052.

Harvey AJ, Speksnijder G, Baugh LR, Morris JA, Ivarie R (2002) Expression of exogenous protein in the egg white of transgenic chickens. Nat Biotechnol 20:396–399.

Hotta A, Kamihira M, Itoh K, Morshed M, Kawabe Y, Ono K, Matsumoto H, Nishijima K, Iijima S (2004) Production of anti-CD2 chimeric antibody by recombinant animal cells. J Biosci Bioeng 98:298–303.

Houdebine LM (2009) Production of pharmaceutical proteins by transgenic animals. Comp Immunol Microbiol Infect Dis 32:107–121.

Ivarie R (2003) Avian transgenesis: progress towards the promise. Trends Biotechnol 21:14–19.

Kamihira M, Kawabe Y, Shindo T, Ono K, Esaka K, Yamashita T, Nishijima K, Iijima S (2009) Production of chimeric monoclonal antibodies by genetically manipulated chickens. J Biotechnol 141:18–25.

Kamihira M, Ono K, Esaka K, Nishijima K, Kigaku R, Komatsu H, Yamashita T, Kyogoku K, Iijima S (2005) High-level expression of single-chain Fv-Fc fusion protein in serum and egg white of genetically manipulated chickens by using a retroviral vector. J Virol 79:10864–10874.

Kawabe Y, Kamihira M, Ono K, Kyogoku K, Nishijima K, Iijima S (2006a) Production of scFv-Fc fusion protein using genetically manipulated quails. J Biosci Bioeng 102:297–303.

Kawabe Y, Naka T, Ando-Noumi N, Matsumoto H, Ono K, Nishijima K, Kamihira M, Iijima S (2006b) Transport of human immunoglobulin G and Fc-fusion proteins to chicken egg yolk. J Biosci Bioeng 102:518–523.

Kawabe Y, Naka T, Komatsu H, Nishijima K, Iijima S, Kamihira M (2008) Retroviral gene transduction into chicken embryo gonads through blood circulation. J Biosci Bioeng 106:598–601.

Kitaguchi K, Osada K, Horio F, Murai A (2008) Exclusion of polymeric immunoglobulins and selective immunoglobulin Y transport that recognizes its Fc region in avian ovarian follicles. Vet Immunol Immunopathol 121:290–299.

Kodama D, Nishimiya D, Iwata K, Yamaguchi K, Yoshida K, Kawabe Y, Motono M, Watanabe H, Yamashita T, Nishijima K, Kamihira M, Iijima S (2008) Production of human erythropoietin by chimeric chickens. Biochem Biophys Res Commun 367:834–839.

Köhler G, Milstein C (1975) Continuous cultures of fused cells secreting antibody of predefined specificity. Nature 256:495–497.

Köhler G (1980) Immunoglobulin chain loss in hybridoma lines. Proc Natl Acad Sci USA 77:2197–2199.

Koo BC, Kwon MS, Lee H, Kim M, Kim D, Roh JY, Park YY, Cui XS, Kim NH, Byun SJ, Kim T (2010) Tetracycline-dependent expression of the human erythropoietin gene in transgenic chickens. Transgenic Res 19:437–447.

Kues WA, Niemann H (2004) The contribution of farm animals to human health. Trends Biotechnol 22:286–294.

Kuwana T (1993) Migration of avian primordial germ cells toward the gonadal anlage. Dev Growth Differ 35:237–243.

Kwon SC, Choi JW, Jang HJ, Shin SS, Lee SK, Park TS, Choi IY, Lee GS, Song G, Han JY (2010) Production of biofunctional recombinant human interleukin 1 receptor antagonist (rhIL1RN) from transgenic quail egg white. Biol Reprod 82:1057–1064.

Kwon MS, Koo BC, Choi BR, Park YY, Lee YM, Suh HS, Park YS, Lee HT, Kim JH, Roh JY, Kim NH, Kim T (2008) Generation of transgenic chickens that produce bioactive human granulocyte-colony stimulating factor. Mol Reprod Dev 75:1120–1126.

Kyogoku K, Yoshida K, Watanabe H, Yamashita T, Kawabe Y, Motono M, Nishijima K, Kamihira M, Iijima S (2008) Production of recombinant tumor necrosis factor receptor/Fc fusion protein by genetically manipulated chickens. J Biosci Bioeng 105:454–459.

Lavial F, Pain B (2010) Chicken embryonic stem cells as a non-mammalian embryonic stem cell model. Dev Growth Differ 52:101–114.

Lavine G (2009) FDA approves first biological product derived from transgenic animal. Am J Health Syst Pharm 66:518.

Lee SH, Gupta MK, Han DW, Han SY, Uhm SJ, Kim T, Lee HT (2007) Development of transgenic chickens expressing human parathormone under the control of a ubiquitous promoter by using a retrovirus vector system. Poult Sci 86:2221–2227.

Lillico SG, McGrew MJ, Sherman A, Sang HM (2005) Transgenic chickens as bioreactors for protein-based drugs. Drug Discov Today 10:191–196.

Lillico SG, Sherman A, McGrew MJ, Robertson CD, Smith J, Haslam C, Barnard P, Radcliffe PA, Mitrophanous KA, Elliot EA, Sang HM (2007) Oviduct-specific expression of two therapeutic proteins in transgenic hens. Proc Natl Acad Sci USA 104:1771–1776.

Loeken MR, Roth TF (1983) Analysis of maternal IgG subpopulations which are transported into the chicken oocyte. Immunology 49:21–28.

Love J, Gribbin C, Mather C, Sang H (1994) Transgenic birds by DNA microinjection. Biotechnology 12:60–63.

Massoud M, Attal J, Thépot D, Pointu H, Stinnakre MG, Théron MC, Lopez C, Houdebine LM (1996) The deleterious effects of human erythropoietin gene driven by the rabbit whey acidic protein gene promoter in transgenic rabbits. Reprod Nutr Dev 36:555–563.

McGrew MJ, Sherman A, Ellard FM, Lillico SG, Gilhooley HJ, Kingsman AJ, Mitrophanous KA, Sang H (2004) Efficient production of germline transgenic chickens using lentiviral vectors. EMBO Rep 5:728–733.

Mizuarai S, Ono K, Yamaguchi K, Nishijima K, Kamihira M, Iijima S (2001) Production of transgenic quails with high frequency of germ-line transmission using VSV-G pseudotyped retroviral vector. Biochem Biophys Res Commun 286:456–463.

Mizuguchi H, Xu Z, Ishii-Watabe A, Uchida E, Hayakawa T (2000) IRES-dependent second gene expression is significantly lower than cap-dependent first gene expression in a bicistronic vector. Mol Ther 1:376–382.

Mohammed SM, Morrison S, Wims L, Trinh KR, Wildeman AG, Bonselaar J, Etches RJ (1998) Deposition of genetically engineered human antibodies into the egg yolk of hens. Immunotechnology 4:115–125.

Morrison SL, Mohammed MS, Wims LA, Trinh R, Etches R (2002) Sequences in antibody molecules important for receptor-mediated transport into the chicken egg yolk. Mol Immunol 38:619–625.

Mozdziak PE, Borwornpinyo S, McCoy DW, Petitte JN (2003) Development of transgenic chickens expressing bacterial β-galactosidase. Dev Dyn 226:439–445.

Murray KM, Dahl SL (1997) Recombinant human tumor necrosis factor receptor (p75) Fc fusion protein (TNFR:Fc) in rheumatoid arthritis. Ann Pharmacother 31:1335–1338.

Penno CA, Kawabe Y, Ito A, Kamihira M (2010) Production of recombinant human erythropoietin/Fc fusion protein by genetically manipulated chickens. Transgenic Res 19: 187–195.

Pollock DP, Kutzko JP, Birck-Wilson E, Williams JL, Echelard Y, Meade HM (1999) Transgenic milk as a method for the production of recombinant antibodies. J Immunol Methods 231: 147–157.

Raju TS, Briggs JB, Borge SM, Jones AJ (2000) Species-specific variation in glycosylation of IgG: evidence for the species-specific sialylation and branch-specific galactosylation and importance for engineering recombinant glycoprotein therapeutics. Glycobiology 10:477–486.

Rapp JC, Harvey AJ, Speksnijder GL, Hu W, Ivarie R (2003) Biologically active human interferon alpha-2b produced in the egg white of transgenic hens. Transgenic Res 12:569–575.

Reichert JM, Valge-Archer VE (2007) Development trends for monoclonal antibody cancer therapeutics. Nat Rev Drug Discov 6:349–356.

Robertson G, Garrick D, Wu W, Kearns M, Martin D, Whitelaw E (1995) Position-dependent variegation of globin transgene expression in mice. Proc Natl Acad Sci USA 92: 5371–5375.

Sang H (2004) Prospects for transgenesis in the chick. Mech Dev 121:1179–1186.

Sherman A, Dawson A, Mather C, Gilhooley H, Li Y, Mitchell R, Finnegan D, Sang H (1998) Transposition of the *Drosophila* element *mariner* into the chicken germ line. Nat Biotechnol 16:1050–1053.

Shukla AA, Thömmes J (2010) Recent advances in large-scale production of monoclonal antibodies and related proteins. Trends Biotechnol 28:253–261.

Song G, Park TS, Kim TM, Han JY (2010) Avian biotechnology: insights from germ cell-mediated transgenic systems. J Poult Sci 47:197–207.

Steurer W, Nickerson PW, Steele AW, Steiger J, Zheng XX, Strom TB (1995) Ex vivo coating of islet cell allografts with murine CTLA4/Fc promotes graft tolerance. J Immunol 155: 1165–1174.

Tajima A, Hayashi H, Kamizumi A, Ogura J, Kuwana T, Chikamune T (1999) Study on the concentration of circulating primordial germ cells (cPGCs) in early chick embryos. J Exp Zool 284:759–764.

Takahashi N, Ishii I, Ishihara H, Mori M, Tejima S, Jefferis R, Endo S, Arata Y (1987) Comparative structural study of the *N*-linked oligosaccharides of human normal and pathological immunoglobulin G. Biochemistry 26:1137–1144.

Takahashi K, Tanabe K, Ohnuki M, Narita M, Ichisaka T, Tomoda K, Yamanaka S (2007) Induction of pluripotent stem cells from adult human fibroblasts by defined factors. Cell 131:861–872.

Takahashi K, Yamanaka S (2006) Induction of pluripotent stem cells from mouse embryonic and adult fibroblast cultures by defined factors. Cell 126:663–676.

Tsumoto K, Shinoki K, Kondo H, Uchikawa M, Juji T, Kumagai I (1998) Highly efficient recovery of functional single-chain Fv fragments from inclusion bodies overexpressed in Escherichia coli by controlled introduction of oxidizing reagent – application to a human single-chain Fv fragment. J Immunol Methods 219:119–129.

van de Lavoir MC, Diamond JH, Leighton PA, Mather-Love C, Heyer BS, Bradshaw R, Kerchner A, Hooi LT, Gessaro TM, Swanberg SE, Delany ME, Etches RJ (2006) Germline transmission of genetically modified primordial germ cells. Nature 441:766–769.

Wheeler MB (2007) Agricultural applications for transgenic livestock. Trends Biotechnol 25: 204–210.

Whitelaw CB (2004) Transgenic livestock made easy. Trends Biotechnol 22:157–159.

Wilson C, Bellen HJ, Gehring WJ (1990) Position effects on eukaryotic gene expression. Annu Rev Cell Biol 6:679–714.

Wurm FM (2004) Production of recombinant protein therapeutics in cultivated mammalian cells. Nat Biotechnol 22:1393–1398.

Yu F, Ding LJ, Sun GB, Sun PX, He XH, Ni LG, Li BC (2010) Transgenic sperm produced by electrotransfection and allogeneic transplantation of chicken fetal spermatogonial stem cells. Mol Reprod Dev 77:340–347.

Zhu L, van de Lavoir MC, Albanese J, Beenhouwer DO, Cardarelli PM, Cuison S, Deng DF, Deshpande S, Diamond JH, Green L et al. (2005) Production of human monoclonal antibody in eggs of chimeric chickens. Nat Biotechnol 23:1159–1169.

Chapter 7
Production of Antibodies in Plants

Alejandro Sarrion-Perdigones, Paloma Juarez, Antonio Granell, and Diego Orzaez

Abstract Plants are specially suited organisms for production of high amounts of carbon-reduced molecules at low cost. Is it possible to take advantage of this privileged position for the manufacturing of value-added proteins as antibodies? Three decades ago the assembly of functional antibodies in plants was first demonstrated. Since then, many examples of plant-made antibodies in different formats and for different applications have been reported. It was not until improvements in transient expression technologies were introduced in recent years that plant-made antibody yields became similar to those of competing platforms. Transient expression systems are adapting well to the latest market trends of speed and quality, and therefore successful products are expected to reach the market propelled by this technology wave. New developments will probably come from optimized platforms, designed according to synthetic biology standards, to integrate in a single plant chassis a number of traits specifically engineered for recombinant protein production. Platform optimization will include advances on subcellular targeting, transcriptional control, protein degradation, glyco-engineering and downstream processing, all of which are reviewed here, together with other adaptations to the agricultural scale as gene containment or identity preservation.

7.1 Introduction

Plants represent an advantageous platform for large-scale production of antibodies because they are economically sustainable, scalable and are not easily contaminated with human pathogens (Stoger et al., 2002, Ma et al., 2003, Ko et al., 2009). Recombinant proteins have been mainly produced in *E.coli* and yeast but today eukaryote systems are in the lead of the market. Even when Chinese Hamster Ovary (CHO) (Wurm, 2004) or baculovirus-infected insect cells (Berger et al., 2004) are the current preferred systems, the number and types of antibodies expressed in plants have increased incessantly since the first reports in 1989 (Hiatt et al., 1989), illustrating the versatility of plants as a production system. In the last years, some of these

D. Orzaez (✉), A. Sarrion-Perdigones, P. Juarez, and A. Granell
Instituto de Biología Molecular y Celular de Plantas (IBMCP), UPV-CSIC, 46022 Valencia, Spain
e-mail: dorzaez@ibmcp.upv.es

Alejandro Sarrion-Perdigones and Paloma Juarez contributed equally to this book chapter.

M. Al-Rubeai (ed.), *Antibody Expression and Production*, Cell Engineering 7,
DOI 10.1007/978-94-007-1257-7_7, © Springer Science+Business Media B.V. 2011

Plant-made antibodies (PMAbs) have been presented as promising therapeutic solutions. For instance, Bayer Innovation GMBH's Non-Hodgkin's lymphoma IgG has recently got the approval to start a Phase I study by the FDA, which represents a milestone for plant-manufactured antibodies.

Antibodies have been produced in different formats in various dicot and monocot plants, in cell cultures and different plant organs, either by transient or stable transformation. Competitive yields, in comparison to those achieved with CHO cells, have been reported not only in a laboratory scale (Giritch et al., 2006, Petruccelli et al., 2006) but also in prototype industrial setups (Vezina et al., 2009, Bendandi et al., 2010). Despite these high production levels, there are still many technical aspects to deal with, such as the efficient targeting to appropriate subcellular localizations, the understanding of how antibody degradation occurs and the way to tackle this problem, the steps taken towards improved glycosylation patterns and the optimization of downstream processing. This chapter analyzes these technical aspects by briefly describing and discussing some of the most relevant experimental work carried out in each particular aspect, with special attention paid to most recent developments as those concerning transient expression systems. The authors' view on the perspectives of the field is also discussed. The aim of this chapter is to provide a general view on the *state of the art* of plant-made antibodies rather than to focus in certain aspects of it. For those readers interested in deepening their knowledge in a particular subject (e.g. antibody formats, glycosilation, viral vectors, etc), the text includes a number of references to excellent recent reviews, which can be consulted for more specific information.

7.2 Diversity of PMAb Formats

The field of PMAbs came into light in 1989 by Hiatt et al. In this first work, tobacco plants expressing single gamma or kappa immunoglobulin chains were sexually crossed to yield progeny which produced a complete functional antibody. Hiatt's work was soon followed by other pioneering groups (Hein et al., 1991, Düring et al., 1990, De Neve et al., 1993) who robustly established that the plant cell machinery is competent to synthesise, fold and secrete functional antigen-binding full-size IgG antibodies.

To this date, the great majority of mAb produced in plants are human IgG. The simplicity of downstream processing methods which are already well established and its widespread use for immunotherapy are two of the main reasons for this choice. IgA, the most abundant isotype in humans has also captured the attention of plant biotechnologists due to its relevance for passive immunotherapeutic approaches. Chimeric IgG/A antibodies both in its monomeric and in its secretory (sIgA/G) forms, have been shown to properly fold and assembly in tobacco and rice and full sIgAs have been produced in corn see (Wycoff, 2005) for a review.

In addition to full-size antibodies, a numbr of a engineered antibody fragments and fusion proteins have been successfully produced in plants (see (Conrad and Floss, 2010) for a recent review). Single-chain variable fragments (scFv) consisting of two variable regions VH and VL of an antibody artificially linked by a flexible polypeptide, have been successfully synthesized not only in different plant species (tobacco, tomato, pea) and plant organs (leaves, fruits, seed) but also in plant cell cultures (Fiedler et al., 1997, Stoger et al., 2002). Although the reported production levels are highly variable, in general scFvs are better expressed as ER-retained proteins, levels ranging from almost undetectable to up to 6.8% total soluble protein (TSP) (Fiedler et al., 1997).

In 1993, a class of antibodies devoided of light chain was discovered in camels and llamas (Hamers-Casterman et al., 1993). The variable domain, named V_HH or nanobody, acts as the real antibody and can itself bind an antigen. V_HH have a molecular weight of 14–15 kDa and a structure that is highly similar to the human family III VH domain (Muyldermans, 2001). Nanobodies can be easily produced in prokaryotic or eukaryotic host organisms (as they do not require glycosylation) and their unique biophysical and pharmacological characteristics render these molecules as ideal candidates for drug development, as a part of a biosensor to diagnose infections, and to treat diseases like cancer or trypanosomosis (Van Bockstaele et al., 2009, Muyldermans et al., 2009). Teh and Kavanagh (2010) demonstrated the potential of plants as a nanobody production system by producing at an exceptional high level an anti-hen egg white lysozyme nanobody in *Nicotiana* leaves. They got post-purification yields of 0.1–0.22 mg g^{-1} fresh weight, only comparable to those achieved by Giritch et al. (2006) and Huang et al. (2010) when producing full-size IgGs.

In order to improve plant-made antibodies stability, downstream processing or functions, several structural modifications have been proposed. As a first example, a fusion with elastin-like peptides (ELP) facilitated the anti-HIV-1 antibody 2G12 downstream processing, increasing the final yields (Floss et al., 2008, 2009). Obregon et al. (2006) reported a new strategy for plant-based protein production. They engineered a human immunodeficiency virus-1 (HIV-1) p24—immunoglobulin A (IgA) antigen—antibody fusion molecule for therapeutic purposes. HIV-p24 antigen was expressed as a genetic fusion with the α2-3 regions from human IgG heavy chain, and targeted to the endomembrane system. The expression of this fusion protein was obtained with yields 13-fold higher than HIV p24 expressed alone, probably due to stabilization of the antigen production. Structures where a certain antibody is fused to its own antigen have been first reported by Chargelegue et al. (2005). They created a structure with potential immunomodulating activity consisting of a tetanus toxin C fragment-specific monoclonal antibody fused with the tetanus toxin C fragment, presenting with it a new vaccine model.

7.3 Diversity of Antibody Functions

The production of antibodies in plants has pursued multiple goals, from the immunomodulation of plant endogenous functions to the production of veterinarian or human therapeutic proteins. Many consideration including the choice for host species (GRAS or otherwise), the expected yields, the cell compartment for accumulation, etc., are strongly dependent on its intended use. Here we review the most relevant applications described to date.

7.3.1 Antibody-Mediated Modulation of Plant Endogenous Targets

PMAbs have been employed to obtain disease resistance against plant pathogens. For instance, Van Engelen et al. (1994) studied the expression of genes encoding antibodies in roots of transgenic plants in order to obtain resistance against plant root pathogens. They used a model IgG that binds to a fungal cutinase. Moreover, Voss et al. (1995) developed a specific antibody against Tobacco Mosaic Virus (anti-TMV) to control plant infection. In another example, a functional IgM monomer specific to stylet secretions of the root-knot nematode without J Chain was produced (Baum et al., 1996). They performed a similar approach to the previous described, and achieved to assemble a monomeric IgM specific to stylet secretions of the root-knot nematode. Boonrod et al. (2004) reported a strategy to achieve virus resistance based on the expression of scFvs against a conserved domain in a plant viral RNA-dependent RNA polymerase (RdRp), which are essential for replication, conferring resistance to different plant viruses.

The unique binding characteristics of antibodies make them also powerful tools for the modulation of plant input and output traits. PMAbs have been used to facilitate the immunomodulation of metabolic pathways (Nolke et al., 2006). Jobling et al. (2003) targeted a VHH against potato starch branching enzyme A (SBE A) to the chloroplast and reported a more efficient inhibition of the enzyme activity and a stronger phenotype than previously seen using antisense technology. It has also been reported the effect of an ornithine decarboxylase (ODC) specific murine scFV in tobacco plants, that resulted in an inhibition up to 90% of endogenous ODC activity, detecting a significant reduction in putrescine, spermidine and spermine levels (Nolke et al., 2005). Several other metabolic applications have been reported such as the control of gibberellins-regulated plant growth using an antigibberellin scFV fragment in the ER of tobacco plants (Shimada et al., 1999) or the subcellular distribution of ABA that was also influenced in transgenic potato plants constitutively expressing an anti-ABA scFV (Strauss et al., 2001).

7.3.2 Anti-microbial Antibodies

Passive immunization with plant-made neutralizating antibodies is among the most promising application of PMAb, particularly considering that GRAS species may have lower purification requirements for topic and mucosal applications. Accordingly, several examples in the literature demonstrate the potential of PMAb as anti-microbial agents.

Zeitlin et al. (1998) compared a humanized anti-herpes simplex virus 2 (HSV-2) mAb expressed in mammalian cell culture with its counterpart expressed in soybean, proving not only the similarity in their stability in mucosal secretions of the human reproductive tract, but also efficacy for prevention of vaginal HSV-2 infection in mouse. Ko et al. (2003) produced a neutralizing anti-rabies virus IgG in tobacco. In a more recent example, Ramessar et al. (2008) developed a mAb based vaginal microbicide to prevent HIV transmission and proved that the neutralization capability was equal to or superior to its counterpart produced in CHO cells.

An interesting isotype for anti-microbial activity is IgA, the most abundant antibody in humans. Moreover, its secretory form (sIgA) is the most abundant form of immunoglobulin in mucosal secretions and it is involved, among other function in passive protection against mucosal pathogens (see (Corthesy, 2002, Corthesy and Spertini, 1999) for reviews). Using tobacco as recombinant platform, Ma and coworkers assembled a Secretory IgA-G which recognized the streptococcal antigen (SA) I/II cell surface adhesion molecule of *Streptococcus mutans* and *S. sobrinus*, conceived for passive mucosal immunotherapy (Ma et al., 1995). Three years later (Ma et al., 1998), the same authors showed that the secretory antibody produced in tobacco, had a higher functional affinity than its parent murine IgG antibody Guy's 13, probably due to its dimeric structure, and provided specific protection in humans against oral streptococcal colonization for at least 4 months. This IgA-G based product, CaroRX, became the first plant-made medical device approved in Europe. More recent examples are the plant assembly of avian chicken sIgA (Wieland et al., 2006) and the use of viral vectors to express IgA molecules able to neutralize TGEV infections both in vitro and in vivo (Alamillo et al., 2006).

7.3.3 Immuno-Modulation and Anti Tumor Activity

Several PMAbs have been produced for immunotherapeutic purposes, mainly but not exclusively represented by anti-tumor activities. Kathuria et al. (2002) transiently expressed in tobacco leaves a chimeric IgG against Human Chorionic Gonadotropin (HCG). Passive immunization with these antibodies may have clinical utility such as contraceptive measures and immunotherapy of cancers. There are other research groups which have reported inhibition of tumour growth by PMAbs (Ko and Koprowski, 2005, Brodzik et al., 2006). McCormick et al. (2008) designed individualized idiotype vaccines for the treatment of non-Hodgkin's lymphoma, which were administered to lymphoma patients in a phase I safety and immunogenicity clinical trial. These vaccines consisted of idiotypic scFv derived from each patient's tumor. Every individualized scFv vaccine was well tolerated by all patients and was shown to induce tumor-specific immune response in 75% of the cases. However, the expression levels of the scFv were highly variable and downstream processing of these fragments lacked an established purification method. Bendandi et al. (2010) reported new individualized idiotype vaccines of whole immunoglobulins G in plants for non-Hodgkin's lymphoma and achieved efficient expression levels with magnifection process (see below). The antigen purification with protein A affinity capture provided a product with pharmaceutical-grade purity.

7.3.4 *Affinity Purification*

An interesting application for PMAb was exemplified by the work of Ramirez et al. (2003) and Yano et al. (2004) who reported the expression in transgenic tobacco and posterior purification of an anti-hepatitis B virus surface antigen mouse IgG1 monoclonal antibody (anti-HBsAg). This recombinant antibody is currently used for the industrial purification of the recombinant vaccine antigen.

7.4 Practical Considerations on PMAb Design

The design of a plant biofactory for antibody manufacturing requires multiple considerations involving not only the expression levels of the antibody but also additional aspects as subcellular targeting, protein degradation, glycosilation patterns and downstram strategies, all of them influencing the yield, quality and cost of the final product. To date, most of these aspects have been addressed separately, mainly on a empirical basis. However, future optimizations will probably require designs that integrate all of them following a systems biology approach. Here we describe some of the main factors affecting antibody production in plants.

7.4.1 *Subcellular Localization*

Although extracellular secretion is the natural route for antibodies in mammals, targeting antibody chains to specific compartments in the plant cell can result in advantages in terms of stability, yield or downstream processing (De Muynck et al., 2010).

Different compartments have been tested as destination for recombinant antibodies, including cytosol (De Jaeger et al., 1999), chloroplast (Düring et al., 1990, Jobling et al., 2003), plasma membrane (Vine et al., 2001), vacuole (Petruccelli et al., 2006), ER (De Wilde et al., 1998, De Muynck et al., 2009), ER-derived protein bodies (PBs) (Frigerio et al., 2000) and protein storage vacuoles (PSV) (Petruccelli et al., 2006). Amongst them, the secretory pathway appears to be the most convenient route for a correct antibody folding and assembly due to the oxidizing environment, the low abundance of proteases and the presence of molecular chaperones found in the endoplasmic reticulum (Ma et al., 2003).

Antibody chains are targeted to the secretory pathway using an appropriate N-terminal signal peptide, either native (Hiatt et al., 1989, Sainsbury et al., 2008a) or replaced by a plant one (Düring et al., 1990). Once there, they can either be efficiently retrieved from the cis-Golgi back to the ER using a C-terminal H/KDEL retention signal or secreted to the apoplast, following the secretory pathway (Petruccelli et al., 2006, De Muynck et al., 2009). Although several antibodies have been reported as strictly apoplastic (Düring et al., 1990, De Wilde et al., 1998, De Muynck et al., 2009), retention in the ER not only seems a possibility for yield improvement but also for avoiding complex plant Golgi-derived glycosylation patterns that could cause immunogenicity in target organisms.

The most important results in yield improvement using ER retention are those involving several groups producing single-chain antibodies. Both anti-cutinase and anti-oxazolone ScFVs produced in *Nicotiana tabacum* by Schouten et al. (1996) and Fiedler et al. (1997), increased the TSP by adding a KDEL motif (100-fold increase from 0.01 to 1% TSP in the first case and a 10- to 20-fold increase, up to 6.8% TSP on the second). De Jaeger et al. (1999) reported a 30-fold increase, up to 3% TSP, in anti-dihydroflavonol 4-reductase ScFV production in *Petunia hybrid* petals.

In contrast, targeting to the ER does not improve full-antibody yield as much as it does for single-chain antibody. The addition of the KDEL sequence to light and heavy chains of the 14D9 monoclonal IgG1 produced by Petrucelli et al. (2006) in transgenic tobacco leaves and seeds led to a TSP 4—11-fold increase when chains were expressed individually. The accumulation of the complete antibody was higher than individual chains but the increase was somewhat lower, from 2.90% TSP for the secreted antibody to 5.20% TSP for the KDEL-antibody (twofold increase). Vaquero et al. (2002) reported a tenfold increase in their anti-carcinoembyonic diabody when expressed in transgenic *Nicotiana tabacum* (0.9 mg kg^{-1} for the apoplast diabody and 9 mg kg^{-1} for the ER targeted one). Although the majority of KDEL-antibodies are localized in the expected compartment, part of the chains can escape the ER and accumulate in other subcellular compartment. Petrucelli et al. (2006) described that their KDEL-tagged IgG1 was efficiently retained in the ER but it was partially secreted and sorted to PSV in tobacco seeds where it undergoes proteolytic cleavage. This *detour* effect was previously reported by Torres et al. (2001) that found KDEL-scFv in storage compartments in rice endosperms. These effects should be further investigated because they may led to heterogeneous glycosylation patterns.

Regarding apoplast secretion, Hadlington et al. (2003) compared the accumulation of two isotypes of anticaries Guy's 13 antibody in tobacco leaves (IgG and a sIgG/A hybrid form). Murine IgG antibody was efficiently secreted to the apoplast and expressed at 1% TSP but sIgA accumulated at higher levels (8% TSP) and was not fully secreted to the apoplast as described by Frigerio et al. (2000). The authors described a cryptic sorting signal in the tailpiece of the IgA/G heavy chain that led to vacuolar delivery. It was proved that the deletion of this signal led to hybrid IgA/G secretion at levels comparable to IgG secretion. This tailpiece signal can be useful for intracellular targeting and accumulation as it seems to increase mAb productivity in plants, but this retention was accompanied by discrete degradation as described by the authors. It has also been shown in the case of sIgA that the subcellular localization of this antibody depends on its assembly status (Nicholson et al., 2005). The nonassembled light chain, heavy chain and secretory component accumulated predominantly within endoplasmic reticulum-derived protein bodies, while the assembled antibody, with antigen-binding function, accumulated specifically in protein storage vacuoles.

7.4.2 Antibody Degradation

An important problem associated with PMAb production is the presence of partially assembled or partially degraded antibody fragments. Gathering together results from various host species and different antibodies, a vast collection of fragments of different sizes have been reported, ranging from partially assembled complexes H2L or H2 showing molecular weights between 100 and 150 kDa, to smaller ones as partially degraded un-associated heavy chains (44 kDa), Fab, (Fab)2 and FC fragments around 44 kDa (De Neve et al., 1993), unassembled light chains and eventually, smaller degradation fragments difficult to assign.

Partially assembled antibody complexes may appear as result of unbalanced expression of the two antibody chains. It has been proposed that a balanced co-expression of heavy and light chain is one of the clue factors for achieving high yield, since unassembled antibody chains which have been retained by the ER-resident chaperone BiP, are often degraded via ERAD. Besides, fragments may result from endogenous protease activities located either in the ER, the vacuole or the apoplast (Van Engelen et al., 1994) and/or from activity of peptidases released during tissue homogenization.

The problem of antibody degradation fragments was systematically addressed by Sharp and Doran (2001) who showed that the antibody fragments produced by a secreted IgG1 (Guy's 13) in a root expression system were not caused by degradation during grinding, but generated in the apoplast and also between the ER and Golgi. Using gelatine and human IgG as substrate, Muynck et al. (De Muynck et al., 2009) also showed, a correlation between the decrease in the amount of antibody produced in suspension leaf cells, and the increase of peptidase activity. Nevertheless it has been proved that ER-retained antibodies are susceptible to degradation when expressed in plants (Sainsbury and Lomonossoff, 2008b). This fact is possibly due to the antibody escaping from the ER (De Muynck et al., 2010) or even though correctly retained, later transported to PSV to be then cleaved (Petruccelli et al., 2006).

The prevention of unintended proteolysis would require an in-depth understanding of the plant protease activities in the different PMAb platforms. To date, several genetic strategies can be used to deal with this quality drawback. For instance, gene knockout or silencing of plant peptidases could be a tool if there is a single or only a few target peptidases which are not essential for the plant growth (De Muynck et al., 2010). Other approaches are blocking the secretory pathway to avoid extracellular peptidase activity (Sharp and Doran, 2001), the use of tissue-specific promoters to confine transgene expression to compartments with reduced metabolic activity, targeting proteins to specific cellular organelles (Benchabane et al., 2008), fusing stabilizing partners (De Muynck et al., 2010) or addition of gelatine as competitor substrate for peptidases (Wongsamuth and Doran, 1997).

7.4.3 Glycosylation

The glycosylation of the constant regions of antibodies is a determinant feature for important antibody properties such as resistance to protein degradation, half life in serum and also for mAb-based immunotherapy (Ko et al., 2009).

One of main advantages of plants is that, as eukaryotic platforms, they are able to express glycoproteins. However, final glycopatterns differ between plant and animal cells (Fig. 7.1). Whereas ER glycosylation patterns are shared, a number of differences occur at the level of the Golgi apparatus. In the early steps in the ER, certain Asn residues are decorated with an oligosaccharide rich in mannose which is later partially removed by ER-resident manosidase. Still in the ER, an *N*-acetylglucosamine transferase adds terminal N-acetylglucosamine residues and the formed biantennary complex finally enters the Golgi apparatus. It is at this point where differences in the complex *N*-glycans formation start occurring. In the plant Golgi, *N*-linked glycan complexes are decorated with β(1,2)-xylose and α(1,3)-fucose residues, whereas the human *N*-glycan contain α(1,6)-fucose. Plants also seldom carry galactose and lack the typical human decoration with sialic acid (Ko et al., 2009).

Although it is quite controversial and has been briefly studied, it seems convenient to "humanize" plant-made mAb glycosylation patterns to avoid potential side effects during immunotherapy. Antigenicity of non-mammalian *N*-glycan residues on a plant-derived mAb has been observed in some experimental animals. Both xylose and α(1,3)-fucose residues were able to elicit specific antibodies in rabbits after parenteral application (Jin et al., 2008) however, this cannot be generalized to every species.

The therapeutic importance of glycoproteins has driven the recent interest in the glycoengineering of production platforms and plants have not been an exception. The main trend consists in mimicking the human glycosylation by means of genetic engineering (humanization) (Fig. 7.1), although there is the possibility of glycoengineering for novel (non-native) glycopatterns in order to improve farmacokinetics and/or effector functions (biobetters). To date, there are two ways to tackle the humanization problem, which summed up are close to make this specific sugar decoration possible: the inactivation of endogenous glycotransferases and/or expression of heterologous glycotransferases (De Muynck et al., 2010). A first approach to humanize plant glycosylation-patterns was performed by Koprivova et al. (2004) in the moss *Physcomitrella patens.* By means of homologous recombination the genes for α1,3-fucosyltransferase (FT) and β1,2-xylosyltransferase (XT) were disrupted. Two years later Cox et al. (2006) expressed a human mAb in the aquatic plant *Lemna minor* silenced in both FT and XT, using an RNA interference construct. These plant glycoengineered antibodies showed greater binding affinity for human Fc receptors compared to its counterpart produced in CHO.

The next step to approach human decorated antibodies consisted of a chimeric form of the human β1,4-galactosyltransferase (GalT). This was first attempted by Bakker et al. (2001) and later improved by Strasser et al. (2009) who reported the production of mAbs with a homogeneous galactose structure. This latter humanized mAb was achieved by eliminating plant-specific N-glycosylation (FT/XT knocked-out plants) and targeting GalT to a final stage of the glycosylation pathway.

To achieve the complete human pattern there was still a modification to sum up, the addition of terminal sialic residues. The first approach in this direction consisted in the engineering of the full mammalian *N*-acetylneuraminic acid (Neu5Ac) biosynthesis pathway into *Arabidopsis thaliana* (Castilho et al., 2008). Following a similar approach, the sialylation of the recombinant monoclonal antibody 2G12 retaining full neutralization activity was recently reported in *Nicotiana benthamiana*

(Castilho et al., 2010). Moreover, the sialylated antibody was free of xylose and fucose, as it was produced in mutants that lacked plant-specific *N*-glycan residues. Although many of the transgenes involved in this work were expressed transiently, this latter step represents an important milestone in the humanization of plant glycosilation patterns.

7.4.4 Downstream Processing

Downstream processing is an important bottleneck in PMAb platforms, as extraction and purification costs represent an important part of total manufacturing costs. Therefore, the optimization of downstream processing has received much attention in recent years.

The optimal extraction method changes with the subcellular localization of the target antibody. Hassan et al. (2008) studied several extraction methods of monoclonal antibodies targeted to different subcellular compartments in transgenic *N. tabacum* plants: apoplast, ER and membrane-bound. Grinding was necessary for the optimal extraction of ER-retained and membrane-bound antibodies with the later also requiring the use of detergent in the extraction buffer. However, secreted antibodies could be easily recovered just by freeze-thawing before adding an extraction buffer.

The possibility of choosing Generally Recognized as Safe (GRAS) plant tissues makes the purification a not always necessary-step, especially for mucosal immunotherapy, which could mean a drastic reduction of final costs. For other applications, the development of antibody purification methodology is necessary. An interesting option is PEG/salt aqueous two-phase partition system used, for instance, by Platis and Labrou (2009) for the recovery of monoclonal antibodies from unclarified transgenic tobacco extract. This system avoids centrifugation and filtration steps and therefore reduces final costs (Azevedo et al., 2009a, b).

Besides, to achieve a more efficient purification it is possible to use an affinity purification protocol with protein G sepharose columns or protein A agarose columns (De Muynck et al., 2009, Ko et al., 2009), with the drawback that it is not possible to avoid previous extract clarification steps (Valdes et al., 2003). These approaches have been applied particularly for IgG purification but, currently there is a lack of established protocols for purification of other isotypes such as IgA. However, recently the Staphylococcal Superantigen-Like Protein 7 (SSL7) (Langley et al., 2005, Wines et al., 2006, Ramsland et al., 2007) has been used for affinity chromatography purification of plant-made IgA (Juárez et al. unpublished results).

7.5 Transient Expression Systems

In opposition to stable transformants, transient expression systems offer as main advantage the speed in which new recombinant proteins can be obtained. Transcriptionally-active foreign coding sequences can be transiently introduced in plants by different methods, the most widely used being (i) pressure infiltration of cultures (agroinfiltration) carrying binary vectors, (ii) recombinant viral vectors, or (iii) a combination of both, consisting in the agroinfiltration of a viral infective clone inserted in a binary vector. See (Lico et al., 2008) for a review.

Agrobacterium-mediated transient expression has traditionally been used for construct-testing or to quickly produce small amounts of proteins. Some mAbs have been expressed by direct agroinjection in leaves (Vaquero et al., 1999) or fruits (Orzaez et al., 2006) at yields up to 20 mg kg^{-1} fresh weight (less than 0.4% TSP). *Agrobacterium* suspension cultures are infiltrated, either using a syringe and pressing the leaf (D'Aoust et al., 2009) or using more complex vacuum systems (Kapila et al., 1997), into the intercellular space of leaf mesophyll. Bacteria transfer its engineered T-DNA to the plant cell nucleus and the encoded genes are actively transcribed and translated in a temporal window that covers 2–7 days post-inoculation. Silencing suppressors can increase the expression of the protein of interest in leaves, indicating that transient expression is limited by silencing rather than by T-DNA degradation as it was described by Voinnet et al. (2003) using the tomato bushy stunt virus p19 protein. Both the mentioned p19 protein and the HcPro suppressor from potato virus Y (Brigneti et al., 1998) have been used to increase yields of antibody production (Sainsbury et al., 2009, Vezina et al., 2009, De Muynck et al., 2010).

Viral-based vectors have also made their way through stable transformation. They offer a useful tool for large-scale production not only for antibodies but also for other proteins with industrial importance. Verch et al. (1998) first reported the use of a plant virus vector to express and assemble a full-size antibody. They used a tobacco mosaic virus-based vector to express mAb CO17-1A, directed against a colon cancer antigen, and co-infiltrated *Nicotiana benthamiana* leaves to assemble the full-length antibody. This kind of vectors, named *First-generation* vectors, led in recent years to *Second*-generation vectors, including a deconstructed virus (Gleba et al., 2007). Magnifection system, developed by the Icon Genetics group (Gleba et al., 2005), is based on two TMV-related viruses (TVCV and CrTMV) with a replicative unit deconstructed in two pieces (Fig. 7.2a). Each segment is flanked by a homologous recombination site and introduced into separate binary vectors. Introduction of these two vectors (together with a third one containing a recombinase) within the cell via agroinfiltration leads to the reconstruction of a functional viral replicon. This RNA reassembled unit contains a polymerase and a movement protein so replication and cell-to-cell movement is ensured. Icon vectors were engineered to minimize the undesired posttranslational events like RNA misprocessing in cryptic splicing sites (Marillonnet et al., 2004). Introns were added and cryptical splicing sequences removed, so the reassembled RNA could efficiently led to a high protein production. These vectors were tested using GFP as a recombinant gene, leading to yields up to 80% of TSP (or 5 g kg^{-1} of fresh weight biomass) (Marillonnet et al., 2004). They have also successfully been used for an anti-TNF-VHH production with a yield up to 0.33 g Kg^{-1} of fresh *N.benthamiana* leaves, 9 days after the inoculation (Giersberg et al., 2009).

Antibody production requires the co-expression of two ORFs (antibody chains) within a single cell. This represents a challenge for viral vectors, as Giritch et al.

(2006) reported that agroinfection with two TMV-based clones harbouring two different fluorescent proteins rapidly led to a spatial separation of the two distinct TMV populations in the infiltrated tissues. As a strategy to effectively produce two proteins in the same cell, two non-competitive viruses, TMV and potato virus X (PVX), were magnicon-adapted to highly express both IgH and IgL respectively. It was proved that the noncompeting viral vectors led to high co-expression levels of both antibody chains and resulted in higher yields than standard transient expression cassettes, up to 0.5 g of IgG per kg of fresh weight in early un-optimized experiments. Upon optimization, Bendandi et al. (2010) reported the production of 20 different individualized IgGs for the treatment of Non-Hodgkins lymphoma. The speed achieved by transient expression was demonstrated in this report, where less than 2 weeks were required for the expression and purification of the antibodies, whereas the whole development from biopsy to vaccine took less than 12 weeks.

Another impressive expression system was recently constructed with elements from the bipartite RNA cowpea mosaic virus (CPMV) (Fig. 7.2b). Despite being virus-based, this system is devoid of the virus replicase. It consists of a very simple collection of so-called pEAQ vectors, ready for classical restriction/ligation and/or Gateway cloning. The pEAQ vectors incorporate mutagenized 5′ and 3′ UTR regions of the virus flanking the gene of interest (Sainsbury and Lomonossoff, 2008b). Controlled by a strong plant promoter, the presence of the viral untranslated sequences is sufficient to confer high expression levels when co-expressed with p19. The expression of the monoclonal IgG 2G12 from a single plasmid (Sainsbury et al., 2009) represents the highest reported yield of recombinant antibody from plant tissue infiltrated with a single *Agrobacterium* culture, up to 0.325 g kg^{-1} of fresh weight tissue.

In another interesting development, Huang et al. (2009) engineered a three-component replicon system derived from bean yellow darf virus (BeYDV) which led to a rapid production of single recombinant proteins in plants (Fig. 7.2c). The three components were put together in a latter version of the system. The performance of the system was demonstrated with the production of anti-Ebola virus GP1 protective 6D8 IgG with a yield of 0.5 g of per gram of leaf fresh weight in only 4 days (Huang et al., 2010). This level is comparable to those reached by Giritch et al. (2006) but in this case a single construct instead of six is needed. Future perspectives include the incorporation of some of the enhancer elements from pEAQ system (Sainsbury et al., 2009) and to exploring the effect of p19 in the final yield of the system.

Those transient strategies based on viruses are not the only ones reaching high yields. It was postulated that promoters driving the expression of photosynthesis-related genes would be useful for transient expression as they would provide high transcription rates. Vézina et al. (2009) compared the double 35S promoter, a ribulose-1,5-biphosphate carboxylase/oxigenase promoter and a plastocyanin promoter, both from *Medicago sativa sp.* As a result of this comparison, it was concluded that plastocyanin promoter was driving the highest expression levels. When plastocyanin promoter was used, together HcPro silencing suppressor, in the agroinfiltration of the murine anti-human C5-1 IgG, maximum expression levels of 1.5 g kg^{-1} FW (25% of TSP) and average values between 9.3 and 12.6% TSP were reached for a secreted and an ER-retained version of the antibody respectively. These values exceeded the maximum accumulation level reported for antibodies in plants with any other expression system (transgenic plants or virus-based system).

Although most of the experiments described here were performed in a small scale (syringe-infiltrated leaves), some of them have been escalated using vacuum-based agroinfiltration systems similar to the one described by Kapila et al. (1997). A couple of examples selected from the literature, Vézina et al. (2009) produced in a small pilot unit more than 1 g of their C5-1 IgG per day and Bendandi et al. (2010) reported up to 4.8 g of a IgG per kg of leaf biomass (>70% TSP) in their prototype industrial process, transfecting around 5 kg plant each time.

Comparing the yields obtained by both transient and stable expression, it is clear that in addition to speeding up production, higher yields can be obtained by transient antibody expression. Full antibodies can be now produced approximately in a week, a great achievement for antibody molecular farming.

7.6 Future Perspectives

Plant-made antibodies have already reached the market in the form of custom antibodies for specific applications, as it is the case for the IgG against hepatitis B surface antigen manufactured in Cuba for affinity purification of hepatitis vaccine. An important milestone in the popularization of this technology will come from the first commercialization of a plant-made recombinant therapeutic product. The enzyme glucocerebridase, produced by Protalix for the treatment of Gaucher's disease is close to reach this point (Aviezer et al., 2009). It is likely that a first successful story will pave the way to new plant-made products including antibodies, as might be the case of IgGs for the individualized therapeutic antibodies against Non-Hodgkin's lymphoma developed by former Icon Genetics and Bayer Innovation.

Although entering the therapeutic market may bring confidence in plant platforms, much more effort will be required to reach a situation where the manufacturing potential of plants is fully developed. Production levels are not, at least temporarily, a major driving force in the current recombinant antibodies market, as manufacturing companies have rapidly adapted their production capacities to growing world demand. In this scenario, the once proposed plant-made alternative to CHO cells might have been overoptimistic.

As the current trend is moving towards product improvement and speeding up the production timeline for individualized medicine, so are doing PMAb platforms, strongly adopting transient expression technologies and making effort in glyco-engineering for biosimilars and biobetters. However it has to be noticed that competing platforms as mammalian cell lines, *Pichia pastoris*, or baculovirus-infected insect cells are also rapidly following the same trend, there is a risk exists that plants platforms, despite being evolutionary well adapted for mass-recombinant production, keep lagging behind due to insufficient technological development.

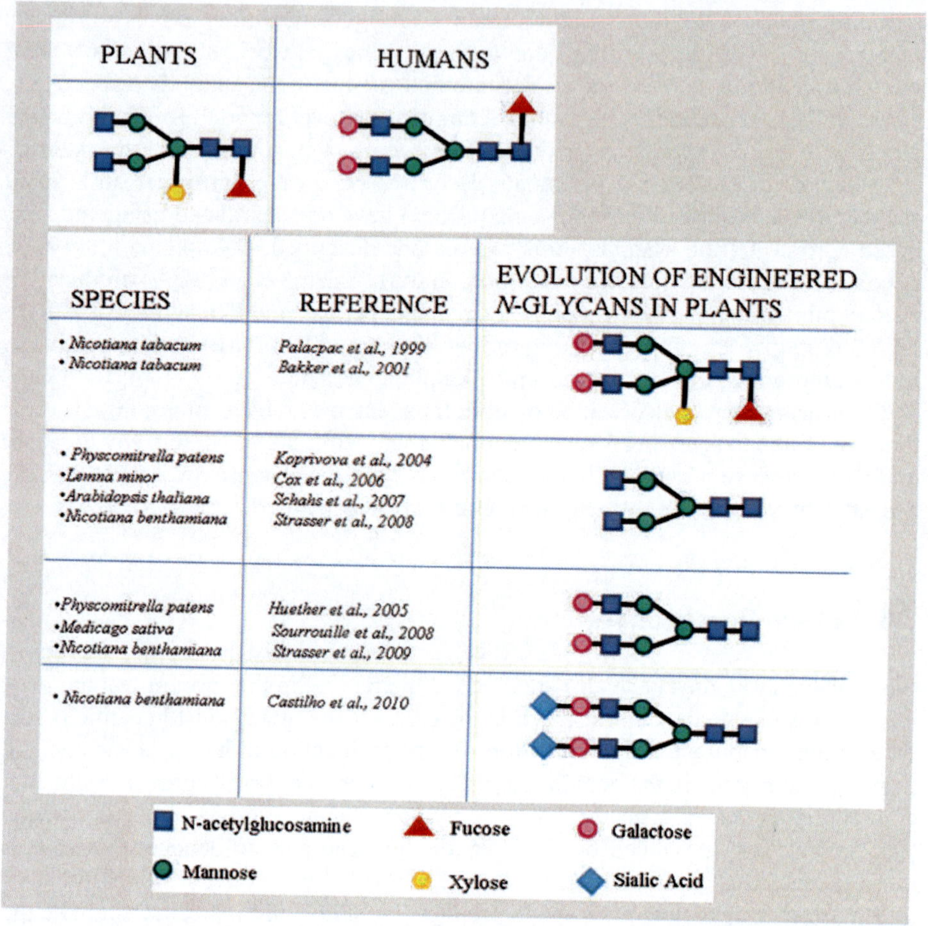

Fig. 7.1 Schematic representation of the steps towards humanized *N*-glycosylation patterns in glycoengineered plants. A comparison with native human and plant glycosylation patterns is also depicted

Fig. 7.2 Viral Vector systems. (**a**) Magnifection. The variable regions of the heavy (IgG1, *green*) and light (Igλ, *blue*) chains of the were subcloned in both tobacco mosaic virus (TMV, *dark blue*) and potato virus X (PVX, *orange*) vectors. The viral pro-vector is *in planta* assembled in the nucleus and then spliced and exported to the cytoplasm. The viral vector replicates and can move either systemically or cell-to cell. Proteins are produced in high levels and full IgGs assembled. (**b**) CPMV based vector. Expression cassettes containing the endoplasmic reticulum-retained heavy chain (HC, *green*) and light chain (HL, *blue*) of the human monoclonal anti-human immunodeficiency virus antibody 2G12 were inserted into the pEAQexpress-GFP-HT vector, containing the 5′ and 3′ UTR from the CPMV (*red*). mRNAs are exported to the cytoplasm as usual and proteins are expressed, in presence of p19 (*grey*), at high yields. (**c**) *BeYDV based vector.* Expression

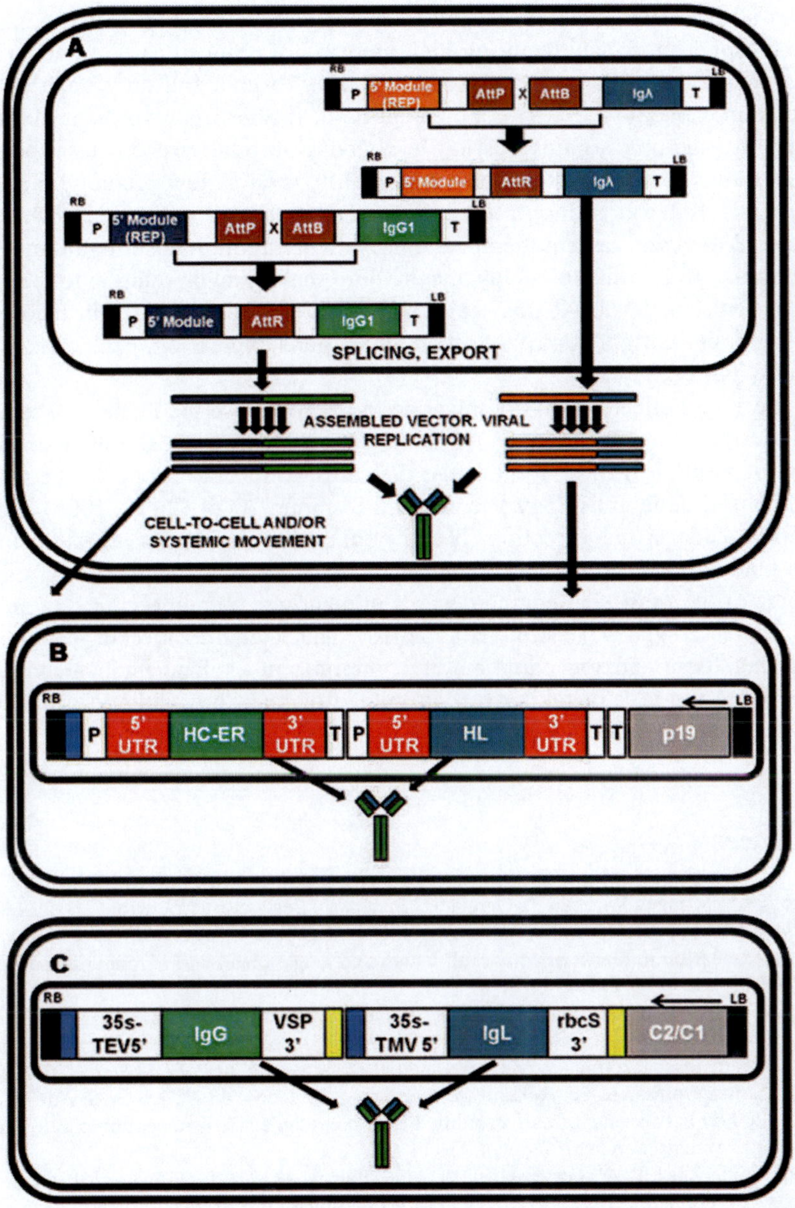

Fig. 7.2 (Continued) cassettes containing the endoplasmic reticulum-retained heavy chain (IgG, *green*) and light chain (IgL, *blue*) of the human anti-Ebola virus GP1 protective 6D8. Schematic representation of the T-DNA regions of the vectors include CaMV 35S promoter with tobacco etch virus and tobacco mosaic virus 5′UTRs (both in *white*, 35S/TEV and 35S/TMV respectively), VSP and rbcS 3′UTR (both in *white*, soybean vspB and pea rbcS 3′UTR). Other elements in the cassetes are LIR (*blue box*, long intergenic region of BeYDV genome) and SIR (*yellow box*, short intergenic region of BeYDV genome) and C2/C1 (*grey*, BeYDV ORFs C1 and C2, encoding replication initiation protein Rep and RepA)

After all plants are, from an evolutionary perspective, the best suited organisms on the biosphere for manufacturing high amounts of reduced carbon molecules at low metabolic cost (Thomas and Sadras, 2001). To gain full advantage of plant biosynthetic capacity, the next goal should be to further improve plant platforms, not only by pursuing high expression levels but also other aspects described here that contribute to reach better yields and quality, such as low proteolytic activity, proper sub cellular targeting, *a la carte* glycosilation, and optimized downstream strategies. Moreover, certain therapeutic uses involving high dose requirements, as topical or mucosal immunotherapy, may require scaling up production to the greenhouse/agricultural level. At this scale, plant platforms should ideally incorporate additional traits as tight control of recombinant gene expression, gene containment or identity preservation.

As it can be noticed, plant optimization in the way described here will require a careful design of traits, following a systems approach, through simultaneous introduction of multiple genes, a challenge that falls within the scope of the modern discipline of Synthetic Biology (Benner and Sismour, 2005, Check, 2005). Despite the efforts made in this direction (Naqvi et al., 2009), multiple gene transformation in plants is still an imprecise and often laborious task, as proved by the fact that some of the most challenging designs in plant glyco-engineering rely on transiently expressed genes (Castilho et al., 2010). The standardization of plant genetic engineering tools and the adoption of common and exchangeable technologies will facilitate the task of taking full advantage of plant biosynthetic capacity for recombinant antibody production.

References

Alamillo, J. M., Monger, W., Sola, I., Garcia, B., Perrin, Y., Bestagno, M., Burrone, O. R., Sabella, P., Plana-Duran, J., Enjuanes, L., Lomonossoff, G. P. & Garcia, J. A. (2006) Use of virus vectors for the expression in plants of active full-length and single chain anti-coronavirus antibodies. *Biotechnol J*, 1, 1103–1111.

Aviezer, D., Brill-Almon, E., Shaaltiel, Y., Hashmueli, S., Bartfeld, D., Mizrachi, S., Liberman, Y., Freeman, A., Zimran, A. & Galun, E. (2009) A plant-derived recombinant human glucocerebrosidase enzyme – a preclinical and phase I investigation. *PLoS One*, 4, e4792.

Azevedo, A. M., Gomes, A. G., Rosa, P. A., Ferreira, I. F., Pisco, A. M. M. O. & Aires-Barros, M. R. (2009a) Partitioning of human antibodies in polyethylene glycol-sodium citrate aqueous two-phase systems. *Sep Purif Technol*, 65, 14–21.

Azevedo, A. M., Rosa, P. A., Ferreira, I. F. & Aires-Barros, M. R. (2009b) Chromatography-free recovery of biopharmaceuticals through aqueous two-phase processing. *Trends Biotechnol*, 27, 240–247.

Bakker, H., Bardor, M., Molthoff, J. W., Gomord, V., Elbers, I., Stevens, L. H., Jordi, W., Lommen, A., Faye, L., Lerouge, P. & Bosch, D. (2001) Galactose-extended glycans of antibodies produced by transgenic plants. *Proc Natl Acad Sci USA*, 98, 2899–2904.

Baum, T. J., Hiatt, A., Parrott, W. A., Pratt, L. H. & Hussey, R. S. (1996) Expression in tobacco of a functional monoclonal antibody specific to stylet secretions of the root-knot nematode. *Mol Plant Microbe Interact*, 5, 382–387.

Benchabane, M., Goulet, C., Rivard, D., Faye, L., Gomord, V. & Michaud, D. (2008) Preventing unintended proteolysis in plant protein biofactories. *Plant Biotechnol J*, 6, 633–648.

Bendandi, M., Marillonnet, S., Kandzia, R., Thieme, F., Nickstadt, A., Herz, S., Frode, R., Inoges, S., Lopez-Diaz De Cerio, A., Soria, E., Villanueva, H., Vancanneyt, G., McCormick, A., Tuse, D., Lenz, J., Butler-Ransohoff, J. E., Klimyuk, V. & Gleba, Y. (2010) Rapid, high-yield production in plants of individualized idiotype vaccines for non-Hodgkin's lymphoma. *Ann Oncol*, doi: 10.1093/annonc/mdq256.

Benner, S. A. & Sismour, A. M. (2005) Synthetic biology. *Nat Rev Genet*, 6, 533–543.

Berger, I., Fitzgerald, D. J. & Richmond, T. J. (2004) Baculovirus expression system for heterologous multiprotein complexes. *Nat Biotechnol*, 22, 1583–1587.

Boonrod, K., Galetzka, D., Nagy, P. D., Conrad, U. & Krczal, G. (2004) Single-chain antibodies against a plant viral RNA-dependent RNA polymerase confer virus resistance. *Nat Biotechnol*, 22, 856–862.

Brigneti, G., Voinnet, O., Li, W. X., Ji, L. H., Ding, S. W. & Baulcombe, D. C. (1998) Viral pathogenicity determinants are suppressors of transgene silencing in Nicotiana benthamiana. *EMBO J*, 17, 6739–6746.

Brodzik, R., Glogowska, M., Bandurska, K., Okulicz, M., Deka, D., Ko, K., Van Der Linden, J., Leusen, J. H., Pogrebnyak, N., Golovkin, M., Steplewski, Z. & Koprowski, H. (2006) Plant-derived anti-Lewis Y mAb exhibits biological activities for efficient immunotherapy against human cancer cells. *Proc Natl Acad Sci USA*, 103, 8804–8809.

Castilho, A., Pabst, M., Leonard, R., Veit, C., Altmann, F., Mach, L., Glossl, J., Strasser, R. & Steinkellner, H. (2008) Construction of a functional CMP-sialic acid biosynthesis pathway in Arabidopsis. *Plant Physiol*, 147, 331–339.

Castilho, A., Strasser, R., Stadlmann, J., Grass, J., Jez, J., Gattinger, P., Kunert, R., Quendler, H., Pabst, M., Leonard, R., Altmann, F. & Steinkellner, H. (2010) In planta protein sialylation through overexpression of the respective mammalian pathway. *J Biol Chem*, 285, 15923–15930.

Chargelegue, D., Drake, P. M., Obregon, P., Prada, A., Fairweather, N. & Ma, J. K. (2005) Highly immunogenic and protective recombinant vaccine candidate expressed in transgenic plants. *Infect Immun*, 73, 5915–5922.

Check, E. (2005) Synthetic biology: designs on life. *Nature*, 438, 417–418.

Conrad, U. & Floss, D. (2010) Expression of antibody fragments in transgenic plants. *Antibody Eng*, III, 377–386.

Corthesy, B. (2002) Recombinant immunoglobulin A: powerful tools for fundamental and applied research. *Trends Biotechnol*, 20, 65–71.

Corthesy, B. & Spertini, F. (1999) Secretory immunoglobulin A: from mucosal protection to vaccine development. *Biol Chem*, 380, 1251–1262.

Cox, K. M., Sterling, J. D., Regan, J. T., Gasdaska, J. R., Frantz, K. K., Peele, C. G., Black, A., Passmore, D., Moldovan-Loomis, C., Srinivasan, M., Cuison, S., Cardarelli, P. M. & Dickey, L. F. (2006) Glycan optimization of a human monoclonal antibody in the aquatic plant Lemna minor. *Nat Biotechnol*, 24, 1591–1597.

De Jaeger, G., Buys, E., Eeckhout, D., De wilde, C., Jacobs, A., Kapila, J., Angenon, G., Van Montagu, M., Gerats, T. & Depicker, A. (1999) High level accumulation of single-chain variable fragments in the cytosol of transgenic Petunia hybrida. *Eur J Biochem*, 259, 426–434.

De Muynck, B., Navarre, C. & Boutry, M. (2010) Production of antibodies in plants: status after twenty years. *Plant Biotechnol J*, 8, 529–563.

De Muynck, B., Navarre, C., Nizet, Y., Stadlmann, J. & Boutry, M. (2009) Different subcellular localization and glycosylation for a functional antibody expressed in Nicotiana tabacum plants and suspension cells. *Transgenic Res*, 18, 467–482.

De Neve, M., De Loose, M., Jacobs, A., Van Houdt, H., Kaluza, B., Weidle, U., Van Montagu, M. & Depicker, A. (1993) Assembly of an antibody and its derived antibody fragment in Nicotiana and Arabidopsis. *Transgenic Res*, 2, 227–237.

De Wilde, C., De Rycke, R., Beeckman, T., De Neve, M., Van Montagu, M., Engler, G. & Depicker, A. (1998) Accumulation pattern of IgG antibodies and Fab fragments in transgenic Arabidopsis thaliana plants. *Plant Cell Physiol*, 39, 639–646.

Düring, K., Hippe, S., Kreuzaler, F. & Schell, J. (1990) Synthesis and self-assembly of a functional monoclonal antibody in transgenic Nicotiana tabacum. *Plant Mol Biol*, 15, 281–293.

D'Aoust, M. A., Lavoie, P. O., Belles-Isles, J., Bechtold, N., Martel, M. & Vezina, L. P. (2009) Transient expression of antibodies in plants using syringe agroinfiltration. *Methods Mol Biol*, 483, 41–50.

Fiedler, U., Phillips, J., Artsaenko, O. & Conrad, U. (1997) Optimization of scFv antibody production in transgenic plants. *Immunotechnology*, 3, 205–216.

Floss, D. M., Sack, M., Arcalis, E., Stadlmann, J., Quendler, H., Rademacher, T., Stoger, E., Scheller, J., Fischer, R. & Conrad, U. (2009) Influence of elastin-like peptide fusions on the quantity and quality of a tobacco-derived human immunodeficiency virus-neutralizing antibody. *Plant Biotechnol J*, 7, 899–913.

Floss, D. M., Sack, M., Stadlmann, J., Rademacher, T., Scheller, J., Stoger, E., Fischer, R. & Conrad, U. (2008) Biochemical and functional characterization of anti-HIV antibody-ELP fusion proteins from transgenic plants. *Plant Biotechnol J*, 6, 379–391.

Frigerio, L., Vine, N. D., Pedrazzini, E., Hein, M. B., Wang, F., Ma, J. K. & Vitale, A. (2000) Assembly, secretion, and vacuolar delivery of a hybrid immunoglobulin in plants. *Plant Physiol*, 123, 1483–1494.

Giersberg, M., Floss, D. M., Kipriyanov, S., Conrad, U. & Scheller, J. (2009) Covalent dimerization of camelidae anti-human TNF-alpha single domain antibodies by the constant kappa light chain domain improves neutralizing activity. *Biotechnol Bioeng*, 106, 161–166.

Giritch, A., Marillonnet, S., Engler, C., Van Eldik, G., Botterman, J., Klimyuk, V. & Gleba, Y. (2006) Rapid high-yield expression of full-size IgG antibodies in plants coinfected with noncompeting viral vectors. *Proc Natl Acad Sci USA*, 103, 14701–14706.

Gleba, Y., Klimyuk, V. & Marillonnet, S. (2005) Magnifection – a new platform for expressing recombinant vaccines in plants. *Vaccine*, 23, 2042–2048.

Gleba, Y., Klimyuk, V. & Marillonnet, S. (2007) Viral vectors for the expression of proteins in plants. *Curr Opin Biotechnol*, 18, 134–141.

Hadlington, J. L., Santoro, A., Nuttall, J., Denecke, J., Ma, J. K., Vitale, A. & Frigerio, L. (2003) The C-terminal extension of a hybrid immunoglobulin A/G heavy chain is responsible for its Golgi-mediated sorting to the vacuole. *Mol Biol Cell*, 14, 2592–2602.

Hamers-Casterman, C., Atarhouch, T., Muyldermans, S., Robinson, G., Hamers, C., Songa, E. B., Bendahman, N. & Hamers, R. (1993) Naturally occurring antibodies devoid of light chains. *Nature*, 363, 446–448.

Hassan, S., Van Dolleweerd, C. J., Ioakeimidis, F., Keshavarz-Moore, E. & Ma, J. K. (2008) Considerations for extraction of monoclonal antibodies targeted to different subcellular compartments in transgenic tobacco plants. *Plant Biotechnol J*, 6, 733–748.

Hein, M. B., Tang, Y., McLeod, D. A., Janda, K. D. & Hiatt, A. (1991) Evaluation of immunoglobulins from plant cells. *Biotechnol Prog*, 7, 455–461.

Hiatt, A., Cafferkey, R. & Bowdish, K. (1989) Production of antibodies in transgenic plants. *Nature*, 342, 76–78.

Huang, Z., Chen, Q., Hjelm, B., Arntzen, C. & Mason, H. (2009) A DNA replicon system for rapid high-level production of virus-like particles in plants. *Biotechnol Bioeng*, 103, 706–714.

Huang, Z., Phoolcharoen, W., Lai, H., Piensook, K., Cardineau, G., Zeitlin, L., Whaley, K. J., Arntzen, C. J., Mason, H. S. & Chen, Q. (2010) High-level rapid production of full-size monoclonal antibodies in plants by a single-vector DNA replicon system. *Biotechnol Bioeng*, 106, 9–17.

Huether, C. M., Lienhart, O., Baur, A., Stemmer, C., Gorr, G., Reski, R. & Decker, E. L. (2005) Glyco-engineering of moss lacking plant-specific sugar residues. *Plant Biol (Stuttg)*, 7, 292–299.

Jin, C., Altmann, F., Strasser, R., Mach, L., Schahs, M., Kunert, R., Rademacher, T., Glossl, J. & Steinkellner, H. (2008) A plant-derived human monoclonal antibody induces an anti-carbohydrate immune response in rabbits. *Glycobiology*, 18, 235–241.

Jobling, S. A., Jarman, C., Teh, M. M., Holmberg, N., Blake, C. & Verhoeyen, M. E. (2003) Immunomodulation of enzyme function in plants by single-domain antibody fragments. *Nat Biotechnol*, 21, 77–80.

Kapila, J., De Rycke, R., Van Montagu, M. & Angenon, G. (1997) An Agrobacterium-mediated transient gene expression system for intact leaves. *Plant Sci*, 122, 101–108.

Kathuria, S., Sriraman, R., Nath, R., Sack, M., Pal, R., Artsaenko, O., Talwar, G. P., Fischer, R. & Finnern, R. (2002) Efficacy of plant-produced recombinant antibodies against HCG. *Hum Reprod*, 17, 2054–2061.

Ko, K., Brodzik, R. & Steplewski, Z. (2009) Production of antibodies in plants: approaches and perspectives. *Curr Top Microbiol Immunol*, 332, 55–78.

Ko, K. & Koprowski, H. (2005) Plant biopharming of monoclonal antibodies. *Virus Res*, 111, 93–100.

Ko, K., Tekoah, Y., Rudd, P. M., Harvey, D. J., Dwek, R. A., Spitsin, S., Hanlon, C. A., Rupprecht, C., Dietzschold, B., Golovkin, M. & Koprowski, H. (2003) Function and glycosylation of plant-derived antiviral monoclonal antibody. *Proc Natl Acad Sci USA*, 100, 8013–8018.

Koprivova, A., Stemmer, C., Altmann, F., Hoffmann, A., Kopriva, S., Gorr, G., Reski, R. & Decker, E. L. (2004) Targeted knockouts of Physcomitrella lacking plant-specific immunogenic N-glycans. *Plant Biotechnol J*, 2, 517–523.

Langley, R., Wines, B., Willoughby, N., Basu, I., Proft, T. & Fraser, J. D. (2005) The staphylococcal superantigen-like protein 7 binds IgA and complement C5 and inhibits IgA-Fc alpha RI binding and serum killing of bacteria. *J Immunol*, 174, 2926–2933.

Lico, C., Chen, Q. & Santi, L. (2008) Viral vectors for production of recombinant proteins in plants. *J Cell Physiol*, 216, 366–377.

Ma, J. K. C., Drake, P. M. W. & Christou, P. (2003) The production of recombinant pharmaceutical proteins in plants. *Nat Rev Genet*, 4, 794–805.

Ma, J. K., Hiatt, A., Hein, M., Vine, N. D., Wang, F., Stabila, P., Van Dolleweerd, C., Mostov, K. & Lehner, T. (1995) Generation and assembly of secretory antibodies in plants. *Science*, 268, 716–719.

Ma, J. K., Hikmat, B. Y., Wycoff, K., Vine, N. D., Chargelegue, D., Yu, L., Hein, M. B. & Lehner, T. (1998) Characterization of a recombinant plant monoclonal secretory antibody and preventive immunotherapy in humans. *Nat Med*, 4, 601–606.

Marillonnet, S., Giritch, A., Gils, M., Kandzia, R., Klimyuk, V. & Gleba, Y. (2004) In planta engineering of viral RNA replicons: efficient assembly by recombination of DNA modules delivered by Agrobacterium. *Proc Natl Acad Sci USA*, 101, 6852–6857.

McCormick, A. A., Reddy, S., Reinl, S. J., Cameron, T. I., Czerwinkski, D. K., Vojdani, F., Hanley, K. M., Garger, S. J., White, E. L., Novak, J., Barrett, J., Holtz, R. B., Tuse, D. & Levy, R. (2008) Plant-produced idiotype vaccines for the treatment of non-Hodgkin's lymphoma: safety and immunogenicity in a phase I clinical study. *Proc Natl Acad Sci USA*, 105, 10131–10136.

Muyldermans, S. (2001) Single domain camel antibodies: current status. *J Biotechnol*, 74, 277–302.

Muyldermans, S., Baral, T. N., Retamozzo, V. C., De Baetselier, P., De Genst, E., Kinne, J., Leonhardt, H., Magez, S., Nguyen, V. K., Revets, H., Rothbauer, U., Stijlemans, B., Tillib, S., Wernery, U., Wyns, L., Hassanzadeh-Ghassabeh, G. & Saerens, D. (2009) Camelid immunoglobulins and nanobody technology. *Vet Immunol Immunopathol*, 128, 178–183.

Naqvi, S., Farre, G., Sanahuja, G., Capell, T., Zhu, C. & Christou, P. (2009) When more is better: multigene engineering in plants. *Trends Plant Sci*, 15(1), 48–56.

Nicholson, L., Gonzalez-Melendi, P., Van Dolleweerd, C., Tuck, H., Perrin, Y., Ma, J. K., Fischer, R., Christou, P. & Stoger, E. (2005) A recombinant multimeric immunoglobulin expressed in rice shows assembly-dependent subcellular localization in endosperm cells. *Plant Biotechnol J*, 3, 115–127.

Nolke, G., Fischer, R. & Schillberg, S. (2006) Antibody-based metabolic engineering in plants. *J Biotechnol*, 124, 271–283.

Nolke, G., Schneider, B., Fischer, R. & Schillberg, S. (2005) Immunomodulation of polyamine biosynthesis in tobacco plants has a significant impact on polyamine levels and generates a dwarf phenotype. *Plant Biotechnol J*, 3, 237–247.

Obregon, P., Chargelegue, D., Drake, P. M., Prada, A., Nuttall, J., Frigerio, L. & Ma, J. K. (2006) HIV-1 p24-immunoglobulin fusion molecule: a new strategy for plant-based protein production. *Plant Biotechnol J*, 4, 195–207.

Orzaez, D., Mirabel, S., Wieland, W. H. & Granell, A. (2006) Agroinjection of tomato fruits. A tool for rapid functional analysis of transgenes directly in fruit. *Plant Physiol*, 140, 3–11.

Palacpac, N. Q., Yoshida, S., Sakai, H., Kimura, Y., Fujiyama, K., Yoshida, T. & Seki, T. (1999) Stable expression of human beta1,4-galactosyltransferase in plant cells modifies N-linked glycosylation patterns. *Proc Natl Acad Sci USA*, 96, 4692–4697.

Petruccelli, S., Otegui, M. S., Lareu, F., Tran Dinh, O., Fitchette, A. C., Circosta, A., Rumbo, M., Bardor, M., Carcamo, R., Gomord, V. & Beachy, R. N. (2006) A KDEL-tagged monoclonal antibody is efficiently retained in the endoplasmic reticulum in leaves, but is both partially secreted and sorted to protein storage vacuoles in seeds. *Plant Biotechnol J*, 4, 511–527.

Platis, D. & Labrou, N. E. (2009) Application of a PEG/salt aqueous two-phase partition system for the recovery of monoclonal antibodies from unclarified transgenic tobacco extract. *Biotechnol J*, 4, 1320–1307.

Ramessar, K., Rademacher, T., Sack, M., Stadlmann, J., Platis, D., Stiegler, G., Labrou, N., Altmann, F., Ma, J., Stoger, E., Capell, T. & Christou, P. (2008) Cost-effective production of a vaginal protein microbicide to prevent HIV transmission. *Proc Natl Acad Sci USA*, 105, 3727–3732.

Ramirez, N., Rodriguez, M., Ayala, M., Cremata, J., Perez, M., Martinez, A., Linares, M., Hevia, Y., Paez, R., Valdes, R., Gavilondo, J. V. & Selman-Housein, G. (2003) Expression and characterization of an anti-(hepatitis B surface antigen) glycosylated mouse antibody in transgenic tobacco (Nicotiana tabacum) plants and its use in the immunopurification of its target antigen. *Biotechnol Appl Biochem*, 38, 223–230.

Ramsland, P. A., Willoughby, N., Trist, H. M., Farrugia, W., Hogarth, P. M., Fraser, J. D. & Wines, B. D. (2007) Structural basis for evasion of IgA immunity by Staphylococcus aureus revealed in the complex of SSL7 with Fc of human IgA1. *Proc Natl Acad Sci USA*, 104, 15051–15056.

Sainsbury, F., Lavoie, P. O., D'Aoust, M. A., Vezina, L. P. & Lomonossoff, G. P. (2008a) Expression of multiple proteins using full-length and deleted versions of cowpea mosaic virus RNA-2. *Plant Biotechnol J*, 6, 82–92.

Sainsbury, F. & Lomonossoff, G. P. (2008b) Extremely high-level and rapid transient protein production in plants without the use of viral replication. *Plant Physiol*, 148, 1212–1218.

Sainsbury, F., Thuenemann, E. C. & Lomonossoff, G. P. (2009) pEAQ: versatile expression vectors for easy and quick transient expression of heterologous proteins in plants. *Plant Biotechnol J*, 7, 682–693.

Schahs, M., Strasser, R., Stadlmann, J., Kunert, R., Rademacher, T. & Steinkellner, H. (2007) Production of a monoclonal antibody in plants with a humanized N-glycosylation pattern. *Plant Biotechnol J*, 5, 657–663.

Schouten, A., Roosien, J., Van Engelen, F. A., De Jong, G. A., Borst-Vrenssen, A. W., Zilverentant, J. F., Bosch, D., Stiekema, W. J., Gommers, F. J., Schots, A. & Bakker, J. (1996) The C-terminal KDEL sequence increases the expression level of a single-chain antibody designed to be targeted to both the cytosol and the secretory pathway in transgenic tobacco. *Plant Mol Biol*, 30, 781–793.

Sharp, J. M. & Doran, P. M. (2001) Characterization of monoclonal antibody fragments produced by plant cells. *Biotechnol Bioeng*, 73, 338–346.

Shimada, N., Suzuki, Y., Nakajima, M., Conrad, U., Murofushi, N. & Yamaguchi, I. (1999) Expression of a functional single-chain antibody against GA24/19 in transgenic tobacco. *Biosci Biotechnol Biochem*, 63, 779–783.

Sourrouille, C., Marquet-Blouin, E., D'Aoust, M. A., Kiefer-Meyer, M. C., Seveno, M., Pagny-Salehabadi, S., Bardor, M., Durambur, G., Lerouge, P., Vezina, L. & Gomord, V. (2008) Down-regulated expression of plant-specific glycoepitopes in alfalfa. *Plant Biotechnol J*, 6, 702–721.

Stoger, E., Sack, M., Fischer, R. & Christou, P. (2002) Plantibodies: applications, advantages and bottlenecks. *Curr Opin Biotechnol*, 13, 161–166.

Strasser, R., Castilho, A., Stadlmann, J., Kunert, R., Quendler, H., Gattinger, P., Jez, J., Rademacher, T., Altmann, F., Mach, L. & Steinkellner, H. (2009) Improved virus neutralization by plant-produced anti-HIV antibodies with a homogeneous beta1,4-galactosylated N-glycan profile. *J Biol Chem*, 284, 20479–20485.

Strasser, R., Stadlmann, J., Schahs, M., Stiegler, G., Quendler, H., Mach, L., Glossl, J., Weterings, K., Pabst, M. & Steinkellner, H. (2008) Generation of glyco-engineered Nicotiana benthamiana for the production of monoclonal antibodies with a homogeneous human-like N-glycan structure. *Plant Biotechnol J*, 6, 392–402.

Strauss, M., Kauder, F., Peisker, M., Sonnewald, U., Conrad, U. & Heineke, D. (2001) Expression of an abscisic acid-binding single-chain antibody influences the subcellular distribution of abscisic acid and leads to developmental changes in transgenic potato plants. *Planta*, 213, 361–369.

Teh, Y. H. & Kavanagh, T. A. (2010) High-level expression of Camelid nanobodies in Nicotiana benthamiana. *Transgenic Res*, 19, 575–586.

Thomas, H. & Sadras, V. O. (2001) The capture and gratuitous disposal of resources by plants. *Funct Ecol*, 15, 3–12.

Torres, E., Gonzalez-Melendi, P., Stoger, E., Shaw, P., Twyman, R. M., Nicholson, L., Vaquero, C., Fischer, R., Christou, P. & Perrin, Y. (2001) Native and artificial reticuloplasmins co-accumulate in distinct domains of the endoplasmic reticulum and in post-endoplasmic reticulum compartments. *Plant Physiol*, 127, 1212–1223.

Valdes, R., Gomez, L., Padilla, S., Brito, J., Reyes, B., Alvarez, T., Mendoza, O., Herrera, O., Ferro, W., Pujol, M., Leal, V., Linares, M., Hevia, Y., Garcia, C., Mila, L., Garcia, O., Sanchez, R., Acosta, A., Geada, D., Paez, R., Luis Vega, J. & Borroto, C. (2003) Large-scale purification of an antibody directed against hepatitis B surface antigen from transgenic tobacco plants. *Biochem Biophys Res Commun*, 308, 94–100.

Van Bockstaele, F., Holz, J. B. & Revets, H. (2009) The development of nanobodies for therapeutic applications. *Curr Opin Investig Drugs*, 10, 1212–1224.

Van Engelen, F. A., Schouten, A., Molthoff, J. W., Roosien, J., Salinas, J., Dirkse, W. G., Schots, A., Bakker, J., Gommers, F. J., Jongsma, M. A. et al. (1994) Coordinate expression of antibody subunit genes yields high levels of functional antibodies in roots of transgenic tobacco. *Plant Mol Biol*, 26, 1701–1710.

Vaquero, C., Sack, M., Chandler, J., Drossard, J., Schuster, F., Monecke, M., Schillberg, S. & Fischer, R. (1999) Transient expression of a tumor-specific single-chain fragment and a chimeric antibody in tobacco leaves. *Proc Natl Acad Sci USA*, 96, 11128–11133.

Vaquero, C., Sack, M., Schuster, F., Finnern, R., Drossard, J., Schumann, D., Reimann, A. & Fischer, R. (2002) A carcinoembryonic antigen-specific diabody produced in tobacco. *FASEB J*, 16, 408–410.

Verch, T., Yusibov, V. & Koprowski, H. (1998) Expression and assembly of a full-length monoclonal antibody in plants using a plant virus vector. *J Immunol Methods*, 220, 69–75.

Vezina, L. P., Faye, L., Lerouge, P., D'Aoust, M. A., Marquet-Blouin, E., Burel, C., Lavoie, P. O., Bardor, M. & Gomord, V. (2009) Transient co-expression for fast and high-yield production of antibodies with human-like N-glycans in plants. *Plant Biotechnol J*, 7, 442–455.

Vine, N. D., Drake, P., Hiatt, A. & Ma, J. K. (2001) Assembly and plasma membrane targeting of recombinant immunoglobulin chains in plants with a murine immunoglobulin transmembrane sequence. *Plant Mol Biol*, 45, 159–167.

Voinnet, O., Rivas, S., Mestre, P. & Baulcombe, D. (2003) An enhanced transient expression system in plants based on suppression of gene silencing by the p19 protein of tomato bushy stunt virus. *Plant J*, 33, 949–956.

Voss, A., Niersbach, M., Hain, R., Hirsch, H. J., Liao, Y. C., Kreuzaler, F. & Fischer, R. (1995) Reduced virus infectivity in N. tabacum secreting a TMV-specific full-size antibody. *Mol. Breeding*, 1, 39–50.
Wieland, W. H., Lammers, A., Schots, A. & Orzaez, D. V. (2006) Plant expression of chicken secretory antibodies derived from combinatorial libraries. *J Biotechnol*, 122, 382–391.
Wines, B. D., Willoughby, N., Fraser, J. D. & Hogarth, P. M. (2006) A competitive mechanism for staphylococcal toxin SSL7 inhibiting the leukocyte IgA receptor, Fc alphaRI, is revealed by SSL7 binding at the C alpha2/C alpha3 interface of IgA. *J Biol Chem*, 281, 1389–1393.
Wongsamuth, R. & Doran, P. M. (1997) Production of monoclonal antibodies by tobacco hairy roots. *Biotechnol Bioeng*, 54, 401–415.
Wurm, F. M. (2004) Production of recombinant protein therapeutics in cultivated mammalian cells. *Nat Biotechnol*, 22, 1393–1398.
Wycoff, K. L. (2005) Secretory IgA antibodies from plants. *Curr Pharm Des*, 11, 2429–2437.
Yano, A., Maeda, F. & Takekoshi, M. (2004) Transgenic tobacco cells producing the human monoclonal antibody to hepatitis B virus surface antigen. *J Med Virol*, 73, 208–215.
Zeitlin, L., Olmsted, S. S., Moench, T. R., Co, M. S., Martinell, B. J., Paradkar, V. M., Russell, D. R., Queen, C., Cone, R. A. & Whaley, K. J. (1998) A humanized monoclonal antibody produced in transgenic plants for immunoprotection of the vagina against genital herpes. *Nat Biotechnol*, 16, 1361–1364.

Chapter 8
Production of Antibody Fab Fragments in *Escherichia coli*

Hiroshi Tachibana and Masataka Takekoshi

Abstract A phage-display library is the most broadly used platform for preparation of recombinant human monoclonal antibody Fab fragments. Panning is effective for the selection of immunoglobulin genes from naïve and immune libraries. However, it is possible to bypass the phage display system if human peripheral lymphocytes are obtained from seropositive patients with infectious diseases as a source of immunoglobulin genes. Direct screening of bacterial colonies producing Fab fragments by colony blotting using filter membranes is practical for the isolation of human Fab fragments to major antigens of pathogens. An oligoclonal culture can also be used, and is a partial application of Epstein-Barr virus transformation of peripheral lymphocytes. Using these procedures, neutralizing antibody Fab fragments to various antigens can be obtained with a sufficient level of cloning efficacy. Chain shuffling and site-directed mutagenesis are also useful ways to improve the quality of the cloned antibody Fab fragments.

8.1 Introduction

The immunoglobulin molecule is a complex structure of four polypeptide chains organized as a homodimer of a heterodimer comprising heavy and light chains. The proteolytic enzyme papain cleaves the molecules into two identical Fab (fragment, antigen binding) fragments and one Fc (fragment, crystallization) fragment, since the enzyme cuts above hinge disulfide bond(s) in the heavy chains. Thus, the Fab fragment is composed of a light chain consisting of a variable region (VL) and a constant region (CL) and the Fd region of the heavy chain containing a variable region (VH) and a constant region (CH1). The domains formed by the two variable regions bind the epitope as a specific antigen.

H. Tachibana (✉)
Department of Infectious Diseases, Tokai University School of Medicine, Isehara, Kanagawa 259-1193, Japan
e-mail: htachiba@is.icc.u-tokai.ac.jp

M. Takekoshi
Department of Molecular Life Science, Tokai University School of Medicine, Isehara, Kanagawa 259-1193, Japan

M. Al-Rubeai (ed.), *Antibody Expression and Production*, Cell Engineering 7,
DOI 10.1007/978-94-007-1257-7_8, © Springer Science+Business Media B.V. 2011

Several methods for preparation of human monoclonal and polyclonal antibodies have been developed through advances in molecular biology, with the goal of therapeutic use of antibodies (Hoet et al., 2005; Kuroiwa et al., 2002; Mendez et al., 1997; Tomizuka et al., 1997). Application of a phage display system (Smith, 1985) is an effective method for expression of a huge repertoire of immunoglobulin genes (Barbas et al., 1991; Thie et al., 2008). Using this system, valuable Fab fragments can be selected from the gene library (Barbas et al., 1992; Burton et al., 1991). The methods and protocols for phage display have been described in detail in reviews and books (Barbas et al., 2001). An antibody Fv fragment composed of only the variable regions (VH-VL), and especially a single chain Fv (scFv), can also be prepared using phage display technology (Bird et al., 1988) and is frequently used because of its better tolerance and expression in bacteria in comparison with a Fab fragment (Mondon et al., 2008). However, Fab fragments are more stable than scFv due to the association between the CH1 and CL regions.

In this chapter, we describe the production of recombinant human Fab fragments in *Escherichia coli*. In addition to phage display followed by screening using panning, a method that allows bypassing of the panning process is introduced.

8.2 Phage Display System

Phagemid-based systems are most commonly used among a variety of phage display systems. Phagemids contain origins for replication of *E. coli* and filamentous phage (M13, fl or fd), a group of viruses that infect *E. coli* (Mead and Kemper, 1988). A phage particle consists of a single-stranded DNA (ssDNA) genome. The phage surface has 3–5 copies of phage coat protein-3 (pIII), which is involved in host cell recognition and infection. The major coat protein is phage coat protein-8 (pVIII), which covers the length of the particle with approximately 2,700 copies that contribute to the structural stability of the phage particle (Russel, 1991). Thus, phagemid combines the characteristics of a plasmid (antibiotic resistance and facilitation of replication of double-stranded DNA (dsDNA)) with those of a phage (production and packaging of ssDNA into a phage particle). Light chain and Fd region genes are cloned into appropriate sites. Secretion signal sequences (e.g. pelB, ompA or PhoA) are coupled to both genes, which allow secretion of Fd and light chain fragments into the periplasm fraction of *E. coli*. The Fab fragment is fused to pIII or its C-terminal domain to allow display on the phage surface.

Recombinant phagemids are introduced into competent *E. coli* by electroporation (Fig. 8.1). Transformation efficiency is critical because it has a direct influence on library size; therefore, electrocompetent cells are better than normal competent cells. After transformation, *E. coli* is superinfected with helper phage (e.g. VCS M13) to give a whole phage particle. Helper phages have a slightly defective origin of replication and phagemid DNA is dominantly packaged into the phage particle. Synthesized Fd and light chain fragments are transported to the periplasm and a

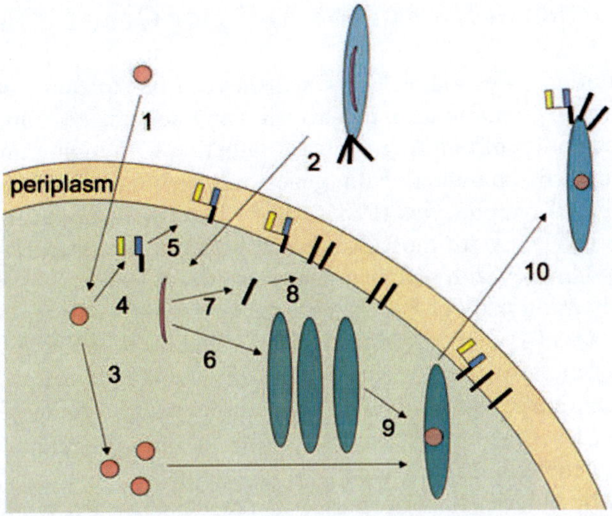

Fig. 8.1 Replication of recombinant phage. *1*, Phagemid is transformed to *E. coli*. *2*, Infection with helper phage. *3*, Phagemid replicates. *4*, Phagemid produces Fd heavy and light chains, one of which is fused to pIII coat protein. *5*, Fd heavy and light chains are transferred to the periplasm and form an inter-chain disulfide bond. *6*, Helper phage produces a virion package. *7*, Normal pIII. *8*, pIII is transferred to the periplasm. *9*, Phagemid is packaged in a virion. *10*, Phagemid moves out of cells with Fab and normal pIII

S-S bond forms between the fragments in the reducing environment. Fab fragments with fused pIII are displayed on the phage surface concomitant with wild-type pIII provided by helper phages. As a result, recombinant phage can infect *E. coli*. Some phagemids utilize pVIII as a Fab fusion protein. However, high valency display libraries generally lead to low affinity Fab fragments due to avidity effects.

Fab-pIII fusion proteins may be toxic to cells, resulting in gene deletion and plasmid instability. Stringent control of fusion protein expression during the propagation steps is critical. In general, *lac* promoter is used to drive gene expression in phagemid vectors; thus, accommodation of leaky expression is needed. Then glucose is added, or *lac* repressor is overexpressed to elevate metabolic activator-regulated repression. A robust terminator upstream of the *lac* promoter also reduces the background expression (Krebber et al., 1996). The *tet* promoter (tetracycline inducible promoter) can be used as an alternative approach (Zahn et al., 1999). With this promoter, gene expression is tightly repressed in the absence of an inducer such as doxycycline or tetracycline.

A phage display system connects the genotype (antibody genes packaged in phage particles) and phenotype (Fab on the surface of phage particles). Affinity selection or biopanning is then needed to isolate Fab fragments against a range of antigens.

8.3 Construction of Human Fab Antibody Gene Libraries

Immunoglobulin genes encoding Fab fragments can be isolated from immunized (immune library) or nonimmunized (naïve library) donors. An immune library is suitable for generation of Fab fragments for targeting with high affinity and specificity. The sources of immunoglobulin genes may be bone marrow, spleen, tonsils or peripheral blood lymphocytes (PBLs). Bone marrow is the ideal source, but is difficult to obtain. PBLs are most commonly used and one-step RT-PCR enables construction of libraries with sufficient amounts of heavy (γ and μ) and light-chain (κ and λ) genes using pairs of Fab specific primers. A naïve library is constructed from IgM mRNA of B cells taken from a nonimmunized donor. Antibodies from naïve libraries generally show weaker affinity, but the library size is larger and the chance of obtaining specific antibodies with affinity against the target is increased (de Haard et al., 1999). A large library is constructed through combination of heavy and light chain genes cloned individually in each library.

Semi-synthetic libraries have been used to merge natural and synthetic diversity. For example, the VH region of the library has been created from semi-synthetic complementarity determining regions (CDRs) 1 and 2 with natural CDR3 from 35 donors with autoimmune diseases and 10 normal donors (Hoet et al., 2005). Synthetic libraries composed of 6 oligonucleotide-derived CDRs for heavy and light chains have diversity that is almost the same as that obtained naturally (Rothe et al., 2008; Shi et al., 2010). Affinities of up to 100 pM have been shown for Fab fragments from these libraries, and semi-synthetic and synthetic libraries are commercially available.

8.4 Screening Procedure by Panning

Selection of phages with Fab fragments reactive to antigen is performed by panning (Fig. 8.2). Phage particles presenting Fab fragments are incubated over an immobilized antigen of choice in ELISA plates or immunotubes (Kang et al., 1991; Marks et al., 1991). Non-binding phage are removed by washing. Phages that bind to the antigen are eluted by changing the binding conditions (e.g. pH change or competitive elution) and amplified by re-infection of *E. coli* cells. The amplified phages are then subjected to further rounds of panning with gradual increases in the frequency and intensity of the washing conditions. The panning process has a combination of positive selection (affinity and specificity) and negative selection (toxicity to bacterial cells and a tendency to delete unnecessary DNA), and 2–3 selection rounds may be preferable. To recover Fabs that are tightly bind to antigen, proteolytic enzyme digestion is used to cut a protease site between the Fab and pIII (Ward et al., 1996).

Some membrane proteins are difficult to produce in soluble form, and protein folding is critical to obtain the epitope conformation. In cells displaying antigen, protein is produced in mammalian cells and displayed on the cell surface. In some cases, antigen-negative cells are used as absorber cells to bind Fab fragments: the absorber cells work to remove sticky clones (de Kruif et al., 1995). Cell sorting

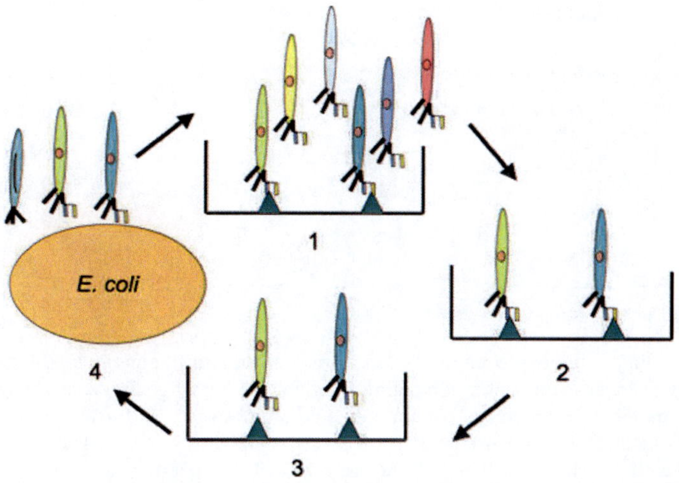

Fig. 8.2 Panning for immobilized antigens. *1*, Binding of recombinant phages with target antigen. *2*, Washing off of free phages. *3*, Elution of bound phages. *4*, Propagation in *E. coli* and next round of panning

enables antigen-specific Fab enrichment (Siegel et al., 1997). Yeast *Saccharomyses cerevisiae* is also used as a cell display platform (van den Beucken et al., 2003), but the transformation efficiency is not as high and the Fab library size is limited. Ribosome display has been developed to expand the library size (Hanes et al., 2000; Hudson and Souriau, 2003), but the antibody is limited to a scFv, rather than a Fab fragment.

8.5 Bypassing Phage Display and Panning

8.5.1 Colony Blot Screening Using Membrane Filters

Bypassing phage display and panning is possible if immunoglobulin gene libraries are constructed from PBLs of immune patients with positive serology. Colony blot screening is a relatively old approach for this purpose (Helfman et al., 1983). The principal of the method is shown in Fig. 8.3. Plasmid vector is used for preparation of combinatorial immunoglobulin gene libraries, instead of use of a phagemid vector. The size of the library is smaller than that of a phage display library, but is sufficient for isolation of Fab fragments to major antigens of pathogens. The number of bacterial colonies screened using filter membranes is also limited to less than 5,000 per 82-mm filter (Fig. 8.4). However, as shown in Table 8.1, positive rates in the first screening by colony blotting were 0.006 to 0.05% to antigens from various pathogens. Plasma recovered during the isolation process of lymphocytes from peripheral blood of the donor can be used for detection of positive clones,

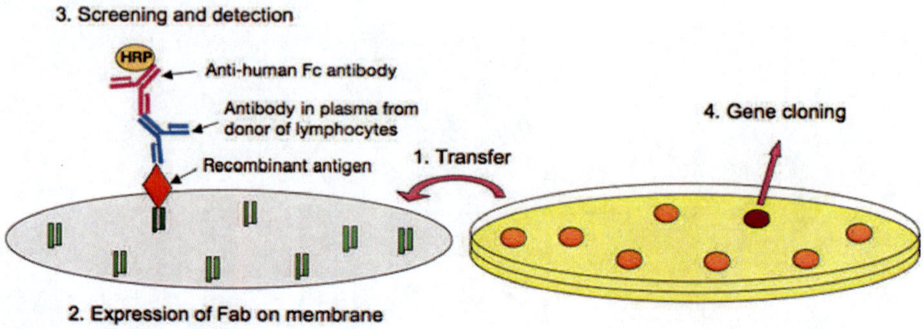

Fig. 8.3 Principle of colony blotting for selection of human Fab fragments. *1*, Bacterial colonies transformed with expression vector containing light and Fd heavy chain genes are transferred to nitrocellulose membranes when the diameter of the colonies reaches 0.1–0.3 mm. *2*, Filters are placed, colony side up, on the surface of fresh plates containing 1 mM IPTG and incubated for expression of Fab fragments. Then colonies are lyzed. *3*, Membranes are reacted serially with blocking solution, target antigen, plasma from donor, and horseradish peroxidase conjugated anti-human IgG Fc antibody and substrate. *4*, Positive colonies are identified on original plates and plasmid DNA is isolated

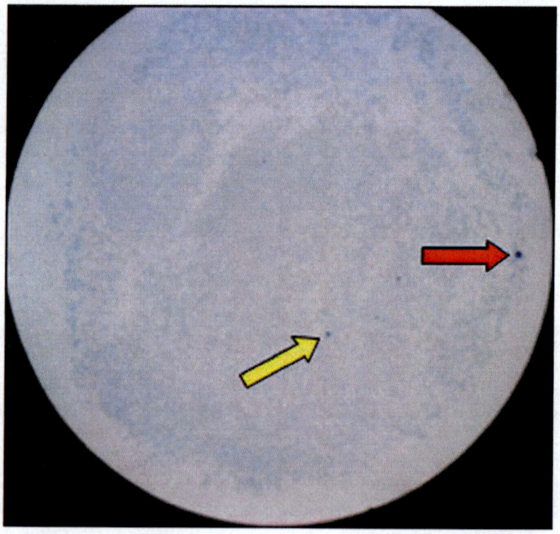

Fig. 8.4 Positive signals (strong, *red arrow*; weak, *yellow arrow*) on a nitrocellulose membrane in screening by colony blotting

instead of the use of monoclonal or polyclonal antibodies from immunized animals, if the donor's antibody titer for the target antigen is sufficient (Cheng et al., 2000). The antibody fraction purified from the plasma is labeled with an enzyme such as horseradish peroxidase and then used for detection. By using a secondary antibody specific for the Fc region of human IgG, direct binding of the second antibody to a human Fab fragment produced in *E. coli* is also avoided in the case of indirect detection (Liu et al., 2006). Although most of the positives in the first screening may become negative in the second and third screenings using ELISA or an indirect

Table 8.1 Efficacy of colony blotting for screening of human Fab fragments from combinatorial immunoglobulin gene libraries derived from patients with infectious diseases

		Positive rate (%)		
Disease in donors of lymphocytes	Target antigen	First screening by colony blot	Final	References
Amebiasis (Liver abscess)	HGL of *Entamoeba histolytica*	0.054*	0.002	Cheng et al. (2000)
Amebiasis (Asymptomatic)	HGL of *E. histolytica*	0.0095*	0.0016	Tachibana et al. (2003)
SARS	Spike protein of SARS coronavirus	0.0063	0.0021	Liu et al. (2006)
Malaria	MSP1-19 of *Plasmodium falciparum*	0.0078	0.00038	Cheng et al. (2007)
Toxoplasmosis	SAG1 of *Toxoplasma gondii*	>0.005	0.00017	Fu et al. (2011)

*Positive rate to crude antigen.
HGL, heavy subunit of galactose- and *N*-acetyl-D-galactosamine-inhibitable lectin; MSP1-19, C-terminal 19 kDa fragment of merozoite surface protein 1; SAG1, surface antigen 1.

immunofluorescein antibody test, colony blotting is a simple and effective procedure for screening for human Fabs with neutralizing activity to pathogens. Screening by colony blotting is especially suitable for relatively small libraries such as chain-shuffled libraries.

Screening of bacterial colonies using two filter membranes is also used (Skerra et al., 1991). Bacteria secreting Fab fragments into the periplasm are grown on a membrane. The secreted Fab fragments are allowed to diffuse to a second membrane coated with anti-globulin, and are probed with antigen. The binding of antigen is detected on the second membrane using enzyme or colloidal gold conjugates. Positive colonies can be grown on the first membrane. The use of a filter membrane for screening is also applicable to phage-expressed antibody libraries (Wu et al., 1998).

8.5.2 Oligoclonal B Lymphoblastoid Cell Culture

In this method, PBLs are collected from healthy human adult volunteers with high antibody titers to the target antigens. The PBLs are infected with Epstein-Barr virus (EBV) strain B95-8 at a dose of 10^5 transformation dose 50 (TD_{50})/mL, and plated on 96-well plates at 10^4 cells per well (Fig. 8.5). Half of the medium is changed every 4 days and the culture is continued for 4 weeks without any treatment, including single isolation. During the process, EBV-transformed B lymphoblastoid cell lines (B-LCL) consisting of several clones are established (Takekoshi et al., 2001). Antibody titers in the culture medium are then checked and the antibody-producing cells are expanded in a 6-cm dish. Cells and medium are harvested and total RNA is

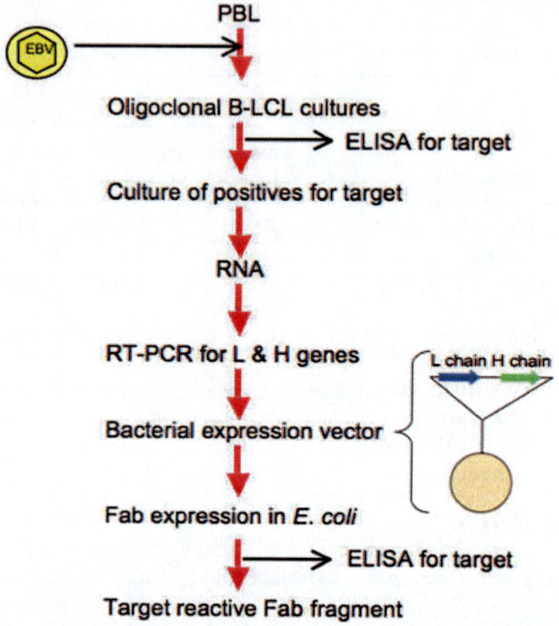

Fig. 8.5 Oligocloning procedure. PBLs are infected with EBV and cultured for 4 weeks without isolation. The EBV-transformed B-LCL is propagated in oligoclonal pools. RNA is extracted from positive clones. Antibody genes are amplified with RT-PCR and cloned into a bacterial expression vector. The vector is transformed into *E. coli* and antibody titers are checked by ELISA. Positive clones are selected

extracted from the cells. The immunoglobulin genes encoding the Fab fragments are amplified by RT-PCR and cloned into the bacterial expression vector. The vector is transformed into *E. coli* and each colony is picked up and incubated for expression of Fab fragments. Screening of 100 clones is generally sufficient to obtain positives in a subsequent assay of the Fab fragments by ELISA.

EBV-transformed B-LCL culture supernatants from a donor were found to be reactive with 71 target antigens (Table 8.2) (Hamatake et al., 2010). There was a tendency to produce autoantibodies, but autoantibody titers in the donor were normal. Using this method, Fab fragments against human cytomegalovirus (Takekoshi et al., 1998), hepatitis B virus s-antigen (Maeda et al., 2005), human tumor necrosis factor-alpha (Takekoshi et al., 2001), ganglioside GM1 (Nagatsuka et al., 2003), and human CD4 (Hamatake et al., 2010) have been obtained. Single isolations of EBV-transformed LCLs frequently cause loss of many clones, which may make it difficult to isolate monoclonal antibodies. However, with oligoclonal B-LCLs, immunoglobulin genes for positive clones are obtained at the semi-cloned level, and there is a low risk of loss of clones. Oligoclonal-LCLs can be stocked in a deep freezer for a long period, and genes for Fab fragments can be rescued by RT-PCR even if there is a loss of cell viability.

Table 8.2 List of antigens recognized by EBV-transformed B-LCL culture supernatants derived from a donor

1. Nuclear antigens	
	Nuclear staining, and SSB and RNP antigens
2. Viral antigens	
	Hepatitis B virus and human cytomegalovirus
3. Bacterial antigens	
	Pseudomonas aeruginosa (13 serotypes), *P. maltophilia, P. cepacia, Staphylococcus aureus, S. epidermidis, S. faecalis, Escherichia coli, Salmonella entertidis, Enterobacter aerogenes, Proteus morganii, P. mirabilis, Klebsiella pneumoniae, K. ozaenae, Serratia marcescens, Shigella dysenteriae, Bacillus subtilis, B. cereus, Listeria monocytogenes, Corynebacterium diphtheriae*, LPS-Rc, and LPS-Re
4. Cytokines and cell surface molecules	
	TNF-alpha, IL-8, monocyte chemotactic protein (MCP)-1, and CD4
5. Glycolipids	
	CMH (GlcCer), CDH (LacCer), CTH, Globoside, Forssman, paragloboside, CPH, Gal-CMH (Galactocerebroside), CSE (Sulfatide), NAGM3, NGGM3 (including N-glycosialic acid or HD antigen), blood group i-type antigen, glycosphingomyelin 1 (GM1), gangliotetraosylceramide 1 (GA1 or asialo-GM1), and GA2
6. Human blood cells	
	B cells, T cells, NK cells, granulocytes, red blood cells, and 5 leukemic cell lines
7. Animal red blood cells	
	Sheep, rabbit, chicken, and guinea pig

8.6 Expression and Purification of Fab Fragments

Fab fragments are produced in the periplasm of *E. coli*, and the expression of recombinant Fab in *E. coli* can vary significantly from antibody to antibody. Severe limitations in yield, folding and functionality are sometimes encountered in bacterial production of Fab fragments. The expression efficacy is affected by vector design and culture conditions; that is, the temperature and concentration of isopropyl-β-D-thiogalactopyranoside (IPTG) (Corisdeo and Wang, 2004; Shibui and Nagahari, 1992). Many cases of successful expression have been shown at relatively low temperature and with gentle induction at a low IPTG concentration. The amount of antibody expressed also depends on its composition. Poor Fab expression is linked to poor intrinsic stability, whereas increases in Fab stability are correlated with higher Fab yields and higher levels of properly folded and functional protein (Demarest et al., 2006). This is also affected by the growth rate of *E. coli* and secretion of Fab fragments into the medium. Limitations on protein expression may be overcome by single amino acid substitutions (Knappik and Pluckthun, 1995). Storage of fragments in periplasmic spaces requires breaking of the outer membrane of the bacteria to release the fragments (Selisko et al., 2004; Takekoshi et al., 1998).

Fab fragments can easily be purified from culture medium or an extracted fraction of bacteria if a tag such as a His-tag is added during the cloning process (Skerra, 1994; Tachibana et al., 2003). The expression level of the light chain is sometimes higher than that of the Fd chain, and light chains also have a natural tendency to form homodimers, whereas Fd fragments do not usually dimerize. Therefore, attachment of a His-tag to the C-terminus of the Fd region is useful to avoid contamination of the purified fraction with light chain dimers. With this system, only functional heterodimers are purified. Affinity chromatography using anti-human Fab or anti-human F(ab')2 antibody is also used for purification of Fab fragments (Cheng et al., 2000; Takekoshi et al., 1998).

8.7 Maturation of Fab Fragments

In the human immune system, affinity of antibodies is matured in a stepwise fashion by incorporation of somatic hypermutation and selection of variants under increasing selective pressures. Various in vitro strategies have been used to maturate the affinity of recombinant Fab fragments (Hoogenboom, 2005). These include chain shuffling, CDR shuffling, and site-directed or random mutagenesis. Site-specific mutagenesis uses modifications based on structural information, whereas random mutagenesis includes use of error-prone PCR (Gram et al., 1992). Shuffling of light or heavy chain genes is effective for finding a better combination of heavy and light chains from the combinatorial library (Hur et al., 2010; Lou et al., 2010; Zhu et al., 2008). When light chain genes are shuffled, the positive rates in colony blot screening are 10- to 20-fold higher than those in shuffling analysis of heavy chain genes (Tachibana et al., 2003). If there is a loss or a partial deletion in the cloned gene, chain shuffling is also useful to find complete genes (Fu et al., 2011; Jia et al., 2008).

Residues in the CDR, and especially in the CDR3 s of the heavy and light chains of the antibody, are thought to be responsible for high-affinity interactions with antigen. Therefore, an increased affinity may occur by mutation if the native residue exhibits a negative effect on the interaction. However, the effect of a mutation is not restricted to contact residues (Winkler et al., 2000). For instance, the CDR3 of the light chain comprises amino acids 89–97 in the Kabat numbering system (Johnson and Wu, 2004; Wu and Kabat, 1970). Although the residue at position 91 may not interact directly with antigenic molecules, it can affect the binding of residue 93 (Hall et al., 1992). Thus, amino acid substitution at position 91 may result in a conformational change that allows redistribution of the neighboring amino acids involved in the antigen-antibody interaction. For site-directed mutagenesis, recombination PCR with high fidelity DNA polymerase is feasible to introduce amino acid substitution and improve the affinity of the original human Fab (Jones and Winistorfer, 1997; Tachibana et al., 2004). The framework in the variable region is also important for establishing the correct conformation of the CDR loops, and mutations in the framework may affect the flexibility of the CDR loops.

It is possible to increase the affinity of antibody Fab fragments using various procedures, and the increased affinity may correlate with biological efficacy. However, if the original antibody already demonstrates high affinity above a threshold, the effect of affinity maturation on biological function may be limited (Hoogenboom, 2005; Tachibana et al., 2003).

8.8 Concluding Remarks

Phage display followed by panning is widely used for preparation of recombinant antibody Fab fragments from large combinatorial immunoglobulin gene libraries from naïve and immune donors. Here, we have shown that bypassing the phage display and panning processes is practical when immunoglobulin genes are derived from seropositive donors. Identification of bacterial colonies using membrane filters and donor plasma is useful for the initial screening of human Fab fragments to major antigenic molecules of pathogens. Preparation of immunoglobulin gene libraries from EB virus-transformed oligoclonal B-LCL can also be used for preparation of human Fab fragments to pathogens and self-antigens. Recently, a new technique bypassing the phage display platform has also been developed. B cells are cultured at near clonal density, culture supernatants are screened, and positives are used to clone authentic antibody genes (Walker et al., 2009). Thus, improvement of technologies for antibody engineering is still in progress to generate high-affinity human antibodies.

References

Barbas CF III, Kang AS, Lerner RA, Benkovic SJ (1991) Assembly of combinatorial antibody libraries on phage surfaces: the gene III site. Proc Natl Acad Sci USA 88:7978–7982

Barbas CF III, Burton DR, Scott JK, Silverman GJ (2001) Phage Display: A Laboratory Manual. Cold Spring Harbor Laboratory Press, New York

Barbas CF III, Bjorling E, Chiodi F, Dunlop N, Cababa D, Jones TM, Zebedee SL, Persson MA, Nara PL, Norrby E et al. (1992) Recombinant human Fab fragments neutralize human type 1 immunodeficiency virus in vitro. Proc Natl Acad Sci USA 89:9339–9343

Bird RE, Hardman KD, Jacobson JW, Johnson S, Kaufman BM, Lee SM, Lee T, Pope SH, Riordan GS, Whitlow M (1988) Single-chain antigen-binding proteins. Science 242:423–426

Burton DR, Barbas CF, Persson MA, Koenig S, Chanock RM, Lerner RA (1991) A large array of human monoclonal antibodies to type 1 human immunodeficiency virus from combinatorial libraries of asymptomatic seropositive individuals. Proc Natl Acad Sci USA 88:10134–10137

Cheng XJ, Hayasaka H, Watanabe K, Tao YL, Liu JY, Tsukamoto H, Horii T, Tanabe K, Tachibana H (2007) Production of high-affinity human monoclonal antibody fab fragments to the 19-kilodalton C-terminal merozoite surface protein 1 of *Plasmodium falciparum*. Infect Immun 75:3614–3620

Cheng XJ, Ihara S, Takekoshi M, Tachibana H (2000) *Entamoeba histolytica*: bacterial expression of a human monoclonal antibody which inhibits in vitro adherence of trophozoites. Exp Parasitol 96:52–56

Corisdeo S, Wang B (2004) Functional expression and display of an antibody Fab fragment in *Escherichia coli*: study of vector designs and culture conditions. Protein Expr Purif 34:270–279

de Haard HJ, van Neer N, Reurs A, Hufton SE, Roovers RC, Henderikx P, de Bruine AP, Arends JW, Hoogenboom HR (1999) A large non-immunized human Fab fragment phage library that permits rapid isolation and kinetic analysis of high affinity antibodies. J Biol Chem 274: 18218–18230

de Kruif J, Terstappen L, Boel E, Logtenberg T (1995) Rapid selection of cell subpopulation-specific human monoclonal antibodies from a synthetic phage antibody library. Proc Natl Acad Sci USA 92:3938–3942

Demarest SJ, Chen G, Kimmel BE, Gustafson D, Wu J, Salbato J, Poland J, Elia M, Tan X, Wong K, Short J, Hansen G (2006) Engineering stability into *Escherichia coli* secreted Fabs leads to increased functional expression. Protein Eng Des Sel 19:325–336

Fu YF, Feng M, Ohnishi K, Kimura T, Itoh J, Cheng XJ, Tachibana H (2011) Generation of a neutralizing human monoclonal antibody Fab fragment to SAG1 of *Toxoplasma gondii* tachyzoites. Infect Immun 79:512–517

Gram H, Marconi LA, Barbas CF III, Collet TA, Lerner RA, Kang AS (1992) In vitro selection and affinity maturation of antibodies from a naive combinatorial immunoglobulin library. Proc Natl Acad Sci USA 89:3576–3580

Hall BL, Zaghouani H, Daian C, Bona CA (1992) A single amino acid mutation in CDR3 of the 3-14-9 L chain abolished expression of the IDA 10-defined idiotope and antigen binding. J Immunol 149:1605–1612

Hamatake M, Komano J, Urano E, Maeda F, Nagatsuka Y, Takekoshi M (2010) Inhibition of HIV replication by a CD4-reactive Fab of an IgM clone isolated from a healthy HIV-seronegative individual. Eur J Immunol 40:1504–1509

Hanes J, Jermutus L, Pluckthun A (2000) Selecting and evolving functional proteins in vitro by ribosome display. Methods Enzymol 328:404–430

Helfman DM, Feramisco JR, Fiddes JC, Thomas GP, Hughes SH (1983) Identification of clones that encode chicken tropomyosin by direct immunological screening of a cDNA expression library. Proc Natl Acad Sci USA 80:31–35

Hoet RM, Cohen EH, Kent RB, Rookey K, Schoonbroodt S, Hogan S, Rem L, Frans N, Daukandt M, Pieters H, van Hegelsom R, Neer NC, Nastri HG, Rondon IJ, Leeds JA, Hufton SE, Huang L, Kashin I, Devlin M, Kuang G, Steukers M, Viswanathan M, Nixon AE, Sexton DJ, Hoogenboom HR, Ladner RC (2005) Generation of high-affinity human antibodies by combining donor-derived and synthetic complementarity-determining-region diversity. Nat Biotechnol 23:344–348

Hoogenboom HR (2005) Selecting and screening recombinant antibody libraries. Nat Biotechnol 23:1105–1116

Hudson PJ, Souriau C (2003) Engineered antibodies. Nat Med 9:129–134

Hur BU, Choi HJ, Song SY, Yoon JB, Liu LK, Cha SH (2010) Development of the dual-vector system-III (DVS-III), which facilitates affinity maturation of a Fab antibody via light chain shuffling. Immunol Lett 132:24–30

Jia L, Yu J, Song H, Liu X, Ma W, Xu Y, Zhang C, Dong S, Li Q (2008) Screening of human antibody Fab fragment against HBsAg and the construction of its dsFv form. Int J Biol Sci 4:103–110

Johnson G, Wu TT (2004) The Kabat database and a bioinformatics example. Methods Mol Biol 248:11–25

Jones DH, Winistorfer SC (1997) Recombination and site-directed mutagenesis using recombination PCR. Methods Mol Biol 67:131–140

Kang AS, Barbas CF, Janda KD, Benkovic SJ, Lerner RA (1991) Linkage of recognition and replication functions by assembling combinatorial antibody Fab libraries along phage surfaces. Proc Natl Acad Sci USA 88:4363–4366

Knappik A, Pluckthun A (1995) Engineered turns of a recombinant antibody improve its in vivo folding. Protein Eng 8:81–89

Krebber A, Burmester J, Pluckthun A (1996) Inclusion of an upstream transcriptional terminator in phage display vectors abolishes background expression of toxic fusions with coat protein g3p. Gene 178:71–74

Kuroiwa Y, Kasinathan P, Choi YJ, Naeem R, Tomizuka K, Sullivan EJ, Knott JG, Duteau A, Goldsby RA, Osborne BA, Ishida I, Robl JM (2002) Cloned transchromosomic calves producing human immunoglobulin. Nat Biotechnol 20:889–894

Liu J, Shao H, Tao Y, Yang B, Qian L, Yang X, Cao B, Hu G, Tachibana H, Cheng X (2006) Production of an anti-severe acute respiratory syndrome (SARS) coronavirus human monoclonal antibody Fab fragment by using a combinatorial immunoglobulin gene library derived from patients who recovered from SARS. Clin Vaccine Immunol 13:594–597

Lou J, Geren I, Garcia-Rodriguez C, Forsyth CM, Wen W, Knopp K, Brown J, Smith T, Smith LA, Marks JD (2010) Affinity maturation of human botulinum neurotoxin antibodies by light chain shuffling via yeast mating. Protein Eng Des Sel 23:311–319

Maeda F, Takekoshi M, Nagatsuka Y, Aotsuka S, Tsukahara M, Ohshima A, Kido I, Ono Y, Ihara S (2005) Production and characterization of recombinant human anti-HBs Fab antibodies. J Virol Methods 127:141–147

Marks JD, Hoogenboom HR, Bonnert TP, McCafferty J, Griffiths AD, Winter G (1991) By-passing immunization. Human antibodies from V-gene libraries displayed on phage. J Mol Biol 222:581–597

Mead DA, Kemper B (1988) Chimeric single-stranded DNA phage-plasmid cloning vectors. Biotechnology 10:85–102

Mendez MJ, Green LL, Corvalan JR, Jia XC, Maynard-Currie CE, Yang XD, Gallo ML, Louie DM, Lee DV, Erickson KL, Luna J, Roy CM, Abderrahim H, Kirschenbaum F, Noguchi M, Smith DH, Fukushima A, Hales JF, Klapholz S, Finer MH, Davis CG, Zsebo KM, Jakobovits A (1997) Functional transplant of megabase human immunoglobulin loci recapitulates human antibody response in mice. Nat Genet 15:146–156

Mondon P, Dubreuil O, Bouayadi K, Kharrat H (2008) Human antibody libraries: a race to engineer and explore a larger diversity. Front Biosci 13:1117–1129

Nagatsuka Y, Hara-Yokoyama M, Kasama T, Takekoshi M, Maeda F, Ihara S, Fujiwara S, Ohshima E, Ishii K, Kobayashi T, Shimizu K, Hirabayashi Y (2003) Carbohydrate-dependent signaling from the phosphatidylglucoside-based microdomain induces granulocytic differentiation of HL60 cells. Proc Natl Acad Sci USA 100:7454–7459

Rothe C, Urlinger S, Lohning C, Prassler J, Stark Y, Jager U, Hubner B, Bardroff M, Pradel I, Boss M, Bittlingmaier R, Bataa T, Frisch C, Brocks B, Honegger A, Urban M (2008) The human combinatorial antibody library HuCAL GOLD combines diversification of all six CDRs according to the natural immune system with a novel display method for efficient selection of high-affinity antibodies. J Mol Biol 376:1182–1200

Russel M (1991) Filamentous phage assembly. Mol Microbiol 5:1607–1613

Selisko B, Cosio G, Garcia C, Becerril B, Possani LD, Horjales E (2004) Bacterial expression, purification and functional characterization of a recombinant chimeric Fab derived from murine mAb BCF2 that neutralizes the venom of the scorpion *Centruroides noxius* Hoffmann. Toxicon 43:43–51

Shi L, Wheeler JC, Sweet RW, Lu J, Luo J, Tornetta M, Whitaker B, Reddy R, Brittingham R, Borozdina L, Chen Q, Amegadzie B, Knight DM, Almagro JC, Tsui P (2010) De novo selection of high-affinity antibodies from synthetic Fab libraries displayed on phage as pIX fusion proteins. J Mol Biol 397:385–396

Shibui T, Nagahari K (1992) Secretion of a functional Fab fragment in *Escherichia coli* and the influence of culture conditions. Appl Microbiol Biotechnol 37:352–357

Siegel DL, Chang TY, Russell SL, Bunya VY (1997) Isolation of cell surface-specific human monoclonal antibodies using phage display and magnetically-activated cell sorting: applications in immunohematology. J Immunol Methods 206:73–85

Skerra A (1994) A general vector, pASK84, for cloning, bacterial production, and single-step purification of antibody Fab fragments. Gene 141:79–84

Skerra A, Dreher ML, Winter G (1991) Filter screening of antibody Fab fragments secreted from individual bacterial colonies: specific detection of antigen binding with a two-membrane system. Anal Biochem 196:151–155

Smith GP (1985) Filamentous fusion phage: novel expression vectors that display cloned antigens on the virion surface. Science 228:1315–1317

Tachibana H, Watanabe K, Cheng XJ, Tsukamoto H, Kaneda Y, Takeuchi T, Ihara S, Petri WA Jr (2003) VH3 gene usage in neutralizing human antibodies specific for the *Entamoeba histolytica* Gal/GalNAc lectin heavy subunit. Infect Immun 71:4313–4319

Tachibana H, Matsumoto N, Cheng XJ, Tsukamoto H, Yoshihara E (2004) Improved affinity of a human anti-*Entamoeba histolytica* Gal/GalNAc lectin Fab fragment by a single amino acid modification of the light chain. Clin Diagn Lab Immunol 11:1085–1088

Takekoshi M, Maeda F, Tachibana H, Inoko H, Kato S, Takakura I, Kenjyo T, Hiraga S, Ogawa Y, Horiki T, Ihara S (1998) Human monoclonal anti-HCMV neutralizing antibody from phage display libraries. J Virol Methods 74:89–98

Takekoshi M, Maeda F, Nagatsuka Y, Aotsuka S, Ono Y, Ihara S (2001) Cloning and expression of human anti-tumor necrosis factor-alpha monoclonal antibodies from Epstein-Barr virus transformed oligoclonal libraries. J Biochem 130:299–303

Thie H, Meyer T, Schirrmann T, Hust M, Dubel S (2008) Phage display derived therapeutic antibodies. Curr Pharm Biotechnol 9:439–446

Tomizuka K, Yoshida H, Uejima H, Kugoh H, Sato K, Ohguma A, Hayasaka M, Hanaoka K, Oshimura M, Ishida I (1997) Functional expression and germline transmission of a human chromosome fragment in chimaeric mice. Nat Genet 16:133–143

van den Beucken T, Pieters H, Steukers M, van der Vaart M, Ladner RC, Hoogenboom HR, Hufton SE (2003) Affinity maturation of Fab antibody fragments by fluorescent-activated cell sorting of yeast-displayed libraries. FEBS Lett 546:288–294

Walker LM, Phogat SK, Chan-Hui PY, Wagner D, Phung P, Goss JL, Wrin T, Simek MD, Fling S, Mitcham JL, Lehrman JK, Priddy FH, Olsen OA, Frey SM, Hammond PW, Kaminsky S, Zamb T, Moyle M, Koff WC, Poignard P, Burton DR (2009) Broad and potent neutralizing antibodies from an African donor reveal a new HIV-1 vaccine target. Science 326:285–289

Ward RL, Clark MA, Lees J, Hawkins NJ (1996) Retrieval of human antibodies from phage-display libraries using enzymatic cleavage. J Immunol Methods 189:73–82

Winkler K, Kramer A, Kuttner G, Seifert M, Scholz C, Wessner H, Schneider-Mergener J, Hohne W (2000) Changing the antigen binding specificity by single point mutations of an anti-p24 (HIV-1) antibody. J Immunol 165:4505–4514

Wu H, Beuerlein G, Nie Y, Smith H, Lee BA, Hensler M, Huse WD, Watkins JD (1998) Stepwise in vitro affinity maturation of Vitaxin, an alphav beta3-specific humanized mAb. Proc Natl Acad Sci USA 95:6037–6042

Wu TT, Kabat EA (1970) An analysis of the sequences of the variable regions of Bence Jones proteins and myeloma light chains and their implications for antibody complementarity. J Exp Med 132:211–250

Zahn G, Skerra A, Hohne W (1999) Investigation of a tetracycline-regulated phage display system. Protein Eng 12:1031–1034

Zhu Z, Bossart KN, Bishop KA, Crameri G, Dimitrov AS, McEachern JA, Feng Y, Middleton D, Wang LF, Broder CC, Dimitrov DS (2008) Exceptionally potent cross-reactive neutralization of Nipah and Hendra viruses by a human monoclonal antibody. J Infect Dis 197:846–853

Chapter 9
Intrabody Expression in Mammalian Cells

Silvia Biocca

Abstract The intracellular expression of antibodies or antibody fragments (intrabodies) in different compartments of mammalian cells allows to block or modulate the function of endogenous molecules. Intrabodies can alter protein folding, protein-protein, protein-DNA, protein-RNA interactions and protein modification. They can induce a phenotypic knockout and work as neutralizing agents by direct binding to the target antigen, by diverting its intracellular traffic or by inhibiting its association with binding partners. They have been largely employed as research tools and are emerging as therapeutic molecules for the treatment of human diseases as viral pathologies, cancer and misfolding diseases. The fast growing bio-market of recombinant antibodies provides intrabodies with enhanced binding specificity, stability and solubility, together with lower immunogenicity, for their use in therapy. This chapter describes the crucial aspects required to express intrabodies in different intracellular compartments of mammalian cells, their various modes of action and gives an update on the applications of intrabodies in human diseases.

9.1 Introduction

The proof of principle that antibodies could be efficiently expressed and targeted to different intracellular compartments in mammalian cells dates to 1990 (Biocca et al. 1990) and led to the concept of exploiting recombinant antibodies to block or modulate the function of target antigens for intercellular and intracellular immunization (Biocca and Cattaneo 1995; Cattaneo and Biocca 1997). Intrabodies have unique advantages comparing to other knockout gene techniques or RNA interference. First, they can target the antigen in different intracellular compartments including extracellular milieu. Secondly, they are highly specific reagents and are very stable in mammalian cells, especially when expressed in the secretory compartment. In addition, intrabodies can hit a variety of possible targets: (i) specific protein domains, (ii) specific protein-protein interaction sites, (iii) post-translational modifications, (iv) multiple conformational isoforms (oligomers, fibrils etc) and (v) even non protein antigens.

S. Biocca (✉)
Department of Neuroscience, University of Rome "Tor Vergata", 00133 Rome, Italy
e-mail: biocca@med.uniroma2.it

M. Al-Rubeai (ed.), *Antibody Expression and Production*, Cell Engineering 7,
DOI 10.1007/978-94-007-1257-7_9, © Springer Science+Business Media B.V. 2011

In order to specifically target intrabodies to the physiological site of the antigen or to new intracellular localisations, dominant and autonomous targeting sequences should be grafted onto antibody chains. Recombinant antibody domains, in particular single-chain Fv (scFv) fragments have been expressed in the cytoplasm (Biocca et al. 1994), the nucleus (Duan et al. 1994; Mhashilkar et al. 1995) and the secretory pathway of mammalian cells (Marasco et al. 1993) and successfully used to inhibit the function of several intracellular antigens. New suitable antibody formats for their intracellular use are now available and functional intrabodies are employed as research tools to define the mechanisms of human pathologies at the molecular level and for a variety of therapeutic applications. For instance, intrabodies have been designed to inhibit single or simultaneously multiple signal transduction pathways (Lener et al. 2000; Tanaka et al. 2007; Jendreyko et al. 2005), inhibition of HIV viral proteins (Lo et al. 2008), inhibition of oncogene products (Williams and Zhu 2006; Griffin et al. 2006), misfolding-prone proteins (Cardinale and Biocca 2008a), receptors of the immune system (Kirschning et al. 2010) and also applied in post-transplantation surgery (Zdoroveac et al. 2008).

9.2 Targeting Intrabodies to Different Intracellular Locations

Antibody fragments can be directed to specific target antigens present in the cytosol, nucleus, endoplasmic reticulum (ER), plasma membrane (PM), mitochondria, peroxisomes and trans-Golgi network (TGN) through in frame fusion with intracellular trafficking sequences (Cardinale et al. 2004). The following is a list of targeting signals successfully used for intracellular expression of antibodies.

The way to target intrabodies throughout the secretory pathway, as secreted proteins, is by exploiting the leader sequence for secretion of the immunoglobulin (Ig) at the N terminus. ER-retained intrabodies are designed with a leader sequence at the N terminus and a retention peptide, KDEL, at the C terminus (Biocca et al. 1995). Intrabodies in the ER behave as intracellular anchors and can be used either to prevent the appearance of receptor proteins on the plasma membrane or to inhibit the secretion of a protein. Similar protein retention in the trans-Golgi has been achieved with a trans-Golgi retention signal (Zhou et al. 1998). Targeting to the plasma membrane has been obtained by fusing a scFv with a receptor transmembrane domain (Chesnut et al. 1996).

Removal of the leader sequence of variable heavy (VH) and variable light (VL) domains which target antibody fragments to the lumen of the endoplasmic reticulum allows the cytoplasmic expression of intrabodies (leader-less). Nuclear targeting can be achieved by adding one or more nuclear localisation sequences (NLS) to the leader-less antibody fragments, such as the PKKKRKV sequence of the large T antigen of SV40, either at the N- and C-terminus (Biocca et al. 1995). Intrabodies have been also targeted to mitochondria. N-terminal presequences are present in most of the nuclear-encoded mitochondrial proteins. These sequences are removed once the protein is translocated through the mitochondrial membrane. The N-terminal presequence of the subunit VIII of human cytochrome oxidase (COX8.21), covering the cleavage junction, can be fused to the scFv fragment. The resulting molecule correctly localizes to mitochondria (Biocca et al. 1995).

To facilitate the expression of scFv fragments as secreted or intracellular proteins, a set of general vectors have been designed. The integrated system of scFvexpress vectors (Persic et al. 1997a) derives from the VHexpress vector, a vector used to produce secretory immunoglobulin heavy chains from cloned IgH regions (Persic et al. 1997b). All the scFvexpress plasmids contain an N- or C-terminal localisation signal that allows the targeting of the antibody fragments to different compartments, including the endoplasmic reticulum (scFvex-ER), the cytoplasm (scFvex-cyt), the nucleus (scFvex-nuc) and the mitochondria (scFvex-mit). The scFvexpress-cyt has no targeting signal (leader-less) and directs the expression of scFv in the cytoplasm, with an N-terminal methionine instead of the leader sequence for secretion. All other targeting vectors are derivatives of the scFvexpress-cyt and were obtained by the insertion of well characterized targeting signals (Biocca et al. 1995) either N- or C-terminal to the scFv, as appropriate.

Besides the specific targeting signals, these vectors contain cassettes that encode for strong promoters and sequences for the resistance to selection antibiotics and a C-terminal myc-tag in frame with the scFv allowing its detection with the monoclonal antibody 9E10 (Persic et al. 1997a).

These vectors can be used for transient or stable transfection of mammalian cells.

9.3 Intrabody Formats

Innovative DNA recombinant technologies have allowed the reformatting of antibody molecules in new smaller fragments with improved properties for their intracellular expression. So far the single chain variable fragment (scFv) has been the recombinant antibody format more widely used for intrabodies. It contains the complete antigen binding site and it consists of the variable domains of the immunoglobulin heavy (VH) and light (VL) chains linked with a flexible polypeptide which prevents dissociation. The resulting molecule is a monovalent antibody fragment, with a molecular weight of about 30 kDa compared to the 150 kDa of the full-length antibody. Other formats that have been successfully expressed inside cells are recombinant bispecific and tetravalent antibody fragments made of two scFvs linked through the second and third heavy chain constant domain, named intradiabodies. These bispecific intrabodies have been designed for simultaneous trapping of two endothelial transmembrane receptors in the same compartment (Jendreyko et al. 2005).

ScFv is not the minimal size for functional antibody fragments. Other fragments made of one variable domain such the single VL and VH domains are the smallest functional fragments derived from immunoglobulin light and heavy chains. The so-called VHH single domain antibody fragments, derived from naturally occurring heavy-chain antibodies devoid of light chain present in the immune system of camelids, retain the antigen specificity of the whole antibody and have excellent properties of solubility, stability and expression in mammalian cells, aside from the absence of intra-domain disulfide bonds (Hamers-Casterman et al. 1993; Wesolowski et al. 2009). They are easily produced as recombinant antibodies, much smaller in size and can be forged into new multispecific and multivalent reagents with enhanced therapeutic efficacy. Due to their smaller size, they possess a great capacity to form long finger-like extensions that can potentially target cryptic epitopes that are difficult for intact antibodies or scFv fragments to reach.

9.4 Selection Strategies to Improve Intrabody Stability and Solubility

Antibody fragments can be potentially targeted to any subcellular compartment, but their folding performance and stability can be limited by the micro-environmental intracellular conditions of the compartment where they are directed to. Thus, a major issue raised in the initial phase of work on intrabodies was the limited half life of antibody and antibody domains and their tendency to aggregate when expressed in the cytoplasm of mammalian cells. Studies on targeting of scFv fragments showed, in fact, that the expression levels of the retargeted antibody domains may vary and the cytoplasm may be considered the worst case (Cattaneo and Biocca 1999). In general, intrabodies expressed in the secretory compartment are more stable than those expressed in the cytoplasm. This is due to the fact that intradomain disulfide bonds, which contribute 4–6 Kcal mol^{-1} to the stability of the antibody domain, do not form in the reducing environment of the cytoplasm (Biocca et al. 1995). As a consequence of the lower stability, some scFvs tend to misfold and aggregate as insoluble proteins. Notwithstanding this fact, it is worth noting that in many cases cytoplasmic-targeted intrabodies bind the antigens and maintain their in vivo functional activity (Cardinale et al. 2001).

A recent comparative study between the cytoplasmic expression in mammalian cells of well characterized single-chain variable fragments and camelid VHHs, selected from antibody libraries based on similar scaffolds, have outlined the physico-chemical determinants that correlate with enhanced intracellular solubility. Soluble expression in the cytoplasm appears to be influenced by the complementary determining regions (CDRs) content and by the overall charge and hydropathicity of the intrabody sequence (Kvam et al. 2010).

This issue is very important and during the last years many selection strategies have been developed to improve stability, solubility and functional properties of antibody fragments under conditions of intracellular expression. Two different strategies have been followed: the knowledge-based and the selection-based approaches. The former relies on the introduction of educated mutations that stabilize pre-existing antibody fragments in order to obtain a "super-framework" (Jung and Plückthun 1997; Wörn et al. 2000; Monsellier and Bedouelle 2006). The second approach exploits the availability of different antibody libraries for selection of the best functional molecules for intracellular expression (Hudson and Souriau 2003). For the expression in the cytoplasm, new human phage antibody libraries have been generated based on single framework optimized for intracellular expression (Philibert et al. 2007). On the other hand, highly stable ribosome-display libraries, based on Escherichia coli SecM translation arrest mechanism, have been used for isolating scFvs that are stable under reducing conditions (Contreras-Martınez and DeLisa 2007). Moreover, antibody-antigen intracellular selection methods have been also developed for the isolation of intrabodies able to efficiently interact with the antigen in vivo. To this aim an in vivo yeast two-hybrid system has been realized, described as intracellular antibody capture (IAC) technology (Visintin et al. 1999). This approach combines a first round of in vitro selection of scFvs with a second round of in vivo screening of selected intrabodies and allows the isolation of antibody domains with no need to use purified antigens. IAC technology has recently allowed the direct in vivo intracellular selection of conformation-sensitive anti-oligomeric scFvs against the Alzheimer's amyloid β peptide (Meli et al. 2009). Another procedure for direct in vivo selection of antigen-specific intrabodies, which utilises a single domain antibody format and is based on a predefined intrabody consensus framework has been described (IAC^2) (Tanaka and Rabbitts 2003) and further developed (IAC^3) (Tanaka and Rabbitts 2010). This new protocol allows the isolation of functional VH and VL domains from different libraries in four steps, including confirmation of functional intrabodies in mammalian cells.

For clinical applications of intrabodies, generation of humanized and/or human-derived antibody domains offers obvious potential advantages. Modern emerging strategies have improved in vitro selection of fully humanized recombinant antibodies directly from human antibody-display libraries, through the creation of large natural or synthetic repertoires of antibody fragments (Hudson and Souriau 2003; Hoogenboom 2005). A direct selection of human phage antibody libraries on tumor cells has been described (Goenaga et al. 2007).

9.5 The Mode of Action of Intrabodies

Intrabodies can mediate their effect inside the cells by neutralizing the target protein through direct binding to the functional domain (Biocca et al. 1994; Cohen et al. 1998), blocking protein-protein interaction (Griffin et al. 2006; Tanaka et al. 2007; Van den Abbeele et al. 2010) or by relocating the antigen to a different intracellular location (Lener et al. 2000; Cardinale et al. 2005; Böldicke 2007). In this last case also non-neutralizing antibodies, according to in vitro biochemical criteria, can be effective when expressed in vivo. This can be obtained by adding to the intrabody a specific targeting signal, as the ER retention signal KDEL, which confers to the antigen-antibody complex the retention in the endoplasmic reticulum. In some described cases, intrabodies divert the antigen to the proteasome degradation pathway (Cardinale et al. 2003; Filesi et al. 2007). As mentioned before, intrabodies have different folding performance and stability particularly when they are expressed in the cytoplasm. In this compartment, intradomain disulfide bonds do not form and intrabodies complexed to their corresponding antigen tend to misfold and aggregate. This has been demonstrated for cytoplasmic, nuclear and secretory intrabodies. In the first case they aggregate and form aggresomes in a perinuclear location, whereas in the secretory compartments the antigen-antibody complexes are retrotranslocated from the ER, ubiquitinated and finally proteasome degraded (Filesi et al. 2007).

In order to define the mode of action of new intrabodies, the solubility properties, the intracellular distribution and the resulting phenotypes should be routinely analysed with ad hoc assays. Protocols for studying the expression, solubility, stability properties, intracellular localisation and for the analysis of the antigen-intrabody complex in vivo have been described (Cardinale et al. 2004 and Cardinale and Biocca 2010).

In summary, intrabodies can be used for preventing or treating human diseases by exploiting their ability to (a) divert the antigen from its functional location, (b) inhibit functional protein-protein interaction, (c) inhibit the functional site of the target antigen, (d) re-route the antigen to the degradation pathway and (e) inhibit different stages of the aggregation process. This latter mode of action can be achieved through stabilization of the native state isoform, inhibition of oligomerization, inhibition of fibril formation and disruption and clearance of preformed aggregates as demonstrated with scFv directed against misfolded prone proteins (Cardinale and Biocca 2008b).

Figure 9.1 summarises the different modes of action of intrabodies observed in vivo. These activities have a great potential in medicine and represent a viable option for different pathologies, including neurodegenerative diseases, infectious diseases and cancer.

9.6 Intrabodies Against Misfolding Diseases

A hallmark of misfolding diseases is the accumulation of amyloid or amyloid-like aggregates deriving from the fibrillization process of a native protein. During the aggregation process of an amyloidogenic protein many different conformers, including misfolded monomers, oligomers and fibrils are generated (Chiti and Dobson 2006). In the neurodegenerative diseases the aggregation occurs in the brain and is accompanied by cognitive decline. The available therapeutic treatments to slow or prevent these devastating disorders are still not effective, so the development of new molecular therapies that target the pathogenic proteins are urgently needed. In particular, there is great interest in searching new molecules able to prevent unfolding and aggregation either by stabilizing the native state of the amyloidogenic precursors or by disassembling amyloid fibrils. Conformational specific antibodies are very promising agents against neurodegenerative disorders, because they can be raised against different structural isoforms of an amyloidogenic antigen, act at the protein level and can be used in an intrabody approach (Fig. 9.1). Interestingly, although there are no obvious structural or sequence identities among proteins that

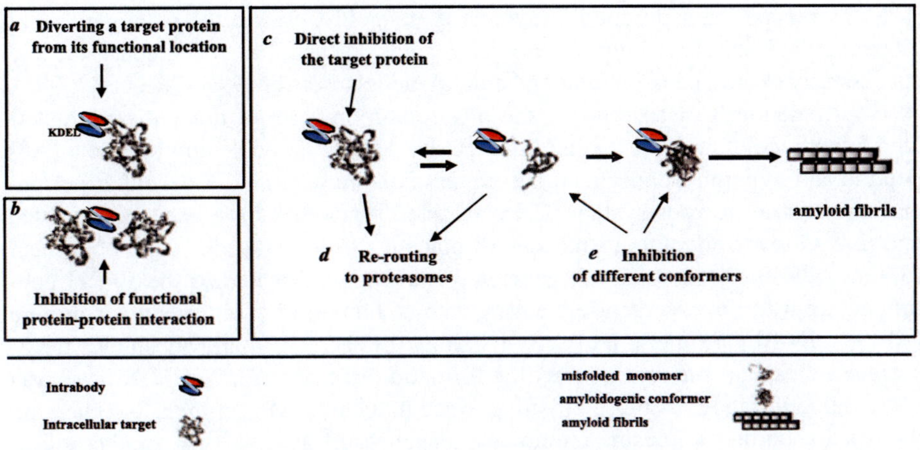

Fig. 9.1 Schematic model for different modes of action of intrabodies. (**a**) Re-routing the antigen from its functional location, in this case by adding the ER-retention signal KDEL at the C-terminus of the intrabody. The retention sequence KDEL is specifically recognized by the ERD2 receptor inside the cis-Golgi. The complex formed by the KDEL-tagged intrabody and the target protein binds to the ERD2 receptors, is transported through the Golgi back to the ER, where it is released. (**b**) Inhibition of functional protein-protein interaction. (**c**) Direct inhibition of the native target protein. (**d**) Re-routing the antigen antibody complex or the prone-misfolded antigen to proteasome degradation. (**e**) Inhibition of different conformers. Intrabodies that distinguish native and misfolded conformations can be generated and used either to re-route the misfolding-prone protein from its site of aggregation and, eventually, divert it to the proteasome pathway or to inhibit different stages of the aggregation process. This latter mode of action can be achieved through (i) stabilization of the monomeric isoform, (ii) inhibition of oligomerization, (iii) neutralization of potentially toxic oligomeric species, (iv) inhibition of fibril formation and (v) clearing of preformed aggregates

cause neurodegenerative diseases, it has been shown that conformation-dependent antibodies, raised against a molecular mimic of Aβ oligomers, react with oligomers derived from different kinds of amyloidogenic proteins, such as α-synuclein, polyglutamine, prions, etc. inhibiting the toxicity associated to these aggregates (Kayed et al. 2003). This finding indicates that the conformational structure rather than the amino acid sequence of these misfolded isoforms is probably the key factor for their neurotoxicity.

To generate conformation-specific intrabodies for therapeutic purposes it can be used either the IAC technology, that does not need the purified antigen (Meli et al. 2009) or it is crucial to isolate specific misfolded isoforms. In this case, atomic force microscopy (AFM) and electron microscopy are methods that allow to visualise any different isoforms even if they are small oligomers. The combination of phage display technology and screening by AFM has been used as a new approach to select morphology-specific intrabodies against different conformers, as reported for Aβ fibrils and α-synuclein oligomers (Emadi et al. 2007; Marcus et al. 2008; Zameer et al. 2008).

Many recombinant intrabodies against proteins involved in the pathogenesis of Alzheimer's, Prion, Huntington's and Parkinson's diseases have been generated and successfully expressed in cellular and animal models (Cardinale and Biocca 2008a).

For Alzheimer's disease (AD), the most common form of dementia characterized by extracellular deposits and intracellular accumulation of amyloid beta (Aβ) peptide and hyperphosphorylated tau protein (Blennow et al. 2006), either peripheral and central nervous system (CNS) targeted antibodies have been largely used and proved to be effective to reduce Aβ placque burden and memory impairment (Steinitz 2009). Paganetti et al. generated intrabodies directed to the EFRH peptide adjacent to the β-secretase cleavage site of human amyloid precursor protein (APP) (scFv-β1). Expression of scFv-β1 along the secretory pathway shields the β-secretase cleavage site and inhibits the formation of toxic Aβ. The KDEL version of the same intrabody is more effective, since it anchors APP in the ER preventing its appearance on the plasma membrane (Paganetti et al. 2005). In another study, the expression of anti-nicastrin intrabodies disrupts the proper folding and glycosylation of the endogenous nicastrin. This protein is required for the stability of the γ-secretase complex. As a result, the anti-nicastrin intrabody suppresses the γ-secretase enzymatic activity in vivo (Hayashi et al. 2009).

Prion diseases are fatal transmissible spongiform encephalopathies affecting humans and animals (Prusiner 1998; Aguzzi et al. 2008). So far, intrabody applications against prion disorders have been targeted to the endogenous prion protein (PrP^C). Anti-prion KDEL-8H4 scFv fragments were generated and stably expressed in a neuronal cell line susceptible to scrapie infection. Its intracellular expression causes a marked impairment of prion maturation and translocation towards the membrane compartment, with a strong reduction of the PrP^C membrane fraction. As a consequence, the pathogenic scrapie isoform (PrP^{Sc}) does not form and accumulate in infected cells (Cardinale et al. 2005). Moreover, mice, intracerebrally injected with a lysate derived from KDEL-8H4 expressing cells infected with scrapie, neither develop scrapie clinical sign nor brain damage, demonstrating effective treatment (Vetrugno et al. 2005). The secretory version of the same intrabody (Sec-8H4), able to recognize PrP^C in the secretory pathway, strongly inhibits PrP^{Sc} accumulation in 139A scrapie strain infected cells. By analysing its mode of action, it was found that PrP^C total level is markedly reduced due to a selective re-routing of PrP^C to the proteasome pathway. Moreover, Sec-8H4 intrabody impairs the secretion of endogenous prion molecules associated to exosomes-like vesicles, a potential spreading route for prion infectivity (Filesi et al. 2007). A drastic reduction of PrP^{Sc} accumulation was also obtained by co-culturing cells secreting anti-prion scFv fragments with chronically scrapie infected neuroblastoma cells (Donofrio et al. 2005).

Huntington's disease (HD) is a genetic disorder associated with a progressive neurodegeneration in the brain areas of cortex and striatum (Walker 2007). It is caused by the aggregation of mutated forms of huntingtin (htt) protein which present abnormally long polyglutamine (poliQ) sequences at the N-terminal. Cytoplasmic intrabodies directed against the poly-proline region, flanking the polyglutamine region at the C-terminal side of htt, inhibit cell death and aggregation in transiently transfected HEK293 cells, while intrabodies to the polyQ region are cytotoxic and ineffective, because they accelerate aggregate formation and apoptotic cell death (Khoshnan et al. 2002). A phage-derived C4 scFv against the N-terminal portion of htt reduces aggregation and toxicity in cell cultures, in neuronal organotypic slice cultures and in a Drosophila model of HD (Murphy and Messer 2004; Wolfgang et al. 2005; McLear et al. 2008). Another potent intrabody against mutant htt, when expressed in the striatum of HD mice via adenoviral infection, reduces neuropil aggregate formation and ameliorates neurological symptoms (Wang et al. 2008). Interaction of this intrabody with mutant huntingtin increases the ubiquitination of cytoplasmic htt and its degradation, indicating that intrabody-mediated re-routing of htt to the proteasome pathway is the mechanism underlying the protective activity (Fig. 9.1). Furthermore, an intrabody against the N-terminal htt, V_L12.3 and another, Happ1, which recognizes the proline rich domain of htt, both prevents htt aggregation and inhibits toxicity in an immortalized striatal cell model of HD by different mechanisms (Southwell et al. 2008). Tested both in brains of five mouse models of HD, V_L12.3 increases severity of phenotype and mortality in two models, while Happ1 treatment ameliorates motor, cognitive and neuropathological symptoms (Southwell et al. 2009).

Parkinson's disease (PD) is the second most prevalent neurodegenerative illness clinically characterized by motor and cognitive dysfunction. The main neuropathological feature is the presence of intracytoplasmic, proteinaceous inclusions termed as Lewy bodies (LB) (Wood-Kaczmar et al. 2006). The protein α-synuclein is a major structural component of LB. Cytoplasmic expression of an anti-monomeric α-synuclein single-chain intrabody rescues the cell adhesion, stabilizes the monomeric isoform and inhibits the formation of high molecular weight insoluble species in a cell model (Zhou et al. 2004). More recently, a novel intrabody against the nonamyloid component of α-synuclein, selected from a yeast surface display library, shows highly significant reduction of aggregation in stably trasfected cellular models (Lynch et al. 2008). Anti-oligomeric scFv fragments block fibril formation in vitro (Emadi et al. 2007) and alleviate toxicity when intracellularly expressed in mammalian cells (Yuan and Sierks 2009).

Oculopharyngeal muscular dystrophy (OPMD) is an autosomal dominant dystrophy clinically characterized by progressive weakening of specific muscles. It is a protein aggregation disorder caused by short expansions of the N-terminal polyalanine tract in the nuclear poly(A)-binding protein 1 (PABPN1). Mutant PABPN1 aggregates in intranuclear inclusions in OPMD patient muscles. In order to inhibit the aggregation of the PABPN1 mutant, a panel of specific single domain antibody fragments (VH) was expressed in cellular and animal OPMD models. One of them was effective in reducing aggregate formation and clearing pre-existing aggregates in cells and was a strong suppressor of muscle degeneration in a characterized Drosophila OPMD model, restoring muscle gene expression (Verheesen et al. 2006; Chartier et al. 2009).

9.7 Intrabodies for Treatment of Infection Diseases

The development of intrabodies for the inhibition of virus structural, regulatory, enzymatic, envelope proteins and receptors on the surface of host cells has been also largely explored, demonstrating their efficacy either at the early and late events of the viral life cycle. In the beginning, they have been used to target proteins of the human immunodeficiency virus 1 (HIV-1) (Lo et al. 2008). Intrabodies equipped with a ER retention signal were employed against the viral coat proteins gp120 and gp41. They caused inhibition of virus replication and syncytial formation (Marasco et al. 1993; Zhou et al. 1998). Inhibition of early and late events of HIV-1 life cycle was also obtained by expression of cytoplasmic intrabodies directed against the matrix protein p17 (Levin et al. 1997) and against proteins involved in replication (reverse transcriptase and Vif). Camelized single-domain intrabodies efficiently interact with HIV-1 Vif protein and neutralize Vif-mediated proviral integration in non permissive cells (Aires da Silva et al. 2004). Moreover, intrabodies that block regulatory proteins, such as Tat and Rev have been used to inhibit HIV-1 replication. A cytosolic anti-Tat scFv fragment modified with a C-terminal human C kappa domain to increase cytoplamic stability inhibits Tat-mediated long terminal repeat (LTR) transactivation and HIV-1 infection in transformed lymphocytes (Mhashilkar et al. 1995; Bai et al. 2003) and a cytosolic anti-Rev scFv fragment inhibits HIV-1 replication in HeLa-T4 cells (Duan et al. 1994). Recently, a llama derived VH domain, selected against the N-terminal α-helical multimerization domain of Rev, was used to block Rev multimerization and prevent HIV replication (Vercruysse et al. 2010).

Intrabodies have been described for treatment of various other viral infections, such as Hepatitis B (HBV) and Hepatitis C virus (HCV) infections, which represent a global health problem. ER-retained single-domain intrabodies (VHHs) targeting the envelope protein of HBV induce more than two log reduction in virion secretion in a HBV mouse model (Serruys et al. 2009) and, in a still preliminary study, cytosolic and nuclear VHH intrabodies targeting the core antigen (HBcAg) were produced and tested in infected cells (Serruys et al. 2010). Intrabodies against the NS3 serine protease, which is necessary for viral replication and innate immune evasion, inhibit HCV replication when expressed in hepatoma cells (Gal-Tanamy et al. 2010).

Intrabodies inhibit Papilloma virus protein function in cervical cancer cells (Griffin et al. 2006), decrease Kaposi sarcoma-associated herpes viral persistence in lymphoma cells (Corte-Real et al. 2005) and reduce transcription and replication of influenza A virus (Mukhtar et al. 2009).

Interestingly, novel disulfide-free proteins that target severe acute respiratory syndrome (SARS) N protein with high affinity and selectivity have been generated by using mRNA display selection and directed evolution. Although these molecules are structurally very different from antibodies as they utilises a discontinuous binding surface, they may represent interesting alternatives to intrabodies. At least for the molecules tested in this study, seven block SARS replication with different efficiency and do not disrupt mammalian cell function (Liao et al. 2009).

9.8 Intrabodies in Cancer

Many tumors are characterized by the presence of mutated or aberrantly expressed proteins involved in the control of vital processes. Most oncogenic proteins are located inside cells and are not available to be targeted by standard antibody-mediated anti-tumor therapies. Intrabodies have clear clinical potential in cancer therapy and have been successfully used as effectors of intracellular cancer targets and pathways associated with tumor cell proliferation, differentiation and invasion. In particular intrabodies have been extensively studied as inhibitors of growth factors receptors or other oncogenic antigens in the secretory compartment, in the cytoplasm and in the nuclei.

Aberrant expression of members of the epidermal growth factor receptor (EGFR) family, such as EGFR and erbB2, has been observed in a variety of human tumors, including breast and ovary carcinoma. Many reports describe the successful phenotypic knockout of the epidermal growth factor receptor (EGFR) and erbB2 achieved by retention of them by ER-retained scFv intrabodies (Graus-Porta et al. 1995; Jannot et al. 1996; Deshane et al. 1997). Down regulation of surface-exposed erbB2 receptors results in induction of apoptosis, cytotoxicity and inhibition of tumor cell proliferation in vitro and in vivo. However, a phase 1 clinical trial using an anti-erbB-2 scFv-encoding adenovirus, carried out to treat erbB-2-overexpressing ovarian cancer, highlighted the need for more efficient gene delivery systems (Alvarez et al. 2000). Reversion of transformed phenotype in ovarian cancer cells was also obtained by the intracellular expression of ER-retained anti-α folate receptor intrabodies (Figini et al. 2003).

To identify and functionally characterize tumor specific markers, a human scFv phage display library was used to select intrabodies able to bind and rapidly internalize into human breast cancers. One of these, the 3GA5, has been proved to knock down the surface display of the CD9 partner 1, a tumor specific receptor antigen, when expressed as an ER-retained intrabody inside cells (Goenaga et al. 2007).

RAS is a guanine nucleotide binding protein which plays a crucial role in the regulation of cell proliferation, oncogenic transformation and differentiation located at the inner surface of the plasma membrane. A panel of non neutralizing anti-RAS scFvs were proved to inhibit cell proliferation by sequestering the antigen and diverting it in cytoplasmic aggresomes (Lener et al. 2000). Intrabody-mediated RAS aggregation led to proteasome dysfunction and apoptosis (Cardinale et al. 2003). In another study, neutralization of RAS promoted apoptosis in human cancer cells and led to tumor regression of a colon carcinoma tumour model in nude mice (Cochet et al. 1998). More recently, a soluble single VH domain that specifically binds to activated GTP-bound RAS inhibited RAS-effector protein interactions with RAS, preventing tumorigenesis in a mouse model (Tanaka et al. 2007).

The tumor suppressor gene p53 is mutated in almost half of human tumors. Restoration of its transcriptional activity may trigger massive apoptosis of cancer cells. Intracellular expression of nuclear targeted anti-p53 scFv fragments in human tumor cells leads to restoration of the p53 mutant deficient transcriptional activity (Caron de Fromentel et al. 1999) and induces tumor regression in an animal model of mice carrying human xenografts, with no apparent deleterious side effects (Orgad et al. 2010).

Functional apoptotic pathways are crucial for cell homeostasis and for the elimination of damaged or transformed cells. A dysregulation of apoptosis is implicated in many pathologies, including neurodegenerative diseases and cancer. Two camelid-derived single VH domains, selected against recombinant Caspase-3 and expressed as intrabodies in a neuroblastoma cell line, show different in vivo apoptotic-modulating effects. Notably, while one of them is an antagonist towards Caspase 3 and protects cells from oxidative-stress-induced apoptosis, the other is able to induce cell death (McGonigal et al. 2009).

9.9 Against Toxins

Other very recent applications are the use of intrabodies as anti-toxin agents. Toxins from microbial and other sources continue to cause substantial human and veterinary pathologies and represent serious biosecurity threats. Currently there is no antidote that can reverse symptoms of the botulinum neurotoxin. Camelid single domain intrabodies (VHHs) specifically selected against Clostridium botulinum neurotoxin (BoNT) protect neuronal cell synaptosomal-associated protein 25 (SNAP25) protein from cleavage, demonstrating its potential as a component of therapeutic agents against botulism intoxication (Tremblay et al. 2010). Some ADP-ribosylation toxins, such as Salmonella SpvB toxin, are secreted directly from the Salmonella-containing vacuole into the cytosol of target cells and, thus, are inaccessible to conventional antibodies. A single-domain antibody expressed as a cytosolic intrabody blocks the actin ADP-ribosylating toxin of Salmonella typhimurium (Alzogaray et al. 2010).

9.10 Concluding Remarks

Targeting antigens not accessible by circulating antibody molecules is the powerful of the intrabody technology. This approach has greatly improved by recent advances in scaffold design, repertoire construction and ad hoc selection methods to improve the stability and solubility of intrabodies inside cells. The camelid derived single domain antibodies, which are the smallest fragments intracellularly expressed, represent the new generation of intrabodies for their capacity to enter small cavities on antigens, their high solubility and good tissue penetration.

There are still major problems to be solved before intrabodies will be clinically approved therapeutic agents. These include, in particular, development of appropriate, efficient and safe delivery strategies, in terms of transduction systems and routes of delivery. Adeno-associated (AAV) viral vectors still represent the best choice of delivery in mammalian tissues although a novel generation of non viral delivery systems (nanoparticles, protein transduction domain peptides or modified liposomes) characterized by low toxicity and immunogenicity are now emerging. The general interest in this issue is so diffuse that promises to bring new improvements in the next future.

Acknowledgements I wish to thank Alessio Cardinale for helpful discussions, suggestions and critical reading of the manuscript.

References

Aguzzi A, Baumann F, Bremer J (2008) The prion's elusive reason for being. Annu Rev Neurosci 31:439–77

Aires da Silva F, Santa-Marta M, Freitas-Vieira A, Mascarenhas P, Barahona I, Moniz-Pereira J, Gabuzda D, Goncalves J (2004) Camelized rabbit-derived VH single-domain intrabodies against Vif strongly neutralize HIV-1 infectivity. J Mol Biol 340:525–42

Alvarez RD, Barnes MN, Gomez-Navarro J, Wang M, Strong TV, Arafat W, Arani RB, Johnson MR, Roberts BL, Siegal GP, Curiel DT (2000) A cancer gene therapy approach utilizing an anti-erbB-2 single-chain antibody-encoding adenovirus (AD21): a phase I trial. Clin Cancer Res 6:3081–7

Alzogaray V, Danquah W, Aguirre A, Urrutia M, Berguer P, García Véscovi E, Haag F, Koch-Nolte F, Goldbaum FA (2010) Single-domain llama antibodies as specific intracellular inhibitors of SpvB, the actin ADP-ribosylating toxin of Salmonella typhimurium. FASEB J 25:526–34

Bai J, Sui J, Zhu RY, Tallarico AS, Gennari F, Zhang D, Marasco WA (2003) Inhibition of Tat-mediated transactivation and HIV-1 replication by human anti-hCyclinT1 intrabodies. J Biol Chem 278:1433–42

Biocca S, Cattaneo A (1995) Intracellular immunization: antibody targeting to subcellular compartments. Trends Cell Biol 5:248–52

Biocca S, Neuberger MS, Cattaneo A (1990) Expression and targeting of intracellular antibodies in mammalian cells. EMBO J 9:101–8

Biocca S, Pierandrei-Amaldi P, Campioni N, Cattaneo A (1994) Intracellular immunization with cytosolic recombinant antibodies. Biotechnology (NY) 12:396–9

Biocca S, Ruberti F, Tafani M, Pierandrei-Amaldi P, Cattaneo A (1995) Redox state of single chain Fv fragments targeted to the endoplasmic reticulum, cytosol and mithocondria. Bio/Technology 13:1110–15

Blennow K, de Leon MJ, Zetterberg H (2006) Alzheimer's disease. Lancet 368:387–403

Böldicke T (2007) Blocking translocation of cell surface molecules from the ER to the cell surface by intracellular antibodies targeted to the ER. J Cell Mol Med 11:54–70

Cardinale A, Biocca S (2008a) The potential of intracellular antibodies for therapeutic targeting of protein-misfolding diseases. Trends Mol Med 14:373–80

Cardinale A, Biocca S (2008b) Combating protein misfolding and aggregation by intracellular antibodies. Curr Mol Med 8(1):2–11

Cardinale A, Biocca S (2010) Expressing intrabodies in mammalian cells. In Antibody Engineering Vol 2. Springer Protocols 161–172

Cardinale A, Filesi I, Biocca S (2001) Aggresome formation by anti-Ras intracellular scFv fragments. The fate of the antigen-antibody complex. Eur J Biochem 268:268–77

Cardinale A, Filesi I, Mattei S, Biocca S (2003) Evidence for proteasome dysfunction in cytotoxicity mediated by anti-Ras intracellular antibodies. Eur J Biochem 270:3389–97

Cardinale A, Filesi I, Mattei S, Biocca S (2004) Intracellular targeting and functional analysis of single-chain Fv fragments in mammalian cells. Methods 34:171–8

Cardinale A, Filesi I, Vetrugno V, Pocchiari M, Sy MS, Biocca S (2005) Trapping prion protein in the endoplasmic reticulum impairs PrP^{C} maturation and prevents PrP^{Sc} accumulation. J Biol Chem 280:685–94

Caron de Fromentel C, Gruel N, Venot C, Debussche L, Conseiller E, Dureuil C, Teillaud JL, Tocque B, Bracco L (1999) Restoration of transcriptional activity of p53 mutants in human tumour cells by intracellular expression of anti-p53 single chain Fv fragments. Oncogene 18:551–7

Cattaneo A, Biocca S (1997) Intracellular Antibodies: Development and Applications. Springer, Berlin, Heidelberg

Cattaneo A, Biocca S (1999) The selection of intracellular antibodies. Trends Biotech 17:115–21

Chartier A, Raz V, Sterrenburg E, Verrips CT, van der Maarel SM, Simonelig M (2009) Prevention of oculopharyngeal muscular dystrophy by muscular expression of Llama single-chain intrabodies in vivo. Hum Mol Genet 18:1849–59

Chesnut JD, Baytan AR, Russell M, Chang MP, Bernard A, Maxwell IH, Hoeffler JP (1996) Selective isolation of transiently transfected cells from a mammalian cell population with vectors expressing a membrane anchored single-chain antibody. J Immunol Methods 193:17–27

Chiti F, Dobson CM (2006) Protein misfolding, functional amyloid, and human disease. Annu Rev Biochem 75:333–66

Cochet O, Kenigsberg M, Delumeau I, Virone-Oddos A, Multon MC, Fridman WH, Schweighoffer F, Teillaud JL, Tocqué B (1998) Intracellular expression of an antibody fragment-neutralizing p21 ras promotes tumor regression. Cancer Res 58:1170–6

Cohen PA, Mani JC, Lane DP (1998) Characterization of a new intrabody directed against the N-terminal region of human p53. Oncogene 17:2445–56

Contreras-Martínez LM, DeLisa MP (2007) Intracellular ribosome display via SecM translation arrest as a selection for antibodies with enhanced cytosolic stability. J Mol Biol 372:513–24

Corte-Real S, Collins C, Aires da Silva F, Simas JP, Barbas CF 3rd, Chang Y, Moore P, Goncalves J (2005) Intrabodies targeting the Kaposi sarcoma-associated herpesvirus latency antigen inhibit viral persistence in lymphoma cells. Blood 106:3797–802

Deshane J, Siegal GP, Wang M, Wright M, Bucy RP, Alvarez RD, Curiel DT (1997) Transductional efficacy and safety of an intraperitoneally delivered adenovirus encoding an anti-erbB-2 intracellular single-chain antibody for ovarian cancer gene therapy. Gynecol Oncol 64:378–85

Donofrio G, Heppner FL, Polymenidou M, Musahl C, Aguzzi A (2005) Paracrine inhibition of prion propagation by anti-PrP single-chain Fv miniantibodies. J Virol 79:8330–8

Duan L, Bagasra O, Laughlin MA, Oakes JW, Pomerantz RJ (1994) Potent inhibition of human immunodeficiency virus type 1 replication by an intracellular anti-Rev single-chain antibody. Proc Natl Acad Sci USA 91:5075–9

Emadi S, Barkhordarian H, Wang MS, Schulz P, Sierks MR (2007) Isolation of a human single chain antibody fragment against oligomeric α-synuclein that inhibits aggregation and prevents α-synuclein-induced toxicity. J Mol Biol 368:1132–44

Figini M, Ferri R, Mezzanzanica D, Bagnoli M, Luison E, Miotti S, Canevari S (2003) Reversion of transformed phenotype in ovarian cancer cells by intracellular expression of anti folate receptor antibodies. Gene Ther 10:1018–25

Filesi I, Cardinale A, Mattei S, Biocca S (2007) Selective re-routing of prion protein to proteasomes and alteration of its vesicular secretion prevent PrP(Sc) formation. J Neurochem 101:1516–26

Gal-Tanamy M, Zemel R, Bachmatov L, Jangra RK, Shapira A, Villanueva RA, Yi M, Lemon SM, Benhar I, Tur-Kaspa R (2010) Inhibition of protease-inhibitor-resistant hepatitis C virus replicons and infectious virus by intracellular intrabodies. Antiviral Res 88(1):95–106

Goenaga AL, Zhou Y, Legay C, Bougherara H, Huang L, Liu B, Drummond DC, Kirpotin DB, Auclair C, Marks JD, Poul MA (2007) Identification and characterization of tumor antigens by using antibody phage display and intrabody strategies. Mol Immunol 44:3777–88

Graus-Porta D, Beerli RR, Hynes NE (1995) Single-chain antibody-mediated intracellular retention of ErbB-2 impairs Neu differentiation factor and epidermal growth factor signaling. Mol Cell Biol 15:1182–91

Griffin H, Elston R, Jackson D, Ansell K, Coleman M, Winter G, Doorbar J (2006) Inhibition of papillomavirus protein function in cervical cancer cells by intrabody targeting. J Mol Biol 355:360–78

Hamers-Casterman C, Atarhouch T, Muyldermans S, Robinson G, Hamers C, Songa EB, Bendahman N, Hamers R (1993) Naturally occurring antibodies devoid of light chains. Nature 363(6428):446–8

Hayashi I, Takatori S, Urano Y, Iwanari H, Isoo N, Osawa S, Fukuda MA, Kodama T, Hamakubo T, Li T, Wong PC, Tomita T, Iwatsubo T (2009) Single chain variable fragment against nicastrin inhibits the gamma-secretase activity. J Biol Chem 284:27838–47

Hoogenboom HR (2005) Selecting and screening recombinant antibody libraries. Nat Biotechnol 23:1105–16

Hudson PJ, Souriau C (2003) Engineered antibodies. Nat Med 9:129–34

Jannot CB, Beerli RR, Mason S, Gullick WJ, Hynes NE (1996) Intracellular expression of a single-chain antibody directed to the EGFR leads to growth inhibition of tumor cells. Oncogene 13:275–82

Jendreyko N, Popkov M, Rader C, Barbas CF 3rd (2005) Phenotypic knockout of VEGF-R2 and Tie-2 with an intradiabody reduces tumor growth and angiogenesis in vivo. Proc Natl Acad Sci USA 102:8293–8

Jung S, Plückthun A (1997) Improving in vivo folding and stability of a single-chain Fv antibody fragment by loop grafting. Protein Eng 10:959–66

Kayed R, Head E, Thompson JL, McIntire TM, Milton SC, Cotman CW, Glabe CG (2003) Common structure of soluble amyloid oligomers implies common mechanism of pathogenesis. Science 300:486–9

Khoshnan A, Ko J, Patterson PH (2002) Effects of intracellular expression of antihuntingtin antibodies of various specificities on mutant huntingtin aggregation and toxicity. Proc Natl Acad Sci USA 99:1002–7

Kirschning CJ, Dreher S, Maass B, Fichte S, Schade J, Köster M, Noack A, Lindenmaier W, Wagner H, Böldicke T (2010) Generation of anti-TLR2 intrabody mediating inhibition of macrophage surface TLR2 expression and TLR2-driven cell activation. BMC Biotechnol 10:31

Kvam E, Sierks MR, Shoemaker CB, Messer A (2010) Physico-chemical determinants of soluble intrabody expression in mammalian cell cytoplasm. Protein Eng Des Sel 23:489–98

Lener M, Horn IR, Cardinale A, Messina S, Nielsen UB, Rybak SM, Hoogenboom HR, Cattaneo A, Biocca S (2000) Diverting a protein from its cellular location by intracellular antibodies. The case of p21Ras. Eur J Biochem 267:1196–205

Levin R, Mhashilkar AM, Dorfman T, Bukovsky A, Zani C, Bagley J, Hinkula J, Niedrig M, Albert J, Wahren B, Göttlinger HG, Marasco WA (1997) Inhibition of early and late events of the HIV-1 replication cycle by cytoplasmic Fab intrabodies against the matrix protein, p17. Mol Med 3:96–110

Liao HI, Olson CA, Hwang S, Deng H, Wong E, Baric RS, Roberts RW, Sun R (2009) mRNA display design of fibronectin-based intrabodies that detect and inhibit severe acute respiratory syndrome coronavirus nucleocapsid protein. J Biol Chem 284:17512–20

Lo AS, Zhu Q, Marasco WA (2008) Intracellular antibodies (intrabodies) and their therapeutic potential. Handb Exp Pharmacol 181:343–73

Lynch SM, Zhou C, Messer A (2008) An scFv intrabody against the nonamyloid component of alpha-synuclein reduces intracellular aggregation and toxicity. J Mol Biol 377:136–47

Marasco WA, Haseltine WA, Chen SY (1993) Design, intracellular expression, and activity of a human anti-human immunodeficiency virus type 1 gp120 single-chain antibody. Proc Natl Acad Sci USA 90:7889–93

Marcus WD, Wang H, Lindsay SM, Sierks MR (2008) Characterization of an antibody scFv that recognizes fibrillar insulin and β-amyloid using atomic force microscopy. Nanomedicine 4:1–7

McGonigal K, Tanha J, Palazov E, Li S, Gueorguieva-Owens D, Pandey S (2009) Isolation and functional characterization of single domain antibody modulators of Caspase-3 and apoptosis. Appl Biochem Biotechnol 157:226–36

McLear JA, Lebrecht D, Messer A, Wolfgang WJ (2008) Combinational approach of intrabody with enhanced Hsp70 expression addresses multiple pathologies in a fly model of Huntington's disease. FASEB J 22:2003–11

Meli G, Visintin M, Cannistraci I, Cattaneo A (2009) Direct in vivo intracellular selection of conformation-sensitive antibody domains targeting Alzheimer's amyloid-beta oligomers. J Mol Biol 387:584–606

Mhashilkar AM, Bagley J, Chen SY, Szilvay AM, Helland DG, Marasco WA (1995) Inhibition of HIV-1 Tat-mediated LTR transactivation and HIV-1 infection by anti-Tat single chain intrabodies. EMBO J 14:1542–51

Monsellier E, Bedouelle H (2006) Improving the stability of an antibody variable fragment by a combination of knowledge-based approaches: validation and mechanisms. J Mol Biol 362: 580–93

Mukhtar MM, Li S, Li W, Wan T, Mu Y, Wei W, Kang L, Rasool ST, Xiao Y, Zhu Y, Wu J (2009) Single-chain intracellular antibodies inhibit influenza virus replication by disrupting interaction of proteins involved in viral replication and transcription. Int J Biochem Cell Biol 41:554–60

Murphy RC, Messer A (2004) A single-chain Fv intrabody provides functional protection against the effects of mutant protein in an organotypic slice culture model of Huntington's disease. Brain Res Mol Brain Res 121:141–5

Orgad S, Dimant H, Dor-On E, Azriel-Rosenfeld R, Benhar I, Solomon B (2010) TAR1, a human anti-p53 single-chain antibody, restores tumor suppressor function to mutant p53 variants. J Immunother 33:146–54

Paganetti P, Calanca V, Galli C, Stefani M, Molinari M (2005) β-site specific intrabodies to decrease and prevent generation of Alzheimer's Aβ peptide. J Cell Biol 168:863–8

Persic L, Righi M, Roberts A, Hoogenboom HR, Cattaneo A, Bradbury A (1997a) Targeting vectors for intracellular immunization. Gene 187:1–8

Persic L, Righi M, Roberts A, Hoogenboom HR, Cattaneo A, Bradbury A (1997b) An integrated vector system for the eukaryotic expression of antibodies or their fragments after selection from phage display libraries. Gene 187:9–18

Philibert P, Stoessel A, Wang W, Sibler AP, Bec N, Larroque C, Saven JG, Courtête J, Weiss E, Martineau P (2007) A focused antibody library for selecting scFvs expressed at high levels in the cytoplasm. BMC Biotechnol 7:81

Prusiner SB (1998) Prions. Proc Natl Acad Sci USA 95:13363–83

Serruys B, Van Houtte F, Farhoudi-Moghadam A, Leroux-Roels G, Vanlandschoot P (2010) Production, characterization and in vitro testing of HBcAg-specific VHH intrabodies. J Gen Virol 91:643–52

Serruys B, Van Houtte F, Verbrugghe P, Leroux-Roels G, Vanlandschoot P (2009) Llama-derived single-domain intrabodies inhibit secretion of hepatitis B virions in mice. Hepatology 49:39–49

Southwell AL, Khoshnan A, Dunn DE, Bugg CW, Lo DC, Patterson PH (2008) Intrabodies binding the proline-rich domains of mutant huntingtin increase its turnover and reduce neurotoxicity. J Neurosci 28:9013–20

Southwell AL, Ko J, Patterson PH (2009) Intrabody gene therapy ameliorates motor, cognitive, and neuropathological symptoms in multiple mouse models of Huntington's disease. J Neurosci 29:13589–602

Steinitz M (2009) Immunotherapy for Alzheimer's disease. Immunotherapy 1:461–9

Tanaka T, Rabbitts TH (2003) Intrabodies based on intracellular capture frameworks that bind the RAS protein with high affinity and impair oncogenic transformation. EMBO J 22:1025–35

Tanaka T, Rabbitts TH (2010) Protocol for the selection of single-domain antibody fragments by third generation intracellular antibody capture. Nat Protoc 5:67–92

Tanaka T, Williams RL, Rabbitts TH (2007) Tumour prevention by a single antibody domain targeting the interaction of signal transduction proteins with RAS. EMBO J 26:3250–9

Tremblay JM, Kuo CL, Abeijon C, Sepulveda J, Oyler G, Hu X, Jin MM, Shoemaker CB (2010) Camelid single domain antibodies (VHHs) as neuronal cell intrabody binding agents and inhibitors of Clostridium botulinum neurotoxin (BoNT) proteases. Toxicon 56:990–8

Van den Abbeele A, De Clercq S, De Ganck A, De Corte V, Van Loo B, Soror SH, Srinivasan V, Steyaert J, Vandekerckhove J, Gettemans J (2010) A llama-derived gelsolin single-domain antibody blocks gelsolin-G-actin interaction. Cell Mol Life Sci 67:1519–35

Vercruysse T, Pardon E, Vanstreels E, Steyaert J, Daelemans D (2010) An intrabody based on a llama single-domain antibody targeting the N-terminal alpha-helical multimerization domain of HIV-1 rev prevents viral production. J Biol Chem 285:21768–80

Verheesen P, de Kluijver A, van Koningsbruggen S, de Brij M, de Haard HJ, van Ommen GJ, van der Maarel SM, Verrips CT (2006) Prevention of oculopharyngeal muscular dystrophy-associated aggregation of nuclear polyA-binding protein with a single-domain intracellular antibody. Hum Mol Genet 15:105–11

Vetrugno V, Cardinale A, Filesi I, Mattei S, Sy MS, Pocchiari M, Biocca S (2005) KDEL-tagged anti-prion intrabodies impair PrP lysosomal degradation and inhibit scrapie infectivity. Biochem Biophys Res Commun 338:1791–7

Visintin M et al. (1999) Selection of antibodies for intracellular function using a two-hybrid in vivo system. Proc Natl Acad Sci USA 96:11723–8

Walker FO (2007) Huntington's disease. Lancet 369:218–28

Wang CE, Zhou H, McGuire JR, Cerullo V, Lee B, Li SH, Li XJ (2008) Suppression of neuropil aggregates and neurological symptoms by an intracellular antibody implicates the cytoplasmic toxicity of mutant huntingtin. J Cell Biol 181:803–16

Wesolowski J, Alzogaray V, Reyelt J, Unger M, Juarez K, Urrutia M, Cauerhff A, Danquah W, Rissiek B, Scheuplein F, Schwarz N, Adriouch S, Boyer O, Seman M, Licea A, Serreze DV, Goldbaum FA, Haag F, Koch-Nolte F (2009) Single domain antibodies: promising experimental and therapeutic tools in infection and immunity. Med Microbiol Immunol 198:157–74

Williams BR, Zhu Z (2006) Intrabody-based approaches to cancer therapy: status andprospects. Curr Med Chem 13:1473–80

Wolfgang WJ, Miller TW, Webster JM, Huston JS, Thompson LM, Marsh JL, Messer A (2005) Suppression of Huntington's disease pathology in Drosophila by human single-chain Fv antibodies. Proc Natl Acad Sci USA 102:11563–8

Wood-Kaczmar A, Gandhi S, Wood NW (2006) Understanding the molecular causes of Parkinson's disease. Trends Mol Med 12:521–8

Wörn A, Auf der Maur A, Escher D, Honegger A, Barberis A, Plückthun A (2000) Correlation between in vitro stability and in vivo performance of anti-GCN4 intrabodies as cytoplasmic inhibitors. J Biol Chem 275:2795–803

Yuan B, Sierks MR (2009) Intracellular targeting and clearance of oligomeric alpha-synuclein alleviates toxicity in mammalian cells. Neurosci Lett 459(1):16–18

Zameer A, Kasturirangan S, Emadi S, Nimmagadda SV, Sierks MR (2008) Anti-oligomeric Abeta single-chain variable domain antibody blocks Abeta-induced toxicity against human neuroblastoma cells. J Mol Biol 384:917–28

Zdoroveac A, Doebis C, Laube H, Brösel S, Schmitt-Knosalla I, Volk HD, Seifert M (2008) Modulation of graft arteriosclerosis in a rat carotid transplantation model. J Surg Res 145:161–9

Zhou C, Emadi S, Sierks MR, Messer A (2004) A human single-chain Fv intrabody blocks aberrant cellular effects of overexpressed a-synuclein. Mol Ther 10:1023–31

Zhou P, Goldstein S, Devadas K, Tewari D, Notkins AL (1998) Cells transfected with a non-neutralizing antibody gene are resistant to HIV infection: targeting the endoplasmic reticulum and trans-Golgi network. J Immunol 160:1489–96

Chapter 10
Engineering Antibodies for Cancer Therapy

Lisa E. Goldsmith and Matthew K. Robinson

Abstract Over the past 20 years monoclonal antibodies (mAbs) have evolved to become a major class of therapeutics for the treatment of a variety of indications, including cancer. The evolution of mAbs into front-line cancer therapies required significant advances in the strategies used to both isolate and optimize these agents. Here we discuss development of the steps that facilitated this evolution and the criteria that drive current development of next-generation mAb-based cancer therapies.

10.1 Introduction

In the early twentieth century, Paul Ehrlich dreamed of treating cancer with a "magic bullet" that would specifically target a therapeutic agent to cancer cells, while leaving healthy cells unharmed. Advances in antibody engineering over the past few decades have transported the field of antibody-based therapy from its infancy of using rodent derived antibodies to today when increasingly sophisticated approaches have resulted in the production and clinical testing of over 66 fully human antibodies in a variety of disease settings, including cancer. Together with antibody engineering, the development of modern molecular biology-based approaches for monoclonal antibody (mAb) production described throughout this volume is allowing Ehrlich's vision of the magic bullet for the treatment of cancer to be realized. Monoclonal antibodies now comprise the fastest growing segment of the pharmaceutical industry. The United States Food and Drug Administration has approved over 20 mAbs for the treatment of multiple diseases, including 10 for the treatment of cancer. Together these represented $20 billion in annual revenue in 2009.

10.2 Antibody Structure

Monoclonal antibody-based cancer drugs are large (150 kDa) multidomain proteins whose in vivo behavior, their pharmacokinetics (PK) and pharmacodynamics (PD), are dictated by a number of variables that must be accounted for during the

L.E. Goldsmith and M.K. Robinson (✉)
Developmental Therapeutics Program, Fox Chase Cancer Center, Philadelphia, PA 19111, USA
e-mail: lisa.goldsmith@fccc.edu; matthew.robinson@fccc.edu

M. Al-Rubeai (ed.), *Antibody Expression and Production*, Cell Engineering 7,
DOI 10.1007/978-94-007-1257-7_10, © Springer Science+Business Media B.V. 2011

development process. At the most basic level, interactions with the targeted tumor and the patient's immune system must be optimized. Currently all antibody-based agents approved for the treatment of cancer are based on the naturally occurring Immunoglobulin-G (IgG) structure. IgG are the major antibody isotype produced by B cells in response to antigen challenge. The bilaterally symmetric structure of an IgG is shown in Fig. 10.1. Two heterodimers, each comprised of one heavy and one light chain, dimerize to form the canonical Y-shape structure associated with an IgG. Enzymatic cleavage of the IgG results in release of three domains, two identical antigen-binding fragments (Fab) and a constant fragment (Fc). Both the Fab and Fc domains play critical functions in defining the in vivo behavior of an intact mAb and therefore each must be taken into account when designing a therapeutic mAb.

The Fab are subdivided into variable and constant domains. The variable domains of both the heavy (V_H) and light (V_L) chains comprise the antigen-binding site, with specificity defined by the amino acid composition of the six complementarity determining regions (CDRs). Although each of the six CDRs, three from each of the light and heavy chains, have the potential to contribute to the binding process, CDR3 of the heavy chain often plays a dominant role (Komissarov et al., 1996). The Fc portion of the molecule, comprised of the CH2 and CH3 domains, mediates both the biological half-life of the IgG and its ability to direct immunologic effector functions, such as targeted cellular killing and immune complex clearance. Biological half-life is regulated through interactions with the neonatal Fc receptor

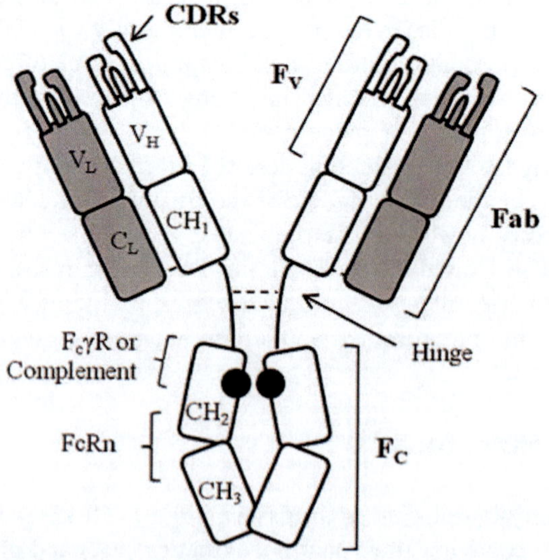

Fig. 10.1 Structure of an IgG. An IgG is a bilaterally symmetric protein complex comprised of two identical heavy (*white*) and two identical light (*grey*) chains that form a dimer of heterodimers held together by disulphide bonds (*dashed lines*). Asn297 (*black dot*) of each heavy chain is a site for post-translational carbohydrate modification. The major domains and sites of interaction with components of the immune system are noted and defined in the text

(FcRn), which is found on the surface of endothelial cells. This receptor was originally identified for its role in transfer of IgG across the placental barrier but is now known to function in adults as a salvage recepetor for IgG in circulation (Ghetie et al., 1997). The Fc domain is also critical for promoting antibody-dependent cellular cytotoxicity (ADCC) and triggering the complement cascade (Natsume et al., 2009). IgG isotypes (e.g. IgG1, IgG2, etc) are defined by structural differences associated with the Fc domain and these isotypes differ dramatically in their ability to elicit immune effector functions.

An increased understanding of the factors that influence in vivo behavior of antibodies coupled with technological advances in antibody engineering has allowed the field of antibody-based therapies to mature.

10.3 Reducing Antibody Immunogenicity

The human immune system has evolved adept mechanisms to recognize and eradicate foreign proteins that it detects within the body. This ability lies in direct opposition to the use of biologic therapies, such as mAbs, to treat diseases like cancer where high levels of the drugs must be maintained in serum in order to elicit the desired therapeutic effect. A major emphasis over the past few decades has been the development of approaches to circumvent detection by the patient's immune system.

10.3.1 Murine Antibodies: The Beginning

Early attempts at antibody-based cancer therapy involved treating patients with polyclonal antisera raised in animals against fragments of the patient's own tumours (Currie, 1972). In 1895, Hericourt and Richet (1895) reported improvement in the condition of 50 patients with advanced cancers treated with antisera. However, life threatening side effects arose after a few treatments with little overall benefit.

In 1975, Kohler and Milstein (1975) developed the technology to create a hybrid cell line, or hybridoma. Antibody-producing B cells isolated from spleens of immunized mice were immortalized through fusion with myeloma cells. This seminal work laid the foundation for production of monospecific, or monoclonal, antibodies (mAb) against defined tumor-associated antigens and their use in the clinical setting (Table 10.1). Because murine antibodies are comprised of foreign protein sequences, those that entered clinical trials were almost universally recognized by the patient's humoral immune system leading to production of human antimouse antibody (HAMA) responses. HAMA responses are classified as either anti-isotypic or anti-idiotypic (Saldanha, 2009). Anti-isotypic responses are the most common, and are antibodies directed against the Fc region of the therapeutic murine antibody. Anti-idiotypic antibodies are directed against the variable regions of the therapeutic antibody, and may or may not compete for binding to the therapeutic target.

Table 10.1 FDA-approved antibodies for the treatment of cancer

Generic name	Trade name	Antibody description[a]	Antigen[b]	Approved indications[c]	Proposed mechanism[d]	Year
Rituximab	Rituxan/ MabThera	Chimeric (IGg1κ)	CD20	Relapsed/refractory NHL	ADCC/CDC; apoptosis by receptor cross-linking; sensitizes to chemo	1997
Trastuzumab	Herceptin	Humanized (IgG1κ), consen	HER2	Metastatic breast cancer	Receptor binding, internalization; cell cycle arrest; apoptosis; ADCC; sensitizes to chemo	1998
Gemtuzumab ozogamicin	Mylotarg	Humanized (IgG4κ), fixed, immunotoxin (calicheamicin)	CD33	Acute myeloid leukemia	Toxin breaks double-stranded DNA, causes cell death	2000
Alemtuzumab	Campath	Humanized (IgG1κ), fixed	CD52	CLL	ADCC/CDC	2001
^{90}Y-Ibritumomab tiuxetan	Zevalin	Murine (IgG1κ)	CD20	Relapsed/refractory NHL	Cell death by radiation; apoptosis	2002
^{131}I-Tositumomab	Bexxar	Murine (IgG2aλ)	CD20	NHL refractory to rituximab	Cell death by radiation; apoptosis; ADCC/CDC	2003
Cetuximab	Erbitux	Chimeric (IGg1κ)	EGFR	EGFR-positive mCRC	Receptor binding, antagonism; cell cycle arrest; apoptosis; sensitizes to chemo; ADCC	2004
Bevacizumab	Avastin	Humanized (IgG1), consen	VEGF	mCRC; NSCLC	Ligand binding, receptor antagonism; inhibits angiogenesis, metastasis	2004
Panitumumab	Vectibix	Human (IgG2κ), transgenic mouse	EGFR	mCRC	Receptor binding, antagonism, down-regulation; cell cycle arrest; apoptosis	2006
Ofatumumab	Arzerra	Human (IgG1), transgenic mouse	CD20	CLL	ADCC/CDC	2009

[a]Consen = Human consensus framework; Fixed = Fixed framework approach.
[b]HER2 = Human epidermal growth factor receptor 2; EGFR = Epidermal growth factor receptor; VEGF = Vascular endothelial growth factor.
[c]NHL = Non-Hodgkin's lymphoma; CLL = Chronic lymphocytic leukemia; mCRC = Metastatic colorectal cancer; NSCLC = Non-small cell lung cancer.
[d]ADCC = Antibody dependent cell-mediated cytotoxicity; CDC = Complement-dependent cytotoxicity.

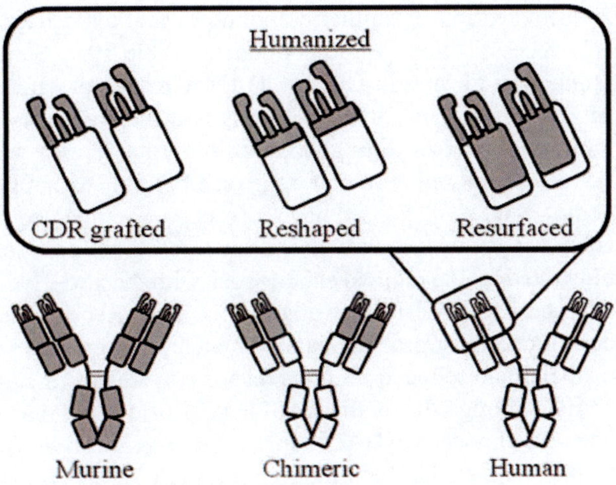

Fig. 10.2 Evolution from murine to fully human antibodies. A schematic representation of murine (*grey*) and human (*white*) components of different classes of therapeutic mAbs

The development of HAMA responses varies widely depending on the type of cancer, the presence of pre-existing anti-globulin antibodies in patient serum (Kricka, 1999), as well as the therapeutic dose and schedule (Schroff et al., 1985, Shawler et al., 1985). Recently, it was found that elevated HAMA responses in patients with B cell malignancy correlated with longer survival (Azinovic et al., 2006). For the most part however, HAMA responses are associated with decreased clinical response. HAMA responses result in the neutralization and clearance of the therapeutic murine antibodies from the patient's system (Badger et al., 1987, Khazaeli et al., 1994). In addition they can be associated with infusion related hypersensitivity reactions that compromise safety and limit the number of times an agent can be administered (Hwang and Foote, 2005).

Advances in antibody engineering have resulted in increasingly sophisticated approaches to create antibodies comprised of human germline sequences in an effort limit HAMA responses and increase efficacy of antibody-based therapies. The different classes of engineered IgGs are cartooned in Fig. 10.2. The approaches used to generate these molecules and their impact on antibody functions are described below. Excellent summaries of FDA-approved recombinant antibodies as well as those currently under investigation are provided elsewhere (Albrecht, 2010, Boulianne et al., 1984, Morrison et al., 1984, Wright and Morrison, 1994).

10.3.2 Chimeric Antibodies

The first approach developed to limit HAMA responses takes advantage of the domain-based structure of antibodies that makes it possible to genetically fuse human constant domain genes with antigen-specific murine variable domain genes (Boulianne et al., 1984, Morrison et al., 1984). The resulting chimeric antibodies (Fig. 10.2) are approximately 70% human in content (Almagro and Fransson,

2008). Creation of chimeric antibodies marked a major breakthrough in the clinical success of therapeutic antibodies (Table 10.1; Saldanha, 2009) by limiting the production of an anti-isotypic HAMA response while both retaining the intrinsic anti-cancer activity of the antibody and increasing its ability to induce anti-tumor immune responses through high affinity interactions with Fc receptors found on the surface of immune effector cells (see below). Although human anti-chimeric antibody (HACA) responses have also been documented, they are more rare than the HAMA responses seen with murine hybridoma-produced antibodies. This is nicely illustrated by the clinical experiences with the anti-glycoprotein IIb/IIIa 7E3 murine Fab and its chimeric derivative abciximab (Reopro) that is FDA-approved for use during coronary procedures as a platelet aggregation inhibitor. Conversion of 7E3 to abciximab reduced immune responses rates from 17 to 1% (Knight et al., 1995).

Even though the incidence of HACA is low and may not interfere with treatment, previous exposure to HACA antibodies may still desensitize patients to further treatment. The fact that the binding specificity of an IgG is dictated by the amino acid composition of the CDRs, which comprise only 5–10% of the total protein content of an IgG, provides an opportunity to further optimize human germline composition of therapeutic IgG through production of "humanized" antibodies.

10.3.3 Humanized Antibodies

Four humanized antibodies are currently approved for the treatment of cancer (Table 10.1). Multiple strategies have been developed to humanize antibodies in part related to intellectual property concerns. These strategies result in essentially the same outcome as the CDR-grafting approach originally described by Winter and colleagues (Jones et al., 1986). In this strategy the CDRs of the murine mAb are grafted onto a human IgG framework, at the gene level, to create a humanized antibody that is comprised of ~90–95% human germline sequences (Fig. 10.2). In designing humanized, CDR-grafted antibodies, there are three considerations: (1) determining the specificity-defining regions of the murine CDRs to be grafted, (2) selecting an appropriate human framework, and (3) identifying human framework residues outside the CDRs which can be "back-mutated" to their murine counterparts in order to restore or improve antibody affinity, a process known as "reshaping". Almagro and Fransson (2008) discuss these decisions in detail, but we review them here briefly.

10.3.3.1 CDR Selection

Antibody-antigen crystal structures, if available, enable easy selection of amino acids involved in binding. In the absence of crystal data, consensus studies of the primary sequences of antibodies (Kabat, 1978) and the solved crystal structures of CDR loops in limited sets of 17 (Chothia and Lesk, 1987) or fewer than 60 (Martin and Thornton, 1996) led to methods for both locating the primary sequence of the CDR regions and predicting the structure of the CDR loops to be grafted. Martin and

Allen (2007) provide a table comparing these and other overlapping CDR numbering schemes and definitions. Recently, Dunbrack and colleagues (North et al., 2011) have refined the definitions of both CDR location and predicted structures through incorporation of 300 crystal structure data sets in the analysis. These refinements have the potential to improve results from CDR grafting.

10.3.3.2 Framework Selection

Well-characterized human germline antibody frameworks (Gonzales et al., 2004), such as those available in the ImMunoGeneTics (IMGT) database (Lefranc et al., 2005), are used as scaffolds for CDR grafting. The "fixed" and "best-fit" approaches represent two general strategies for framework selection. The "fixed" approach (Jones et al., 1986, Riechmann et al., 1988), focuses only on the physical characteristics of the human framework. This has been refined to the "best fit" framework approach which uses human sequences with the highest homology to the original murine sequence, even in the absence of crystal structures (Co et al., 1992, Lo, 2004). In a number of cases, the best-fit approach has been more successful than the fixed framework approach (Gorman et al., 1991, Graziano et al., 1995). Regardless of the framework selection strategy employed, there is no doubt that the scaffold structure is important for conferring antibody activity (Bourne et al., 2004), and protocols for CDR grafting of murine antibodies onto human frameworks using both approaches have been detailed elsewhere (Lo, 2004).

10.3.3.3 Reshaping Antibody CDR Loops

CDR-grafted antibodies often exhibit diminished affinity as a result of incompatibilities between the murine CDRs and the human framework. In order to fine-tune ("reshape") the conformation of the antibody CDR loops, additional human residues within the variable domain framework (outside the CDRs) can be "back-mutated" to their chimeric murine counterparts (Foote and Winter, 1992). Although inclusion of additional murine residues increases the risk of an immunogenic response, these "back-mutations" can drastically restore or improve antibody affinity, specificity, and expression (Riechmann et al., 1988, Saldanha et al., 1999). In addition to providing structural support, these framework residues may also be involved in antigen binding (Mian et al., 1991). Selection of which framework residues to back-mutate can be guided by structural models of the antibody (Foote and Winter, 1992) and is nicely exemplified by the design of trastuzumab (Carter et al., 1992). Grafting of the mu4D5 CDRs onto a human IgG1 framework to create hu4D5-1 resulted in both an approximate 80-fold loss of affinity and loss of anti-proliferative effect as compared to mu4D5. Back-mutating of a series of seven framework residues to their murine identity restored biologic activity and increased affinity 250-fold over hu4D5-1 and three-fold over the parent mu4D5.

The process of designing, producing, and testing each engineered mutant for improved affinity can be tedious. Phage display, however, offers an efficient alternative. A combinatorial library of antibody variable domains containing both the murine and human residues at desired positions can be created and rapidly screened for mutants with increased affinity over the original humanized antibody (Baca et al., 1997, Kim et al., 2010, Rosok et al., 1996).

10.3.3.4 Resurfacing the Variable Domains

Padlan (1991) proposed that only the murine residues exposed on the surface of the chimeric antibody, and not those buried in the hydrophobic regions of the folded protein, cause HACA responses. To prevent B cell activation and subsequent humoral response against the murine variable domains, Padlan employed an alternative method to CDR grafting, termed "resurfacing" or "veneering". In this approach, only the solvent-exposed surface residues of the chimeric antibody's murine variable domains are humanized (Fig. 10.2). A resurfaced antibody should provide the necessary structural support to the CDR loops, increasing the chance of obtaining an active antibody with reduced immunogenicity. A general protocol for resurfacing of antibodies is described elsewhere (Desmet et al., 2010). In comparisons of CDR-grafted and resurfaced antibodies in vitro, results have been similar (Roguska et al., 1996). Although no clinical trial data of resurfaced antibodies exists currently to substantiate this theory, resurfaced antibodies have been made with decreased reactivity toward the serum of patients with HAMA response (Zhu et al., 2009). The resurfaced therapeutic antibody is therefore expected to have reduced immunogenicity in patients as well.

Even with a resurfaced antibody, there is a possibility that the murine sequences in the core of the resurfaced antibody may contain T-cell epitopes (discussed in further detail below), which would be presented after the antibody is catabolised in the cell, causing an immune response.

10.3.4 Removal of T Cell Epitopes

Even with the human germline content maximized, immune responses may still arise (Hwang and Foote, 2005) due to the presence of short linear peptides or conformational motifs in the humanized antibody that are presented in the context of histocompatibility complex class II (MHC-II) molecules on the surface of antigen presenting cells (APCs). Presentation of these peptides by APCs can activate a cellular (T-cell) response against the mAb. Multiple, *in silico*-based approaches have been developed to identify T-cell epitopes, in an effort to "deimmunize" mAbs. One such approach is that described by Lazar et al. (2007) known as human string content (HSC) optimization. In this method, humanness is measured as the proportion of conserved 9 amino acid peptide strings within a given mAb as compared to those found in the IMGT database of human germline antibodies. It is hypothesized that optimizing with this set of germline antibodies will decrease immunogenicity since the database is comprised of antibodies that are minimally immunogenic to the human immune system. To ensure the identified mutations do not compromise the structural integrity of the antibody, the HSC method employs a sequence- and structure-based scoring method known as analogous contact environments (ACE) (Desjarlais, 2005). HSC and ACE were applied to four murine mAbs with clinically relevant, engineered derivatives: the anti-CD30 mAb AC10 (SGN-30), the anti-EGFR mAb 225 (cetuximab), the anti-HER2/neu mAb 4D5 (trastuzumab), and the anti-EpCAM mAb 17-1A (c17-1A). The newly created antibodies had improved HSC versus their chimeric or humanized counterparts and some even exhibited higher affinity for their antigens than the parental antibodies, without the need for affinity maturation.

Previously, removing potential immunogenic T-cell epitopes within the antibody required generation of large peptide libraries that were systematically probed for binding (Tangri et al., 2005). Advances in modeling however, such as the approaches described above, have allowed for efficient identification of T-cell epitopes, and their subsequent removal by amino acid substitutions. This rational engineering approach has been used to develop three fully human antibodies targeting CD25, VEGF, and TNF, based on murine sequences (Bernett et al., 2010). These human antibodies were shown to have sequence and binding properties equivalent to those isolated from transgenic mice and phage display.

In the design of a humanized antibody, the commonly used "best fit" approach for framework selection of CDR-grafted antibodies, as discussed above, is based on the global similarity of the human acceptor framework and murine or chimeric donor framework sequences. Modelling approaches to measure the degree of "humanness" of a prospective antibody, by determining how typical its sequence is within the human repertoire, are currently being developed (Abhinandan and Martin, 2007). To date, the success of these approaches has been limited by the inability to define a correlation between "humanness" and observed immune responses in the clinic. However, as our understanding of the immune system continues to evolve, efforts to design new modelling approaches are justified.

10.3.5 Human Antibodies

Shortly after the introduction of the initial methods to humanize murine antibodies, technologies were developed that facilitated the isolation of antibodies derived directly from human germline sequences. These technologies can be broadly classified into two distinct approaches: in vitro assembly of recombinant human antibody libraries for use in phage-display type approaches, and in vivo generation of human antibodies using transgenic mice engineered to encode the human immunoglobulin heavy and light chain loci. FDA-approved fully human IgG molecules have been developed using both types of approaches, validating their clinical utility. A third approach, human hybridoma technology, represents an additional promising approach for isolation of fully human therapeutic antibodies.

10.3.5.1 Phage Display

In addition to aiding in reshaping of CDR loops of humanized antibodies (as discussed above), phage display has emerged as a powerful tool to create fully human antibodies, with affinities even higher than can be obtained in the immune system (for review see Bradbury and Marks, 2004). The basis for this approach was demonstrated by George Smith when he expressed peptides on the surface of filamentous phage by fusing DNA encoding the peptides in frame with the phage gene that encodes viral capsid proteins (Smith, 1985). By coupling phenotype (expression of a novel protein on the phage surface) with genotype (the ability to isolate the DNA encoding the novel protein), this system enables large scale screening approaches to antibody isolation. Within 4 years, the first functional antibodies were isolated from a phage display library derived from an immunized mouse (Huse et al., 1989). Human V genes are amplified from hybridomas or B cells from the peripheral blood of donors using polymerase chain reaction (PCR), resulting in the creation of large libraries of randomly combined V_H and V_L segments. The libraries are classified as either naïve (from non-immunized donors) (McCafferty et al., 1990) or immune (from immunized donors, biasing the library toward antibodies of a certain specificity) (Persson et al., 1991). Subsequently, phage carrying V genes express the antibody V domains on the surface of the phage as single chain Fv (sc-Fv) followed by panning for antibody binding using antigen (Marks et al., 1991). Selected clones can be amplified and expressed as soluble scFv by infecting *E. coli* with the phage (Marks et al., 1992). Single-chain Fv isolated as lead candidates in initial selections are often low to moderate affinity and require rounds of in vitro affinity maturation to obtain desired binding affinity. This is accomplished through a variety of mutagenesis and selection strategies including CDR mutagenesis, chain shuffling, and yeast display. A protocol for constructing human scFv gene libraries and antibody selection is provided by Schirrmann and Hust (2010).

Adalimumab, originally D2E7 (Kempeni, 1999, Salfeld et al., 2000), was the first FDA-approved fully human monoclonal antibody (Weinblatt et al., 2003) created by phage display technology (Salfeld et al., 1998). Adalimumab is an IgG1 antibody used for the treatment of rheumatoid arthritis. It binds to ($K_D = 6 \times 10^{-10}$ M) and inhibits the activity of tumor necrosis factor alpha (TNFα).

Phage display has also been extended to the production of antibody Fab fragments (Barbas and Lerner, 1991, Hoogenboom et al., 1991). In addition, new display platforms have emerged, such as yeast (Boder and Wittrup, 1997, Feldhaus et al., 2003), lymphocyte (Alonso-Camino et al., 2009), *E. coli* (Mazor et al., 2007), and ribosomes (Hanes et al., 1998). Fully synthetic *Hu*man *C*ombinatorial *A*ntibody *L*ibraries (HuCAL) (Morphosys, Martinsried, Germany) have also been created to mimic human V_H and V_L subfamilies that are seen frequently in immune responses. Originally, these synthetic libraries only contained randomized CDR3 encoding regions (Knappik et al., 2000), but have now been extended to all six CDRs (Rothe et al., 2008). Several HuCAL antibodies are now in clinical trials, including BHQ880 for multiple myeloma (Novartis), CNTO888 for metastatic prostate cancer (Centocor), and BAY79-4620 for advanced solid tumors (Bayer).

10.3.5.2 Transgenic Mice

It was initially suggested by Alt and colleagues that transgenic mice, engineered to encode unrearranged human immunoglobulin loci, could be valuable tools for generating human antibodies (Alt et al., 1985). Bruggemann et al. (1989) achieved an initial step toward this goal with their creation of a transgenic mouse carrying a human heavy-chain minigene locus comprised of unrearranged V, D, J segments fused to a common *C*μ region. Antigen challenge and hybridoma formation demonstrated that this transgenic mouse line was capable of mounting a transgene-encoded immune response. Following this, two groups reported the creation of mice harbouring targeted disruption of the murine heavy- and κ light-chain genes along with insertion of human transgenes comprising large portions of the human immunoglobulin repertoire (Lonberg et al., 1994, Green et al., 1994). These transgenic strains supported VDJ joining, somatic mutation, and class switching and led directly to the UltiMAb (Medarex, Princeton, NJ) and XenoMouse (Amgen, Thousand Oaks, CA) strains that have been been used to isolate clinically validated fully human monoclonal antibodies. It has been speculated (Lonberg, 2008) that the natural affinity maturation process that occurs in the transgenic mouse system provides an advantage over in vitro approaches by eliminating the need for subsequent affinity maturation of lead antibodies that is often required of antibodies isolated by techniques such as phage-display.

The anti-EGFR mAb panitumumab (Hecht et al., 2007, Weiner et al., 2008) derived using the XenoMouse technology represents the first FDA-approved mAb (Giusti et al., 2007) created from a transgenic mouse platform. Although direct comparison of immune responses between clinical trials is complicated by variables such as assay design, immunogenicity does appear to be decreased compared to cetuximab. No human anti-human antibody (HAHA) responses were detected in a 231 patient phase II trial (Van Cutsem et al., 2007) with panitumumab as compared to approximately a 4% HACA response seen with cetuximab (Tan et al., 2006). Perhaps a more significant indication is the decreased rate of infusion reactions associated with panitumumab (5%, no grade 3 or 4 adverse events) versus cetuximab (7.5%, 1.7% grade 3 or 4 events). The overall improvement associated with fully human mAbs as compared to either chimeric or humanized mAbs will be more clear from the clinical experiences associated with the extensive list of additional transgenic mouse-derived antibodies, from both platforms, that are currently in various stages of clinical development (Lonberg, 2008).

10.3.5.3 Human Hybridoma

Fully human antibodies produced by human hybridomas, particularly ones developed directly from the patient, would be ideal for eliminating deleterious immune responses raised against mAb-based therapies. Obviously, significant ethical and practical considerations limit the capability of developing hybridomas in a manner analogous to the original murine counterparts; however, alternate methods are being developed. Antigen-specific B cells collected from peripheral blood, which is readily accessible and may be sampled repeatedly, have been utilized to create monoclonal antibodies (Houghton et al., 1983, Olsson et al., 1984, Borrebaeck et al., 1988). Additional methods have been developed to eliminate the need for human immunization. These include the in vitro immunization of human splenocytes (Boerner et al., 1991) as well as *ex vivo* immunization of cryopreserved B cells from healthy donors and subsequent immortalization by electrofusion to K6H6/B5 cells (Li et al., 2006b).

Several patient-derived immortal cell lines have been developed for creation of human antibodies using hybridoma technology (Karpas et al., 2001, Olsson et al., 1984, Sikora et al., 1983). The main setback to creation of human hybridomas, however, has been the difficulty in obtaining or creating an immortal human B cell line. Early work with human hybridomas fused antibody-producing B cells from peripheral blood with the immortal Epstein Barr Virus (EBV) (Steinitz et al., 1977, Cole et al., 1984), a cancer-causing virus of the herpes family. These early attempts suffered from low transformation efficiency, unstable growth, and loss of human chromosomes (Glassy and Ferrone, 1982). Improvements to fusion efficiency and yield could be accomplished by cross-linking fusion partners using an avidin-biotin conjugation scheme (Kozbor and Roder, 1981), or by fusing EBV-transformed B lymphocytes with a heteromyeloma (mouse-human) cell line (Teng et al., 1983, Kudo et al., 1988). Activity, titer, and growth rates of sub-optimal hybridomas could be enhanced using morphogenics (Nicolaides et al., 1995), a proprietary process of Morphotek (Exton, PA), which transiently regulates DNA mismatch repair and increases the phenotypic diversity of the cell line.

10.4 Optimizing the Intrinsic Antitumor Properties of mAbs

The mechanisms-of-action for the majority of therapeutic antibodies in clinical use rely on intrinsic properties of the antibody, such as the ability of the mAb to redirect the patient's immune system against tumor cells (Fig. 10.3a) or to inhibit signaling (Fig. 10.3b). Antibody and cell engineering technologies can be employed to alter the pharmacokinetic and pharmacodynamic properties of mAbs in effort to both increase efficacy and introduce defined biologic activities.

10.4.1 Regulating Serum Half-Life and Tumor Targeting

10.4.1.1 FcRn and Serum Persistance

The neonatal Fc receptor (FcRn) is a critical regulator of IgG half-life in serum and contributes to the prolonged (~21 day) half-life of therapeutic IgG1 molecules (Carter, 2006). FcRn is a dimer comprised of a 50 kDa protein with homology to the major histocompatibility complex α-chain and a 15 kDa nonpolymorphic β_2-microglobulin. First discovered in mice (Junghans and Anderson, 1996), FcRn is responsible for transport of IgG across the placenta from a mother to a growing fetus. In adults, FcRn is expressed in vascular endothelial cells and regulates homeostasis of IgG (for review see (Ghetie and Ward, 2000)) and albumin (Chaudhury et al., 2003) each binding to distinct sites on FcRn (Chaudhury et al., 2006). IgG in serum is sampled by fluid phase pinocytosis and binds to FcRn at the slightly acidic pH (pH 6.0–6.5) found in the endosomes resulting in the complex being recycled

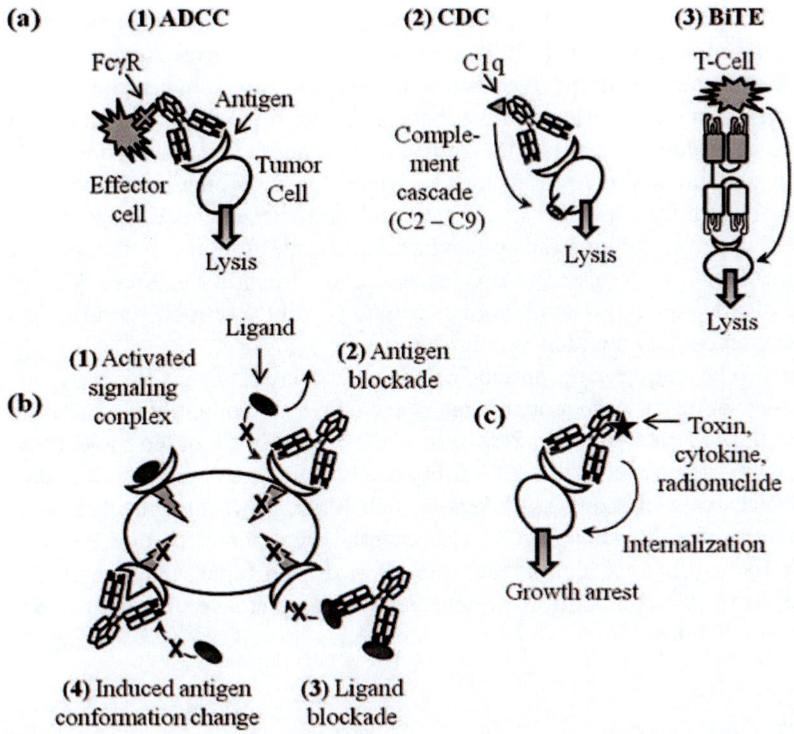

Fig. 10.3 Mechanisms of action of therapeutic antibodies. (**a**) Antibody-based therapies can focus the cytotoxic effects of immune system components to kill tumor cells via (1) antibody-dependent cellular cytotoxicity (ADCC) and (2) complement dependent cytoxicity (CDC). The Fc domain of standard mAb-based therapies is critical for tumor cell killing via these mechanisms. (3) Antibody-engineering facilitates the generation of novel structures such as BiTEs (Micromet) that are capable of recruiting defined classes of immune effector cells without an Fc domain. (**b**) Antibody-based therapies can inhibit signal transduction through a series of mechanisms. (**c**) Antibodies that lack intrinsic activity can target cytotoxic agents to tumor

back to the surface of the cell where the IgG is released into the serum at physiological pH (pH 7.0–7.4), avoiding lysosomal degradation (Ober et al., 2004a, b). Damaged IgG or IgG of the wrong subclass are not recognized by FcRn, and are therefore left for degradation by lysosomes. In addition, when the FcRn is saturated, excess IgG is degraded, thus regulating the amount of IgG in circulation.

Control of IgG half-life can be accomplished by engineering site-directed mutations into the IgG CH2-CH3 interface, the region of Fc domain responsible for binding FcRn (Fig. 10.1). Such mutations can either increase or decrease the affinity for FcRn for the IgG Fc. Several residues, conserved across species, have been identified as being important for binding of FcRn to Fc, including His^{310}, His^{435}, and Asn^{434} (Ghetie et al., 1997). Together with His^{250} and His^{251} of FcRn, these Fc domain residues impart a pH-dependence to the FcRn-IgG interaction (Shields et al., 2001). Several antibodies have been developed with attenuated FcRn binding at pH 6.0 but unchanged binding properties at neutral pH, resulting in increased IgG half-life (Hinton et al., 2004, 2006, Petkova et al., 2006, Dall'Acqua et al., 2006). In addition to increased serum half-life, increased FcRn-Fc interaction has recently been correlated with enhanced therapeutic efficacy in preclinical models. Zalevsky et al. have created a variant of the anti-vascular endothelial growth factor (VEGF) antibody bevacizumab (Avastin) termed Xtend-VEGF which exhibited an 11-fold increase in affinity for FcRn at pH 6.0, and half-life extended up to five-fold in transgenic mice expressing the human FcRn and three-fold in cynomolgus monkeys at longer dosing intervals (Zalevsky et al., 2010). In addition, Xtend-VEGF and a variant of the anti-EGFR antibody cetuximab (Erbitux) termed Xtend-EGFR both showed therapeutic benefit in xenograft mice.

Conversely, engineering antibodies to have decreased half-life may be desirable in the setting of immunoconjugates (see below) or for antibody-based imaging applications (Kenanova et al., 2005). In addition, antibody-based molecules called Abdegs, engineered to have high affinity for FcRn, can compete with endogenous IgGs (Vaccaro et al., 2005). Abdegs, which block FcRn and result in increased degradation of endogenous IgGs, are potential treatments for diseases involving autoreactive antibodies such as multiple sclerosis and lupus. Such antibodies may also be useful for competing with and increasing clearance of immunotoxins and immunoconjugates.

10.4.1.2 Glycosylation

Protein glycosylation is a natural covalent post-translational modification that aids in proper protein folding, protects the protein from proteolytic degradation (Vegarud and Christensen, 1975), and enhances stability (Sola and Griebenow, 2010, Sola et al., 2007). In addition, the persistence of therapeutic antibodies in circulation depends, in part, on the structure and identity of the N-linked oligosaccharide attached to Asn297 within the Fc region (Fig. 10.4). Human IgGs isolated from normal serum contain multiple glycoforms that differ in both the number of potentially

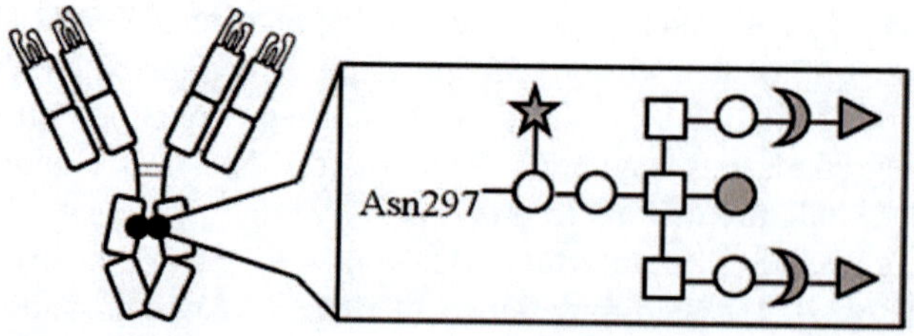

Fig. 10.4 Structure of IgG oligosaccharide. IgG are postranslationally modified at Asn297 of the heavy chain by the addition of a complex N-linked oligosaccharide. The components of the core (*white*) and variable (*grey*) regions of the oligosaccharide are detailed. *Circle* = N-acetylglucosamine, *Square* = mannose, *Moon* = galactose, *Triangle* = sialic acid, *Star* = fucose

glycosylated N- and O-linked sites as well as in the heterogeneity of the oligosaccharide structure. N-linked glycosylation is also observed in both the V_H and V_L domains of IgG from normal human serum (up to 30%) due to random hypermutations leading to generation of N-linked glycosylation motifs (Abes and Teillaud, 2010).

The biological importance of glycosylation as it relates to both half-life and effector function (see below) of therapeutic mAbs requires that it be taken into account during development and production. Analogous to the concerns with the primary amino acid sequence of mAbs that led to the development of humanized antibodies (see above), non-human glycosylation patterns are associated with infusion related hypersensitivities. Therefore, optimizing mAb glycosylation provides an opportunity to further enhance efficacy and safety.

Glycosylation of recombinant therapeutic mAbs depends on several factors including (1) the type of expression system, (2) the presence and activity of specific enzymes including glycosyltransferases to build to the oligosaccharide to the terminal mannose moieties (Kornfeld and Kornfeld, 1985), and glycosidases to break them down, and (3) the availability of appropriate sugar-nucleotide donors in the culture media. The mammalian cell lines commonly used to express therapeutic antibodies are of non-human origin and therefore glycosylate in patterns significantly different than those seen in humans (see Chapter 13 of this volume and also Raju, 2003, Brooks, 2004). Chinese hamster ovary (CHO) cells produce glycosylation patterns differing from human by the type of sialic acid linkage and lack of bisecting GlcNAc residues (Raju, 2003). Cell engineering strategies have been employed to develop cells that express the β(1,4)-N-acetylglucosaminyltransferase III enzyme required to add the bisecting GlcNAc residues (Umana et al., 1999). Such "glycoengineering" of expression systems to mimic natural human glycosylation patterns may allow for lot-to-lot standardization of glycosylation as required by regulatory agencies.

Alternative production systems such as insect cells and yeast produce glycosylation patterns that differ more dramatically from those seen with mammalian expression systems. High-mannose glycosylation by yeast decreases half-life by binding of the antibody oligosaccharide to macrophage mannose receptors, leading to degradation of the antibodies (Beck et al., 2008a). In addition, fucose and xylene found in plant-produced glycoproteins can be toxic. In contrast, antibody glycoforms with high sialic acid content (negative charges) can prolong circulation (Elliott et al., 2003, Byrne et al., 2007). Analogous to engineering of CHO cells to express β(1,4)-N-acetylglucosaminyltransferase III, the endogenous yeast glycosylation pathways of *Pichia pastoris* have been replaced with a synthetic in vivo glycosylation pathway including eukaryotic mannosidases I and II and N-acetylglucosaminyl transferases I and II (Hamilton et al., 2003, Jacobs et al., 2009). These modifications allow for production of human glycoproteins with uniform complex N-glycosylation (Li et al., 2006a). More recently, *E. coli* have also been successfully glycoengineered (Schwarz et al., 2010); a promising development for production of future therapeutic antibodies with enhanced serum half-life.

10.4.1.3 Antibody Size and Affinity

Antibody-based therapeutics direct the killing of a tumor cell by binding to antigens on its surface. Therefore, in order to eradicate all cells within a tumor, antibodies must penetrate into the tumor, away from their site of extravasation from the vasculature. Tumors are characterized by fenestrated vasculature and a lack of draining lymphatics, which combine to result in high interstitial fluid pressure within the tumor. Upon extravasation of macromolecules from the vasculature, this pressure is predicted to oppose their diffusion into the tumor in a manner inversely proportional to the cube root of the molecular weight (Jain, 1987, 1990). In addition, target antigens are typically expressed at high levels on tumor cells. Weinstein and colleagues postulated that high levels of antigen near the vasculature would act as a "binding site barrier" to antibody diffusion (Fujimori et al., 1990). Wittrup and colleagues extended this binding-site-barrier model to account for the role of internalization and catabolism of antibodies once they engage antigen at the cell surface (Schmidt and Wittrup, 2009). As detailed above, the pharmacokinetic behavior of IgG mAbs have made them the favored structure for antibody-based therapeutics. However, the large size and high functional affinity of therapeutic IgGs are predicted to exacerbate the pressure gradient and binding-site-barrier, thereby limiting the ability of these agents to penetrate into tumor. When coupled with heterogeneous antigen expression these effects potentially account for the non-uniform distribution of systemically administered antibodies that is generally observed in biopsy specimens of solid tumors. Although size and intrinsic affinity of engineered antibodies are independent characteristics they must be evaluated in combination when designing an optimized therapeutic. The interplay between molecular size and affinity and how they impact in vivo behavior was elegantly modeled by Schmidt et al. (2009).

The inverse correlation between antibody size and rate of diffusion into the tumor has led to the prediction that antibody fragments smaller than intact IgG molecules (150 kDa) would be more effective at diffusing away from the tumor vasculature. In turn this has led to the creation and testing of the tumor targeting properties of a wide variety of antibody-based structures, some of which are shown in Fig. 10.5. The most basic of these structures is the 25 kDa single-chain variable fragment (scFv), comprised of the V_H and V_L domains connected by a peptide linker. Single-chain Fv serve as the building blocks for the creation of $(scFv)_2$ and diabodies, as well as bispecific single-chain antibodies (bs-scFv) and BiTEs (bispecific T-cell engaging antibodies, see below). Additional structures not depicted in Fig. 10.5, such as triabodies (90 kDa) and tetrabodies (120 kDa) have also been created (Todorovska et al., 2001). The impact of molecular size on tumor penetration was examined with a series of engineered antibodies (scFv, Fab', $F(ab)_2$, and IgG) binding to tumor associated glycoprotein 72 (TAG-72) in LS174T human colon carcinoma xenografts. Data from this study was consistent with the hypothesis that tumor penetration is inversely related to the size of the antibody molecules (Yokota et al., 1992).

Antibodies that are less than 50–65 kDa in size undergo rapid, first-pass renal clearance (Wochner et al., 1967). Although the scFv and $(scFv)_2$ would be expected

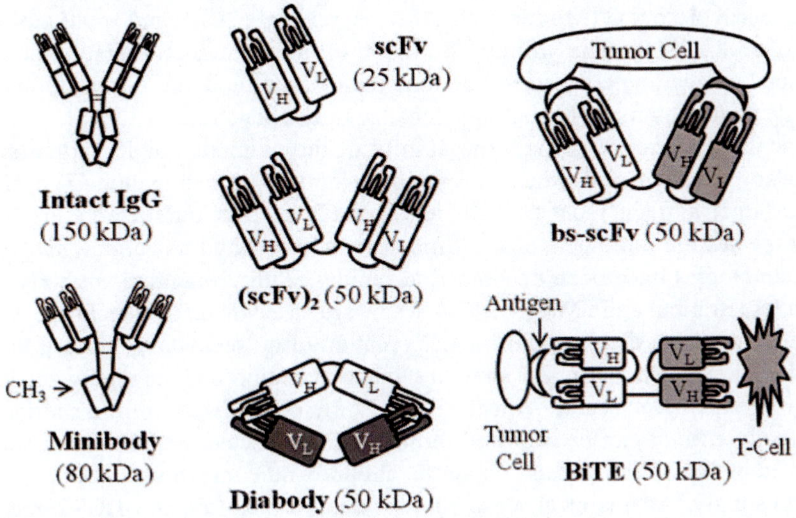

Fig. 10.5 Structures of engineered antibodies. The 25 kDa single-chain variable fragment (scFv), consisting of the V_H and V_L linked by a peptide, is the basic building block for a variety of engineered antibody constructs, as detailed in the figure

to have increased tumor penetration than their larger counterparts, these proteins are rapidly removed from circulation, limiting the overall level of tumor uptake that can be achieved. A number of strategies are being investigated to circumvent first-pass clearance and increase the half-life of promising therapeutic scFv. Fusion of polyethylene glycol (PEG) to the scFv (for review see Constantinou et al., 2010) has improved antibody stability and tumor targeting without increasing toxicity (Harris and Chess, 2003, Pasut and Veronese, 2009). For example, conjugating PEG3400 to the anti-TAG-72 diabody (Li et al., 2010) or anti-CEA diabody (Li et al., 2006c) confers an apparent molecular size to the protein complexes of ~75 kDa, equivalent to that of a minibody (80 kDa), with improved tumor-to-liver ratios. It has been hypothesized that PEGylated diabodies such as these, or PEGylated scFv such as the 4D5-PEG20 (50 kDa) (Kubetzko et al., 2006), could be useful for imaging or delivery of cytotoxic payloads.

Similar to IgGs, the half-life of albumin in serum is regulated through interaction with the FcRn (see above). Multiple strategies are being investigated to exploit albumin as a mechanism to prolong the half-lives of scFv. Creation of albumin gene fusions is one such approach. For example, an anti-CEA scFv-albumin fusion protein has shown 2.5 times improvement in tumor retention over the naked scFv due to increased serum persistence (Yazaki et al., 2008). Another protein, MM-111, a bispecific anti-Her2/HER3 scFv-albumin fusion protein currently in clinical trials was developed using this approach (Denlinger et al., 2010, Huhalov et al., 2010). Alternative strategies employ fusion of albumin-binding moieties to the therapeutic antibody. These have included peptides (Dennis et al., 2002, Nguyen et al., 2006, Dennis et al., 2007), single domain antibodies (Holt et al., 2008), domains of streptococcal protein G (Muller et al., 2007, Hopp et al., 2010) and small molecules (Trussel et al., 2009). The affinity of the non-covalent interaction with albumin in these methods can be tailored to fine tune the PK and facilitate dissociation of the antibody from albumin to potentially enhance tumor penetration.

In addition to antibody size, the affinity of the antibody for its target antigen must also be considered when designing an optimized therapeutic. The affinity for the target antigen must be sufficient to both support tumor targeting and to impart the desired biological effect. Similar to antibody humanization, many effective technologies have been developed to rapidly affinity maturate antibody-based molecules (Bradbury and Marks, 2004). Many are based on display technology, such as phage and yeast display, and involve creation of a "secondary" library through diversification of the V genes, and subsequent selection and screening for higher affinity variants (Boder and Wittrup, 2000, Colby et al., 2004). Further refinement of antibody affinity can be achieved through CDR-directed modelling methods, as discussed in the section of this chapter on antibody humanization.

Interestingly, Adams et al. have shown, using a panel of anti-HER2/neu scFv with affinities ranging from 1.6×10^{-6} M to 1.5×10^{-11} M, that the highest affinity antibodies may in fact be suboptimal for tumor targeting (Adams et al., 2001). Consistent with the binding-site-barrier hypothesis the authors found that a low affinity scFv distributed diffusely through the vascularized regions of the tumors within 24 h post-injection, whereas a scFv with 10,000-fold higher affinity was only detected within a fell cell diameters of the blood vessels. The clinical significance of affinity may be addressed, at least in part, by clinical comparisons between the low affinity anti-EGFR mAb nimotuzumab (YM Biosciences, Ontario, Canada) and either cetuximab or panitumumab.

10.4.2 Improving the Effector Function of Antibodies

Analogous to the opsonization of pathogens, binding of therapeutic antibodies to the surface of tumor cells has the potential to direct antibody-dependent cell-mediated cytotoxicity (ADCC) against the tumor via Fc-dependent interactions with Fcγ receptors found on the surface of immune effector cells (e.g. natural killer cells and CD8+ T cells). Likewise, tumor-bound IgG can kill tumor cells via complement-dependent cytotoxicity (CDC). These processes are depicted in Fig. 10.3a. The clinical success of several therapeutic mAbs is thought to depend, in part, on their ability to elicit ADCC (such as trastuzumab; Gennari et al., 2004), CDC (alemtuzumab; Zent et al., 2008), or both (such as rituximab; van Meerten et al., 2006). IgG framework selection, Fc engineering, and novel antibody-based structures represent approaches to increase recruitment of an anti-tumor immune response.

10.4.2.1 IgG Isotypes

Whether designing therapeutic chimeric or human antibodies the choice of the Fc domain IgG isotype is important, as it is the most direct way to control ADCC and CDC (Salfeld, 2007). The IgG isotypes share 95% homology of the Fc region, but differ in amino acid composition and hinge region structure. The IgG1 isotype is the most commonly used framework for therapeutic antibodies, as it maximizes ADCC and CDC (Liu et al., 1987). IgG3 also exhibit enhanced CDC over IgG1 (Bruggemann et al., 1987), however they are rarely used because their long hinge regions are more prone to proteolysis (Carter, 2006). The IgG2 isotype has low affinity for Fc receptors and the C1q component of complement cascade, and is therefore not effective at eliciting either ADCC or CDC (Bruggemann et al., 1987). Therefore, IgG2 isotype antibodies are useful for situations in which stimulating the patient's immune system is unwanted, as in the use of antibodies to deliver cytotoxic drugs to the tumor or when the antibody is itself therapeutic, or unnecessary, as for the neutralization of soluble antigens. IgG4 have been used for similar purposes, however with half-lives approximately two times smaller than IgG1 (Salfeld, 2007).

10.4.2.2 Altering Interaction with Fc Gamma Receptor Through Mutations

The FcγR family is comprised of three classes (type I, II, and III) that are further divided into subclasses (IIa/IIb and IIIa/IIIb). FcγRI has a much higher affinity for IgG than FcγRII or FcγRIII. Signalling through type I, IIa, and IIIa receptors causes activation of effector cells and ADCC, due to associated immunoreceptor tyrosine-based activation motifs (ITAM). Signalling through type IIb receptors, however, inhibits cell activation through associated immunoreceptor tyrosine-based inhibitory motifs (ITIM) (Raghavan and Bjorkman, 1996).

Significant effort has focused on modifying the Fc domain of IgGs to optimize engagement of subclasses of FcγR for induction of ADCC (Fig. 10.1). The major interaction sites of FcγR with the IgG are in the Fc CH2 (Shields et al., 2001) and hinge regions (Morgan et al., 1995). Shields et al. identified several mutations to the general Fc IgG structure that increased binding to FcγR IIIa, but either had no effect or lowered binding to IIb, enhancing ADCC. More recently, approaches that combine computational structure-based protein design methods coupled with high throughput screening have identified mutations which optimize the FcγR binding capacity of two clinically relevant antibodies, the anti-CD52 Ab alemtuzumab and the anti-HER2 Ab trastuzumab (Lazar et al., 2006). Antibodies with increased ADCC activity have also been generated against CD-30 (XmAb5574, Horton et al., 2008) and CD-19 (XmAb2513), which is now in clinical trials for the treatment of Hodgkin lymphoma or anaplastic large cell lymphoma (Blum et al., 2009). In addition, modification of Fc residues that specifically contact the carbohydrate moiety can also modulate recognition by FcγR (Lund et al., 1995).

The binding of C1q to the antibody Fc region is the first step in activating CDC, and the cascading response depends on the intensity of this first step. Therefore, researchers have investigated increasing CDC by facilitating binding to C1q. This can be accomplished by mutating residues in the Fc CH2 (Idusogie et al., 2001) or hinge regions of the antibody (Dall'Acqua et al., 2006). Using a proprietary approach based on imparting the intrinsic complement fixation activity of the

IgG3 isotype to an IgG1 backbone, Natsume et al. demonstrated that engineered anti-CD20 showed increased CDC over the parental IgG1 antibody, without changing its ADCC properties (Natsume et al., 2009, 2008).

10.4.2.3 Altering Interaction with Fc Receptor Thru Glycan Modifications

Human IgG molecules are glycosylated in CH2 (Fig. 10.1) at Asn297, and in IgG1, the oligosaccharide is of the biantennary complex type (Fig. 10.4), composed of a mannosyl-chitobiose core structure containing a core fucose, a bisecting N-acetylglucosamine (GlcNAc), and a terminal galactose and sialic acid (Natsume et al., 2009). As mentioned above, the structure and identity of the N-linked oligosaccharide greatly influences the stability and half-life of the IgG in circulation, as well as the ability of the IgG to bind to FcγR and elicit effector functions. Truncations of the carbohydrate group have been shown to reduce the binding affinity of FcγRI for IgG1 by four- to six-fold (Wright and Morrison, 1994). In addition, antibodies in which the core fucose has been removed have exhibited dramatic increases in binding to FcγRIIIa and enhanced ADCC (Shields et al., 2002, Kanda et al., 2006a).

Despite the apparent advantage of non-fucosylated antibodies to improve ADCC, current licensed therapeutic antibodies contain a mixture of fucosylated (~90%) and non-fucosylated IgG (Natsume et al., 2009). This is likely due to the difficulty of finding suitable hosts to produce large quantities of non-fucosylated antibodies (Kanda et al., 2006b). However, recent genetic engineering of the fucosylation pathway in antibody-producing CHO cells using siRNA may help to drive the future production of non-fucosylated antibodies (Imai-Nishiya et al., 2007).

10.4.2.4 Bispecific Antibodies and BiTES

An alternative to increasing effector function by modifying the Fc region of mAbs is to create bispecific antibodies (bsAbs) that recognize both a tumor associated antigen and a "trigger antigen" or receptor present on the surface of an immune effector cell (Figs. 10.3 and 10.5). Simultaneous engagement of both antigens can redirect the cytotoxic potential of effector cells against the tumor (Weiner et al., 1993, Keler et al., 1997). There are several advantages to using bispecific antibodies instead of mAbs to recruit the immue system. First, selection and affinity maturation of bispecific antibodies can be performed using bacterial or yeast surface display methods (as discussed in the section on human antibodies), allowing for custom tailoring of the affinity of the bsAb to match effector cell characteristics (Carter, 2001). Second, bsAbs can be selected to bind epitopes on FcγR that are distinct from the epitopes involved in the FcγR-Fc interaction. This allows recruitment of effector function in the presence of excess IgG (Weiner et al., 1993), as expected to occur in vivo. Third, the specificity of the bsAb for its oncogenic target can be used to direct the cytotoxic potential of any immune effector cell to the tumor, including T cells (Liu et al., 1985), which are not normally recruited by IgG molecules.

BiTE (Bispecific T-cell Engager) antibodies (Figs. 10.3 and 10.5) are emerging as a versatile platform for specifically recruiting T cells for cancer therapy (Mack et al., 1997, Wolf et al., 2005). BiTE-activated T cells have been shown to kill cancer cells by membrane perforation using perforin, and subsequent release of granzyme B, inducing apoptosis (Haas et al., 2009). There are several BiTEs in pre-clinical and clinical development that show promise. A BiTE targeting EGFR has shown in vitro and in vivo activity against colorectal cancer cells with mutated KRAS and BRAF oncogenes that are resistant to Cetuximab (Lutterbuese et al., 2010). Also a BiTE for the therapy of colorectal cancer patients previously treated with conventional chemotherapy, called MEDI-565, which targets carcinoembryonic antigen (CEA) and CD3, was able to engage a patient's own T cells to attack the cancerous cells in vitro (Osada et al., 2010). Blinatumomab, a BiTE against CD19/CD3 is now in clinical trials for patients with relapsed non-Hodgkin's lymphoma (Bargou et al., 2008, Baeuerle and Reinhardt, 2009). MT110, a BiTE targeting EpCAM/CD3 in the setting of advanced lung and gastrointestinal tumors is also in clinical trials (Brischwein et al., 2006, Fiedler et al., 2010). Bispecific and BiTE antibodies are reviewed in greater detail in Chapter 12 of this volume.

10.4.3 Polyclonal Therapeutics

As mentioned above, preparations of polyclonal antibodies (pAb) derived from the serum of immunized animals have been used as therapeutics. However, limited efficacy and significant safety concerns have severely hampered their utility. It has been hypothesized that in vitro generation of an antigen-specific pAb mixture would have benefits over a standard mAb-based therapy targeting the same antigen. Since the immune system normally responds to antigen challenge by the formation of a pAb response, pAb-based therapeutics may be ideally suited to recruit immune effector cells to kill tumor cells. Additionally, pAbs could more effectively clear activating ligands from the tumor microenvironment or inactivate cell surface receptors than their mAb-based counterparts. The advent of recombinant antibody technology has allowed this hypothesis to be tested.

In principle, in vitro generation of a pAb therapeutic could be accomplished either through mixing or simultaneous administration of mAbs that were individually expressed and purified by standard techniques (Fig. 10.6a–c). Clinical testing could then follow the standard combination therapy approaches currently in use. This strategy has been employed with the anti-HER2 mAbs trastuzumab and pertuzumab (Baselga et al., 2010) with promising results. Metastatic breast cancer patients that had previously progressed on trastuzumab therapy showed an objective response rate of 24.2% and a clinical benefit rate of 50% in a phase II trial.

Nielsen et al. (2010) describe an alternative method for generation of a polyclonal therapeutic that is depicted in Fig. 10.6a, d, e. In this approach, mixtures of engineered cell lines, each with defined growth and IgG expression properties, are created and used to express and purify pAb preparations. This approach, which forms the basis of the proprietary Sympress technology (Symphogen, Denmark),

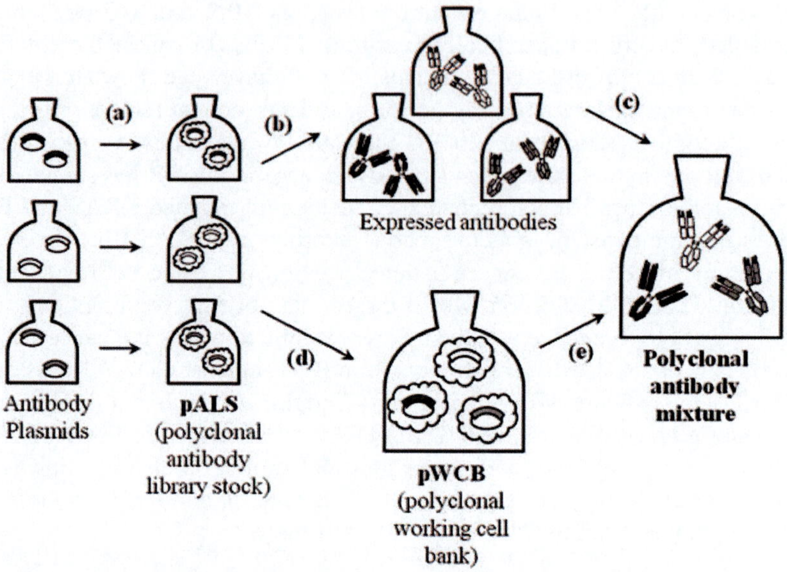

Fig. 10.6 Recombinant polyclonal antibodies. Polyclonal antibody library stocks (pALS) (**a**) are created by stable transfection of cells with individual antibody clones isolated by techniques such as phage display. Traditionally recombinant polyclonal antibody mixtures are created by expressing (**b**) and then mixing (**c**) individual clones. An emerging method involves combining stable cell lines (**d**) to obtain a polyclonal working cell bank (pWCB), and expressing large batches of the polyclonal antibody mixture (**e**) under controlled conditions

is reported to allow for cost-effective single-batch manufacturing of pAb preparations of up to six different IgG. Concerns related to cell engineering aspects of this approach include lot-to-lot variability of products that could result from variations in growth or expression characteristics in the production runs. These concerns are addressed, in part, by a multi-stage production scheme that involves generation of polyclonal antibody library stocks (pALS, Fig. 10.6a) and polyclonal working cell banks (pWCB, Fig. 10.6d), frozen before and after mixing of stable antibody-producing cell lines, respectively. Clinical validation of this approach is currently ongoing.

10.5 Arming Antibodies for Therapy

The majority of FDA-approved mAbs function via mechanisms-of-action that are intrinsic to the antibody. By binding to their target antigens they either alter signalling through receptor-mediated pathways that drive tumor formation and progression or direct tumor cell killing by the immune system. However, mAbs that lack intrinsic anti-cancer activity can still be exploited as delivery vehicles for cytotoxic payloads (Fig. 10.3c) such as chemotherapy, radionuclides, toxins, and cytokines (for review see Carter, 2006). Engineering of these classes of mAb-based drugs requires that additional variables be taken into consideration during the design process.

10.5.1 Immunodrug Conjugates

Gemtuzumab ozogamicin is a humanized IgG4 anti-CD33 antibody conjugated to the cytotoxic antibiotic calicheamicin. It is used for the treatment of patients with relapsed CD33-positive acute myeloid leukaemia and represents the only FDA-approved immunodrug conjugate to date. Although recent data has called into question its safety (Petersdorf et al., 2009), gemtuzumab ozogamicin laid the groundwork for what is now a burgeoning field of immunodrug conjugate development. The hypothesis driving this field is that the tumor targeting properties of mAb can focus the cytotoxic effects of chemotherapy to the tumor, allowing for use of anti-cancer agents that are otherwise too toxic for systemic use. Calicheamicin and maytansinoid derivatives represent the two classes of chemotherapies with the most clinical validation to date. Trastuzumab-DM1 is representative of this expanding class of cancer therapies (Lewis Phillips et al., 2008). It has shown significant single-agent activity in phase II trials (Krop et al., 2009, Vogel et al., 2009) in the setting of HER2-positive breast cancer that had progressed on HER2-targeted therapies (i.e. trastuzumab or lapatinib).

Linker design, the chemistry used to conjugate the cytotoxic compound to the mAb, is a critical aspect of immunodrug conjugate development. In broad terms two classes of linkers, cleavable (e.g. disulphide bond) and noncleavable (e.g. thioether), are used and selection between the two types is driven, at least in part, by the underlying biology of the target antigen. This is exemplified by the preclinical studies supporting both trastuzumab-DM1 (Lewis Phillips et al., 2008) and the anti-CanAg immunodrug conjugate cantuzumab mertansine (Tolcher et al., 2003, Erickson et al., 2006). Trastuzumab-DM1 utilizes a thioether-based linkage to couple trastuzumab to the DM1 maytansinoid based on preclinical data demonstrating a better PK profile and therapeutic window than seen with a cleavable disulphide-based linker (Lewis Phillips et al., 2008). In contrast, Erickson and colleagues (Erickson et al., 2006) demonstrated that a cleavable linker improved the efficacy of cantuzumab mertansine. The differences seen in these systems are due to the impact of internalization and lysosomal degradation of the immunoconjugates and the subsequent drug metabolites that are released. Additional considerations include the location and number of drugs conjugated to each mAb and how these alter properties of the mAb (solubility, PK, etc). These considerations are nicely addressed in a recent review by Carter and Senter (2008).

10.5.2 Immunotoxins

Bacterial- and plant-derived toxins represent another class of highly toxic compounds being investigated as immunoconjugates for cancer therapy (Reiter, 1998). Toxins can be broadly classified into two main categories. The first is catalytic toxins, which include the bacterial proteins diptheria toxin (DT) and *Pseudomonas* exotoxin (PE) (Govindan and Goldenberg, 2010), and the ribosome inactivating proteins (Bagga et al., 2002) and ricin (Santanche et al., 1997) from plants. The catalytic nature of these proteins makes them sufficiently toxic that internalization of a single molecule into a cell is sufficient to kill that cell (Kreitman, 2006). The second class is the superantigens that are highly potent immune modulatory proteins and can be used to induce T cell infiltration into the tumor, eliciting a focused cytotoxic T cell response (for review see Robinson et al., 2010).

Unlike with immunodrug conjugates, immunotoxins are typically developed as gene fusions, with an engineered antibody fragment replacing the toxin's natural translocation domain. This is hypothesized to limit normal tissue toxicities and facilitate internalization into the tumor. Although promising preclinical results have been generated with this approach, clinical successes have been limited and associated with significant toxicities (Baluna and Vitetta, 1997).

Immunogenicitiy of current toxins is a major concern that limits both safety and efficacy (Posey et al., 2002). Analogous to the immunogenicity seen with murine mAb, it is necessary to "deimmunize" the toxins to make them amenable for use in patients. Therefore, steps to identify and remove B and T cell epitopes represent a major focus in the field of immunotoxin development. This effort is driven by experiments designed to identify epitopes that react with serum from patients involved in clinical trials. For example, antibodies isolated from the serum of patients treated with PE38 immunoconjugates identified seven B cell epitope groups within PE38 (Roscoe et al., 1997, Onda et al., 2006). Elimination or mutation of these large hydrophilic residues to small nonpolar residues decreased immunogenicity of the immunotoxin (Onda et al., 2008). The antibody and protein engineering steps undertaken in the development of Napatumomab estafenatox, an antibody-superantigen fusion protein, exemplifies the processes necessary to overcome both the toxicity and immunogenicity issues associated with this class of agent (for review see Robinson et al., 2010). Pegylation has been investigated as an additional strategy to reduce the immunogenicity of immunotoxins and has the added benefit of increasing the serum half-life of these molecules (Tsutsumi et al., 2000, Youn et al., 2005). Although promising, the clinical utility of this approach is not yet validated. A table of current clinical trials using immunotoxins is provided elsewhere (Risberg et al., 2010).

Finally, the use of endogenous proteins is expected to limit immunogenicity. Ribonucleases (RNases) are one class of proteins being investigated. Like toxins, RNAses can act catalytically to degrade intracellular RNAs, leading to cell death. Ranpirnase (marketed as Onconase) is an amphibian RNase with demonstrated activity in patients with unresectable malignant mesothelioma (Lee and Raines, 2008, Beck et al., 2008b). Despite its origin, ranpirnase could be administered repeatedly with limited immunogenicity (Mikulski et al., 2001), and is currently awaiting FDA approval. These results have led to development of anti-CD22 (Newton et al., 2001), anti-CD30 (Menzel et al., 2008), and anti-Trop2 (Chang et al., 2010) immunoRNases for use in a variety of clinical settings. In addition, the success of ranpirnase has lead researchers to develop endogenous human

RNases, such as human pancreatic RNase1 and the plasma protein angiogenin for similar purposes (Mathew and Verma, 2009).

10.6 Conclusions

Advances in antibody engineering and production over the past few decades have matured the field of antibody-based therapy from its infancy of using rodent derived antibodies to today, when over 66 fully human mAbs have been tested clinically in a variety of disease settings, including cancer. Reducing the immunogenicity of therapeutic antibodies has been a major factor in their clinical success. In addition to model-based approaches to antibody humanization, new production platforms such as phage display, transgenic mice, and human hybridomas have allowed generation of fully human antibodies. The majority of antibodies in clinical use rely on intrinsic properties of the antibody, such as the ability of the mAb to inhibit signalling or to redirect the patient's immune system against tumor cells. Antibody and cell engineering technologies can therefore be employed to tailor the serum persistence, size, and affinity of antibodies or to introduce new functionalities, allowing control of their pharmacokinetic and pharmacodynamic properties. In addition, mAbs that lack intrinsic anti-cancer activity can still be exploited as delivery vehicles for cytotoxic payloads such as chemotherapy, radionuclides, toxins, and cytokines. Together with advances in modern molecular biology, the innovative approaches to antibody engineering, described in this chapter, and to antibody production, described in this entire volume, are allowing Ehrlich's vision of the magic bullet for the treatment of cancer to be realized.

References

Abes, R. & Teillaud, J. (2010) Imapct of glycosylation on effector functions of therapeutic IgG. *Pharmaceuticals,* 3, 146–157.

Abhinandan, K. R. & Martin, A. C. R. (2007) Analyzing the "degree of humanness" of antibody seqences. *Journal of Molecular Biology,* 369, 852–862.

Adams, G. P., Schier, R., Mccall, A. M., Simmons, H. H., Horak, E. M., Alpaugh, R. K., Marks, J. D. & Weiner, L. M. (2001) High affinity restricts the localization and tumor penetration of single-chain fv antibody molecules. *Cancer Research,* 61, 4750–4755.

Albrecht, H. (2010) Role of antibodies in cancer treatment (an overview). In Hayat, M. A. (Ed.) *Methods of Cancer Diagnosis, Therapy, and Prognosis.* Houten, Netherlands, Springer.

Almagro, J. C. & Fransson, J. (2008) Humanization of antibodies. *Frontiers in Bioscience,* 13, 1619–1633.

Alonso-Camino, V., Sanchez-Martin, D., Compte, M., Sanz, L. & Alvarez-Vallina, L. (2009) Lymphocyte display: A novel antibody selection platform based on T cell activation. *Public Library of Science,* 4, e7174.

Alt, F. W., Blackwell, T. K. & Yancopoulos, G. D. (1985) Immunoglobulin genes in transgenic mice. *Trends in Genetics,* 1, 231–236.

Azinovic, I., Denardo, G. L., Lamborn, K. R., Mirick, G., Goldstein, D., Bradt, B. M. & Denardo, S. J. (2006) Survival benefit associated with human anti-mouse antibody (HAMA) in patients with B-cell malignancies. *Cancer Immunology and Immunotherapy,* 55, 1451–1458.

Baca, M., Presta, L. G., O'Connor, S. J. & Wells, J. A. (1997) Antibody humanization using phage display. *Journal of Biological Chemistry,* 272, 10678–10684.

Badger, C. C., Anasetti, C., Davis, J. & Bernstein, I. D. (1987) Treatment of malignancy with unmodified antibody. *Pathology and Immunopathology Research,* 6, 419–434.

Baeuerle, P. A. & Reinhardt, C. (2009) Bispecific T-cell engaging antibodies for cancer therapy. *Cancer Research,* 69, 4941–4944.

Bagga, S., Seth, D. & Batra, J. K. (2002) The cytotoxic activity of ribosome-inactivating protein saporin-6 is attributed to its rRNA N-glycosidase and internucleosomal DNA fragmentation activities. *Journal of Biological Chemistry,* 287, 4813–4820.

Baluna, R. & Vitetta, E. S. (1997) Vascular leak syndrome: A side effect of immunotherapy. *Immunopharmacology,* 37, 117–132.

Barbas, C. F. & Lerner, R. A. (1991) Combinatorial immunoglobulin libraries on the surface of phage (Phabs): Rapid selection of antigen-specific fabs. *Methods,* 2, 119–124.

Bargou, R., Leo, E., Zugmaier, G., Klinger, M., Goebeler, M., Knop, S., Noppeney, R., Viardot, A., Hess, G., Schuler, M., Einsele, H., Brandl, C., Wolf, A., Kirchinger, P., Klappers, P., Schmidt, M., Riethmuller, G., Reinhardt, C., Baeuerle, P. A. & Kufer, P. (2008) Tumor regression in cancer patients by very low doses or a T cell-engaging antibody. *Science,* 321, 974–977.

Baselga, J., Gelmon, K. A., Verma, S., Wardley, A., Conte, P., Miles, D., Bianchi, G., Cortes, J., Mcnally, V. A., Ross, G. A., Fumoleau, P. & Gianni, L. (2010) Phase II trial of pertuzumab and trastuzumab in patients with human epidermal growth factor receptor 2 – Positive metastatic breast cancer that progressed during prior trastuzumab therapy. *Journal of Clinical Oncology,* 28, 1138–1144.

Beck, A. K., Pass, H. I., Carbone, M. & Yang, H. (2008b) Ranpirnase as a potential antitumor ribonuclease treatment for mesothelioma and other malignancies. *Future Oncology,* 4, 341–349.

Beck, A., Wagner-Rousset, E., Bussat, M., Lokteff, M., Klinguer-Hamour, C., Haeuw, J., Goetsch, L., Wurch, T., Can Dorsselaer, A. & Corvaia, N. (2008a) Trends in glycosylation, glycoanalysis and glycoengineering of therapeutic antibodies and Fc-fusion proteins. *Current Pharmaceutical Biotechnology,* 9, 482–501.

Bernett, M. J., Karki, S., Moore, G. L., Leung, I. W. L., Chen, H., Pong, E., Nguyen, D. T., Jacinto, J., Zalevsky, J., Muchhal, U. S., Desjarlais, J. R. & Lazar, G. A. (2010) Engineering fully human monoclonal antibodies from murine variable regions. *Journal of Molecular Biology,* 396, 1474–1490.

Blum, K. A., Smith, M., Fung, H., Zalevsky, J., Combs, D., Ramies, D. A. & Younes, A. (2009) Phase I study of an anti-CD30 Fc engineered humanized monoclonal antibody in Hodgkin lymphoma (G+HL) or anaplastic large cell lymphoma (ALCL) patients: Safety, pharmacokinetics (PK), immunogenicity, and efficacy. *2009 ASCO Annual Meeting*. Abstract #8531, Orlando, FL.

Boder, E. T. & Wittrup, K. D. (1997) Yeast surface display for screening combinatorial polypeptide libraries. *Nature Biotechnology,* 15, 553–557.

Boder, E. T. & Wittrup, K. D. (2000) Yeast surface display for directed evolution of protein expression, affinity, and stability. *Methods in Enzymology,* 328, 430–444.

Boerner, P., Lafond, R., Lu, W., Brams, P. & Royston, I. (1991) Production of antigen-specific human monoclonal antibodies from in vitro-primed human splenocytes. *Journal of Immunology,* 147, 86–95.

Borrebaeck, C. A. K., Danielsson, L. & Moller, S. A. (1988) Human monoclonal antibodies produced by primary in vitro immunization of peripheral blood lymphocytes. *Proceedings of the National Academy of Sciences of the USA,* 85, 3995–3999.

Boulianne, G. L., Hozumi, N. & Shulman, M. J. (1984) Production of functional chimaeric mouse/human antibody. *Nature,* 312, 643–646.

Bourne, P. C., Terzyan, S. S., Cloud, G., Landolfi, N. F., Casquez, M. & Edmundson, A. B. (2004) Three-dimensional structures of a humanized anti-IFN-gamma Fab (HuZAF) in two crystal structures. *Acta Crystallographica,* D60, 1761–1769.

Bradbury, A. R. M. & Marks, J. D. (2004) Antibodies from phage antibody libraries. *Journal of Immunological Methods,* 290, 29–49.

Brischwein, K., Schlereth, B., Guller, B., Steiger, C., Wolf, A., Lutterbuese, R., Offner, S., Locher, M., Urbig, T., Raum, T., Kleindienst, P., Wimberger, P., Kimmig, R., Fichtner, I., Kufer, P., Hofmeister, R., Da Silva, A. J. & Baeuerle, P. A. (2006) MT110: A novel bispecific single-chain antibody construct with high efficacy in eradicating established tumores. *Molecular Immunology,* 43, 1129–1143.

Brooks, S. A. (2004) Appropriate glycosylation of recombinant proteins for human use. *Molecular Biotechnology,* 23, 241–255.

Bruggemann, M., Caskey, H. M., Teale, C., Waldmann, H., Williams, G. T., Surani, M. A. & Neuberger, M. S. (1989) A repertoire of monoclonal antibodies with human heavy chains from transgenic mice. *Proceedings of the National Academy of Sciences of the USA,* 86, 6709–6713.

Bruggemann, M., Willimas, G. T., Bindon, C. I., Clark, M. R., Walker, M. R., Jefferis, R., Waldmann, H. & Neuberger, M. S. (1987) Comparison of the effector functions of human immunoglobulins using a matched set of chimeric antibodies. *Journal of Experimental Medicine,* 166, 1351–1361.

Byrne, B., Donohoe, G. G. & O'Kennedy, R. (2007) Sialic acids: Carbohydrate moieties that influence the biological and physical properties of biopharmaceutical proteins and living cells. *Drug Discovery Today,* 12, 319–326.

Carter, P. (2001) Bispecific human IgG by design. *Journal of Immunological Methods,* 248, 7–15.

Carter, P. J. (2006) Potent antibody therapeutics by design. *Nature Reviews Immunology,* 6, 343–357.

Carter, P., Presta, L., Gorman, C. M., Ridgway, J. B., Henner, D., Wong, W. L., Rowland, A. M., Kotts, C., Carver, M. E. & Shepard, H. M. (1992) Humanization of an anti-p185HER2 antibody for human cancer therapy. *Proceedings of the National Academic of Sciences of the USA,* 89, 4285–4289.

Carter, P. J. & Senter, P. D. (2008) Antibody-drug conjugates for cancer therapy. *Cancer Journal,* 14, 154–169.

Chang, C., Gupta, P., Michel, R., Loo, M., Wang, Y., Cardillo, T. M. & Gondenberg, D. M. (2010) Ranpirnase (frog RNase) targeted with a humanized internalizing, anti-Trop-2 antibody has potent cytotoxicity against diverse epithelial cancer cells. *Molecular Cancer Therapeutics,* 9, 2276–2286.

Chaudhury, C., Brooks, C. L., Carter, D. C., Robinson, J. M. & Anderson, C. L. (2006) Albumin binding to FcRn: Distinct from the FcRn-IgG interaction. *Biochemistry,* 45, 4983–4990.

Chaudhury, C., Mehnaz, S., Robinson, J. M., Hayton, W. L., Pearl, D. K., Roopenian, D. C. & Anderson, C. L. (2003) The major histocompatibilty complex-related Fc receptor for IgG (FcRn) binds albumin and prolongs its lifespan. *Journal of Experimental Medicine,* 197, 315–322.

Chothia, C. & Lesk, A. M. (1987) Canonical structures for the hypervariable regions of immunoglobulins. *Journal of Molecular Biology,* 196, 901–917.

Co, M. S., Avdalovic, N. M., Caron, P. C., Avdalovic, M. V., Scheinberg, D. A. & Queen, C. (1992) Chimeric and humanized antibodies with specificity for the CD33 antigen. *Journal of Immunology,* 148, 1149–1154.

Colby, D. W., Kellogg, B. A., Graff, C. P., Yeung, Y. A., Swers, J. S. & Wittrup, K. D. (2004) Engineering antibody affinity by yeast surface display. *Methods in Enzymology,* 388, 348–358.

Cole, S. P., Campling, B. G., Louwman, I. H., Kozbor, D. & Roder, J. C. (1984) A strategy for the production of human monoclonal antibdies reactive with lung tumor cell lines. *Cancer Research,* 44, 2750–2753.

Constantinou, A., Chen, C. & Doeonarain, M. P. (2010) Modulating the pharmacokinetics of therapeutic antibodies. *Biotechnology Letters,* 32, 609–622.

Currie, G. A. (1972) Eighty years of immunotherapy: A review of immunological methods used for the treatment of human cancer. *British Journal of Cancer,* 26, 141–153.

Dall'Acqua, W. F., Kiener, P. A. & Wu, H. (2006) Properties of human IgG1s engineered for enhanced binding to the neonatal Fc receptor (FcRn). *Journal of Biological Chemistry,* 281, 23514–23524.

Denlinger, C. S., Beeram, M., Tolcher, A. W., Goldstein, L. J., Slichenmyer, W. J., Murray, J., Mcdonagh, C. F., Andreas, K. & Moyo, V. M. (2010) A phase I/II and pharmacologic study of MM-111 in patients with advanced, refractory HER2-positive (HER2+) cancers. *Journal of Clinical Oncology,* 28, TPS169.

Dennis, M. S., Jin, H., Dugger, D., Yang, R., Mcfarland, L., Ogasawara, A., Williams, S., Cole, M. J., Ross, S. & Schwall, R. (2007) Imging tumors with an albumin-binding Fab, a novel tumor-targeting agent. *Cancer Research,* 67, 254–261.

Dennis, M. S., Zhang, M., Meng, Y. G., Kadkhodayan, M., Kirchhofer, D., Combs, D. & Damico, L. A. (2002) Albumin binding as a general strategy for improving the pharmacokinetics of proteins. *Journal of Biological Chemistry,* 277, 35035–35043.

Desjarlais, J. R. (2005) Protein engineering with analogous contact environments. In Wipo (Ed.) *WO/2005/057486*. G06F 19/00 (2006.01) ed. USA.

Desmet, J., Vanhoorelbeke, K. & Deckmyn, H. (2010) Humanization by resurfacing. In Kontermann, R. & Dubel, S. (Eds.) *Antibody Engineering*, 2nd ed. Heidelberg, Springer.

Elliott, S., Lorenzini, T., Asher, S., Aoki, K., Brankow, D., Buck, L., Busse, L., Chang, D., Fuller, J., Grant, J., Hernday, N., Hokum, M., Hu, S., Knudten, A., Levin, N., Komorowski, R., Martin, F., Navarro, R., Osslund, T., Rogers, G., Rogers, N., Trail, G. & Egrie, J. (2003) Enhancement of therapeutic protein in vivo activities through glycoengineering. *Nature Biotechnology,* 21, 414–421.

Erickson, H. K., Park, P. U., Widdison, W. C., Kovtun, Y. V., Garrett, L. M., Hoffman, K., Lutz, R. J., Goldmacher, V. S. & Blattler, W. A. (2006) Antibody-maytansinoid conjugates are activated in targeted cancer cells by lysosomal degradation and linker-dependent intracellular processing. *Cancer Research,* 66, 4426–4433.

Feldhaus, J. M., Siegel, R. W. & Wittrup, K. D. (2003) *Yeast Display scFv Antibody Library User's Manual.* Richland, WA, Pacific Northwest National Laboratory.

Fiedler, W. M., Ritter, B., Seggewiss, R., Bokemeyer, C., Fettes, P., Klinger, M., Vieser, E., Reuttinger, D., Kaubitzsch, S. & Wolf, M. (2010) Phase I safety and pharmacology study of the EPCAM/CD3-bispecific antibody MT110 in patients with metastatic colorectal, gastric, or lung cancer. *Journal of Clinical Oncology,* 28, 2573.

Foote, J. & Winter, G. (1992) Antibody framework residues affecting the conformation of the hypervariable loops. *Journal of Molecular Biology,* 224, 487–499.

Fujimori, K., Covell, D. G., Fletcher, J. E. & Weinstein, J. N. (1990) A modeling analysis of monoclonal antibody percolation through tumors: A binding site barrier. *The Journal of Nuclear Medicine,* 31, 1191–1198.

Gennari, R., Menard, S., Fagnoni, F., Ponchio, L., Scelsi, M., Tagliabue, E., Castiglioni, F., Villani, L., Magalotti, C., Gibelli, N., Oliviero, B., Ballardini, B., Da Prada, G., Zambelli, A., & Costa, A. (2004) Pilot study of the mechanism of action of preoperative trastuzumab in patients with primary operable breast tumors overexpressing HER2. *Clinical Cancer Research,* 10, 5650–5655.

Ghetie, V., Popov, S., Borvak, J., Radu, C., Matesoi, D., Medesan, C., Ober, R. J. & Ward, E. S. (1997) Increasing the serum persistence of an IgG fragment by random mutagenesis. *National Biotechnology,* 15, 637–640.

Ghetie, V. & Ward, E. S. (2000) Multiple roles for the major histocompatibility complex class I-related receptor FcRn. *Annual Review of Immunology,* 18, 739–766.

Giusti, R. M., Shastri, K. A., Cohen, M. H., Keegan, P. & Pazdur, R. (2007) FDA approval summary: Panitumumab (Vectibix). *Oncologist,* 12, 577–583.

Glassy, M. C. & Ferrone, S. (1982) Differential segregation patters of human chromosomes in somatic cell hybrids constructed with human B-lymphocytes and human melanoma cells. *Cancer Research,* 42, 3971–3973.

Gonzales, N. R., Padlan, E. A., De Pascalis, R., Schuck, P., Schlom, J. & Kashmiri, S. V. S. (2004) SDR grafting of a murine antibody using multiple human germline templates to minimize its immunogenicity. *Molecular Immunology,* 40, 863–872.

Gorman, S. D., Clark, M. R., Routledge, E. G., Cobbold, S. P. & Waldmann, H. (1991) Reshaping a therapeutic CD4 antibody. *Proceedings of the National Academic of Sciences of the USA,* 88, 4181–4185.

Govindan, S. V. & Goldenberg, D. M. (2010) Immunoconjugate anticancer therapeutics. In Reddy, L. H. & Couvreur, P. (Eds.) *Macromolecular Anticancer Therapeutics*. New York, NY, Springer.

Graziano, R. F., Tempest, P. R., White, P., Keler, T., Deo, Y., Ghebremariam, H., Coleman, K., Pfefferkorn, L. C., Fanger, M. W. & Guyre, P. M. (1995) Construction and characterization of a humanized anit-gamma-Ig receptor type I monoclonal antibody. *Journal of Immunology,* 155, 4996–5002.

Green, L. L., Hardy, M. C., Maynard-Currie, C. E., Tsuda, H., Louie, D. M., Mendez, M. J., Abderrahim, H., Noguchi, M., Smith, D. H., Zeng, Y., David, N. E., Sasai, H., Garza, D., Brenner, D. G., Hales, J. F., Mcguinness, R. P., Capon, D. J., Klapholz, S. & Jakobovits, A. (1994) Antigen-specific human monoclonal antibodies from mice engineered with human Ig heavy and light chain YACs. *Nature Genetics,* 7, 13–21.

Haas, C., Krinner, E., Brischwein, K., Hoffmann, P., Lutterbuse, R., Schlereth, B., Kufer, P. & Baeuerle, P. A. (2009) Mode of cytotoxic action of T-cell engaging BiTE antibody MT110. *Immunobiology,* 214, 441–453.

Hamilton, S. R., Bobrowicz, P., Bobrowicz, B., Davidson, R. C., Li, H., Mitchell, T., Nett, J. H., Rausch, S., Stadheim, T. A., Wischnewski, H., Wildt, S. & Gerngross, T. U. (2003) Production of complex human glycoproteins in yeast. *Science,* 301, 1244–1246.

Hanes, J., Jermutus, L., Weber-Bornhauser, S., Bosshard, H. R. & Pluckthun, A. (1998) Ribosome display efficiently selects and evolves high-affinity antibodies in vitro from immune libraries. *Proceedings of the National Academy of Sciences of the USA,* 95, 14130–14135.

Harris, J. M. & Chess, R. B. (2003) Effect of pegylation on pharmaceuticals. *Nature Reviews Drug Discovery,* 214, 214–221.

Hecht, J. R., Patnaik, A., Berlin, J., Venook, A., Malik, I., Tchekmedyian, S., Navale, L., Amado, R. G. & Meropol, N. J. (2007) Panitumumab monotherapy in patients with previously treated metastatic colorectal cancer. *Cancer,* 10, 980–988.

Hericourt, J. & Richet, C. (1895) Traitement d'un cas de sarcome par la serotherapie. *Comptes Rendus Hebdomadaires des Seances de l Academie des Sciences,* 120, 948–950.

Hinton, P. R., Johlfs, M. G., Xiong, J. M., Hanestad, K., Ong, K. C., Bullock, C., Keller, S., Tang, M. T., Tso, J. Y., Vasquez, M. & Tsurushita, N. (2004) Engineered human IgG antibodies with longer serum half-lives in primates. *Journal of Biological Chemistry,* 279, 6213–6216.

Hinton, P. R., Xiong, J. M., Johlfs, M. G., Tang, M. T., Keller, S. & Tsurushita, N. (2006) An engineered human IgG1 anitbody with longer serum half-life. *Journal of Immunology,* 176, 346–356.

Holt, L. J., Basran, A., Jones, K., Chorlton, J., Jespers, L. S., Brewis, N. D. & Tomlinson, I. M. (2008) Anti-serum albumin domain antibodies for extending the half-lives of short lived drugs. *Protein Engineering, Design & Selection,* 21, 283–288.

Hoogenboom, H. R., Griffiths, A. D., Johnson, K. S., Chiswell, D. J., Hudson, P. J. & Winter, G. (1991) Multi-subunit proteins on the surface of filamentous phage: Methodologies for displaying antibody (Fab) heavy and light chains. *Nucleic Acids Research,* 19, 4133–4137.

Hopp, J., Hornig, N., Zettlitz, K. A., Schwarz, A., Fu, N., Muller, D. & Kontermann, R. E. (2010) The effects of affinity and valency of an albumin-binding domain (ABD) on the half-life of a single-chain diabody-ABD fusion protein. *Protein Engienering, Design & Selection,* 23, 827–834.

Horton, H. M., Bernett, M. J., Pong, E., Peipp, M., Karki, S., Chu, S. Y., Richards, J. O., Vostiar, I., Joyce, P. F., Repp, R., Desjarlais, J. R. & Zhukovsky, E. A. (2008) Potent in vitro and in vivo activity of an Fc-engineered anti-CD19 monoclonal antibody against lymphoma and leukemia. *Cancer Research,* 68, 8049–8057.

Houghton, A. N., Brooks, H., Cote, R. J., Taormina, M. C., Oettgen, H. F. & Old, L. J. (1983) Detection of cell surface and intracellular antigens by human monoclonal antibodies:

Hybrid cell lines derived from lymphocytes of patients with malignant melonoma. *Journal of Experimental Medicine,* 158, 53–65.

Huhalov, A., Adams, S., Paragas, V., Oyama, S., Overland, R., Luus, L., Gibbons, F., Zhang, B., Nguyen, S., Nielsen, U. B., Niyikiza, C., Mcdonagh, C. F. & Kudla, A. J. (2010) MM-111, an ErbB2/ErbB3 bispecific antibody with potent activity in ErbB2-overexpressing cells, positively combines with trastuzumab to inhibit growth of breast cancer cells driven by the ErbB2/ErbB3 oncogenic unit. *Proceedings of the AACR*, Washington DC, Abstract 3485.

Huse, W. D., Sastry, L., Iverson, S. A., Kang, A. S., Alting-Mees, M., Burton, D. R., Benkovic, S. J. & Lerner, R. A. (1989) Generation of a large combinatorial library of the immunoglobulin repertoire in phage lambda. *Science,* 246, 1275–1281.

Hwang, W. Y. K. & Foote, J. (2005) Immunogenicity of engineered antibodies. *Methods,* 36, 3–10.

Idusogie, E. E., Wong, P. Y., Presta, L. G., Gazzano-Santoro, H., Totpal, K., Ultsch, M. & Mulkerrin, M. G. (2001) Engineered antibodies with increased activity to recruit complement. *Journal of Immunology,* 166, 2571–2575.

Imai-Nishiya, H., Mori, K., Inoeu, M., Wakitani, M., Iida, S., Shitara, K. & Satoh, M. (2007) Double knockdown of alpha1,6-fucosyltransferase (FUT8) and GDP-mannose 4,6-dehytratase (GMD) in antibody-producing cells: A new strategy for generating fully non-fucosylated therapeutic antibodies with enhanced ADCC. *BMC Biotechnology,* 7:84.doi:10.1186/1472-6750-7-84.

Jacobs, P. P., Geysens, S., Vervecken, W., Contreras, R. & Callewaert, N. (2009) Engineering complex-type N-glycosylation in Pichia pastoris using GlycoSwitch technology. *Nature Protocols,* 4, 58–70.

Jain, R. K. (1987) Transport of molecules in the tumor interstitum: A review. *Cancer Research,* 47, 3039–3051.

Jain, R. K. (1990) Physiological barriers to delivery of monoclonal antibodies and other macromolecules in tumors. *Cancer Research,* 50, 814–819.

Jones, P. T., Dear, P. H., Foote, J., Nueuberger, M. S. & Winter, G. (1986) Replacing the complementarity-determining regions in a human antibody with those from a mouse. *Nature,* 321, 522–525.

Junghans, R. P. & Anderson, C. L. (1996) The protection receptor for IgG catabolism is the B2-microglobulin-containing neonatal intestinal transport receptor. *Proceedings of the National Academy of Sciences of the USA,* 93, 5512–5516.

Kabat, E. A. (1978) The structural basis for antibody complementary. *Advances in Protein Chemistry,* 32, 1–75.

Kanda, Y., Yamada, T., Mori, K., Okazaki, A., Inoeu, M., Kitajima-Miyama, K., Kuni-Kamochi, R., Nakano, R., Yano, K., Kakita, S., Shitara, K. & Satoh, M. (2006a) Comparison of biological activity among nonfucosylated therapeutic IgG1 antibodies with three differnt N-linked Fc oligosaccharides: The high-mannose, hybrid, and complex types. *Glycobiology,* 17, 104–118.

Kanda, Y., Yamane-Ohnuki, N., Sakai, N., Yamano, K., Nakano, R., Inoeu, M., Misaka, H., Iida, S., Wakitani, M., Konno, Y., Yano, K., Shitara, K., Hosoi, S. & Satoh, M. (2006b) Comparison of cell lines for stable production of fucose-negative antibodies with enhanced ADCC. *Biotechnology and Bioengineering,* 94, 680–688.

Karpas, A., Dremucheva, A. & Czepulkowski, B. H. (2001) A human myeloma cell line suitable for the generation of human monoclonal antibodies. *Proceedings of the National Academy of Sciences of the USA,* 98, 1799–1804.

Keler, T., Graziano, R. F., Mandal, A., Wallace, P. K., Fisher, J., Guyre, P. M., Fanger, M. W. & Deo, Y. M. (1997) Bispecific antibody-dependent cellular cytotoxicity of HER2/new-overexpressing tumor cells by FCgamma receptor type I-expressing effector cells. *Cancer Research,* 57, 4008–4014.

Kempeni, J. (1999) Preliminary results of early clinical trials with the fully human anti-TNFalpha monoclonal antibody D2E7. *Annals of the Rheumatic Diseases,* 58(Suppl I), I70–I72.

Kenanova, V., Olafsen, T., Crow, D. M., Sundaresan, G., Subbarayan, M., Carter, N. H., Ikle, D. N., Yazaki, P. J., Chatziioannou, A. F., Gambhir, S. S., Williams, L. E., Shively, J. E., Colcher, D., Raubitschek, A. A. & Wu, A. M. (2005) Tailoring the pharmacokinetics and positron emission tomography imaging properties of anti-carcinoembryonic antigen single-chain Fv-Fc antibody fragments. *Cancer Research,* 65, 622–631.

Khazaeli, M. B., Conry, R. M. & Lobuglio, A. F. (1994) Human immune-response to monoclonal-antibodies. *Journal of Immunotherapy,* 15, 42–52.

Kim, S. J., Park, I. & Hong, H. J. (2010) Human antibodies by guided selection. In Kontermann, R. & Dubel, S. (Eds.) *Antibody Engineering*, 2nd ed. Heidelberg, Springer.

Knappik, A., Ge, L., Honegger, A., Pack, P., Fischer, M., Wellnhofer, G., Hoess, A., Wolle, J., Pluckthun, A. & Virnekas, B. (2000) Fully synthetic human combinatorial antibody libraries (HuCAL) based on modular consenses frameworks and CDRs randomized with trinucleotides. *Journal of Molecular Biology,* 296, 57–86.

Knight, D. M., Wagner, C., Jordan, R., Mcaleer, M. F., Derita, R., Fass, D. N., Coller, B. S., Weisman, H. F. & Ghrayeb, J. (1995) The immunogenicity of the 7E3 murine monoclonal Fab antibody fragment variable region is dramatically reduced in humans by substitution of human for murine constant regions. *Molecular Immunology,* 32, 1271–1281.

Kohler, G. & Milstein, C. (1975) Continuous cultures of fused cells secreting antibody of predefined specificity. *Nature,* 256, 495–497.

Komissarov, A. A., Calcutt, M. J., Marchbank, M. T., Peletskaya, E. N. & Deutsher, S. L. (1996) Equilibrium binding studies of recombinant anti-single-stranded DNA Fab. Role of heavy chain complementarity-determining regions. *Journal Biological Chemistry,* 271, 12241–12246.

Kornfeld, R. & Kornfeld, S. (1985) Assembly of asparagine-linked oligosaccharides. *Annual Review of Biochemistry,* 54, 631–664.

Kozbor, D. & Roder, J. C. (1981) Requirements for the establishment of high-titered human monoclonal antibodies against tetanus toxoid using the epstein-barr virus technique. *Journal of Immunology,* 127, 1275–1280.

Kreitman, R. J. (2006) Immunotoxins for targeted cancer therapy. *The AAPS Journal,* 8, E532–E551.

Kricka, L. J. (1999) Human anti-animal antibody interferences in immunological assays. *Clinical Chemistry,* 45, 942–956.

Krop, I., Lorusso, P., Miller, K. D., Modi, S., Yardley, D., Rodriguez, G., Agresta, S., Zheng, M., Amler, L. & Rugo, H. (2009) A phase II study of trastuzumab-DM1 (T-DM1), a novel HER2 antibody-drug conjugate, in patients with HER2+ metastatic breast cancer who were previously treated with an anthracyline, a taxane, capecitabine, lapatinib, and trastuzumab. *San Antonio Breast Cancer Symposium*, San Antonio, TX.

Kubetzko, S., Balic, E., Wiaibel, R., Zangemeister-Wittke, U. & Pluckthun, A. (2006) PEGylation and multimerization of the anti-p185 HER-2 single chain Fv fragment 4D5. *Journal of Biological Chemistry,* 281, 35186–35201.

Kudo, T., Asao, A. & Tachibana, T. (1988) Highly efficient procedure for production of human monoclonal antibodies: Establishment of hybrids between Epstein-Barr virus-transformed B lymphocytes and hetermyeloma cells by use of GIT culture medium. *Tohoku Journal of Experimental Medicine,* 154, 345–355.

Lazar, G. A., Dang, W., Karki, S., Vafa, O., Peng, J. S., Hyun, L., Chan, C., Chung, H. S., Eivazi, A., Yoder, S. C., Vielmetter, J., Carmichael, D. F., Hayes, R. J. & Dahiyat, B. I. (2006) Engineered antibody Fc variants with enhanced effector funciton. *Proceedings of the National Academy of Sciences of the USA,* 103, 4005–4010.

Lazar, G. L., Desjarlais, J. R., Jacinto, J., Karki, S. & Hammond, P. W. (2007) A molecular immunology approach to antibody humanization and functional optimization. *Molecular Immunology,* 44, 1986–1998.

Lee, J. E. & Raines, R. T. (2008) Ribonucleases as novel chemotherapeutics: The ranpirnase example. *BioDrugs,* 22, 53–58.

Lefranc, M., Giudicelli, V., Kaas, Q., Duprat, E., Jabado-Michaloud, J., Scaviner, D., Ginestoux, C., Clement, O., Chaume, D. & Lefranc, G. (2005) IGMT, the international ImMunoGeneTics information system. *Nucleic Acids Research,* 33, D593–D597.

Lewis phillips, G. D., Li, G., Dugger, D. L., Crocker, L. M., Parsons, K. L., Mai, E., Blattler, W. A., Lambert, J. M., Chari, R. V., Lutz, R. J., Wong, W. L., Jacobson, F. S., Koeppen, H., Schwall, R. H., Kenkare-Mitra, S. R., Spencer, S. D. & Sliwkowski, M. X. (2008) Targeting HER2-positive breast cancer with trastuzumab-DM1, an antibody-cytotoxic drug conjugate. *Cancer Research,* 68, 9280–9290.

Li, J., Sai, T., Berger, M., Chao, W., Davidson, D., Deshmukh, G., Drozdowski, B, Ebel, W., Harley, S., Henry, M., Jacob, S., Yao, J., Zhou, Y., Kavuru, M., Bonfield, T., Thomassen, M. J., Sass, P. M., Nicolaides, N. C. & Grasso, L. (2006b) Human antibodies for immunotherapy development generated via a human B cell hybridoma technology. *Proceedings of the National Academy of Sciences of the USA,* 103, 3557–3562.

Li, H., Sethuraman, N., Stadheim, T. A., Zha, D., Prinz, B., Ballew, N., Bobrowicz, P., Choi, B., Cook, W. J., Cukan, M., Houston-Cummings, N. R., Davidson, R., Gong, B., Hamilton, S. R., Hoopes, J. P., Jiang, Y., Kim, N., Mansfield, R., Nett, J. H., Rios, S., Strawbridge, R., Wildt, S. & Gerngross, T. U. (2006a) Optimization of humanized IgGs in glycoengineered Pichia pastoris. *Nature Biotechnology,* 24, 210–215.

Li, L., Turatti, F., Crow, D., Bading, J. R., Anderson, A., Poku, E., Yazaki, P. J., Williams, L. E., Tamvakis, D., Sanders, P., Leong, D., Raubitschek, A., Hudson, P. J., Colcher, D. & Shively, J. E. (2010) Monodispersed DOTA-PEG-conjugated anti-TAG-72 diabody has low kidney uptake and high tumor-to-blood ratios resulting in improved 64Cu PET. *Journal of Nuclear Medicine,* 51, 1139–1146.

Li, L., Yazaki, P. J., Anderson, A., Crow, D., Colcher, D., Wu, A. M., Williams, L. E., Wong, J. Y. C., Raubitschek, A. & Shively, J. E. (2006c) Improved biodistribution and radioimmunoimaging with poly(ethylene glycol)-DOTA-conjugated anti-CEA diabody. *Bioconjugate Chemistry,* 17, 68–76.

Liu, M. A., Kranz, D. M., Kurnick, J. T., Boyle, L. A., Levy, R. & Eisen, H. N. (1985) Heteroantibody duplexes target cells for lysis by cytotoxic T lymphocytes. *Proceedings of the National Academy of Sciences of the USA,* 82, 8648–8652.

Liu, A. Y., Robinson, R. R., Hellstrom, K. E., Murray, E. D., Chang, C. P. & Hellstrom, I. (1987) Chimeric mouse-human IgG1 antibody that can mediate lysis of cancer cells. *Proceedings of the National Academy of Sciences of the USA,* 84, 3439–3443.

Lo, B. K. C. (2004) Antibody humanization by CDR grafting. In Lo, B.K.C. (Ed.) *Methods in Molecular Biology.* Totowa, NJ, Humana Press Inc, 248, 135–159.

Lonberg, N. (2008) Fully human antibodies from transgenic mouse and phage display platforms. *Current Opinion in Immunology,* 20, 450–459.

Lonberg, N., Taylor, L. D., Harding, F. A., Trounstine, M., Higgins, K. M., Schramm, S. R., Kuo, C., Mashayekh, R., Wymore, K., Mccabe, J. G., Munoz-O'Regan, D., O'Donnell, S. L., Lapachet, E. S. G., Bengoechea, T., Fishwild, D. M., Carmack, C. E., Kay, R. M. & Huszar, D. (1994) Antigen-specific human antibodes from mice comprising four distinct genetic modifications. *Nature,* 368, 856–859.

Lund, J., Takahashi, N., Pound, J. D., Goodall, M., Nakagawa, H. & Jefferis, R. (1995) Oligosaccharide-protein interactions in IgG can modulate recognition by Fc gamma receptors. *Federation of American Societies for Experimental Biology,* 9, 115–119.

Lutterbuese, R., Raum, T., Kischel, R., Hoffmann, P., Mangold, S., Rattel, B., Friedrich, M., Thomas, O., Lorenczewski, G., Rau, D., Schaller, E., Herrmann, I., Wolf, A., Urbig, T., Baeuerle, P. A. & Kufer, P. (2010) T cell-engaging BiTE antibodies specific for EGFR potently eliminate KRAS- and BRAF-mutated colorectal cancer cells. *Proceedings of the National Academy of Sciences of the USA,* 107, 12605–12610.

Mack, M., Riethmuller, G. & Kufer, P. (1997) A small bispecific antibody construct expressed as a functional single-chain molecule with high tumor cell cytotoxicity. *Proceedings of the National Academy of Sciences of the USA,* 92, 7021–7025.

Marks, J. D., Hoogenboom, H., Bonnert, T. P., Mccafferty, J., Griffiths, A. D. & Winter, G. (1991) By-passing immunization: Human antibodies from V-gene libraries displayed on phage. *Journal of Molecular Biology,* 222, 581–597.

Marks, J. D., Hoogenboom, H. R., Griffiths, A. D. & Winter, G. (1992) Molecular evolution of proteins on filamentous phage. *Journal of Biological Chemistry,* 267, 16007–16010.

Martin, A. C. R. & Allen, J. (2007) Bioinformatics tools for antibody engineering. In Dubel, S. (Ed.) *Handbook of Therapeutic Antibodies.* Weinheim, Wiley-VCH.

Martin, A. C. & Thornton, J. M. (1996) Structural families in loops of homologous proteins: Automatic classification, modelling and application to antibodies. *Journal Molecular Biology,* 263, 800–815.

Mathew, M. & Verma, R. S. (2009) Humanized immunotoxins: A new generation of immunotoxins for targeted cancer therapy. *Cancer Science,* 100, 1359–1365.

Mazor, Y., Van Blarcom, T., Mabry, R., Iverson, B. L. & Georgiou, G. (2007) Isolation of engineered, full-length antibodies from libraries expressed in Escherichia coli. *Nature Biotechnology,* 25, 563–565.

Mccafferty, J., Griffiths, A. D., Winter, G. & Chiswell, D. (1990) Phage antibodies: Filamentous phage displaying antibody variable domains. *Nature,* 348, 552–554.

Menzel, C., Schirrmann, T., Hust, M., Konthur, Z., Jostock, T. & Dubel, S. (2008) Human antibody RNase fusion protein targeting CD30+ lymphomas. *Blood,* 111, 3830–3837.

Mian, I. S., Bradwell, A. R. & Olson, A. J. (1991) Structure, function and properfies of antibody binding sites. *Journal of Molecular Biology,* 217, 133–151.

Mikulski, S. M., Costanzi, J. J., Vogelzang, N. J., Mccachren, S., Taub, R. N., Chun, H., Mittelman, A., Panella, T., Puccio, C., Fine, R. & Shogen, K. (2001) Phase II trial of a single weekly intravenous dose of ranpirnase in patients with unresectable malignant mesothelioma. *Journal of Clinical Oncology,* 20, 274–281.

Morgan, A., Jones, N. D., Nesbitt, A. M., Chaplin, L., Bodmer, M. W. & Emtage, J. S. (1995) The N-terminal end of the CH2 domain of chimeric human IgG1 anti-HLA-DR is necessary for C1q, FcgammaRI and FcgammaRIII binding. *Immunology,* 86, 319–324.

Morrison, S. L., Johnson, M. J., Herzenberg, L. A. & Oi, V. T. (1984) Chimeric human antibody molecules: Mouse antigen-binding domains with human constant region domains. *Proceedings of the National Academy of Sciences of the USA,* 81, 6851–6855.

Muller, D., Karle, A., Meiburger, B., Hofig, I., Stork, R. & Kontermann, R. E. (2007) Improved pharmacokinetics of recombinant bispecific antibody molecules by fusion to human serum albumin. *Journal of Biological Chemistry,* 282, 12650–12660.

Natsume, A., In, M., Takamura, H., Nakagawa, T., Shimizu, Y., Kitajima, K., Wakitani, M., Ohta, S., Satoh, M., Shitara, K. & Niwa, R. (2008) Engineered antibodies of IgG1/IgG3 mixed isotype with enhanced cytotoxic activities. *Cancer Research,* 68, 3863–3872.

Natsume, A., Niwa, R. & Satoh, M. (2009) Improving effector functions of antibodies for cancer treatment: Enhancing ADCC and CDC. *Drug Design, Development and Therapy,* 3, 7–16.

Newton, D. L., Hansen, H. J., Mikulski, S. M., Goldenberg, D. M. & Rybak, S. M. (2001) Potent and specific antitumor effects of an anti-CD22-targeted cytotoxin ribonuclease: Potential for the treatment of non-Hodgkin lymphoma. *Blood,* 97, 528–535.

Nguyen, A., Reyes, A. E., 2nd, Zhang, M., Mcdonald, P., Wong, W. L., Damico, L. A. & Dennis, M. S. (2006) The pharmacokinetics of an albumin-binding Fab (AB.Fab) can be modulated as a function of affinity for albumin. *Protein Engineering, Design & Selection,* 19, 291–7.

Nicolaides, N. C., Ebel, W., Kline, B., Chao, Q., Routhier, E., Sass, P. M. & Grasso, L. (1995) Morphogenics as a tool for target discovery and drug development. *Annals of the New York Academy of Sciences,* 1059, 86–96.

Nielsen, L. S., Baer, A., Müller, C., Gregersen, K., Mønster, N. T., Rasmussen, S. K., Weilguny, D. & Tolstrup, A. B. (2010) Single-batch production of recombinant human polyclonal antibodies. *Molecular Biotechnology,* 45, 257–266.

North, B. H., Lehmann, A. & Dunbrack, R. L. (2011) A new clustering of antibody loop conformations. *Journal Moleculer Biology*, 406, 228–256. doi:10.1016/j.jmb.2010.10.030. PMID 21035459.

Ober, R. J., Martinez, C., Lai, X., Zhou, J. & Ward, E. S. (2004a) Exocytosis of IgG as mediated by the receptor, FcRn: An analysis at the single-molecule level. *Proceedings of the National Academic of Sciences of the USA,* 101, 11076–11081.

Ober, R. J., Martinez, C., Vaccaro, C., Zhou, J. & Ward, E. S. (2004b) Visualizing the site and dynamics of IgG salvage by the MHC class I-related receptor, FcRn. *Journal Immunology,* 172, 2021–2029.

Olsson, L., Andreasen, R. B., Ost, A., Christensen, B. & Biberfeld, P. (1984) Antibody producing human-human hybridomas II: Derivation and characterization of an antibody specific for human leukemia cells. *Journal of Experimental Medicine,* 159, 537–550.

Onda, M., Beers, R., Xiang, L., Nagata, S., Wang, Q. & Pastan, I. (2008) An immunotoxin with greatly reduced immunogenicity by identification and removal of B cell epitopes. *Proceedings of the National Academy of Sciences of the USA,* 105, 11311–11316.

Onda, M., Nagata, S., Fitzgerald, D. J., Beers, R., Fisher, R. J., Vincent, J. J., Lee, B., Nakamura, M., Hwang, J., Kreitman, R. J., Hassan, R. & Pastan, I. (2006) Characterization of the B cell epitopes associated with a truncated form of pseudomonas exotoxin (PE38) used to make immunotoxins for the treatment of cancer patients. *Journal of Immunology,* 177, 8822–8834.

Osada, T., Hsu, D., Hammond, S., Hobeika, A., Devi, G., Clay, T. M., Lyerly, H. K. & Morse, M. A. (2010) Metastatic colorectal cancer cells from patients previously treated with chemotherapy are sensitve to T-cell killing mediated by CEA/CD3-bispecific T-cell-engaging BiTE antibody. *British Journal of Cancer,* 102, 124–133.

Padlan, E. A. (1991) A possible procedure for reducing the immunogenicity of antibody variable domains while preserving their ligand-binding properties. *Molecular Immunology,* 28, 489–498.

Pasut, G. & Veronese, F. M. (2009) PEG conjugates in clinical development or use as anticancer agents: An overview. *Advanced Drug Delivery Reviews,* 61, 1177–1188.

Persson, M. A. A., Caothien, R. H. & Burton, D. R. (1991) Generation of diverse high-affinity human monoclonal antibodies by repertoire cloning. *Proceedings of the National Academy of Sciences of the USA,* 88, 2432–2436.

Petersdorf, L., Kopecky, K., Stuart, R. K. & Al., E. (2009) Preliminary results of southwest oncology group study S0106L an international intergroup phase 3 randomized trial comparing the addition of gemtuzumab ozogamicin to standard induction therapy versus standard induction therapy followed by a second randomization to post-consolidation gemtuzumab ozogamicin versus no additiona therapy for previously untreated acute myeloid leukemia. *Blood,* 114, 790.

Petkova, S. B., Akilesh, S., Sproule, T. J., Christianson, G. J., Khabbaz, H. A., Brown, A. C., Presta, L. G., Meng, Y. G. & Roopenian, D. C. (2006) Enhanced half-life of genetically engienered human IgG1 antibodies in a humanized FcRn mouse model: Potential application in humorally mediated autoimmune disease. *International Immunology,* 18, 1759–1769.

Posey, J. A., Khazaeli, M. B., Bookman, M. A., Nowrouzi, A., Grizzle, W. E., Thornton, J., Carey, D. E., Loenz, J. M., Sing, A. P., Siegall, C. B., Lobuglio, A. F. & Saleh, M. N. (2002) A phase I trial of the single-chain immunotixin SGN-10 (BR96 scFv-PE40) in patients with advanced solid tumors. *Clinical Cancer Research,* 8, 3092–3099.

Raghavan, M. & Bjorkman, P. J. (1996) Fc receptors and their interactions with immunoglobulins. *Annual Review of Cell and Developmental Biology,* 12, 181–220.

Raju, T. S. (2003) Glycosylation variations with expression systems and their impact on biological activity of therapeutic immunoglobulins. *Bioprocess International,* 1, 44–53.

Reiter, Y. & Pastan, I. (1998) Recombinant Fv immunotoxins and Fv fragments as novel agents for cancer therapy and diagnosis. *Trends in Biotechnology,* 16, 513–520.

Riechmann, L., Clark, M., Waldmann, H. & Winter, G. (1988) Reshaping human antibodies for therapy. *Nature,* 332, 323–327.

Risberg, K., Fodstad, O. & Andersson, Y. (2010) Immunotoxins: A promising treatment modality for metastatic melanoma? *Ochsner Journal,* 10, 193–199.

Robinson, M. K., Alpaugh, R. K. & Borghaei, H. (2010) Naptumomab estafenatox: A new immunoconjugate. *Expert Opinion on Biological Therapy,* 10, 273–9.

Roguska, M. A., Pedersen, J. T., Henry, A. H., Searle, S. M. J., Roja, C. M., Avery, B., Hoffee, M., Cook, S., Lamber, J. M., Blattler, W. A., Rees, A. R. & Guild, B. C. (1996) A comparison of two murine monoclonal antibodies humanized by CDR-grafting and variable domain resurfacing. *Protein Engineering,* 9, 895–904.

Roscoe, D. M., Pai, L. H. & Pastan, I. (1997) Identification of epitopes on a mutant form of Pseudomonas exotoxin using serum from humans treated with Pseudomonas exotoxin containing immunotoxins. *European Journal of Immunology,* 27, 1459–1468.

Rosok, M. J., Yelton, D. E., Harris, L. J., Bajorath, J., Hellstrom, K. E., Hellstrom, I., Cruz, G. A., Kristensson, K., Lin, H., Huse, W. D. & Glaser, S. M. (1996) A combinatorial library strategy for the rapid humanization of anticarcinoma BR96 Fab. *Journal Biology Chemistry,* 271, 22611–22618.

Rothe, C., Ulrlinger, S., Lohning, C., Prassler, J., Stark, Y., Jager, U., Hubner, B., Bardroff, M., Pradel, I., Boss, M., Bittlingmaier, R., Bataa, T., Frisch, C., Brocks, B., Honegger, A. & Urban, M. (2008) The human combinatorial antibody library HuCAL GOLD combines diversification of all six CDRs according to the natural immune system with a novel display method for efficient selection of high-affinity antibodies. *Journal of Molecular Biology,* 376, 1182–1200.

Saldanha, J. W. (2009) *Humanization of Recombinant Antibodies.* New York, NY, Cambridge University Press.

Saldanha, J. W., Martin, A. C. & Leger, O. J. (1999) A single backmutation in the human kIV framework of a previously unsuccessfully humanized antibody restores the binding activity and increases the secretion in cos cells. *Molecular Immunology,* 36, 709–719.

Salfeld, J. G. (2007) Isotype selection in antibody egineering. *Nature Biotechnology,* 25, 1369–1372.

Salfeld, J. G., Allen, D. J., Hoogenboom, H. R. J. M., Kaymakcalan, Z., Labkovsky, B., Mankovich, J. A., Cfuinness, B. T., Roberts, A. J., Sakorafas, P., Schoenhaut, D., Vaughan, T. J., White, M. & Wilton, A. J. (2000) Human antibodies that bind human TNFalpha, patent number US 6258562 B1, 10, July 2001.

Salfeld, J. G., Zaymakcalan, Z., Tracey, D., Roberts, A. & Kamen, R. (1998) Generation of fully human anti-TNF antibody D2E7 [Abstract]. *Arthritis and Rheumetism,* 41, S57.

Santanche, S., Bellelli, A. & Brunori, M. (1997) The unusual stability of saporin, a candidate for the synthesis of immunotoxins. *Biochemical and Biophysical Research Communications,* 234, 129–132.

Schirrmann, T. & Hust, M. (2010) Construction of human antibody gene libraries and selection of antibodies by phage display. In Yotnda, P. (Ed.) *Immunotherapy of Cancer.* New York, NY, Humana Press.

Schmidt, M. M. & Wittrup, K. D. (2009) A modeling analysis of the effects of molecular size and binding affinity on tumor targeting. *Molecular Cancer Therapy,* 8, 2861–2871.

Schroff, R. W., Foon, K. A., Beatty, S. M., Oldham, R. K. & Morgan, A. C. (1985) Human anti-murine immunoglobulin in response in patients receiving monoclonal antibody therapy. *Cancer Research,* 45, 879–885.

Schwarz, F., Huang, W., Li, C., Schulz, B. L., Lizak, C., Palumbo, A., Numao, S., Neri, D., Aebi, M. & Wang, L. X. (2010) A combined method for producing homogeneous glycoproteins with eukaryotic N-glycosylation. *National Chemistry Biology,* 6, 264–266.

Shawler, D. L., Bartholomew, R. M., Smith, L. M. & Dillman, R. O. (1985) Human immune response to multiple injections of murine monoclonal IgG1. *Journal of Immunology,* 135, 1530–1535.

Shields, R. L., Lai, J., Keck, R., O'Connell, L. Y., Hong, K., Meng, Y. G., Weikert, S. H. A. & Presta, L. G. (2002) Lack of fucose on human IgG1 N-linked oligosaccharide improves binding to human FcgammaRIII and antibody-dependent cellular toxicity. *Journal of Biological Chemistry,* 277, 36733–36740.

Shields, R. L., Namenuk, A. K., Hong, K., Meng, Y. G., Rae, J., Briggs, J., Xie, D., Lai, J., Stadlen, A., Li, B., Fox, J. A. & Presta, L. G. (2001) High resolution mapping of the binding site on human IgG1 for Fc gamma RI, Fc gamma RII, Fc gamma RIII, and FcRn and design of IgG1 variants with improved binding to the Fc gamma R. *Journal Biology Chemistry,* 276, 6591–6604.

Sikora, K., Alderson, T., Ellis, J., Phillips, J. & Watson, J. (1983) Human hybridomas from patients with malignant disease. *British Journal of Cancer,* 47, 135–145.

Smith, G. P. (1985) Filamentous fusion phage: Novel expression vectors that display cloned antigens on the virion surface. *GP Smith,* 228, 1315–1317.

Sola, R. J. & Griebenow, K. (2010) Glycosylation of therapeutic proteins: An effective strategy to optimize efficacy. *BioDrugs,* 24, 9–21.

Sola, R. J., Rodriguez-Martinez, J. A. & Griebenow, K. (2007) Modulation of protein biophysical properties by chemical glycosylation: Biochemical insights and biomedical implications. *Cellular and Molecular Life Sciences,* 64, 2133–2152.

Steinitz, M., Klein, G., Koskimies, S. & Makel, O. (1977) EB virus-induced B lymphocyte cell lines producing specific antibody. *Nature,* 269, 420–422.

Tan, A. R., Moore, D. F., Hidalgo, M., Doroshow, J. H., Poplin, E. A., Goodin, S., Mauro, D. & Rubin, E. H. (2006) Pharmacokinetics of cetuximab after administration of escalating single dosing and weekly fixed dosing in patients with solid tumors. *Clinical Cancer Research,* 12, 6517–6522.

Tangri, S., Mothe, B. R., Eisenbraun, J., Sidney, J., Southwood, S., Briggs, K., Zinckgraf, J., Newman, M., Chesnut, R., Licalsi, C. & Sette, A. (2005) Rationally engineered therapeutic proteins with reduced immunogenicity. *Journal of Immunology,* 174, 3187–3196.

Teng, N. N. H., Lam, K. S., Riera, F. C. & Kaplan, H. S. (1983) Construction and testing of mouse-human heteromyelomas for human monoclonal antibody production. *Proceedings of the National Academy of Sciences of the USA,* 80, 7308–7312.

Todorovska, A., Roovers, R. C., Dolezal, O., Kortt, A. A., Hoogenboom, H. R. & Hudson, P. J. (2001) Design and application of diabodies, triabodies and tetrabodies for cancer targeting. *Journal of Immunological Methods,* 248, 47–66.

Tolcher, A. W., Ochoa, L., Hammond, L. A., Patnaik, A., Edwards, T., Takimoto, C., Smith, L., De Bono, J., Schwartz, G., Mays, T., Jonak, Z. L., Johnson, R., Dewitte, M., Martino, H., Audette, C., Maes, K., Chari, R. V., Lambert, J. M. & Rowinsky, E. K. (2003) Cantuzumab mertansine, a maytansinoid immunoconjugate directed to the CanAg antigen: A phase I, pharmacokinetic, and biologic correlative study. *Journal Clinical Oncology,* 21, 211–222.

Trussel, S., Dumelin, C., Frey, K., Villa, A., Buller, F. & Neri, D. (2009) New strategy for the extension of the serum half-life of antibody fragments. *Bioconjugate Chemistry,* 20, 2286–2292.

Tsutsumi, Y., Onda, M., Nagata, S., Lee, B., Kreitman, R. J. & Pastan, I. (2000) Site-specific chemical modification with polyethylene glycol of recombinant immunotoxin anti-Tac(Fv)-PE38 (LMB-2) improves antitumor activity and reduces animal toxicity and immunogenicity. *Proceedings of the National Academy of Sciences of the USA,* 97, 8548–8553.

Umana, P., Jean-Mairet, J., Moudry, R., Amstutz, H. & Bailey, J. E. (1999) Engineered glycoforms of an antineuroblastoma IgG1 with optimized antibody-dependent cellular cytotoxic activity. *Nationl Biotechnology,* 17, 176–180.

Vaccaro, C., Zhou, J., Ober, R. J. & Ward, E. S. (2005) Engineering the Fc region of immunoglobulin G to modulate in vivo antibody levels. *National Biotechnology,* 23, 1283–1288.

Van Cutsem, E., Peeters, M., Siena, S., Humblet, Y., Hendlisz, A., Neyns, B., Canon, J. L., Van Laethem, J. L., Maurel, J., Richardson, G., Wolf, M. & Amado, R. G. (2007) Open-label phase III trial of panitumumab plus best supportive care compared with best supportive care alone in patients with chemotherapy-refractory metastatic colorectal cancer. *Journal of Clinical Oncology,* 25, 1658–1664.

Van Meerten, T., Van Rijn, R. S., Hol, S., Hagenbeek, A. & Ebeling, S. B. (2006) Complement-induced cell death by Rituximab depends on CD20 expression level and acts complementary to antibody-dependent cellular toxicity. *Clinical Cancer Research,* 12, 4027–4035.

Vegarud, G. & Christensen, T. B. (1975) The resistance of glycoproteins to proteolytic inactivation. *Acta Chemica Scandinavica B,* 29, 887–888.

Vogel, C. L., Burris, H. A., Limentani, S., Borson, R., O'Shaughnessy, J., Vukelja, S., Agresta, S., Klencke, B., Birkner, M. & Rugo, H. (2009) A phase II study of trastuzumab-DM1 (T-DM1), a HER2 antibody-drug conjugate (ADC), in patients (pts) with HER2+ metastatic breast cancer (MBC): Final results. *Journal Clinical Oncology,* 27, 1017.

Weinblatt, M. E., Keystone, E. C., Furst, D. E., Moreland, L. W., Weisman, M. H., Birbara, C. A., Teoh, L. A., Fischkoff, S. A. & Chartash, E. K. (2003) Adalimumab, a fully human anti-tumor necrosis factor alpha monoclonal antibody, for the treatment of rheumatoid arthritis in patients taking concomitant methotrexate. *Arthritis and Rheumetism,* 48, 35–45.

Weiner, L. M., Blelldegrun, A. S., Crawford, J., Tolcher, A. W., Lockbaum, P., Arends, R. H., Anavale, L., Amado, R. G., Schwab, G. & Figlin, R. A. (2008) Dose and schedule study of Panitumumab monotherapy in patients with advanced solid malignancies. *Clinical cancer Research,* 14, 502–508.

Weiner, L. M., Holmes, M., Richeson, A., Godwin, A., Adams, G. P., Hsieh-Ma, S. T., Ring, D. B. & Alpaugh, R. K. (1993) Binding and cytotoxicity characteristics of the bispecific murine monoclonal antibody 2B1. *Journal of Immunology,* 151, 2877–2886.

Wochner, R. D., Strober, W. & Waldmann, T. A. (1967) The role of the kidney in the catabolism of Bence Jones proteins and immunoglobulin fragments. *Journal of Experimental Medicine,* 126, 207–221.

Wolf, E., Hofmeister, R., Kufer, P., Schlereth, B. & Baeuerle, P. A. (2005) BiTEs: Bispecific antibody constructs with unique anti-tumor activity. *Drug Discovery Today,* 10, 1237–1244.

Wright, A. & Morrison, S. L. (1994) Effect of altered CH2-associated carbohydrate structure on the funcitonal properties and in vivo fate of chimeric mouse-human immunoglobulin G1. *Journal of Experimental Medicine,* 180, 1087–1096.

Yazaki, P. J., Kassa, T., Cheung, C., Crow, D. M., Sherman, M. A., Bading, J. R., Anderson, A. J., Colcher, D. & Raubitschek, A. (2008) Biodistribution and tumor imaging of an anti-CEA single-chain antibody-albumin fusion protein. *Nuclear Medicine and Biology,* 35, 151–158.

Yokota, T., Milenic, D. E., Whitlow, M. & Schlom, J. (1992) Rapid tumor penetration of a single-chain Fv and comparison with other immunoglobulin forms. *Cancer Research,* 52, 3402–3408.

Youn, Y. S., Na, D. H., Yoo, S. D., Song, S. & Lee, K. C. (2005) Carbohydrate-specifically polyethylene glycol-modified ricin A-chain with improved therapeutic potential. *International Journal of Biochemistry and Cell Biology,* 37, 1525–1533.

Zalevsky, J., Chamberlain, A. K., Horton, H. M., Karki, S., Leung, I. W. L., Sproule, T. J., Lazar, G. A., Roopenian, D. C. & Desjarlais, J. R. (2010) Enhanced anitbody half-life improves *in vivo* activity. *Nature Biotechnology,* 28, 157–159.

Zent, C. S., Secreto, C. R., Laplant, B. R., Bone, N. D., Call, T. G., Shanafelt, T. D., Jelinek, D. F., Tschumper, R. C. & Kay, N. E. (2008) Direct and complement dependent cytotoxicity in CLL cells from patients with high risk early stage chronic lyphocytic leukemia (CLL) treated with alemtuzumab and rituximab. *Leukemia Research,* 32, 1849–1856.

Zhu, H., Yang, B., Yang, X., Wang, L., Xu, J., Liao, C., Feng, Q., Tang, H., Hu, L., Chen, Z. & Li, Y. (2009) A novel antibody fragment targeting HAb18G/CD147 with cytotoxicity and decreased immunogenicity. *Cancer Biology & Therapy,* 8, 1035–1044.

Chapter 11
Recombinant Bispecific Antibodies for Cancer Therapy

Dafne Müller and Roland E. Kontermann

Abstract Bispecific antibodies are molecules capable of simultaneously binding to two different antigens. While initially bispecific antibodies have been developed mainly for cellular cancer immunotherapy through retargeting of effector cells to tumor cells, recent developments include also dual targeting strategies and the retargeting of effector molecules, e.g. in radioimmunotherapy. In addition to various applications, a plethora of bispecific antibody formats have been developed, including small recombinant bispecific molecules comprising only the variable domains of two antibodies and tetravalent IgG-like bispecific antibodies. In this chapter we will focus on the most relevant bispecific antibody formats, explaining the therapeutic concepts, the methodology involved in their production and the preclinical and clinical achievements, so far.

11.1 Introduction

Bispecific antibodies are artificially generated molecules capable of binding simultaneously two different epitopes, either on the same antigen or on different antigens. Thus, from a therapeutic point of view, improved targeting with blocking or neutralizing activity as well as retargeting of effector cells or effector molecules is possible, opening up a broad spectrum of therapeutic strategies, especially for cancer therapy (Segal et al., 1999; Müller & Kontermann, 2007; Chames & Baty, 2009). Over the years, seeking for improved efficacy and selectivity, advances in antibody engineering have led to the development of diverse bispecific antibody formats. Although diverse drawbacks left many approaches at the preclinical stage, an increasing number of bispecific antibodies have meanwhile entered clinical trials, indicating a growing interest for these molecules in the field (Müller & Kontermann, 2010).

D. Müller and R.E. Kontermann (✉)
Institut für Zellbiologie und Immunologie, Universität Stuttgart, 70569 Stuttgart, Germany
e-mail: roland.kontermann@izi.uni-stuttgart.de

M. Al-Rubeai (ed.), *Antibody Expression and Production*, Cell Engineering 7,
DOI 10.1007/978-94-007-1257-7_11, © Springer Science+Business Media B.V. 2011

11.2 Bispecific Antibodies by Somatic Hybridization

Initially, bispecific antibodies were generated by somatic hybridization, taking advantage of the hybridoma technology developed by Köhler and Milstein in 1975. Hybridomas are producer cell lines of monoclonal antibodies, generated by the fusion of a B cell and a myeloma cell line, contributing antibody specificity and immortality, respectively (Köhler & Milstein, 1975). Two hybridomas, producing antibodies of different binding specificity can be further fused, generating a hybrid-hybridoma, i.e quadroma (Milstein & Cuello, 1983). In these cells, heavy and light chains of both antibodies are produced, assembling randomly into IgG molecules with an average of 10% being bispecific. The formation of bispecific antibodies can be further increased, taking into account preferential species-restricted heavy/light chain pairing, thus employing hybridomas of different species (mouse/rat). Nevertheless, purification steps in form of e.g. affinity chromatography (Protein A) and ionic exchange chromatography are always required to obtain homogeneous bispecific antibody preparations (Lindhofer et al., 1995).

For cancer therapy, hybrid-hybridoma antibodies were conceived that retarget immune effector cells to tumor cells. TRION Pharma together with Fresenius Biotech developed hybrid mouse/rat IgG molecules with specificity for a tumor-associated antigen (EpCAM, Her2, CD20) and CD3 (Shen & Zhu, 2008; Kiewe & Thiel, 2008; Stanglmaier et al., 2008). Considering that the Fc part constitutes also a functional unit of the molecule, these antibodies are trifunctional (triomabs). In combination with tumor targeting, binding to CD3 enables triggering of T cells bypassing MHC restriction, i.e. recruiting effector cells of the immune system that are normally not involved in an antibody-mediated immune response. In addition, binding of the antibody Fc part to FcγRI, FcγRIIA and FcγRIII activates accessory cells (e.g. NK cells, macrophages and dendritic cells). Thus, diverse effector mechanisms as T cell-mediated lysis, antibody-dependent cell-mediated cytotoxicity (ADCC) and phagocytosis can be elicited, contributing to the tumor destruction. Thereby, anti-tumor effects are also thought to be supported by concomitant cytokine release (e.g. TNF-α, IFN-γ) and the induction anti-tumor immunity (Seimetz et al., 2010; Kiewe & Thiel, 2008).

Catumaxomab (EpCAM x CD3) is the first and so far only bispecific antibody on the market. It has been approved 2009 in the European Union (EU) for the intraperitoneal (i.p.) treatment of malignant ascites in patients with EpCAM-positive carcinomas, in circumstances where standard therapy is not available or no longer feasible. In the pivotal phase II/III clinical trial catumaxomab was administrated after paracentesis to patients diagnosed with malignant ascites due to EpCAM-positive epithelial cancer. Catumaxomab was administered i.p. on days 0, 3, 7 and 10 at escalating doses of 10, 20, 50 and 150 μg, respectively, leading to a significant prolongation of the puncture-free survival. Most common adverse advents observed were the cytokine-release-related-symptoms (CRRS) pyrexia, nausea and vomiting that were mostly classified as mild to moderate and fully reversible. Development of human anti-mouse or anti-rat antibodies (HAMA/HARA) was detected in the majority of the patients after the treatment. Clinical studies with catumaxomab for other indications than malignant ascites, e.g. ovarian and gastric cancer, are in progress (Shen & Zhu, 2008; Seimetz et al., 2010). Other triomabs, targeting Her2 (ertumaxomab) and CD20 (FBTA05) on tumor cells have also reached clinical trials (phase I/II) being tested in metastatic breast cancer and B-cell lymphoma patients, respectively.

11.3 Bispecific Antibodies by Chemical Conjugation

Bispecific antibodies can also be generated by chemical conjugation. For example heteroconjugation of monoclonal antibodies with different specificity can be employed to generate trifunctional antibody constructs composed of two whole immunoglobulins (Lum et al., 2006). Therefore, in a first step, a monoclonal antibody of specificity A is treated with the crosslinker Sulfo-SMCC (sulfosuccinimidyl 4-[N-maleimidomethyl]cyclohexane-1-carboxylate) that reacts with primary amines introducing reactive maleimide groups. In parallel, a monoclonal antibody with specificity B is treated with Traut's reagent which reacts also with primary amines but, introduces sulphydryl (-SH) groups. After purification by chromatography, both antibody preparations are mixed at equimolar ratios and conjugation takes place by the reaction between maleimide and sulfhydryl groups of the antibodies forming stable thioether bonds. Thus, approximately 20–30% active dimer formation was reported, whereby the percentage of monomers and multimers varied considerably (Reusch et al., 2006; Gall et al., 2005). By this means, several bispecific antibody molecules with specificity for a tumor associated antigen (EGFR, Her2, CD20, CA125) and CD3 have been generated (Lum et al., 2006; Grabert et al., 2006; Chan et al., 2006; Gall et al., 2005). Due to the multivalent character (two binding sites for each antigen and two Fc parts per molecule) of these bispecific antibodies, the therapeutic strategy stipulate first the activation and expansion of T cells *ex-vivo*, which are then re-armed with the bispecific antibody in vitro, before an adoptive transfer back into the patient. Thus, the number of targeted effector cells was expected to be increased and the risk of toxicity associated to the systemic administration to be reduced.

In vitro, enhanced tumor cell directed cytotoxicity and cytokine secretion mediated by antibody armed ATCs (activated T cells) in comparison to unarmed ATCs was shown in settings with healthy donor T cells and tumor cell lines, as well as with patient material (i.e. primary tumor and effector cells) (Sen et al., 2001; Reusch et al., 2006; Chan et al., 2006). Furthermore, in an ovarian xenograft mouse model, i.p. treatment with bispecific antibody (Her2 x CD3 or CA125 x CD3) armed ATCs led to reduced tumor burden and increased survival (Chan et al., 2006). Also, for certain colon xenograft mouse model, enhanced tumor growth inhibition was observed if ATCs were armed with a respective bispecific antibody (EGFR x CD3) before i.v. application (Reusch et al., 2006). Currently a clinical phase I trial focuses on the evaluation of the CD20Bi (CD20 x CD3) antibody for the treatment of high-risk, refractory or relapsed $CD20^{+}$ non-Hodgkin's lymphomas. Here, infusion of bispecific antibody re-armed ATCs, following myeloablative therapy and autologous peripheral blood stem cell transplantation is expected to account for immune consolidation.

Alternatively, bispecific antibodies can also be produced by chemical conjugation of defined antibody fragments or modules. For example, Fab' fragments can be generated by pepsin digestion from each monoclonal antibody, which are combined by chemical conjugation in a second step, creating bsF(ab')$_2$ molecules. These antibody constructs are exempt from the Fc part and therefore bifunctional and significantly smaller than a complete IgG molecule. Chemical conjugation can be achieved by using thiol-reactive homobifunctional crosslinkers such as DTNB and o-PDM leading either to disulfide bonds or highly stable thioether bonds, respectively (Graziano & Guptill, 2004).

11.4 Recombinant Bispecific Antibodies

Advances in genetic engineering provided the tools for the generation and defined combination of variable and constant heavy and light chain domains of antibodies, thus enabling the creation of a great variety of recombinant bispecific antibody formats. They can be roughly classified into two categories: (i) formats resulting from the combination of variable regions only and (ii) formats combining variable regions with constant domains.

Main representatives of the first category are tandem scFv (taFv), diabodies (Db) and single-chain diabodies (scDbs), as well as derivatives thereof. They are composed of the V_H and V_L region of two antibodies with specificity A and B, thus reducing the bispecific antibody to its binding units. The arrangement of the domains and the length of the peptide linkers connecting them define the respective format (Fig. 11.1). Tandem scFv (taFv) are generated by connecting two scFv molecules in series. scFvs in turn consist of a V_H and V_L region connected by an linker of approximately 15 amino acids, folding into an independent binding module. scFvs of specificity A and B are connected via a middle linker (linker M) that can vary in length and composition. Here, 3-residue alanine linker (Brandão et al., 2003), hydrophilic 6-residue linker identified by a phage display approach (Korn et al., 2004), glycine-serine-rich linkers (Kufer et al., 1997; McCall et al., 2001), linkers adopting a helical structure (Hayden et al., 1994) and linkers derived from various natural interconnecting sequences from immunoglobulins or immunoglobulin-like molecules (Grosse-Hovest et al., 2004; Ren-Heidenreich et al., 2004) have been reported. In contrast to tandem scFv, which are monomeric molecules, diabodies are heterodimeric molecules. In the diabody format the variable domains are arranged in two chains V_HA–V_LB and V_HB–V_LA (or V_LA–V_HB and V_LB–V_HA), in which the linker connecting the variable domains is reduced to a length of 5 amino acids (e.g. G_4S). Thus, assembling of V_H and V_L from the same chain is sterically precluded, favoring interchain pairing in a head-to-tail orientation, resulting in non-covalently bound dimers (Holliger et al., 1993). Heterodimer formation retrieves active bispecific antibody molecules, nevertheless inactive homodimer formation cannot be excluded as well. Attempts to improve

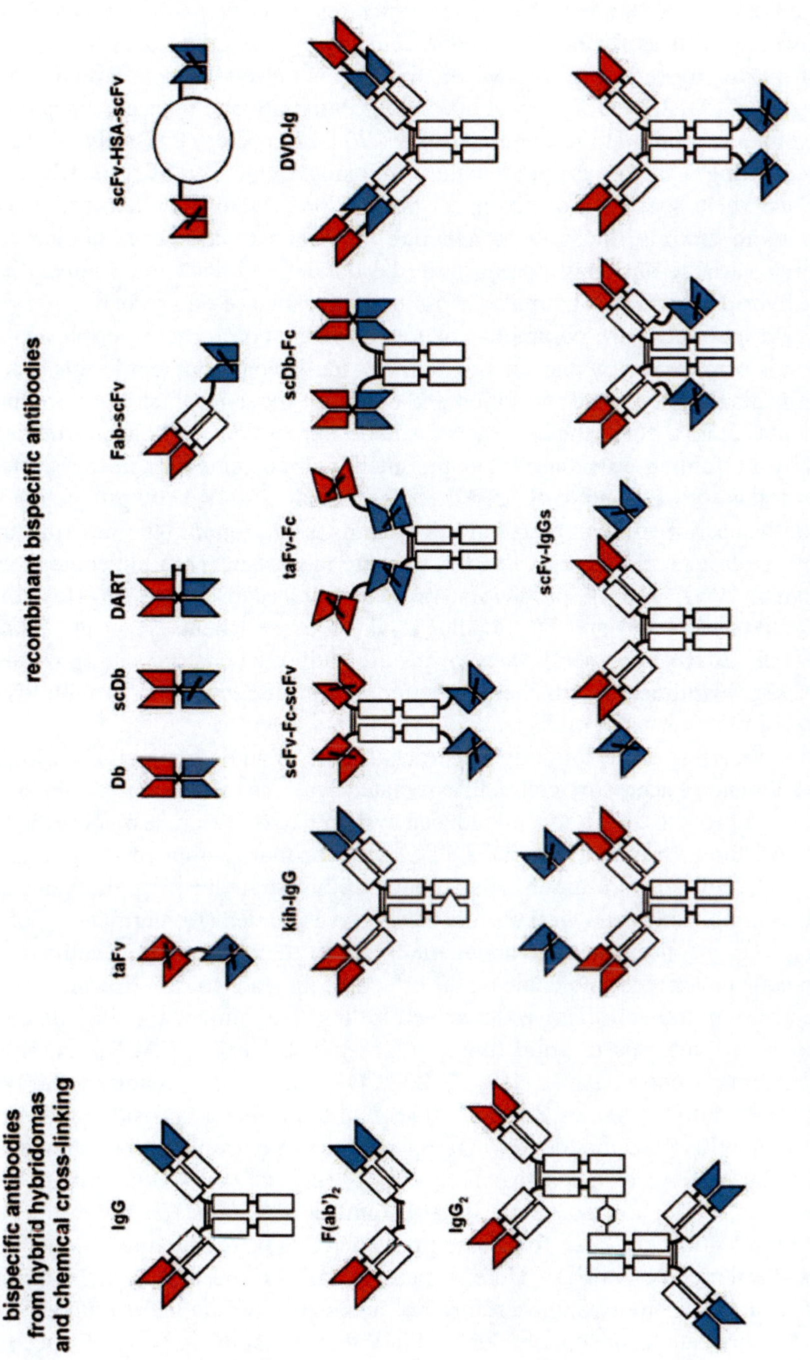

Fig. 11.1 Examples of bispecific antibodies generated from hybrid hybridoma, chemical crosslinking of two antibodies, or by genetic engineering

heterodimer formation include the introduction of interchain disulphide bounds (FitzGerald et al., 1997), knobs-into-holes structures in the V_H–V_L interface (Zhu et al., 1997) as well as fusion of cysteine-containing C-terminal tails to the two chains of a diabody resulting in covalent linkage of the two chains ("dual affinity re-targeting", DART) (Johnson et al., 2010). Alternatively, both chains can be connected by an additional middle linker of 15–20 amino acids (Brüsselbach et al., 1999), generating a monomeric molecule in the single-chain diabody (scDb) format. Db and scDb share a more compact disposition, whereas tandem scFvs are generally more flexible. In order to generate these constructs, genes coding for the required variable antibody domains can be obtained either from sources like cDNA of hybridomas and combinatorial antibody libraries or be synthetically synthesized. Db and scDb are commonly expressed in E. coli, where periplasmatic production is predominantly chosen, due to the oxidative environment in this space that permits disulphide bound formation and therefore the correct antibody folding (Kontermann, 2005). Nevertheless, cytoplasmatic expression as inclusion bodies, followed by a refolding procedure to obtain functional molecules has been reported for diabodies as well (Asano et al., 2002; Hayashi et al., 2004). TaFv production is preferentially performed in mammalian cell systems (secretion into the supernatant) for solubility reasons, although expression in bacteria was successful in some cases (Kontermann, 2005). Lately, production in mammalian cells becomes also more often an option for scDb and Db (Müller et al., 2007; Kashentseva et al., 2009; Johnson et al., 2010). In general, these bispecific antibody constructs are provided with tags (e.g. histidine-tag) for their detection and purification (e.g. immobilized metal ion affinity chromatography).

By generating bispecific antibody constructs devoid of an Fc fragment, heterogeneous recruitment of accessory cells and associated cytokine release was aimed to be avoided and improved selectivity and reduced systemic toxicity, i.e. less side effects, expected. All three formats (taFv, Db, scDb) were primarily applied to strategies retargeting effector cells to tumor cells, although other strategies, e.g. the retargeting of effector molecules and viral vectors have been evaluated (Kontermann, 2005). According to this concept, simultaneous binding to a tumor-associated antigen on the tumor cell and a trigger molecule on the effector cell, leads to site-directed effector cell activation and consecutive tumor cell killing. For tumor targeting, diverse tumor-associated antigens of solid tumors (e.g. EGFR, Her2, CEA, EpCAM) as well as leukemic tumors (e.g. CD19, CD20, CD30) have been reported (Müller & Kontermann, 2007; Chames & Baty, 2009). On the effector cell side, although retargeting of NK cells, monocytes/macrophages and neutrophils was shown to be feasible by binding to respective Fc receptors (e.g. CD16, CD64 and CD89) (Bruenke et al., 2005; Johnson et al., 2010; Ranft et al., 2009; Guettinger et al., 2010), main attention has been focused on the approach of retargeting T cells, the most potent killer cells, via CD3. Here, bypassing MHC restriction, T cells, which are exempt of Fc receptors and therefore not accessible for standard monoclonal antibody therapy can be selectively directed to kill tumor cells. All three bispecific antibody formats (taFv, Db, scDb) turned out to be applicable for this strategy; mediating tumor cell killing by T cells in vitro (Kontermann, 2005). In addition, tumor growth inhibition and increased survival was shown for many of them in xenograft tumor mouse models, in vivo (Müller & Kontermann, 2007). Considering that under physiological conditions T cell activation is a strictly regulated event, requiring appropriate costimulation, T cells were generally preactivated, in order to assure full effector cell response. Later on it was shown, that providing costimulation in form of monoclonal antibodies (CD28 specific) (Cochlovius et al., 2000) or antibody fusion proteins with costimulatory ligands (B7, 4-1BBL) (Blanco et al., 2003; Müller et al., 2007, 2008; Liu et al., 2010) could enhance significantly the effect of bispecific antibodies (TAA x CD3) achieved with prestimulated or even unstimulated PBMCs. Furthermore it turned out, that it is possible to generate bispecific antibodies capable to induce efficient tumor cell killing by T cells obtained from unstimulated PBMCs without additional help (Löffler et al., 2000). Here, the BiTE (Bispecific T-cell engager) class molecules developed by Micromet, which belong to the taFv format, are most advanced. In vitro, target-dependent activation of T cells in a costimulatory-independent manner was achieved with high efficiency at low concentrations (picomolar range) and low effector-to-target cell ratio (Wolf et al., 2005). $CD8^+$ as well as $CD4^+$ T cells could be activated to participate in tumor cell killing, inducing apoptosis via the perforin/granzyme B mechanism (Haas et al., 2009). Furthermore it was shown, that BiTEs can induce serial killing by T cells (Hoffmann et al., 2005). Epitope distance to the target cell membrane and antigen size were identified as factors that determine the BiTE efficiency (Bluemel et al., 2010). In vivo, promising anti-tumor effects were shown in xenograft and syngeneic mouse models and even in a non-human primate model (Dreier et al., 2003; Schlereth et al., 2006a, b; Lutterbuese et al., 2009). Currently a growing set of BiTE molecules retargeting T cells (CD3) to diverse solid (EGFR, Her2, CEA, EpCAM, EphA2 (Eph receptor tyrosine kinase A2), MCSP (melanoma-associated chondroitin sul fate proteoglycan)) and leukemic (CD19, CD33) tumors are being tested (Baeuerle & Reinhardt, 2009). Two of them, MT103 (blinatumomab) targeting CD19 and CD3 and MT110 targeting EpCAM and CD3 entered clinical trials. MT103 was evaluated in a phase I study with patients with relapsed B-cell non Hodgkin lymphoma (NHL) (Bargou et al., 2008). Therefore the BiTE was administered by continuous infusion over 4–8 weeks. Partial and complete response was observed (21–100% patients with tumor regression) at very low doses (0.015–0.06 $mg/m^2/day$). Adverse events frequently observed at the beginning of the treatment included mainly leukopenia and/or lymphopenia, chills, pyrexia and transient elevated levels of liver enzymes. Treatment was discontinued in 7 of 40 patients because of symptoms related to the central nervous system (CNS) (confusion, disorientation and speech disorder), that were however fully reversible. Currently, blinatumomab is being tested in a phase II trial in patients with acute B-cell lymphoblastic leukemia (B-ALL) who have shown minimal residual disease after conventional treatment. Here first results indicate BiTE potential for the eradication of rare disseminated tumor cells in the bone marrow, whereas adverse events of the CNS have not been observed so far (Nagorsen et al., 2009). In addition, MT110, targeting EpCAM and CD3 is being evaluated in a clinical phase I trial in patients with late stage lung or gastrointestinal cancers.

All of these recombinant antibody formats (taFv, Db, scDb, DART) possess a molecular weight of approximately 60 kDa. This small size is expected to favor tumor penetration on one hand, but involves also rapid blood clearance and therewith reduced availability over a longer time period. Therefore, recombinant bispecific antibody variants of enhanced molecular size and avidity have been created, in order to improve the half-life and binding efficiency (see Fig. 11.1).

For example, tetravalent bispecific recombinant antibodies can be generated, favoring homodimerization of unfolded single-chain diabody polypeptide chains in an antiparallel orientation. This can be achieved by adjustments of the linker lengths and the production conditions. Different sizes of middle linker (6–27 amino acids) and flanking linker (2–10 amino acids) in diverse combinations have been described (Kipriyanov et al., 1999; Völkel et al., 2001; Le Gall et al., 2004). Tandem diabodies targeting CD19 and CD3 have been reported to show enhanced stability, plasma retention time and antibody-mediated cytotoxicity (Kipriyanov et al., 1999).

In another approach an additional scFv is fused to a bispecific taFv molecule, generating a bispecific and trivalent triple body (sctb). sctb (CD19 x CD16 x CD19) (Kellner et al., 2008) and (CD33 x CD16 x CD33) (Singer et al., 2010) have been reported, in which the central localized scFv is additionally stabilized by disulphide bounds. In vitro, these sctbs showed in comparison to the respective bispecific taFv molecule, a 3-fold increase in tumor cell binding translating into an approximately 10- to 200-fold gain in cytotoxic activity (depending on the target cell line). In addition, a 2-fold enhancement in plasma half-life was achieved in mice for sctb (CD19 x CD16 x CD19). Alternatively, scFvs targeting different antigens on the same tumor cell were employed for sctb generation (sctb CD123 x CD16 x CD33) (Kügler et al., 2010). Here, evaluating the cytotoxic potential (ADCC) of the constructs on primary leukemia cells, in general dual targeting showed to be as effective as the best performing mono targeting (sctb CD123 x CD16 x CD123 or sctb CD33 x CD16 x CD33) approach in each case. Thus, although antigen expression and density might favor mono targeting in a particular case, dual targeting should be advantageous to address tumors presenting more heterogeneous profiles.

A large variety of recombinant bispecific antibodies have been generated in the past two decades comprising also constant immunoglobulin domains. Examples are shown in Fig. 11.1. Many of these IgG-like bispecific antibodies contain a complete Fc region, thus are capable of excerting Fc-mediated activities such as antibody-dependent cellular cytotoxicity (ADCC) and complement-dependent cytotoxicity (CDC). Furthermore, the Fc part enables recycling by the neonatal Fc receptor (FcRn) leading to extended half-lives in plasma (Kontermann, 2009).

Early attempts to generate recombinant bispecific antibodies were based on the expression of two different heavy and light chains within one cell. Similar to hybrid hybridomas, this leads, however, to the formation of a mixture of bispecific, monospecific and inactive molecules. In order to force heterodimerization of two different heavy chains, the knobs-into-holes strategy was developed. Here, amino acids with large side chains are introduced at the C_H3–C_H3 interface of one heavy chain, while residues with smaller side chains are introduced at the opposite site of the second heavy chain. This hinders homodimerization of two identical heavy chains leading to IgG molecules composed of two different heavy chains. However, the two light chains are still able to bind to both heavy chains resulting also in molecules with inactive binding sites. This obstacle was circumvented by using antibodies as building blocks that utilize the same light chain (Merchant et al., 1998). More recently, similar approaches leading to Fc heterodimer formation were established based on an electrostatic steering effect (Gunasekaran et al., 2010) or creating complementary C_H3 domains through strand-exchanged engineered domains (SEED) derived from human IgG and IgA C_H3 domains (Davis et al., 2010). These Fc heterodimers can also be used as fusion partners for scFv fragments, thus completely avoiding the use of light chains.

A new type of bispecific IgG-like antibody was developed by Abbott. In the dual-variable-domain (DVD) approach, a second V_H domain is fused to the N-terminus of the heavy chain and a complementary second V_L domain to the N-terminus of the light chain. Interestingly, they could show that these leads to molecules capable of simultaneously binding to two different antigens while maintaining the functions of IgG molecules (Wu et al., 2007). However, it was found that the affinity of the two binding sites is influenced by the order of the two specificities and by the linkers connecting the variable domains within each chain (Wu et al., 2009).

A diverse set of bispecific IgG-like antibodies were generated by fusion of small bispecific antibody fragments, e.g. tandem scFv, diabodies and single-chain diabodies to an Fc fragment (Alt et al., 1999; Lu et al., 2005; Li et al., 2010) (Fig. 11.1). This leads to tetravalent molecules, i.e. possessing four binding sites, two for each antigen. Similarly, scFv moieties have been fused to the N- or C-terminus of the heavy or light chain generating also tetravalent and bispecific molecules (Orcutt et al., 2010; Michaelson et al., 2009) (Fig. 11.1). A further simplification of this strategy was realized by Trubion. They fused scFv moieties to the N- and C-terminus of an Fc fragment with the advantage that only a single polypeptide chain needs to be produced.

These tetravalent, bispecific antibodies are especially useful for dual targeting approaches, either aiming at (i) an increased and more selective binding to targeted cells, e.g. tumor cells, or (ii) simultaneously neutralizing two target molecules, e.g. for the treatment of inflammatory diseases but also in cancer therapy (Marvin & Zhu, 2006). Thus, it was shown in vitro that a bispecific diabody-Fc fusion protein targeting EGFR and IGFR was able to block the respective receptor-ligand interactions and down-regulate receptor surface expression therewith inhibiting growth signaling into tumor cells. In addition, direct tumor cell killing by ADCC was shown as well. Furthermore, tumor growth inhibition was reported in two xenograft mouse models (Lu et al., 2005). Tumor growth inhibition mediated by IgG-like, bispecific antibody formats could also be achieved in mouse models, targeting other receptor pairs like TRAIL-R2 and LTβR (Michaelson et al., 2009) or soluble ligands like vascular endothelial growth factor A (VEGF) and osteopontin (OPN) (Kou et al., 2010).

Dual targeting strategies in cancer therapy were also established for the retargeting of cytotoxic effector molecules, including cytokines and bacterial toxins. Thus, bispecific tandem scFv molecules were fused to the catalytic and translocation domains of diptheria toxin (DT) or Pseudomonas eotoxin (ETA) (Marvin & Zhu, 2006). A bispecific DT immunotoxin targeting CD19 and CD22 was shown to exhibit broader and improved cytotoxicity against B cell malignancies (Vallera et al., 2005, 2009). Recently, a deimmunized version of a bispecific ETA(P38) immunotoxin was developed, which should allow for multiple drug treatments due to a strongly reduced immunogenicity of this fusion protein (Vallera et al., 2010).

11.5 Conclusions

During the past two decades a plethora of different recombinant bispecific antibody formats have been developed for cancer therapy. Initially, these molecules were designed for the retargeting of immune effector cells to tumor cells (cellular cancer immunotherapy). In 2009, the first bispecific antibody, catumaxomab, was approved for cancer therapy and several other bispecific antibodies, including small recombinant formats, are currently at varying stages of clinical trials. Recently, developments have shifted towards dual targeting strategies, i.e. the simultaneous binding to two target molecules on cancer cells. This can lead to improved selectivity and binding and, thus, retargeting of associated effector functions. Alternatively, dual targeting and neutralization of two surface receptors expressed on cancer cells can efficiently inhibit signal transduction events responsible for tumor cell proliferation. These examples highlight that bispecific antibodies represent versatile molecules for therapeutic interventions. Importantly, the biochemical, pharmacological and functional properties of these molecules can be modulated and adjusted through genetic engineering to the medical need.

References

Alt M, Müller R, Kontermann RE. (1999) Novel tetravalent and bispecific IgG-like antibody molecules combining single-chain diabodies with the immunoglobulin g1 Fc or CH3 region. FEBS Lett. 454(1–2):90–4.

Asano R, Kudo T, Nishimura Y, Makabe K, Hayashi H, Suzuki M, Tsumoto K, Kumagai I. (2002) Efficient construction of a diabody using a refolding system: anti-carcinoembryonic antigen recombinant antibody fragment. J Biochem. 132(6):903–9.

Baeuerle PA, Reinhardt C. (2009) Bispecific T-cell engaging antibodies for cancer therapy. Cancer Res. 69(12):4941–4.

Bargou R, Leo E, Zugmaier G, Klinger M, Goebeler M, Knop S, Noppeney R, Viardot A, Hess G, Schuler M, Einsele H, Brandl C, Wolf A, Kirchinger P, Klappers P, Schmidt M, Riethmüller G, Reinhardt C, Baeuerle PA, Kufer P. (2008) Tumor regression in cancer patients by very low doses of a T cell-engaging antibody. Science. 321(5891):974–7.

Blanco B, Holliger P, Vile RG, Alvarez-Vallina L. (2003) Induction of human T lymphocyte cytotoxicity and inhibition of tumor growth by tumor-specific diabody-based molecules secreted from gene-modified bystander cells. J Immunol. 171(2):1070–7.

Bluemel C, Hausmann S, Fluhr P, Sriskandarajah M, Stallcup WB, Baeuerle PA, Kufer P. (2010) Epitope distance to the target cell membrane and antigen size determine the potency of T cell-mediated lysis by BiTE antibodies specific for a large melanoma surface antigen. Cancer Immunol Immunother. 59(8):1197–209.

Brandão JG, Scheper RJ, Lougheed SM, Curiel DT, Tillman BW, Gerritsen WR, van den Eertwegh AJ, Pinedo HM, Haisma HJ, de Gruijl TD. (2003) CD40-targeted adenoviral gene transfer to dendritic cells through the use of a novel bispecific single-chain Fv antibody enhances cytotoxic T cell activation. Vaccine. 21(19–20):2268–72.

Bruenke J, Barbin K, Kunert S, Lang P, Pfeiffer M, Stieglmaier K, Niethammer D, Stockmeyer B, Peipp M, Repp R, Valerius T, Fey GH. (2005) Effective lysis of lymphoma cells with a stabilised bispecific single-chain Fv antibody against CD19 and FcgammaRIII (CD16). Br J Haematol. 130(2):218–28.

Brüsselbach S, Korn T, Völkel T, Müller R, Kontermann RE. (1999) Enzyme recruitment and tumor cell killing in vitro by a secreted bispecific single-chain diabody. Tumor Targeting. 4:115–23.

Chames P, Baty D. (2009) Bispecific antibodies for cancer therapy. Curr Opin Drug Discov Devel. 12(2):276–83.

Chan JK, Hamilton CA, Cheung MK, Karimi M, Baker J, Gall JM, Schulz S, Thorne SH, Teng NN, Contag CH, Lum LG, Negrin RS. (2006) Enhanced killing of primary ovarian cancer by retargeting autologous cytokine-induced killer cells with bispecific antibodies: a preclinical study. Clin Cancer Res. 12(6):1859–67.

Cochlovius B, Kipriyanov SM, Stassar MJ, Christ O, Schuhmacher J, Strauss G, Moldenhauer G, Little M. (2000) Treatment of human B cell lymphoma xenografts with a CD3 x CD19 diabody and T cells. J Immunol. 165(2):888–95.

Davis JH, Aperlo C, Li Y, Kurosawa E, Lan Y, Lo KM, Huston JS. (2010) SEEDbodies: fusion proteins based on strand-exchange engineered domain (SEED) CH3 heterodimers in an Fc analogue platform for asymmetric binders or immunofusions and bispecific antibodies. Protein Eng Des Sel. 23(4):195–202.

Dreier T, Baeuerle PA, Fichtner I, Grün M, Schlereth B, Lorenczewski G, Kufer P, Lutterbüse R, Riethmüller G, Gjorstrup P, Bargou RC. (2003) T cell costimulus-independent and very efficacious inhibition of tumor growth in mice bearing subcutaneous or leukemic human B cell lymphoma xenografts by a CD19-/CD3-bispecific single-chain antibody construct. J Immunol. 170(8):4397–402.

FitzGerald K, Holliger P, Winter G. (1997) Improved tumour targeting by disulphide stabilized diabodies expressed in Pichia pastoris. Protein Eng. 10(10):1221–5.

Gall JM, Davol PA, Grabert RC, Deaver M, Lum LG. (2005) T cells armed with anti-CD3 x anti-CD20 bispecific antibody enhance killing of $CD20^{+}$ malignant B cells and bypass complement-mediated rituximab resistance in vitro. Exp Hematol. 33(4):452–9.

Grabert RC, Cousens LP, Smith JA, Olson S, Gall J, Young WB, Davol PA, Lum LG. (2006) Human T cells armed with Her2/neu bispecific antibodies divide, are cytotoxic, and secrete cytokines with repeated stimulation. Clin Cancer Res. 12(2):569–76.

Graziano RF, Guptill P. (2004) Chemical production of bispecific antibodies. Methods Mol Biol. 283:71–85.

Grosse-Hovest L, Müller S, Minoia R, Wolf E, Zakhartchenko V, Wenigerkind H, Lassnig C, Besenfelder U, Müller M, Lytton SD, Jung G, Brem G. (2004) Cloned transgenic farm animals produce a bispecific antibody for T cell-mediated tumor cell killing. Proc Natl Acad Sci USA. 101(18):6858–63.

Guettinger Y, Barbin K, Peipp M, Bruenke J, Dechant M, Horner H, Thierschmidt D, Valerius T, Repp R, Fey GH, Stockmeyer B. (2010) A recombinant bispecific single-chain fragment variable specific for HLA class II and Fc alpha RI (CD89) recruits polymorphonuclear neutrophils for efficient lysis of malignant B lymphoid cells. J Immunol. 184(3):1210–17.

Gunasekaran K, Pentony M, Shen M, Garrett L, Forte C, Woodward A, Ng SB, Born T, Retter M, Manchulenko K, Sweet H, Foltz IN, Wittekind M, Yan W. (2010) Enhancing antibody Fc heterodimer formation through electrostatic steering effects: applications to bispecific molecules and monovalent IgG. J Biol Chem. 285(25):19637–46.

Haas C, Krinner E, Brischwein K, Hoffmann P, Lutterbüse R, Schlereth B, Kufer P, Baeuerle PA. (2009) Mode of cytotoxic action of T cell-engaging BiTE antibody MT110. Immunobiology. 214(6):441–53.

Hayashi H, Asano R, Tsumoto K, Katayose Y, Suzuki M, Unno M, Kodama H, Takemura S, Yoshida H, Makabe K, Imai K, Matsuno S, Kumagai I, Kudo T. (2004) A highly effective and stable bispecific diabody for cancer immunotherapy: cure of xenografted tumors by bispecific diabody and T-LAK cells. Cancer Immunol Immunother. 53(6):497–509.

Hayden MS, Linsley PS, Gayle MA, Bajorath J, Brady WA, Norris NA, Fell HP, Ledbetter JA, Gilliland LK. (1994) Single-chain mono- and bispecific antibody derivatives with novel biological properties and antitumour activity from a COS cell transient expression system. Ther Immunol. 1(1):3–15.

Hoffmann P, Hofmeister R, Brischwein K, Brandl C, Crommer S, Bargou R, Itin C, Prang N, Baeuerle PA. (2005) Serial killing of tumor cells by cytotoxic T cells redirected with a CD19-/CD3-bispecific single-chain antibody construct. Int J Cancer. 115(1):98–104.

Holliger P, Prospero T, Winter G. (1993) "Diabodies": small bivalent and bispecific antibody fragments. Proc Natl Acad Sci USA. 90(14):6444–8.

Johnson S, Burke S, Huang L, Gorlatov S, Li H, Wang W, Zhang W, Tuaillon N, Rainey J, Barat B, Yang Y, Jin L, Ciccarone V, Moore PA, Koenig S, Bonvini E. (2010) Effector Cell recruitment with novel Fv-based dual-affinity re-targeting protein leads to potent tumor cytolysis and in vivo B-cell depletion. J Mol Biol. 399(3):436–49.

Kashentseva EA, Douglas JT, Zinn KR, Curiel DT, Dmitriev IP. (2009) Targeting of adenovirus serotype 5 pseudotyped with short fiber from serotype 41 to c-erbB2-positive cells using bispecific single-chain diabody. J Mol Biol. 388(3):443–61.

Kellner C, Bruenke J, Stieglmaier J, Schwemmlein M, Schwenkert M, Singer H, Mentz K, Peipp M, Lang P, Oduncu F, Stockmeyer B, Fey GH. (2008) A novel CD19-directed recombinant bispecific antibody derivative with enhanced immune effector functions for human leukemic cells. J Immunother. 31(9):871–84.

Kiewe P, Thiel E. (2008) Ertumaxomab: a trifunctional antibody for breast cancer treatment. Expert Opin Investig Drugs. 17(10):1553–8.

Kipriyanov SM, Moldenhauer G, Schuhmacher J, Cochlovius B, Von der Lieth CW, Matys ER, Little M. (1999) Bispecific tandem diabody for tumor therapy with improved antigen binding and pharmacokinetics. J Mol Biol. 293(1):41–56.

Köhler G, Milstein C. (1975) Continuous cultures of fused cells secreting antibody of predefined specificity. Nature. 256(5517):495–7.

Kontermann RE. (2005) Recombinant bispecific antibodies for cancer therapy. Acta Pharmacol Sin. 26(1):1–9.

Kontermann RE. (2009) Strategies to extend plasma half-lives of recombinant antibodies. BioDrugs. 23(2):93–109.

Korn T, Nettelbeck DM, Völkel T, Müller R, Kontermann RE. (2004) Recombinant bispecific antibodies for the targeting of adenoviruses to CEA-expressing tumour cells: a comparative analysis of bacterially expressed single-chain diabody and tandem scFv. J Gene Med. 6(6): 642–51.

Kou G, Shi J, Chen L, Zhang D, Hou S, Zhao L, Fang C, Zheng L, Zhang X, Liang P, Zhang X, Li B, Guo Y. (2010) A bispecific antibody effectively inhibits tumor growth and metastasis by simultaneous blocking vascular endothelial growth factor A and osteopontin. Cancer Lett. 299(2):130–6.

Kufer P, Mack M, Gruber R, Lutterbüse R, Zettl F, Riethmüller G. (1997) Construction and biological activity of a recombinant bispecific single-chain antibody designed for therapy of minimal residual colorectal cancer. Cancer Immunol Immunother. 45(3–4):193–7.

Kügler M, Stein C, Kellner C, Mentz K, Saul D, Schwenkert M, Schubert I, Singer H, Oduncu F, Stockmeyer B, Mackensen A, Fey GH. (2010) A recombinant trispecific single-chain Fv derivative directed against CD123 and CD33 mediates effective elimination of acute myeloid leukaemia cells by dual targeting. Br J Haematol. 150(5):574–86.

Le Gall F, Reusch U, Little M, Kipriyanov SM. (2004) Effect of linker sequences between the antibody variable domains on the formation, stability and biological activity of a bispecific tandem diabody. Protein Eng Design Sel. 17:357–66.

Li B, Zhang X, Shi S, Zhao L, Zhang D, Qian W, Zheng L, Gao J, Wang H, Guo Y. (2010) Construction and characterization of a bispecific anti-CD20 antibody with potent antitumor activity against B-cell lymphoma. Cancer Res. 70(15):6293–302.

Lindhofer H, Mocikat R, Steipe B, Thierfelder S. (1995) Preferential species-restricted heavy/light chain pairing in rat/mouse quadromas. Implications for a single-step purification of bispecific antibodies. J Immunol. 155(1):219–25.

Liu R, Jiang W, Yang M, Guo H, Zhang Y, Wang J, Zhu H, Shi R, Fan D, Yang C, Zhu Z, Xie Y, Xiong D. (2010) Efficient inhibition of human B-cell lymphoma in SCID mice by synergistic antitumor effect of human 4-1BB ligand/anti-CD20 fusion proteins and anti-CD3/anti-CD20 diabodies. J Immunother. 33(5):500–9.

Löffler A, Kufer P, Lutterbüse R, Zettl F, Daniel PT, Schwenkenbecher JM, Riethmüller G, Dörken B, Bargou RC. (2000) A recombinant bispecific single-chain antibody, CD19 x CD3, induces rapid and high lymphoma-directed cytotoxicity by unstimulated T lymphocytes. Blood. 95(6):2098–103.

Lu D, Zhang H, Koo H, Tonra J, Balderes P, Prewett M, Corcoran E, Mangalampalli V, Bassi R, Anselma D, Patel D, Kang X, Ludwig DL, Hicklin DJ, Bohlen P, Witte L, Zhu Z. (2005) A fully human recombinant IgG-like bispecific antibody to both the epidermal growth factor receptor and the insulin-like growth factor receptor for enhanced antitumor activity. J Biol Chem. 280(20):19665–72.

Lum LG, Davol PA, Lee RJ. (2006) The new face of bispecific antibodies: targeting cancer and much more. Exp Hematol. 34(1):1–6.

Lutterbuese R, Raum T, Kischel R, Lutterbuese P, Schlereth B, Schaller E, Mangold S, Rau D, Meier P, Kiener PA, Mulgrew K, Oberst MD, Hammond SA, Baeuerle PA, Kufer P. (2009) Potent control of tumor growth by CEA/CD3-bispecific single-chain antibody constructs that are not competitively inhibited by soluble CEA. J Immunother. 32(4):341–52.

Marvin JS, Zhu Z. (2006) Bispecific antibodies for dual-modality cancer therapy: killing two signaling cascades with one stone. Curr Opin Drug Discov Develop. 9(2):184–93.

McCall AM, Shahied L, Amoroso AR, Horak EM, Simmons HH, Nielson U, Adams GP, Schier R, Marks JD, Weiner LM. (2001) Increasing the affinity for tumor antigen enhances bispecific antibody cytotoxicity. J Immunol. 166(10):6112–17.

Merchant AM, Zhu Z, Yuan JQ, Goodard A, Adams CW, Presta LG, Carter P. (1998) An efficient route to human bispecific IgG. Nat Biotechnol. 16(7):677–81.

Michaelson JS, Demarest SJ, Miller B, Amatucci A, Snyder WB, Wu X, Huang F, Phan S, Gao S, Doern A, Farrington GK, Lugovskoy A, Joseph I, Bailly V, Wang X, Garber E, Browning J, Glaser SM. (2009) Anti-tumor activity of stability-engineered IgG-like bispecific antibodies targeting TRAIL-R2 and LTbetaR. MAbs. 1(2):128–41.

Milstein C, Cuello AC. (1983) Hybrid hybridomas and their use in immunohistochemistry. Nature. 305(5934):537–40.

Müller D, Frey K, Kontermann RE. (2008) A novel antibody-4-1BBL fusion protein for targeted costimulation in cancer immunotherapy. J Immunother. 31(8):714–22.

Müller D, Karle A, Meissburger B, Höfig I, Stork R, Kontermann RE. (2007) Improved pharmacokinetics of recombinant bispecific antibody molecules by fusion to human serum albumin. J Biol Chem. 282(17):12650–60.

Müller D, Kontermann RE. (2007) Recombinant bispecific antibodies for cellular cancer immunotherapy. Curr Opin Mol Ther. 9(4):319–26.

Müller D, Kontermann RE. (2010) Bispecific antibodies for cancer immunotherapy: current perspectives. BioDrugs. 24(2):89–98.

Nagorsen D, Bargou R, Ruttinger D, Kufer P, Baeuerle PA, Zugmaier G. (2009) Immunotherapy of lymphoma and leukemia with T-cell engaging BiTE antibody blinatumomab. Leuk Lymphoma. 50(6):886–91.

Orcutt KD, Ackerman ME, Cieslewicz M, Quiroz E, Slusarczyk AL, Frangioni JV, Wittrup KD. (2010) A modular IgG-scFv bispecific antibody topology. Protein Eng Des Sel. 23(4):221–8.

Ranft K, Thepen T, Fischer R, Barth S, Stöcker M. (2009) Recombinant bispecific single chain antibody fragments induce Fc gamma-receptor-mediated elimination of CD30$^+$ lymphoma cells. Cancer Lett. 282(2):187–94.

Ren-Heidenreich L, Davol PA, Kouttab NM, Elfenbein GJ, Lum LG. (2004) Redirected T-cell cytotoxicity to epithelial cell adhesion molecule-overexpressing adenocarcinomas by a novel recombinant antibody, E3Bi, in vitro and in an animal model. Cancer. 100(5):1095–103.

Reusch U, Sundaram M, Davol PA, Olson SD, Davis JB, Demel K, Nissim J, Rathore R, Liu PY, Lum LG. (2006) Anti-CD3 x anti-epidermal growth factor receptor (EGFR) bispecific antibody redirects T-cell cytolytic activity to EGFR-positive cancers in vitro and in an animal model. Clin Cancer Res. 12(1):183–90.

Schlereth B, Quadt C, Dreier T, Kufer P, Lorenczewski G, Prang N, Brandl C, Lippold S, Cobb K, Brasky K, Leo E, Bargou R, Murthy K, Baeuerle PA. (2006a) T-cell activation and B-cell depletion in chimpanzees treated with a bispecific anti-CD19/anti-CD3 single-chain antibody construct. Cancer Immunol Immunother. 55(5):503–14.

Schlereth B, Kleindienst P, Fichtner I, Lorenczewski G, Brischwein K, Lippold S, da Silva A, Locher M, Kischel R, Lutterbüse R, Kufer P, Baeuerle PA. (2006b) Potent inhibition of local and disseminated tumor growth in immunocompetent mouse models by a bispecific antibody construct specific for Murine CD3. Cancer Immunol Immunother. 55(7):785–96.

Segal DM, Weiner GJ, Weiner LM. (1999) Bispecific antibodies in cancer therapy. Curr. Opin. Immunol. 11:558–62.

Seimetz D, Lindhofer H, Bokemeyer C. (2010) Development and approval of the trifunctional antibody catumaxomab (anti-EpCAM x anti-CD3) as a targeted cancer immunotherapy. Cancer Treat Rev. 36(6):458–67.

Sen M, Wankowski DM, Garlie NK, Siebenlist RE, Van Epps D, LeFever AV, Lum LG. (2001) Use of anti-CD3 x anti-HER2/neu bispecific antibody for redirecting cytotoxicity of activated T cells toward HER2/neu+ tumors. J Hematother Stem Cell Res. 10(2):247–60.

Shen J, Zhu Z. (2008) Catumaxomab, a rat/murine hybrid trifunctional bispecific monoclonal antibody for the treatment of cancer. Curr Opin Mol Ther. 10(3):273–84.

Singer H, Kellner C, Lanig H, Aigner M, Stockmeyer B, Oduncu F, Schwemmlein M, Stein C, Mentz K, Mackensen A, Fey GH. (2010) Effective elimination of acute myeloid leukemic cells by recombinant bispecific antibody derivatives directed against CD33 and CD16. J Immunother. 33(6):599–608.

Stanglmaier M, Faltin M, Ruf P, Bodenhausen A, Schröder P, Lindhofer H. (2008) Bi20 (fBTA05), a novel trifunctional bispecific antibody (anti-CD20 x anti-CD3), mediates efficient killing of B-cell lymphoma cells even with very low CD20 expression levels. Int J Cancer. 123(5): 1181–9.

Vallera D, Chen H, Sichender AR, Panoskaltsis-Mortari A, Taras EP. (2009) Genetic alteration of a bispecific ligand-directed toxin targeting human C19 and CD22 receptors resulting in improved efficacy against systemic B cell malignancies. Leuk Res. 33:1233–42.

Vallera D, OH S, Chen H, Shu Y, Frankel AE. (2010) Bioengineering a unique deimmunized bispecific targetred toxin that simultaneously recognizes human CD22 and CD19 receptors in a mouse model of B-cell malignancies. Mol Cancer Ther. 9(6):1872–83.

Vallera D, Todhunter DA, Kuroki DW, Shu Y, Sichender A, Chen H. (2005) A bispecific recombinant immunotoxin, DT2219, targeting human CD199 and CD22 receptors in mouse xenograft model of B-cell leukemia/lymphoma. Clin Cancer Res. 11(10):3879–88.

Völkel T, Korn T, Bach M, Müller R, Kontermann RE. (2001) Optimized linker sequences for the expression of monomeric and dimeric bispecific single-chain diabodies. Protein Eng. 14: 815–23.

Wolf E, Hofmeister R, Kufer P, Schlereth B, Baeuerle PA. (2005) BiTEs: bispecific antibody constructs with unique anti-tumor activity. Drug Discov Today. 10(18):1237–44.

Wu C, Ying H, Bose S, Miller R, Medina L, Santora L, Ghayur T. (2009) Molecular construction and optimization of anti-human IL-1alpha/beta dual variable domain immunoglobulin (DVD-Ig) molecules. MAbs. 1(4):339–47.

Wu C, Ying H, Grinnell C, Bryant S, Miller R, Clabbers A, Bose S, McCarthy D, Zhu RR, Santora L, Davis-Taber R, Kunes Y, Fung E, Schwartz A, Sakorafas P, Gu J, Tarcsa E, Murtaza A, Ghayur T. (2007) Simultaneous targeting of multiple disease mediators by a dual-variable-domain immunoglobulin. Nat Biotechnol. 25(11):1290–7.

Zhu Z, Presta LG, Zapata G, Carter P. (1997) Remodeling domain interfaces to enhance heterodimer formation. Protein Sci. 6(4):781–8.

Chapter 12
The Role of Glycosylation in Therapeutic Antibodies

Maureen Spearman, Ben Dionne, and Michael Butler

Abstract Monoclonal antibodies (Mabs) are biopharmaceuticals that are used increasingly for the treatment of a wide range of diseases such as cancer and autoimmunity. The effectiveness of therapeutic Mabs, most of which are immunoglobulin G (IgG), is dependent upon their ability to link antigen recognition with an appropriate effector function, to elicit a biological response in vivo that will treat the targeted disease. Studies over the last decade have determined that the effector function of Mabs is highly dependent upon the structure of the N-linked glycan of the Fc domain of the Mab. Total removal of the glycan is highly detrimental to the effector function of the Mab, but subtle differences in the glycan structure, such as the lack of fucose, can improve significantly bioactivity and function of the Mabs. Some Mabs are glycosylated in the variable Fab domain but in many cases the function is not known. The host cellular production system including the bioreactor environment can produce Mabs with very different glycosylation profiles that must be considered in bioprocess development. Cell culture conditions such as dissolved oxygen, nutrient levels, pH and feed strategies can all have considerable influence on the glycosylation of the Mab, which will affect product quality and efficacy. Great improvements have been made in techniques for high resolution and high throughput analysis of glycans such as normal phase-high performance liquid chromatography (HPLC) and mass spectrometry (MS). This has allowed a better understanding of the link between the structure and function, which will in turn lead to the development of safer and more effective Mabs.

Abbreviations

2-AA	2-aminobenzoic acid
2-AB	2-aminobenzaminde
2-PA	2-aminopyridine
ADCC	Antibody-dependent cellular cytotoxicity
APTS	1-aminopyrene-3,6,8-trisulfonate
Asn	Asparagine
C1q	Subcomponent of C1 of the complement cascade
CDC	Complement-dependent cytotoxicity
CGE	Capillary gel electrophoresis

M. Spearman, B. Dionne, and M. Butler (✉)
Department of Microbiology, University of Manitoba, Winnipeg, MB, Canada R3T 2N2
e-mail: butler@cc.umanitoba.ca

M. Al-Rubeai (ed.), *Antibody Expression and Production*, Cell Engineering 7,
DOI 10.1007/978-94-007-1257-7_12, © Springer Science+Business Media B.V. 2011

CH1, CH2, or CH3	Constant regions of the heavy chain of immunoglobulin
CHO	Chinese hamster ovary
CIE	Capillary ion electrophoresis
C_L	Constant region of the light chain of immunoglobulin
Dol-P	Dolichol phosphate
ESI-TOF-MS	Electrospray time-of-flight mass spectrometry
Fab	Antibody binding fragment of immunoglobulin
Fc	Crystalizable fragment of immunoglobulin
FcγR	Fc gamma receptor
FT	Fucosyltransferase
FUT8	α1-6 fucosyltransferase
Fuc	Fucose
Fv	Variable region of immunoglobulin
Gal	Galactose
GalT	Galactosyltransferase
GDP	Guanidine diphosphate
Glc	Glucose
GlcNAc	N-acetylglucosamine
GMD	GDP-mannose 4,6 dehydratase
GU	Glucose units
HACA	Human anti-chimeric antibody
HAMA	Human anti-mouse antibody
HEK	Human embryonic kidney
HILIC	Hydrophobic interaction liquid chromatography
HPAEC-PAD	High performance anion exchange chromatography with pulsed amperometric detection
Ig	Immunoglobulin
LC	Liquid chromatography
LIF	Laser-induced fluorescence
Mab	Monoclonal antibody
MALDI-TOF	Matrix-assisted laser desorption/ionization with time of flight
Man	Mannose
ManNac	N-acetylmannosamine
MS	Mass spectrometry
NeuAc	N-acetylneuraminic acid
Neu5Gc	N-glycolylneuraminic acid
NP-HPLC	Normal phase high performance liquid chromatography
RP-HPLC	Reverse phase HPLC
Ser	Serine
Thr	Threonine
UDP	Uridine diphosphate
V_H	Variable region of the heavy chain of immunoglobulin
V_L	Variable region of the light chain of immunoglobulin
XT	Xylosyltransferase

12.1 Introduction

Antibodies are a major component of the immune system that exist as soluble glycoproteins in biological fluids including the blood stream. They are usually referred to as immunoglobulins (Ig) through their association with the gamma-globulin fraction of blood proteins that separate by electrophoresis. The basic Y-structure of these molecules was elucidated by Porter and Edelman through a series of fragmentation experiments using proteolytic enzymes (Nobel Prize for Medicine 1972). However, superimposed on the basic structure is an enormous heterogeneity that allows binding of specific antibodies to selected antigens through a hypervariable region found in the Fab fragment of the molecule. The importance of this in the immune system of mammals is clear in the prevention of disease and removal from the blood stream of undesirable components.

The biotechnological significance of this molecular heterogeneity and highly specific antigen binding was advanced through the development of techniques for the production of monoclonal antibodies (Mabs). These are homogeneous pool of antibodies capable of binding to a single epitope of an antigen. The initial value of these was recognized in diagnostic testing and purification protocols that exploit the highly specific binding of Mabs. However, it was not until the ability to humanize these Mab structures that they could be used as therapeutic agents. Presently there are >20 approved Mabs for therapeutic use with 240 or so in clinical trials (Walsh 2010). This represents a significant proportion of new approvals and developments in the pharmaceutical industry.

The need to understand the structure and function of the glycan portion of Mabs has increased in importance through these developments in therapeutic usage. The first issue is that any biologic destined for therapeutic use is required to have a consistent structure with minimal variation between batches. The control of structural variability in large-scale production of Mabs is not trivial given the environmental factors that could perturb the cellular glycosylation process. The second issue is to gain an understanding of the therapeutic effects of specific glycoforms of Mabs. This was brought into focus with the discovery that the removal of fucose from the glycan structures of Mabs could improve their efficiency for antibody-dependent cellular cytoxicity (ADCC) by nearly two orders of magnitude (Kanda et al. 2006). Thus the control of glycosylation during production leading to homeogenous single-glycoform structures could be a distinct advantage for the use of Mabs as biopharmaceuticals. The following chapter explores our understanding of the glycosylation of Mabs as well as considering the means of control this process during large-scale production.

12.2 Immunoglobulins

The five classes of immunoglobulins (IgA, IgD, IgE, IgG and IgM) share some similar structural components (Fig. 12.1). A heavy and a light chain are linked by disulphide bridges and then form dimers creating a Y-shaped molecule. A hinge or

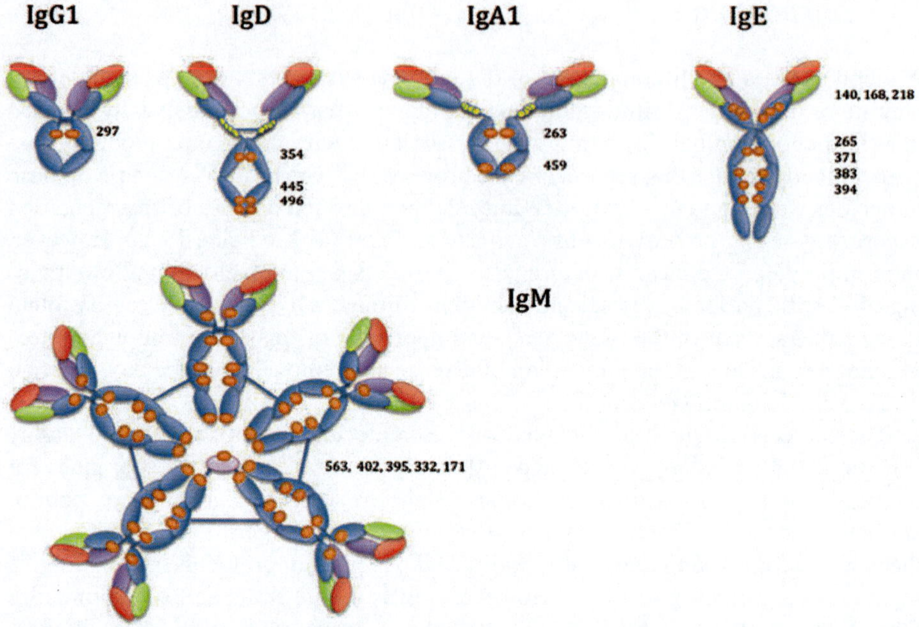

Fig. 12.1 Classes of immunoglobulins with N-linked glycans represented by *orange symbols* and O-linked glycans represented by *yellow symbols*. Domains of the heavy chains are shown in *blue* and *green*. Those of the light chains are shown in *purple* and *red*

linker region, that may be flexible (IgG, IgA, IgD) or rigid (IgM, IgE), separates the Fc (crystalizable fragment) region from the two Fab (antigen binding fragment) regions. The larger IgM is a complex of five heavy and light chains joined. Heavy chains contain three constant regions (CH1, CH2, CH3) and one variable domain (V_H). Light chains contain one variable domain (V_L) and one constant region (C_L).

IgG is the most common class developed for monoclonal antibody production and can be subclassified into 4 isotypes: IgG1, IgG2, IgG3 and IgG4. Each of these subclasses varies in amino acid composition at the hinge region of the molecule lending to different flexibilities for Fab interaction. Other major differences are the ability of IgG2 to dimerize, forming tetravalent molecules, and functional monovalency of IgG4, which has the ability to break intrachain bonds and reform with different half molecules of IgG4 creating a bivalent molecule with bispecificity (Salfeld 2007). IgG1 has thus far been the most popular choice for monoclonal antibodies, and therefore is the most well-characterized with regard to glycosylation, and effector functions, but IgG2 and IgG4 Mabs are also on the market or in development (Beck et al. 2008; Salfeld 2007). IgA is probably the second most important Ig for industrial purposes. It is naturally expressed as a monomer in serum but as a secretory dimer on epithelial cells. IgA's major advantage include increased recruitment of neutrophils, the most abundant type of white blood cell, to attack tumors (Dechant et al. 2007) and to initiate an ADCC response (Zhao et al. 2008). IgM may become a relevant therapeutic in the future due to its potency at complement fixation but little development is currently underway for commercial use. IgM's pentameric nature can be difficult to generate and purify (Mahassni et al. 2009; Tornoe et al. 1997) but if achieved can have potential therapeutic benefit (Bieber et al. 2007; Irie et al. 2004).

12.3 Types of Monoclonal Antibodies

There are several species of monoclonal antibodies available, including murine, chimeric, humanized and fully human. The first clinically tested and approved Mab were of the murine variety which demonstrated good antigen binding characteristics, however several side effects and short half-life in serum have downplayed their importance as therapeutics (Brekke and Sandlie 2003). More recently the focus has been on producing genetically engineering chimeric, humanized and fully human Mabs which are less immunogenic and therefore have better potential as therapeutics.

Chimeric Mabs consist of murine antigen binding (or variable) regions (Fab or Fv) grafted with human Fc regions for an approximate total of 25% murine and 75% human. This human Fc region enables effector function for ADCC activity. The largest concern with murine and chimeric Mabs is that they elicit immune responses resulting in human anti-mouse and anti-chimeric antibodies (HAMA and HACA respectively) which leads to quick clearance of the Mab from the serum thus nullifying any long term effects. Humanized Mabs consist of <10% murine antibody sequences which would include the hypervariable region in the antigen binding domain. These humanized Mabs exhibit increased half-lives, less immunogenicity and increased effector functions. Humanized Mab can be similarly expressed in all major mammalian cell lines with differences mainly observed in the N-glycosylation patterns dependent upon the host cell lines. Fully human Mabs are 100% human in sequence and theoretically exhibit minimal therapeutic side effects. In addition, depending on the host cell line, the glycosylation pattern may be human-like leading to greater effector function. A number of humanized and human Mabs are in development and will continue to lead the way as novel therapeutic Mabs in the foreseeable future.

12.4 Synthesis of Glycans

12.4.1 N-Linked Glycan Synthesis in Mammalian Cells

The addition of N-linked glycans to Mabs, and more generally to protein, occurs as a co-translational modification in the ER, with oligosaccharide chains transferred to the nascent polypeptide chain from a lipid-oligosaccharide precursor, dolichol phosphate (Stanley et al. 2009). Initially, on the cytosolic face of the ER membrane, dolichol phosphate acts as a carrier molecule to build a high mannose precursor molecule consisting of a two N-acetylglucosamine (GlcNAc) molecules with a branched nine mannose (Man) structure and three additional glucose (Glc) residues.

Initial sugars are added in the cytosol using nucleotide sugars (UDP-GlcNAC, GDP-Man) as substrates. Following the addition of the fifth mannose (Dol-P-P-GlcNac$_2$-Man$_5$) the entire dolichol-glycan structure is enzymatically flipped into the endoplasmic reticulum where an additional four Man residues are added through a dolichol-P-Man carrier. Three Glc molecules are then added using dolichol-P-Glc. This high Man/Glc glycan structure is then transferred from the dolichol carrier to the protein by the oligosaccharyltransferase complex.

N-linked oligosaccharides are attached through an "N" of asparagine at specific sites or "sequons" on the protein. Sequons are comprised of a three amino acid sequence of asparagine-X-serine/threonine (Asn-X-Ser/Thr), where X can be any amino acid with a few exceptions, and the third amino acid can be either Ser or Thr. Not all sequons are glycosylated, but typically a specific glycoprotein will have a set of sequons that have a higher percentage of glycosylation. Variation in occupied glycan sequons from one molecule to the next leads to macroheterogeneity within a specific glycoprotein. Glucose residues on the high mannose structure are necessary to influence proper folding the glycoprotein and interact with the calnexin/calreticulin system.

Microheterogeneity, variation in the structure of individual N-linked glycans, occurs during the processing reactions, as the glycoprotein passes through the ER and Golgi (Fig. 12.2). Some oligosaccharides are maintained as high mannose glycans following the removal of the glucose residues in the ER. But for most glycans, glucosidases and mannosidases trim the high Man/Glc structure down to a core molecule containing two GlcNAc (chitobiose core) with five Man residues. In the Golgi, additional sugar residues are added to convert the structures into complex glycans which contain a terminal triplet of sugars consisting of N-acetylglucosamine, galactose (Gal) and sialic acid (N-acetylneuraminic acid, NeuAc or N-acetylglycolylneuraminic acid, Neu5Gc). Fucose (Fuc) can also be added in an α1,6 linkage to the first core GlcNAc residue. These sugar residues are added by a diverse array of glycosyltransferases using nucleotide-sugars as substrates that introduce branching and heterogeneity into the complex glycans. Complex glycans are normally defined by their antennarity with two (biantennary), three (triantennary) or four (tetraantennary) branches. Sugar residues can also be added through different linkages such as sialic acid that can be through an α2,3 or α2,6 linkage. Control of the transferase activity and branching patterns and the ultimate microheterogeneity introduced into the glycans of a specific glycoprotein are not well understood, but pools of intracellular nucleotide sugars, Golgi transit time, pH, and enzyme levels are known to affect the glycosylation (Butler 2006). Figure 12.3 illustrates examples of four common types of oligosaccharide: high mannose, complex, hybrid, and bisected. Key enzymatic control points of sugar additions influence the final structure and lead to variable antennarity of the structure, and extra sugar additions. For example, addition of a bisecting GlcNAc between biantennary arms will prevent the addition of α1,6-fucose to the core GlcNAc adjacent to the sequon (Schachter 1986). As will be discussed, this is an important aspect of Mab glycosylation and influences the effector function of the molecule.

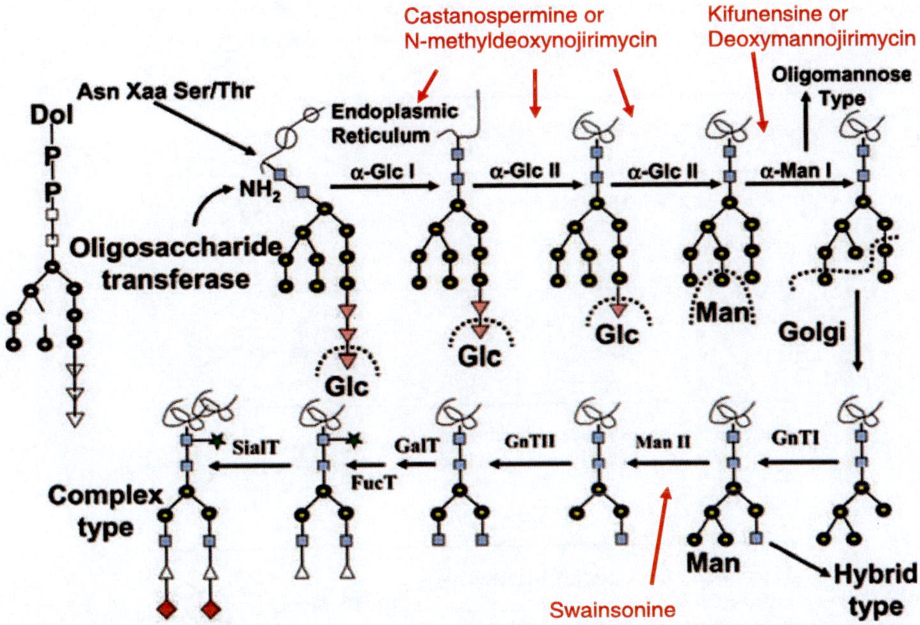

Fig. 12.2 Glycoprotein processing reactions with sites of inhibition of processing inhibitors which can be used to modify IgG glycan structure. Castanospermine and methyldeoxynojirimycin inhibit Glucosidase I and II. Kifunensine and deoxymannojirimycin inhibit Mannosidase I and Swainsonine inhibits Mannosidase II

Fig. 12.3 Types of N-linked glycans: high mannose; complex (with terminal GlcNAc, Gal, NeuAc); hybrid form with one arm high mannose and one arm with complex monosaccharides; and complex with bisecting GlcNAc

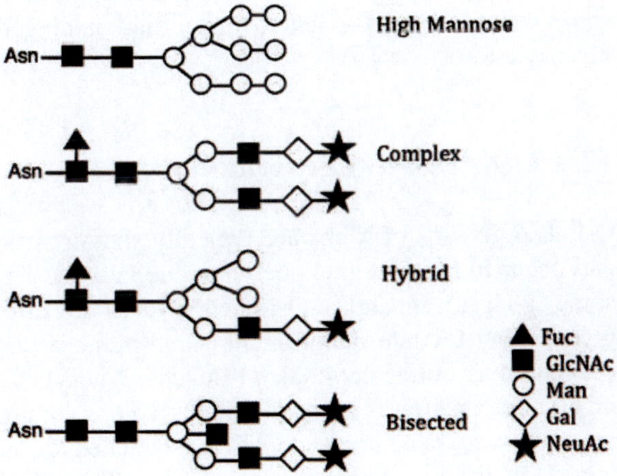

Variations in N linked glycosylation exist between mammalian species and greater variability occurs across eukaryotes (Fig. 12.4). Plants have additional sugars (xylose and α1,3-linked fucose); insects have short mannose glycans (pauci-mannose); yeasts produce predominately high mannose structures. Thus recombinant Mabs produced in these organisms will have very different glycan structures

Cell Type	N-Glycan	Structure
Bacteria	None	---------
Yeast	High Mannose	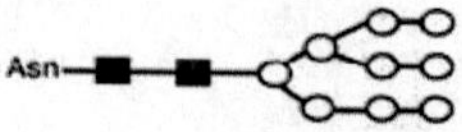
Insect	Fucosylated Core	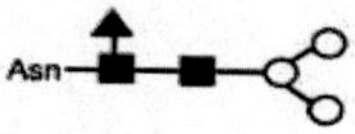
Plant	Xylosylated and α1,3 Fucosylated Core	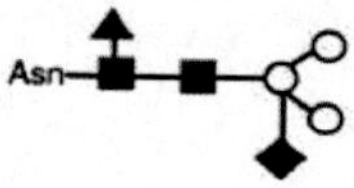
Mammalian	Complex Biantennary	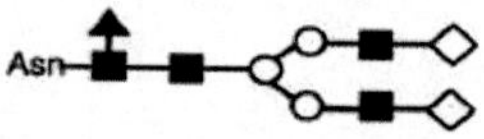

Fig. 12.4 Comparison of N-linked glycosylation in different cell expression systems

compared to those produced in mammalian cells, leading to possible immunogenic events, and have led to the initiation of genetic engineering to humanize glycosylation.

12.4.2 O-Linked Glycosylation

O-linked glycans of Mabs are typically shorter structures than N-linked structures and occur in hinge regions between the Fab and Fc portion of the heavy chain of some Ig (IgA1 and IgD). O-linked glycans are different structures from N-linked glycans, but include many of the same monosaccharides. Whereas N-linked glycosylation is cotranslational, O-linked glycosylation is a posttranslational event involving transferases in the Golgi. O-glycans are divided into several groups based on their core structures and all are attached through an N-acetylgalactosamine (GalNAc) residue to the oxygen of Ser or Thr. Consensus sequences of O-linked glycans are much less well-defined than N-linked glycans, but regions of proline and alanine along with Ser and Thr may influence glycan addition. Glycosyl transferases add individual sugar residues to the protein using nucleotide sugar substrates, unlike the en bloc transfer and processing in N-linked glycosylation. Additional sugars structures such as polylactosamine residues (GlcNac-Gal), can be added to extend the core structure and the terminal sugar residues are commonly sialic acids (NeuAc). Control of the biosynthetic pathway is not well understood, but localization of specific transferases through the Golgi compartments may influence patterns of sugar addition (Hang and Bertozzi 2005).

12.5 Ig Glycosylation

Glycosylation patterns are very diverse across the classes of immunoglobulin (Fig. 12.1). IgG has the simplest glycosylation pattern with one N-linked structure at Asn297 on the Fc region within the CH2 domain. However, a large degree of microheterogeneity exits within the oligosaccharides occupying this site. The majority of structures are complex biantennary glycans but high mannose glycans may also be present. In human IgG a second Fab glycosylation can occur on the kappa (Vκ), lambda (Vλ) or heavy chain (VH) with no consensus sequence (Jefferis 2005b). IgA1 contains two complex glycans (Asn263 and 459) while other Ig's have multiple complex glycans and some have high mannose glycans. Glycosylation can also occur on the Fab region of the heavy chain of IgM, IgE, and IgA2. Clusters of O-linked glycans are found in the hinge regions of IgA1 and IgD.

In the development of Mab's for clinical use, it has become evident that understanding the in vivo role of microheterogeneity and macroheterogeneity of glycosylation is absolutely necessary to provide clinically effective Mabs. A focus of research has been on the most common commercial type, IgG. Four subclasses of IgG (IgG1, IgG2, IgG3 and IgG4) are defined by differences in the Fc polypeptide, disulphide bonds, valency, antigen type, etc. (Salfeld 2007) but glycosylation is maintained at Asn297 of the Fc region in all. The glycan chain is typically a biantennary complex structure of a $GlcNAc_2$-trimannose core with the two arms defined by the branching mannose linkages (α1-3 or α1-6). Complex sugars of GlcNAc and Gal may be added sequentially to the mannose core resulting in structures with no galactose (G0), one galactose (G1) or two galactose (G2). Additional microheterogeneity occurs with additions of fucose (α1-6 to the primary GlcNAc), bisecting GlcNAc (attached to the core mannose) and terminal sialic acids (NeuAc or Neu5Gc added to galactose). High mannose oligosaccharides are also common (reviewed by Arnold et al. 2007). Glycosylation in the Fab region is at a much lower frequency (up to 30%) (Endo et al. 1995; Jefferis 2005b; Jefferis 2007; Rudd et al. 1991) and may result from somatic mutation (Jefferis 2005a).

12.6 Structure/Function of IgG Fc Glycan

The development and use of Mabs as biotherapeutics is rapidly expanding with a wide variety of applications, such as the treatment of rheumatoid arthritis, cancer therapy and many other diseases. And with this is the requirement for the understanding of the intricate role glycosylation plays in the function of the specific Mab. As with many other recombinant biotherapeutic proteins, Mabs require glycosylation for maintaining stability and solubility (Ghirlando et al. 1999) and cellular transport and clearance (Gala and Morrison 2002) but the primary focus is now on the complexities of the relationship between glycan structure and effector function. IgG links the humoral and cellular components of the immune system. The variable region of IgG binds to the target antigen, thus changing the conformation of the Fc region and activating effector mechanisms (Fig. 12.5). There are three groups of human Fcγ receptors (Ia, IIa/b and IIIa/b) that interact with IgG and induce signaling pathways, as well as activation of the complement pathway. Binding of Fc region to C1q of the complement pathway converts the C1q to a protease, thus initiating the complement cascade and resulting in complement-dependent cytotoxicity (CDC) of the target cell. Activation of antibody-dependent cell mediated cytotoxicity (ADCC) occurs through the binding of Fcγ receptors on various leucocytes such as natural killer cells, monocytes, macrophages and neutrophils. Binding of the Fc receptors causes activation of the leucocytes and results in target cell death through the release of granzymes, tumor necrosis factor or lytic enzymes that damage the cell, or through phagocytosis of the cell. FcRn, the neonatal receptor, which is responsible for regulation of serum IgG levels, also binds to the Fc region. The role of glycosylation of Mabs, in particular the Asn297 glycosylation site of the CH2 domain of the Fc region, is of significant importance because it defines the functionality of the Mab by affecting the Fc effector functions.

The necessity for Fc glycosylation in ADCC effector functions has been known for some time, as non-glycosylated IgG has a reduced affinity for the FcγRIII receptor (Sarmay et al. 1992; Tao and Morrison 1989). More recent studies have focused on defining how various oligosaccharide structure(s) affect Fc receptor interactions. Crystallographic studies have shown the α1-6 glycan arm of IgG binds non-covalently with hydrophobic amino acids of the CH2 domain of Fc, whereas the α1-3 branches interact and occupy the space between the CH2 domains, creating a horseshoe shape of the Fc region with the glycan sequestered in between (Arnold

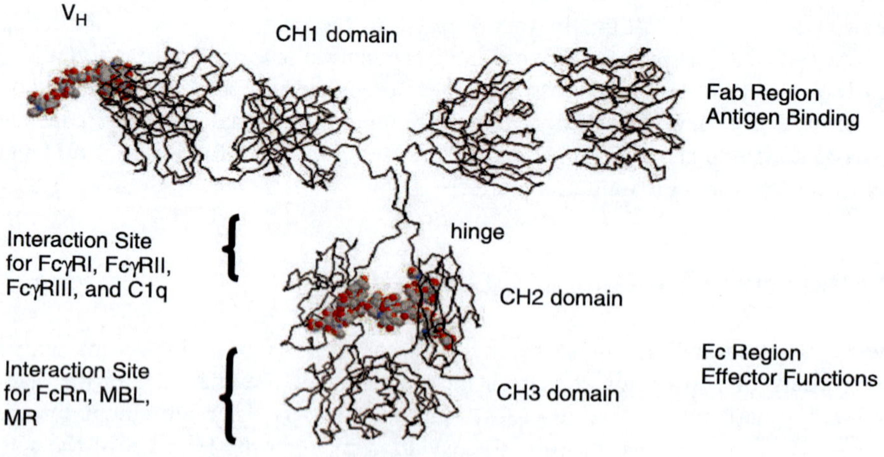

Fig. 12.5 Structure of IgG with Fab (antibody binding) region and Fc region which binds effector molecules. (Courtesy of Max Crispin with modifications)

et al. 2007; Deisenhofer 1981; Padlan 1991). Amino acid replacement of residues established to be important for non-covalent interactions with the two core GlcNAcs reduced complement activation and FcγRI interaction (Lund et al. 1996). As well, amino acid replacement influenced the amount of galactosylation and sialylation of the IgG. Crystallography of an IgG-Fc fragment complexed with the FcγRIII receptor identified some direct interaction of the receptor with the GlcNAc residue of Asn297 glycan (Sondermann et al. 2000) suggesting that the glycan has an intimate role in FcγRIII activation. But others have suggested that the carbohydrate helps stabilize the binding conformation, but does not contact the receptor (Radaev et al. 2001). Truncation of IgG oligosaccharides has shown that the glycan does confer structural stability of the CH2 domain of IgG-Fc for FcγRIIb receptor binding (Mimura et al. 2000). More specifically, the outer-arm GlcNAc is required for the thermal stability of the CH2 domain but the branching mannose residues (α1-3 and α1-6 linked) are required for FcγRIIb binding (Mimura et al. 2001). Removal of the outer GlcNAc and mannose residue induces a large conformational change in the molecule and total removal of the glycan allows a "closing" between the CH2 domains, thus affecting the conformation of the Fc portion of the molecule and causing a reduction in FcγR binding capacity (Ferrara et al. 2006b; Krapp et al. 2003). Sequential removal of outer sugars of the glycan has a minimal effect on FcγRIIIa binding, but cleavage between the inner GlcNAc (chitobiose) residues eliminates receptor binding capacity by inducing conformational alterations in the lower hinge region where the FcγR binding site is located (Yamaguchi et al. 2006). The glycan is therefore required to maintain the Fc portion in the proper protein conformation for receptor binding.

The IgG glycan effect on Fc receptor binding activity has proven to be highly influenced by the presence of core α1,6-fucose. IgG glycans lacking the core fucose have a higher capacity (up to a 50-fold increase) to activate ADCC through the FcγRIIIa receptor but have no effect on binding to FcγR1 and Fcγn or C1q (Shields et al. 2002). Increased levels of bisecting GlcNAc on IgG also increased ADCC through increased FcγIII receptor binding, allowing lower effective therapeutic doses of the Mab, Herceptin (Shinkawa et al. 2003; Umana et al. 1999). This is most likely due to a corresponding reduction in fucosylation with the presence of a bisecting GlcNAc (Schachter 1986; Takahashi et al. 2009; Umana et al. 1999). However, another study has found that low fucosylated (76%) epidermal growth factor Mab had high mononuclear cell-mediated ADCC activity through the FcγRIIIa receptor but a highly fucosylated (95%) Mab showed enhanced tumor killing by polymorphonuclear cells, illustrating that the ADCC response resulting from differences in glycosylation is complex. Only small differences in ADCC responses were observed in whole blood between high and low fucosylated antibodies (Peipp et al. 2008). Barbin et al. (2006) found fucosylated paucimannosylated Mab from insect cells have the same ADCC activity as afucosylated bisected biantennary structures, which raises the question as to how the fucose may or may not affect receptor binding.

A comparison of the crystal structure of fucosylated and non-fucosylated IgG1 glycan indicates only very subtle differences in conformation except for hydration around Tyr296 of IgG1 (Matsumiya et al. 2007). Their model shows that Tyr296 is involved in the interaction with the FcγRIIIa receptor, and the afucosylated form may be more flexible and thus enhance receptor affinity. Analysis by surface plasmon resonance showed that the removal of fucose from the IgG1 glycans, results in increased binding to the FcγRIIIa receptor, due to a corresponding increase in binding enthalpy (Okazaki et al. 2004). Low fucose content may also affect other Fcγ receptor binding. Only a slight increase to binding of FcγRIIb and a polymorphic form of FcγRIIa was observed in IgG lacking core fucose (Shields et al. 2002), but low fucose increased binding to FcγRIIb in another study (Siberil et al. 2006).

The role/effect of sialic acid on antibody activity is somewhat less defined and has proven to be contradictory. Increased sialylation of Mabs via a variety of methods significantly reduced the ADCC activity, either through decreased FcγRIII binding on NK cells, or through reduced affinity to the cell surface antigen (Scallon et al. 2007b). Studies in mice with human IVIG have found that increased sialic acid can play an anti-inflammatory role (Kaneko et al. 2006). However, others have found that sialic acid content has no effect on antibody efficacy (Peipp et al. 2008). These apparently contradictory results may have several explanations such as differences between mouse and human Fc receptors, methods of assay, and antibody preparations (Peipp et al. 2008).

A second effector function of the Fc region is the activation of the complement pathway. IgG produced in Lec 1 mutants lacking GlcNAc transferase I that produced only truncated Man_5 glycans, were unable to activate complement (Wright and Morrison 1994). Terminal galactosylation affects the ability of IgG to initiate CDC activity and can be significantly reduced through in vitro removal of terminal Gal residues from the Fc glycan using galactosidase (Boyd et al. 1995; Hodoniczky et al. 2005; Tsuchiya et al. 1989). However, IgG produced in glycosylation mutant cells lines lacking galactose showed higher complement activation than IgG with Gal present (Wright and Morrison 1998). Therefore the effects of glycan structure on CDC may require further study.

Fc glycan is also important for thermal stability by stabilizing the CH2 domain (Liu et al. 2006) and was shown to increase solubility of an anti-IL3 Mab but did not affect the affinity to IL-13 (Wu et al. 2010). Terminal sugars of the Fc glycan may also increase resistance to proteases (Raju and Scallon 2007). Mabs are cleared from circulation via a mannose binding receptor and asialo-glycan binding receptor in the liver (Wright et al. 2000). IgG with complex glycan (with or without core fucose) has higher serum half-life in mice than IgG with high mannose or hybrid glycan, suggesting the terminal sugar residues (Gal-GlcNAc) contribute to maintaining the Mab in circulation (Kanda et al. 2007b).

Recent clinical trials have shown that polymorphisms in the FcγRIIIa receptor in patients treated with the anti-CD20 Mab, rituximab (Cartron et al. 2002; Weng and Levy 2003) and with cetuximab (Taylor et al. 2009) resulted in a variable Mab response. This suggests that small changes in receptors can have significant impact on antibody interaction. Thus in turn, small changes in glycosylation, such as the presence or absence of fucose on the Fc glycan, can also significantly affect the function of the Mab. Combined, these studies show that the intricate interaction of glycosylated IgG with Fc receptors on various effector cells. This has lead to studies focusing on the modification of Fc glycans by various methods to improve effector function.

12.7 IgG Glycosylation in Normal and Diseased States

Microheterogeneity within the Fc glycan of serum IgG is associated with many immune and pathological states and recent studies of the role of the IgG-Fc glycosylation in normal and diseased states have given insights into how glycosylation of Mabs may affect functionality in treatment. Galactosylation of IgG is reduced in rheumatoid arthritis (Parekh et al. 1985), ovarian cancer (Saldova et al. 2008) and other inflammatory and autoimmune diseases (Abès and Teillaud 2010) (Arnold et al. 2007). Fucosylation is increased in liver disease (Mehta and Block 2008) and decreases in galactosylation are also correlated with age (Yamada et al. 1997) and pregnancy (Arnold et al. 2007; Rook et al. 1991). Also, intravenous immunoglobulin (IVIG), isolated from pools of donor serum, is used to treat patients with immunodeficiencies (Siberil et al. 2007) and autoimmune diseases (Baerenwaldt et al. 2010) and affect responses through Fcγ receptors. Therefore much work has been done to understand the effects of glycosylation on serum IgG, with focus on the Fc glycosylation.

Sialylation of serum IgG has recently been identified as a key control mechanism of effector function. The correlation of high levels of sialylation of serum IgG in mice with an anti-inflammatory state indicates a specific function for terminal sialic acid on IgG glycan. A reduction of sialic acid switches the antigen-specific IgG to a proinflammatory response mimicking rheumatoid arthritis (Kaneko et al. 2006). Thus a high degree of sialylation of IgG may be beneficial when IVIG is used as an anti-inflammatory treatment, but is to be avoided for Mab used for treatments where initiation of an inflammatory pathway (ADCC and C1q) is required. The authors suggest that the sialic acid does not work through a masking effect, but rather by inducing FcγRIIB expression that suppresses inflammation by raising the threshold for activation of other Fcγ receptors. However, differences exist between human and mouse IgG and Fc receptor classes, and therefore further studies are necessary to relate to the human system (Burton and Dwek 2006). Pools of human serum contain much higher concentrations of sialic acid than Mab produced in CHO (Kim et al. 2010), and therefore sialic acid must contribute to an important function of serum IgG.

Linkage specificity of sialic acid also appears to be important in Ig function. Addition of 2,6 and 2,3-sialylated Fc fragments in a mouse arthritis model system found that only 2,6-sialylated Fc reduced inflammation showing that the anti-inflammatory effect is highly linkage specific (Anthony et al. 2008). This may have important consequences to the choice of host cell line, when producing Mab for anti-inflammatory treatment. Other work has found that IgG glycans with one sialic acid residue in the Fc do not bind to a SNA-lectin (*Sambucus nigra* agglutinin) column, whereas two sialic acids within the Fc glycans (either one disialylated or two monosialylated glycans) will bind to the lectin column (Stadlmann et al. 2009). This suggests that glycans with one sialic acid can be sequestered between the heavy chain domains, whereas two sialic results in an exposed sialic acid residue. The authors propose that two sialic acids (with one exposed sialic acid) are required for an anti-inflammatory effect, even though this represents only 1% of the IgG population. MS glycopeptide analysis of the Fc glycan of serum IgG subclasses has also found differences in their glycosylation patterns, with a higher degree of sialylation in IgG1, although the sample size was small (Wuhrer et al. 2007).

A better understanding of the functional interactions of specific glycan structures of serum IgGs with receptors and effector molecules in normal and diseased conditions, will allow the translation of this knowledge into the manipulation of Mab through glycoengineering of glycan microheterogeneity to enhance their clinical efficacy. This also has implications for choices of host cells lines for the production of anti-inflammatory IgG, as many nonhuman host cells such as CHO, do not have the α2,6-sialyltransferase (see below). Most other animal species also add N-acetylglycolylneuraminic acid (Neu5Gc) as well as N-acetyl neuraminic acid (NeuAc) as the terminal sialic acid, which has recently been proven to be immunogenic (Ghaderi et al. 2010). A novel structure of a Gal attached to a bisecting GlcNAc has recently been identified by two different studies in human serum (Harvey et al. 2008; Takegawa et al. 2005). The functional significance of these novel structures has yet to be defined.

12.8 IgG Glycan Modification

12.8.1 Remodeling with Enzymes and Genetic Engineering

Since the discovery of the effect of glycosylation on Mab activity, there have been numerous studies that have used in vitro remodeling with glycosyltransferases (Raju et al. 2001; Warnock et al. 2005), production in glycosylation mutant host cells (Kanda et al. 2006), glycoengineering of host cells (Ferrara et al. 2006a; Imai-Nishiya et al. 2007; Kanda et al. 2007b; Natsume et al. 2005; Natsume et al. 2006; Niwa et al. 2004; Yamane-Ohnuki et al. 2004) and siRNA knockout (Mori et al. 2004) to modify Fc glycans of IgG to study the effect of glycan structure on function. A CHO knockout cell line for FUT8 (α1-6 fucosyltransferase) was more effective in producing non-fucosylated IgG (Yamane-Ohnuki et al. 2004) than lectin-resistant CHO cell lines (Kanda et al. 2006). In CHO cells, a double knockout of FUT8 and GMD (GDP-mannose 4,6 dehydratase, an enzyme required for the production of GDP-fucose) using an siRNA expression vector allowed the production of fully non-fucosylated antibodies (Imai-Nishiya et al. 2007). However, GMD knockout alone can also produce non-fucosylated Mab (Kanda et al. 2007a). The production of low or non-fucosylated Mab may result in high efficacy in vivo and with even low dosages result in high cellular cytotoxicity of tumor cells and some initial trials have shown the effectiveness of low doses of nonfucosylated antibody therapies (Yamane-Ohnuki and Satoh 2009). One study has suggested that non-fucosylated IgG1 can compete with the inhibition of plasma IgG of ADCC, thus resulting in a more effective product (Iida et al. 2006).

Overexpression of GlcNAc transferase III (GnT-III) enzyme with an altered localization domain produced IgG with a high percentage of non-fucosylated bisected complex structures (Umana et al. 1999). Higher expression levels led to hybrid glycans that were afucosylated and bisected. The co-expression of GnT-III with an α-mannosidase II enzyme allows further modifications to increase the amount of complex bisected structure (Ferrara et al. 2006a). The afucosylated bisected hybrid was also found to inhibit complement dependent-cytotoxicity (CDC) (Ferrara et al. 2006a). This suggests that the composition of the sugars linked to the α-6 Man arm are important in maintaining conformation of the Fc region for C1q binding.

Affinity studies with glycoengineered IgG have shown that the Fc glycan with bisecting GlcNAc interacts with the glycan (Asn 162) of the FcγRIIIa receptor, and is reliant on the lack of at least one of the fucose residues on the two Fc glycans (Ferrara et al. 2006b). In another study, remodeling of recombinant Mab, Rituxan and Herceptin, using recombinant GlcNac transferase III to add a bisecting β1,4-linked GlcNAc to Fc glycans increased the ADCC response by 10-fold, with little effect on the CDC response (Hodoniczky et al. 2005). However, a significant percentage of the Mab contained core fucosylation, suggesting that the presence of bisecting GlcNAc can "override" the presence of fucose in promoting ADCC activity, but does not increase binding to the same degree as removal of the core fucose.

The specificity of the ADCC response to glycan structure is possibly more than just a response to afucosylated glycans. Chimeric CD19 antibodies were produced in Sf21 insect cells, resulting in fucosylated paucimannosidic N-linked glycosylation, and in a human 293 T cells, producing primarily fucosylated biantennary glycosylation (Barbin et al. 2006). The antibody with fucosylated paucimannosidic glycans elicited a stronger concentration-dependent ADCC response than antibody produced in the human cell line with fucosylated complex glycan. Glyco-engineering of the human cell line with enhanced GnTIII activity increased the amounts of bisecting GlcNAc from 9.5 to 78.7% and increased the proportion of afucosylated glycan with a bisecting GlcNAc from 0.5 to 15.9% and resulted in an increase of the ADCC activity to the level of the paucimannosylated antibody. However, the presence of the core fucose on the highly active paucimannosylated antibody and the lack of a bisecting GlcNAc suggest that the effect of the glycan structure on ADCC activity is more complex than just afucosylation and bisecting GlcNAc structures. The authors do not exclude the possibility that the antibody from the insect cell may contain small amounts α1,3-linked fucose that may enhance the ADCC response.

12.8.2 Remodelling with Glycosidase Inhibitors

An alternate method of modifying glycan structure of Mab is through the addition of glycoprotein processing inhibitors to the culture media during production, which can result in alternative glycan structures such as high mannose (with or without glucose), or hybrid structures containing one arm of mannose and one arm with complex sugars (GlcNAc, Gal, NeuAc). Modification of glycosylation using glucosidase inhibitors (castanospermine and methydeoxynojirimycin) that results in $GlcNAc_2$-$Man_{8-9}Glc_{1-3}$ glycans, a mannosidase I inhibitor (deoxymannojirimycin) that results in high mannose structures ($GlcNAc_2$-$Man_{8\ \mathrm{or}\ 9}$) and swainsonine, a mannosidase II inhibitor that results in a hybrid glycans, all had no affect on secretion of IgG (Hashim and Cushley 1987) (Fig. 12.2). Early studies on functional relationships of altered glycan structures, found the glucosidase inhibitors castanospermine and methydeoxynojirimycin, and a mannosidase inhibitor, deoxymannojirimycin, significantly enhanced ADCC (Rothman et al. 1989a). IgG1 produced in the presence of kifunensin, another mannosidase I inhibitor also had high ADCC and FcγIIIa binding (Kanda et al. 2007b; van Berkel et al. 2010; Zhou et al. 2008) and reduced C1q binding, but with an apparently unaltered in vivo half-life in mice (Zhou et al. 2008). X-ray crystallography identified a more open conformation of the Fc region with kifunensin-treated high mannose glycosylation that is proposed to facilitate receptor binding (Crispin et al. 2009). A decrease in the core fucosylation in the presence of the mannosidase I and glucosidase inhibitors may explain the increased ADCC response. However, IgG produced with 33% fucosylation still had a 6 fold increase (near maximal activity) in ADCC response (van Berkel et al. 2010), again suggesting that total reduction in core fucosylation is not required for ADCC enhancement which is consistent with other reports (Kanda et al. 2006; Scallon et al. 2007a). The effect of kifunensine is also observed in IgG produced via transient transfection of CHO and HEK293 cells (van Berkel et al. 2010). Swainsonine, a mannosidase II inhibitor producing hybrid glycans, did not enhance ADCC which may be partially due to the presence of fucose on the hybrid glycan (Rothman et al. 1989a).

Currently there are several second generation glycoengineered anti-CD20 Mabs in development which have low or no fucose and all have increased ADCC activation and increased binding to FcγRIIIa (Abès and Teillaud 2010). Hopefully this will translate into better efficacy in patients, without an increase in adverse reactions. However, based on studies with fucosylated paucimannose and high mannose glycans and their ability to increase ADCC, more work is required to understand the explicit involvement of the glycan in Fcγ receptor interactions, and also to elucidate the mechanism of glycan control over effector function of native IgG. Determining specific glycosylation patterns affecting activation of these pathways will allow "glyco-customization" of Mabs to activate or down regulate specific effector functions and thus increase the efficacy of the Mab for the specific clinical function.

12.9 IgG Fab Glycosylation

Glycosylation of the variable regions (V_L and V_H of Fab) of monoclonal antibodies (Wallick et al. 1988; Wright et al. 1991), as well as a significant percentage (up to 30%) of serum IgG (Jefferis 2005a; Jefferis 2007; Stadlmann et al. 2009), is well established. Glycosylation sites can be germ line or introduced through somatic mutation, as in follicular lymphoma (Zhu et al. 2002). Although the presence of Fab glycosylation has been identified some time ago, the structure/function relationship is not as well studied as Fc glycosylation, but can have significant consequences on the effects of serum IgG and Mab. Fab glycans are more heavily galactosylated and sialylated than Fc glycans, possibly due to greater accessibility to transferase enzymes (Jefferis 2007). Analysis of glycans from Fab glycosylation found predominantly mono-sialylated structures (Lim et al. 2008; Stadlmann et al. 2009). These glycans are more readily bound by SNA-lectin than the Fc glycan and therefore may be more exposed on the IgG than Fc glycans. LC/MS analysis identified the Fv glycan structure of a humanized IgG1 as a core-fucosylated, biantennary complex structure, similar to structures found in the Fc region, but the Fv glycan has a higher sialic acid content overall (1.6–2.5 mols sialic acid/mol antibody) and contained Neu5Gc instead of NeuAc (Huang et al. 2006). The functional role of Fab sialylation has yet to be determined. Glycosylation of the Fv region did not have a strong correlation with antibody clearance, but loss of galactose slightly increased clearance rates (Huang et al. 2006). Variable sialylation of the Fv region of a monoclonal IgG1 and lack of glycosylation had no effect on clearance rates of the antibody in mice (Millward et al. 2008). However, introducing glycosylation sites into the single chain Fv fragment of a bispecific single-chain diabody prolonged the circulation time by 2–3 fold (Stork et al. 2008). Some have suggested that because of their sialic acid content Fv glycans may be involved in the anti-inflammatory effect (Abès and Teillaud 2010). N-linked glycosylation of the V_H region of a human Mab for factor VIII (FVIII) did not affect the affinity of the antibody but was found to enhance its FVIII neutralizing activity (Jacquemin et al. 2006). Molecular modeling with the primary structure (a fucosylated biantennary structure with two sialic acids) suggests that the Fv glycan may sterically hinder the active site of FVIII.

Fab regions also have unusual attachment sites for glycosylation. A complex biantennary fucosylated oligosaccharide has been found on a non-consensus glutamine residue (Gln-106) in the V_L domain of a human IgG2 antibody produced in CHO cells (Valliere-Douglass et al. 2010). Interestingly, low levels (0.5–2.0%) of a glycosylated non-consensus sequence has also been identified on the C_H1 (heavy chain) of a human IgG2 antibody produced in CHO, as well as IgG1 in human serum (Valliere-Douglass et al. 2009). Additionally, O-linked mannosylation with one mannose residue has been identified in the V_L chain of a human IgG2 monoclonal antibody produced in CHO and COS cells (Martinez et al. 2007). IgG produced by follicular lymphoma have additional Fv region glycosylation sites. However, glycans on the Fv region are predominantly oligomannose structures, while maintaining normal complex glycosylation on the Fc region (Radcliffe et al. 2007). Production of scFv derived from tumor cells add oligomannose to the introduced sequons and fail to glycosylate the natural glycan site suggesting an important biological role for altered glycosylation of the Fv region in tumor cells (McCann et al. 2008). The biological significance of these alternate glycosylation sites has not been determined and not yet been identified on other Mabs, but with better methods of identification with MS analysis of glycopeptides, more aberrant glycosylation sites and types may be identified, and perhaps linked to alternate functions.

12.10 Production of Mabs

12.10.1 Bacterial Systems

Prokaryotes (bacteria) are useful in the production of small recombinant proteins. However, the limitation is that prokaryotes do not have chaperone proteins that ensure correct folding of larger recombinant proteins or Mab (Schillberg et al. 2003). But much more limiting is that bacteria do not have N-linked glycosylation machinery, therefore not allowing the production of glycosylated recombinant proteins. However, recent identification of N-glycosylation patterns in *Campylobacter jejeuni* and the transfer of this capacity to *E. coli* through genetic engineering has raised the possibility of glycosylated recombinant proteins and Mab production in prokaryotic organisms, although the sequence of glycan is very different than mammalian cells (Chiba and Jigami 2007). Lack of glycosylation in Mab (IgG) produced in prokaryotes limits their ability to carry out the ADCC and complement activity. However, recent work has generated aglycosylated variants of IgG expressed in *E. coli* which are capable of eliciting an ADCC response via binding to FcγRI on dendritic cells (Jung et al. 2010).

For Mabs in which neither ADCC or complement activity is required, bacterial production systems may prove to be useful and much less costly for production.

12.10.2 Yeast and Fungus

Yeast and fungal organisms are potentially a low cost production system and have been utilized for the production of antibodies and antibody fragments (reviewed in Gasser and Mattanovich 2007). They are capable of N-linked glycosylation, disulfide bond formation and proper protein folding, but their glycans are predominantly high mannose structures that contain extended mannose chains, not typically found in mammalian cells, and therefore are potentially immunogenic and at risk of higher clearance rates due to mannose receptors or lectins (Chiba and Jigami 2007). This has necessitated the "humanization" of yeast glycosylation for the production of recombinant proteins and Mabs.

The first humanized mutant of *Sacchromyces cerevisiae* was a mutant lacking the α1,6-mannosyltransferase (OCH1) that is responsible for the addition of extended mannose residues to the Man_8-$GlcNAc_2$ glycan. A subsequent knockout mutant lacking three genes (OCH1, MNN1 and MNN4) has the ability to produce a Man_5-$GlcNAc_2$ structure which is identical to human intermediate mannose structure (Chiba et al. 1998). Full-length antibodies have also been produced in other host systems such as *Aspergillus niger*, but the glycans were larger than human glycoforms (Ward et al. 2004).

A methylotrophic yeast species, *Pichia pastori,* has also been engineered to produce biantennary glycan structures on recombinant proteins that are consistent with human glycosylation. The addition of several genes necessary for the glycosylation machinery were introduced from combinatorial libraries, producing mutants with a repertoire of enzymes and nucleotide-sugar substrates which are able to mimic human glycosylation (Chiba and Akeboshi 2009; Choi et al. 2003; Hamilton et al. 2003). The added advantage of glycoengineered yeast production is the ability to control the population of glycoforms, which is much smaller than in mammalian cells (Gerngross 2004). In 2006 full length Mabs of rituximab with humanized glycosylation were produced using several glycoengineered cell lines expressing heavy and light chains, followed by enzymatic glycan processing of the products (Li et al. 2006). Antibody with several types of glycoforms was generated, but antibody containing non-fucosylated G0 and G2 glycan structures displayed high affinity for the FcγRIIIa receptor (both the normal and low affinity polymorphic form), and high activity in B-cell depletion assays (Li et al. 2006). Surprisingly, over 100-fold increase in binding to the low affinity polymorphic variant of FcγRIIIa compared to the commercial rituximab was observed. This was a major step in the development of yeast expression systems for commercial purposes. Humanized glycans on yeast-produced IgG1-Fc have also been synthesized chemoenzymatically, through removal of yeast glycans with Endo H, followed by enzymatic transglycosylation of a galactose-containing glycan (Wei et al. 2008). This modified glycan with a $GlcNAc_2Man_3Gal_2$ structure lacking the two internal GlcNAc residues was able to bind to the FcγRIIIa receptor 1.4 fold more effectively than CHO-generated IgG-Fc. The authors suggest that this method may be useful in determining modified glycan structures that may be beneficial in developing Mabs for patients with low binding receptor polymorphisms.

12.10.3 Plants

Production of antibodies in plants is another potentially low cost system and has recently been reviewed (Ko et al. 2009; Orzaez et al. 2009). There are many advantages of plant production such as large capacity, ease of manipulation, and avoidance of animal pathogens. Other considerations are increased downstream processing for antibody purification, and environmental and biosafety issues with regard to contamination of food crops. But one of the major hurdles to overcome is the differences in N-linked glycosylation between plants and mammalian cells (Ko et al. 2009). Plant glycans contain β1,2-xylose, α1,3-fucose (instead of α1,6-fucose in mammals) and a β1,3-Gal-α1,4-Fuc (also known as Lewis a), and therefore have the potential of generating immunogenic glycans (Saint-Jore-Dupas et al. 2007). Additionally, plants do not have high levels of sialic acid and galactose (Bakker et al. 2001). Although sialic acid is not generally a requirement for therapeutic Mabs, terminal galactosylation is necessary for efficacy.

Aglycosylated Mabs produced in plants in the presence of tunicamycin (preventing N-linked glycan synthesis) or through genetic mutation of N-glycosylation sites, eliminates possible immunogenicity but retains proper folding of Mabs with no effect on Fab binding activity (Nuttall et al. 2005). However, aglycosylated antibodies have reduced binding to Fc receptors, which reduce their effectiveness in targeted therapies. The second alternative is to induce modifications to the glycan to reduce immunogenicity. Two different methods used have been (1) attaching a KDEL or HDEL C-terminal signal which results in retention of the glycoprotein in the ER, eliminating Golgi processing and resulting in the production of only high mannose oligosaccharides (Sriraman et al. 2004; Tekoah et al. 2004; Triguero et al. 2005) and (2) glycoengineering used to eliminate immunogenic sugars, and increase galactose and sialic acid. The first approach increases the clearance of antibody in vivo through the high-mannose receptor (Ko et al. 2003) thus reducing its efficacy. In the second approach, the knockout of α1,3 fucosyltransferase (FT) and β1,2 xylosyltransferase (XT) using RNA interference has proven effective in the production of Mabs with more humanized glycosylation in several plant species (Cox et al. 2006; Schahs et al. 2007; Strasser et al. 2008). Expression of a chimeric form of β1,4 galactosyltransferase (GalT) containing a medial Golgi anchor portion in tobacco plants allowed high yields of antibody with small amounts of fucose and no detectable xylose (Bakker et al. 2006). The GalT is thought to compete with the FT and XT reducing the content of the potentially immunogenic sugars.

Low yields in stably transfected plants have lead to the use of transient expression of Mabs in plant systems which have higher levels of Mab production. Transient expression of monoclonal antibody along with a chimeric GalT containing a GNT1 membrane anchor portion resulted in high yields of Mab with human-like glycosylation (Vezina et al. 2009). Recent work has transiently transfected six genes necessary for the synthesis, transport and sialylation with NeuAc along with a recombinant antibody into *Nicotiana benthamiana* (Castilho et al. 2010). The monoclonal antibody product contains a highly sialylated Fc and are fully functional. Although sialylation is not required for activation of ADCC and complement pathways in monoclonal antibody therapy, sialylated Mabs have been shown to be efficient in reducing inflammation and therefore this may be a useful and cost effective way of producing Mabs for anti-inflammatory treatments. Production of Mabs in plants with fully humanized glycosylation is now possible and may produce Mabs that are equally, if not more effective, than mammalian cell produced counterparts, with more homogeneous glycosylation.

12.10.4 Insects

Insect cell lines have also been useful as a host for the production of recombinant proteins and more recently for therapeutic Mab production. Stable transfected clones can be generated, but the baculovirus-lepidopteran insect cell transient expression system has been widely used for low cost production of large amounts of N-glycosylated recombinant proteins in serum-free media with minimal downstream processing (Ahn et al. 2008; Tomiya et al. 2004). Other advantages are proper protein folding and the avoidance of human pathogens. However, insect glycosylation produces N-linked glycans that are much shorter than mammalian glycans, typically a paucimannosidic ($Man_{3-6}GlcNAc_2$) structure and α(1,6) linked fucose residues and can also contain one or two terminal GlcNAc residues (Altmann et al. 1999; Tomiya et al. 2004). Some insect cells (e.g. High Five cells) attach α1,3 linked fucose to the core GlcNAc which is potentially immunogenic (Hsu et al. 1997). Glyco-engineering of some insect lines has been successful in producing elongated, more human-like glycosylation (Jarvis 2003; Tomiya et al. 2004).

Transient expression of antibodies in insect cells has been useful in understanding the structure/function relationship of Mab glycans. A comparison of chimeric CD19 antibodies produced in insect cells (Sf21), human 293T and 293T glyco-engineered with GnTIII found the Mab produced in insect cells had an ADCC response (tumor lysis) much higher than the 293T produced Mab and comparable to the Mab produced in the glycoengineered cell line (Barbin et al. 2006). Hence, these unexpected results show that Mabs with paucimannosidic glycans ($GlcNAc_2$-(Fuc)-Man_3) are capable of eliciting ADCC responses through the appropriate receptor. The authors suggest that the smaller glycan structure may allow a closer conformation of the CH2 domains, allowing enhanced Fc receptor binding. Recent work has produced a functional anti-colorectal cancer Mab that contains insect-like N-glycosylation of the Fc site, and is able to bind human colorectal cancer cells and FcγR1 (Song et al. 2010). Besides the paucimannosidic glycans with $Man_{5\text{-}6}GlcNAc_2$ structures, $GlcNAc_2$-Man_3-$GlcNAc_2$ structures were also present. Another chimeric antibody produced with the baculovirus/insect expression system has high ADCC and CDC activity (Shen et al. 2009), but glycosylation analysis was not reported. These studies suggest that functional binding and some effector function of Mabs is not dependent on full mammalian glycosylation patterns, but can occur with glycans lacking terminal galactose. However, N-linked glycans with a high content of mannose can be cleared efficiently with the mannose binding receptor, and thus the efficacy of insect-produced Mab's in in vivo treatment must be demonstrated. An Fab fragment produced in Drosophila S2 cells was found to be glycosylated but to a lesser degree than glycosylation by a mammalian hybridoma of the full Mab (Backovic et al. 2010). Thus glycosylation of the Fab region and variations in production must be considered, even though the functionality of this glycosylation site has not been determined.

12.10.5 Transgenic Animals

The production of recombinant proteins in transgenic animals has several advantages and disadvantages, but optimization of systems for production and acceptance by regulatory agencies is still in the early stages (Houdebine 2009). Ruminant animals (goats, sheep, cattle) have the greatest potential for low cost production of recombinant proteins in their milk. However, besides other disadvantages such as long maturation time and possibility of prion contamination, they also have lower levels of glycosylation. An alternate production system is transgenic chickens with isolation of recombinant proteins from the eggs. A monoclonal antibody produced in chimeric chicken eggs was glycosylated on both the Fc and Fab regions (Zhu et al. 2005). MS analysis of total glycan revealed higher levels of mannose glycans than in CHO, but also complex (branched with GlcNac and some Gal) and hybrid types were present. Complex structures had low fucose and sialic acid content.

Despite the glycosylation differences between the CHO and chicken-produced Mab, similar binding affinities were observed, but chicken derived Mabs had a shorter half-life in mice. The lack of fucose corresponded to an increase in ADCC activity. Interestingly, the chicken Mab had a higher stability (higher melting point) than CHO-derived Mab that was attributed to the different glycosylation patterns. Transgenic silkworm larvae and pupae have also proven to be a cost effective system for the production of recombinant proteins containing terminal GlcNAc and Gal (Kato et al. 2010), and are another potential production systems for monoclonal antibodies.

12.10.6 Mammalian Cell Lines

The optimum cellular production systems for recombinant therapeutic Mabs, with regard to glycosylation and other posttranslational modifications, are mammalian cell lines. Common Mab cell line production platforms include hybridoma, CHO (Chinese hamster ovary), NS0 (murine myeloma), SP2/0 (murine), HEK 293 (human embryonic kidney) and PER.C6 (human cell lines). The glycosylation patterns of Mab produced in these cell lines reflect the species specific glycosylation (Raju et al. 2000).

Major differences in glycosylation of the same antibody expressed in different cells lines have been observed and affect the biological function of the Mab (Lifely et al. 1995). An anti-HIV antibody produced in hybridoma was non-neutralizing to the virus, but the Mab produced in CHO cells was able to neutralize the virus (Miranda et al. 2007). Differences in glycosylation patterns with increases in sialic acid, fucose and N-acetylglucosamine in glycans on the Fc region in CHO-produced antibody were attributed to the increased neutralization by the Mab.

Some mammalian cell lines produce glycans with potentially immunogenic sugar residues, thus reducing their usefulness for therapeutic antibody production. NS0 cells produce glycans with Galα1,3Gal terminal structures, which are immunogenic (Galili 2004; Sheeley et al. 1997). Cetuximab, a monoclonal produced in SP2/0 mouse hybridoma cells, has resulted in an IgE-induced anaphylactic response in some patients due to the presence of the Galα1,3Gal structure in the Fab region (Chung et al. 2008). Although addition of sialic acid is usually low, most Mabs produced in nonhuman (rodent) cell lines add Neu5Gc, as well as NeuAc (Raju et al. 2000). Recent work has shown that Neu5Gc is present in some commercial Mabs and is immunogenic in humans (Ghaderi et al. 2010). Mab produced in mouse myeloma had a greater percentage of Neu5Gc of the total sialic acid than antibody produced in CHO cells. The addition of NeuAc to the media was shown to reduce Neu5Gc content. In addition, CHO cell lines normally only attach sialic acid in an α2,3 linkage, as the α2,6 sialyltransferase is absent. However, a CHO cell with 2,6 sialyltransferase activity has been engineered (Bragonzi et al. 2000). This has implications for anti-inflammatory IgG that requires a α2-6 sialic acid on the Fc glycan (Anthony et al. 2008). Per.C6, an adenovirus-transformed human retinal cell line, has the advantage of human glycosylation patterns. Glycans of IgG produced in Per.C6 are fucosylated biantennary chains with 50% of chains mono Gal (G1), 25% di-Gal (G2) and 25% agalactosylated (G0). Sialic acid and bisecting GlcNAc are at very low levels (Jones et al. 2003).

HEK 293 is another human cell line which has proven useful for the transient expression and production of recombinant proteins for rapid screening of products (Thomas and Smart 2005). Transient IgG transfection of HEK 293 was used in the screening of altered glycosylation of IgG grown in the presence of kifunensine to increase high mannose content and reduce core fucosylation (van Berkel et al. 2010).

12.11 Effect of Culture Conditions on Mab Glycosylation

The choice of expression cell line for production of Mab has a predominant effect on the glycosylation pattern of the product, and these can be further refined with glycoengineering and glycosylation processing inhibitors (above). However, modification of the glycosylation can also occur with variations to the culture conditions. As novel platforms, production systems and feed strategies are investigated for enhanced production, their effect on glycosylation must be monitored to ensure product quality and efficacy. Dissolved oxygen (Restelli et al. 2006), pH (Borys et al. 1993), nutrient concentrations (Andersen et al. 2000; Baker et al. 2001), and ammonia (Andersen and Goochee 1994; Hayter et al. 1992; Yang and Butler 2000) are all known to affect glycosylation of recombinant proteins and similarly may affect Mab glycosylation. Manufacturers are now adopting quality by design (QbD) concepts in their process development and validation, which uses statistical analysis to assess how changes in culture parameters affect product critical quality attributes (CQA) (Horvath et al. 2010). Glycosylation and its effect on CDC and ADCC activity are included in these parameters.

IgG2a produced in a hybridoma had a decrease in galactosylation with elevated osmolality, but galactose content decreased with increased CO_2 concentration (Schmelzer and Miller 2002). Galactosylation of Mab in hybridomas has also been found to be greater with slight increases in pH (to 7.4), and also to enhance incorporation of NeuAc over Neu5Gc (Muthing et al. 2003). Others have also observed modest changes in glycosylation of IgG in hybridomas under different pH conditions (Rothman et al. 1989b). A loss of galactosylation with a reduction of the G2 glycan (30–12%) has been observed with decreasing DO (100–10%) (Kunkel et al. 1998). This is suggested to be due to a decrease in UDP-Gal, or possibly by steric hindrance by the earlier formation of the interchain disulfide bond in low DO conditions. Early sequestering of the Asn297 glycan chain in the pocket between the heavy chains may reduce the exposure to Golgi galactosyltransferase (Kunkel et al. 1998).

Culture methods, media formulations and additives can all significantly affect glycosylation patterns of Mabs. IgG produced in serum-free media had a higher degree of sialylation than IgG produced in serum-containing media or ascites (Patel et al. 1992). Also static and spinner cultures of IgG hybridoma had higher galactosylation than IgG produced in ascites and hollow-fiber systems (Cabrera et al. 2005). Higher yield systems had greater variability in galactosylation but serum did not affect glycosylation in spinner flasks and static cultures. In another study, antibody produced in hybridomas grown in serum-free media (SFM) have a much higher percentage (58%) of G0 glycans (lacking galactose) compared to Mab produced with serum-containing (28% G0) or chemically defined media (CDM) (32% G0) (Serrato et al. 2007). This was attributed to a higher β-galactosidase activity in the media. But increased fucosylation was also found in Mab produced in the SFM and CD media over serum-containing media, which would have consequences on ADCC activity. Decreased sialylation was also observed in the SFM and CD media. As the general trend is for production in serum-free and chemically defined media, careful evaluation of effects of different media on glycosylation of individual Mab must be determined, as alternative media formulations may have very different effects.

Fed-batch and perfusion cultures are commonly used to improve productivity and longevity over batch cultures. Feed strategies should consider the effect on glycosylation of the product, especially in the addition of sugars (galactose, glucose, glucosamine) or other components (glutamine) with known effects on glycosylation of recombinant proteins (Butler 2006). The addition of glucosamine to increase UDP-HexNac in NSO cells, significantly reduced the galactosylation of the Fc glycan (Hills et al. 2001). However, galactosylation was only slightly increased with the addition of galactose. Consistency of glycosylation can be maintained with different feed strategies, as a chemically defined feed strategy found similar glycosylation of a Mab compared to a fed batch with a yeast extract (Hermes and Castro 2010). The type of bioreactor system may also affect glycosylation patterns. The extent of galactosylation of murine IgG was consistently lower in a hollow fiber system compared to batch and continuous cultures (Majid et al. 2007). The authors suggest that high cell densities in the hollow fiber system may result in conditions of low DO, which has been shown to reduce galactosylation (Kunkel et al. 1998). Others have also reported decreased galactosylation of IgG in hybridomas cultured in hollow fiber and membrane bioreactors (Cabrera et al. 2005).

Clonal variation can also affect glycosylation. A high degree of variability in galactosylation, non-core-fucosylation and Man5 content was found in six different IgGs produced in 105 stable CHO-K1SV clones (van Berkel et al. 2009). The authors suggest that these variations may be large enough to affect complement activation, ADCC effector function and clearance by high mannose receptors. Some variations in galactosylation and core-fucosylation were also observed in scale-up from shaker flasks to fed batch bioreactors, indicating that other factors such as feed strategies and dissolved O_2 might have an effect.

A recent report of misincorporation of serine for asparagine in Mab during the growth phase of a fed-batch CHO culture (Khetan et al. 2010), may have significant consequences on product quality with regard to occupancy of glycans. The study did not analyze changes in glycosylation macroheterogeneity, or specifically identify changes to asparagine in known sequons.

The stability of glycosylation of Mab in long term cultures has also been analyzed. An antibody/interleukin 2 fusion protein showed very similar glycosylation patterns over 62 generations in culture (Cruz et al. 2002), indicating that glycosylation can remain consistent. However, this comparison was for static cultures. Repeated batch cultures have been investigated as a means of increasing large scale production of antibody, rather than producing seed cultures from frozen stocks. Although batch CHO cultures have shown long term stability, effect on product integrity with regard to glycosylation was not analyzed (Kaneko et al. 2010), but others have found no significant changes in glycosylation over 60 generations of CHO cultures (van Berkel et al. 2009).

Age of the culture significantly affects glycosylation patterns. Increases in fucosylated G0 glycans of a recombinant IgG4 were observed from day 8–14 fed-batch culture, with a decrease in fucosylated G1 and G2 glycans emphasizing the importance of time of harvest (Reid et al. 2010). The authors suggest changes in the productivity may affect efficiency of glycosylation pathway enzymes. However, other factors such as nutrient levels might also affect glycosylation. The same study has found an increase in glycosylated half antibodies (one heavy + one light chain) during different processing of Mabs. Cell lysis due to shear may cause reduction of disulfide bonds, and decrease product quality. Hydrodynamic stress can also significantly increase the amount of galactosylation in Mabs produced in CHO cells, although the tested stress levels were suggested to be much higher than the normal range in bioreactors (Godoy-Silva et al. 2009). Overexpression of a transcription X-box binding protein (XBP-1) was found to increase the ER and yield of Mab produced in a fed-batch CHO culture, but maintained the glycosylation profile (Becker et al. 2008).

Transient transfection for rapid production of Mab can produce glycosylation patterns consistent with stable clones of CHO (van Berkel et al. 2010; Ye et al. 2009). However, transient transfection of HEK293EBNA cells was much different than CHO with higher levels of galactosylation with increased G1F and G2F relative to G0F (Ye et al. 2009). Transient production of a Mab in CHO using reduced temperature (32°C) and recombinant insulin-like growth factor yielded glycosylation patterns on the Mab consistent with stably transfected CHO (Galbraith et al. 2006).

12.12 Glycan Structural Analysis

The demonstrated relationship between Mab glycan structure and function is largely due to the development and refinement of glycan structural analysis. Consequently the need for consistent glycosylation of clinical Mabs and the optimization of glycosylation for activity has further driven the exploration of new techniques in structural analysis of IgG glycan. Analytical techniques for IgG glycosylation and their applications have been extensively reviewed (Huhn et al. 2009). The techniques to isolate, fluorescently label and then identify glycans through various types of chromatography such as high performance liquid chromatography (HPLC), high performance anion exchange chromatography with pulsed amperometric detection (HPAEC-PAD), and capillary gel electrophoresis (CGE) and capillary ion electrophoresis (CIE) are common methods of analysis. Liquid chromatography and CGE of labeled glycans are often used in combination with mass spectroscopy (MS) analysis, or native glycans can also be directly analyzed by MS. Fragmentation patterns using tandem mass spectroscopy (MS/MS) can give detail information on oligosaccharide sequence and branching (Butler and Perreault 2010; Wada et al. 2007). Alternatively, MS analysis of glycopeptides from proteolytically digested glycoprotein can determine the glycan structure at different glycosylation sites of the Mab, particularly in the Fab region. A more recent focus is on the automation of analysis (or high through-put analysis) to eliminate labour intensive techniques and increase analytical output.

Protein A or G purification is a common method for isolation of IgG for analysis. Recent work has found that oligosaccharide structure can affect the Protein A or G separation of antibody due to changes in structure around the CH2–CH3 domain interface of the Fc region at the low pH required for separation (Gaza-Bulseco et al. 2009). Therefore, care must be taken in purification to ensure that selective isolation of glycosylated Mabs does not skew oligosaccharide analysis. Alternatively, the authors suggest that this may be useful in enrichment for glycosylated populations of Mab.

12.12.1 Released Glycan Analysis

A standard quantitative method for glycoprotein glycan analysis using glycan release, fluorescent labeling and identification using normal phase HPLC (NP-HPLC) with HILIC (hydrophobic interaction liquid chromatography) separation has been developed and is widely accepted (Domann et al. 2007; Guile et al. 1996; Royle et al. 2007) and a database of structures based on retention times is available (Campbell et al. 2008) (Fig. 12.6). The first step of oligosaccharide analysis is removal of the glycan from the protein either chemically (hydrazinolysis) or by using a broad specificity glycosidase such as PNGase F (except α1-3 linked fucose-containing glycans from plants). Endo H removes only high mannose glycans and can be used to analyze subpopulations of glycans. PNGase F can also be applied to an in-gel release method following separation on SDS-PAGE. Recent work has identified low level amounts of artifact isomers of some glycan species which result from epimerization of GlcNAc to ManNac (N-acetylmannosamine) with PNGase F release of Mab glycans (Liu et al. 2009). This can be eliminated by reducing the pH to 5.5 for the digestion.

Glycans can be analyzed in their native state, but more commonly are labeled with a fluorescent tag at the reducing end with 2-AB (2-aminobenzaminde), 2-AA (2-aminobenzoic acid), or 2-PA (2-aminopyridine) dependent upon further separation techniques. NP-HPLC separation of 2-AB labeled N-glycan on a TSK-amide column establishes glucose unit (GU) values to compare to the standard database of glycans (Glycobase) (Guile et al. 1996; Royle et al. 2007) and can be combined with

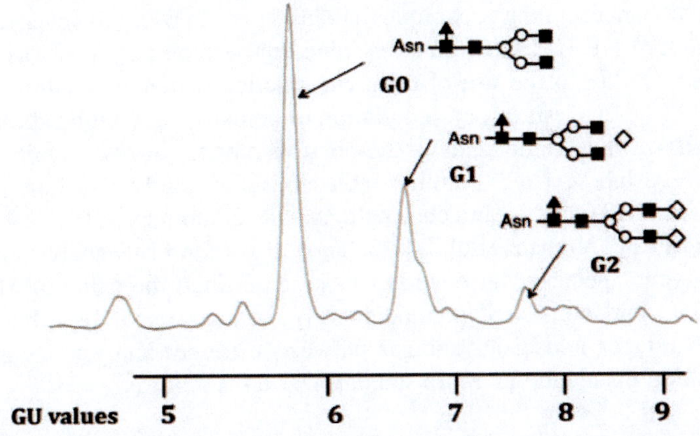

Fig. 12.6 NP-HPLC analysis of 2-AB labeled glycans from IgG produced in NSO cells. Predominant glycans are G0 (no galactose), G1 (isomers of one galactose on either arm), G2 (two galactose)

exoglycosidase digestion for determination of linkages and MS analysis for verification of structures (Kuster et al. 1997). NP-HPLC analysis is the "gold standard" of analysis and has several advantages over other current techniques for glycan analysis including: quantitative values at the femtomole range with stoichiometric 2-AB labeling; ability to analyze for neutral and sialic acid containing glycans at the same time; provides information on sequence and linkage specificity (Royle et al. 2008). New methods using smaller particle size (1.7 μm) amide columns with a UPLC system can reduce run times from 180 to 30 min or less, which is a major advantage in higher throughput analysis (Ahn et al. 2010; Clarke et al. 2009). This is online compatible with MS, and can separate some isomers (G1F and Man6), which are often hard to resolve (Clarke et al. 2009). CGE analysis of fluorescence labeled glycans with APTS (1-aminopyrene-3,6,8-trisulfonate) coupled with a LIF (laser-induced fluorescence) detector also has the advantage of very fast analysis (Domann et al. 2007), and can be directly coupled to ESI-TOF-MS (electrospray time-of-flight) (Gennaro and Salas-Solano 2008).

Native (non-labeled) N-glycan can be directly analyzed using HPAEC-PAD and a new procedure was able to separate and identify neutral and sialic acid-containing glycans with results consistent with MALDI-TOF analysis (Grey et al. 2009). For greater sensitivity with HPAEC-PAD, glycans can be labeled with 2-AA (Dhume et al. 2008) or APTS. MALDI-TOF and ESI-Q-TOF coupled with HILIC fractionation can also be used for analysis of native glycans (Harvey et al. 2008). Graphitized carbon HPLC coupled to ESI-MS of native glycans has been useful in resolving differences between α2,3 and α2,6 linked sialic acid residues and β1,4 and β1,3-linked galactose residues (Stadlmann et al. 2008).

A recent multi-institutional analysis (20 laboratories) compared released glycan and glycopeptide analysis of transferrin and serum IgG from donors using standard chromatographic techniques (NP-HPLC, ESI-MS and MALDI-MS) and found only small variance among techniques (Wada et al. 2007). However, larger variance in standard NP-HPLC chromatographic analysis among labs was noted, and suggested to be due to the use of different reaction protocols leading to variable labeling. MALDI analysis of permethylation of glycans can establish branching and linkages and stabilize sialic acids in the positive mode, however, again variance in results between labs existed. Another inter-laboratory study also found high variability between MS analysis and chromatographic techniques of PNGase F released oligosaccharides (Thobhani et al. 2009). They also found MS analysis could identify more glycan species. These studies have underlined the value of MS analysis as an efficient and consistent method for glycan analysis of IgG, but also indicate the advantages and disadvantages between different analysis techniques and instrumentation exist and that more standardization is required.

12.12.2 Glycopeptide Analysis

Glycopeptide analysis with MS also has the advantage of distinguishing between the more common Fc glycosylation site and the less common Fab glycosylation sites and can differentiate if more than one glycan site exists within the Fab region. LC separation methods coupled with ESI (electro spray ionization) or MALDI-MS have been useful in identifying Mab glycopeptides (reviewed in Huhn et al. 2009) and are useful in distinguishing N-linked glycan from Fc and Fab regions (Lim et al. 2008). Although digestion with papain and reduction of disulphide bonds is required, the method gives a quick qualitative (although not quantitative) structural representation at different glycosylation sites.

12.12.3 Glycoprotein Analysis

With the development of better analytical instrumentation, intact glycosylated IgG can now be analyzed for differences in glycosylation patterns using a number of different methods. CE-LIF and MS analysis are the most useful for rapid analysis and can generate detailed information for quality control in Mab production through the identification of different isoforms. The development of new MS spectrometers (ESI hybrid qTOF and MALDI hybrid qTOF) have allowed the analysis of larger protein masses with better accuracy (Srebalus Barnes and Lim 2007). A comparison of three types of analysis (intact IgG analysis by ESI-Q-TOF MS; MALDI-TOF MS of released glycans; anion exchange separation of 2-AB labeled glycans) found that ESI-Q-TOF MS yielded a quantitative analysis of major galactosylated glycan species with a 1 day preparation time (Siemiatkoski et al. 2006). However, some agalactosyl species were not resolved and species of equal mass could not be distinguished. But MALDI-TOF MS had better sensitivity in distinguishing glycan with smaller agalactosyl glycans when the glycan was removed from IgG. A method using RP-HPLC (reverse phase-HPLC) coupled with ESI-TOF-MS could separate and identify several glycoforms on intact IgG, but high column temperatures with low pH may remove sialic acid residues from glycans (Dillon et al. 2006). In another technique separation, of light and heavy chains through disulphide bond reduction, or fragmentation using protease digestion (e.g. papain or pepsin) can first be done to generate smaller fragments. One study has compared LC/ESI-MS analysis with protease digestion and/or disulfide bond reduction of whole IgG to the standard NP-HPLC analysis of fluorescent labeled glycans (Sinha et al. 2008). They found that generating a Fc/2 fragment (constant region of the single heavy chain) with protease digestion and reduction allowed the best resolution and also detection of low intensity peaks, and gives quantitative glycosylation results comparable to standard normal phase HPLC separation techniques of 2-AB labeled oligosaccharide. However, the limitation of using MS analysis is that it is not able to distinguish between some isomers. Weak cation exchange chromatography can also separate Mab with different macroheterogeneity of glycosylated forms for a intact antibody characterization but for detailed glycan analysis, glycans must be removed by PNGase F (Gaza-Bulseco et al. 2008).

12.12.4 High Throughput Analysis

The focus is now on the establishment of high throughput techniques that will allow the fast and reliable analysis for quality control, to ensure consistent glycosylation patterns in Mab products from batch to batch. But these may also be useful in the future for diagnosis where IgG glycosylation patterns are markers for disease. The current bottleneck in analysis for many techniques is the separate process of removal of glycan from Mab, usually followed by fluorescent labeling. New methods that overcome this or speed the process will contribute to the next level of high throughput analysis. Automation of a NP-HPLC based analysis using a 96-well microtitre plate increases the throughput capabilities and links the analysis to a oligosaccharide database (Glycobase) for automation of identification (Royle et al. 2008). However, processing and analysis of samples still takes several days. Recent reports use microwave-assisted PNGaseF deglycosylation that reduce glycan release to under 30 min (Prater et al. 2009; Sandoval et al. 2007). However, time-consuming fluorescent labeling and cleanup is still required. New MS analysis of intact IgG with better resolving power may overcome these challenges, or novel preparation steps which allow larger scale analysis. A high throughput method utilizing protein A and reverse phase solid phase beads in microtiter plates allows sample preparation of glycopeptides in 1 day with MALDI-TOF analysis and give quantitative results comparable to the standard NP-HPLC method (Bailey et al. 2005). Hansen et al. (2010) has developed a rapid analysis technique (5 h) combining small-scale ultrafiltration, PNGase F glycan removal, protein precipitation and desalting, followed by MALDI-TOF (positive mode) analysis. A technique using phenylhydrazine labeling allows the detection of oligosaccharides in the presence of peptides using MALDI-TOF MS (Lattova et al. 2006) (Fig. 12.7). After PNGase F detachment from glycopeptides, glycans detection is usually suppressed by stronger

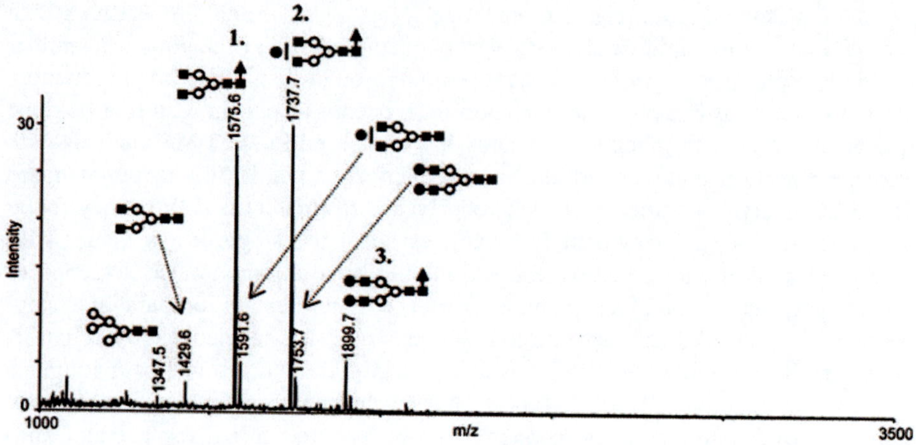

Fig. 12.7 MALDI-QqTOF mass spectrum of Herceptin IgG N-glycans released by PNGase F digestion from glycopeptides and labeled with phenylhydrazine. Symbols: ▲ Fuc; ● Gal; ○ Man; ■ GlcNAc (Lattova et al. 2010). Reprinted with permission from J. Proteome Res., copyright 2010 American Chemical Society

$[M+H]^+$ ions of the peptides. Derivitization with phenylhydrazine increases the detection of the glycans and allows analysis of peptide occupancy and glycan profiles together. An additional step allows the release of oligosaccharide from the glycopeptides using PNGase F and visualization directly on the target matrix (2-aza-2-thiothymine/phenylhydrazine hydrochloride) which shortens the reaction time, eliminates the lengthy labeling cleanup process, prevents loss of sample (Lattova et al. 2007). The drawback of MS analysis is that sialic acid is not easily detected in positive mode. Also, linkage specificity cannot be determined with MS alone.

12.13 Conclusion

Systems biology has led to significant advances in genomics and proteomics which are both based on the analysis of template-based polymers. Because of the structural diversity of carbohydrates and their non-template based structures advances in glycomics have been much slower (Marino et al. 2010). However, analytical techniques for profiling and sequencing glycans have advanced rapidly within the last few years. This has occurred with the realisation of the functional importance of glycans in relation to the therapeutic activity of biopharmaceuticals. High-throughput techniques in mass spectrometry and liquid chromatography now enable high precision analysis of glycoconjugates. These techniques are important not only for the growing list of licensed biopharmaceuticals produced to treat unmet medical needs but also for the rapid production of biosimilars (Kawasaki et al. 2009). The structural validation of these biosimilars is becoming an important element in deciding on clinical equivalence with original products (Schellekens 2009) and so the techniques and developments described in this chapter will form an integral part in these future advances.

Acknowledgements We thank Erika Lattova for her comments on the manuscript. We acknowledge the Natural Science and Engineering Council (NSERC) of Canada for funding our activities in protein glycosylation.

References

Abès R, Teillaud J. 2010. Impact of glycosylation on effector functions of therapeutic IgG. Pharmaceuticals 3(1):146–157.

Ahn J, Bones J, Yu YQ, Rudd PM, Gilar M. 2010. Separation of 2-aminobenzamide labeled glycans using hydrophilic interaction chromatography columns packed with 1.7 microm sorbent. J Chromatogr B Analyt Technol Biomed Life Sci 878(3–4):403–8.

Ahn M-H, Song M, Oh E-Y, Jamal A, Kim HS, Ko K, Choo Y-K, Kim B-J, Ko K. 2008. Production of therapeutic proteins with baculovirus expression system in insect cell. Entomol Res 38:S71–S78.

Altmann F, Staudacher E, Wilson IB, Marz L. 1999. Insect cells as hosts for the expression of recombinant glycoproteins. Glycoconj J 16(2):109–23.

Andersen DC, Bridges T, Gawlitzek M, Hoy C. 2000. Multiple cell culture factors can affect the glycosylation of Asn-184 in CHO-produced tissue-type plasminogen activator. Biotechnol Bioeng 70(1):25–31.

Andersen DC, Goochee CF. 1994. The effect of cell-culture conditions on the oligosaccharide structures of secreted glycoproteins. Curr Opin Biotechnol 5(5):546–9.

Anthony RM, Nimmerjahn F, Ashline DJ, Reinhold VN, Paulson JC, Ravetch JV. 2008. Recapitulation of IVIG anti-inflammatory activity with a recombinant IgG Fc. Science 320(5874):373–6.

Arnold JN, Wormald MR, Sim RB, Rudd PM, Dwek RA. 2007. The impact of glycosylation on the biological function and structure of human immunoglobulins. Annu Rev Immunol 25: 21–50.

Backovic M, Johansson DX, Klupp BG, Mettenleiter TC, Persson MA, Rey FA. 2010. Efficient method for production of high yields of Fab fragments in Drosophila S2 cells. Protein Eng Des Sel 23(4):169–74.

Baerenwaldt A, Biburger M, Nimmerjahn F. 2010. Mechanisms of action of intravenous immunoglobulins. Expert Rev Clin Immunol 6(3):425–34.

Bailey MJ, Hooker AD, Adams CS, Zhang S, James DC. 2005. A platform for high-throughput molecular characterization of recombinant monoclonal antibodies. J Chromatogr B Anal Technol Biomed Life Sci 826:177–87.

Baker KN, Rendall MH, Hills AE, Hoare M, Freedman RB, James DC. 2001. Metabolic control of recombinant protein N-glycan processing in NSO and CHO cells. Biotechnol Bioeng 73(3):188–202.

Bakker H, Bardor M, Molthoff JW, Gomord V, Elbers I, Stevens LH, Jordi W, Lommen A, Faye L, Lerouge P, et al. 2001. Galactose-extended glycans of antibodies produced by transgenic plants. Proc Natl Acad Sci USA 98(5):2899–904.

Bakker H, Rouwendal GJ, Karnoup AS, Florack DE, Stoopen GM, Helsper JP, van Ree R, van Die I, Bosch D. 2006. An antibody produced in tobacco expressing a hybrid beta-1,4-galactosyltransferase is essentially devoid of plant carbohydrate epitopes. Proc Natl Acad Sci USA 103(20):7577–82.

Barbin K, Stieglmaier J, Saul D, Stieglmaier K, Stockmeyer B, Pfeiffer M, Lang P, Fey GH. 2006. Influence of variable N-glycosylation on the cytolytic potential of chimeric CD19 antibodies. J Immunother 29(2):122–33.

Beck A, Wagner-Rousset E, Bussat MC, Lokteff M, Klinguer-Hamour C, Haeuw JF, Goetsch L, Wurch T, Van Dorsselaer A, Corvaia N. 2008. Trends in glycosylation, glycoanalysis and glycoengineering of therapeutic antibodies and Fc-fusion proteins. Curr Pharm Biotechnol 9(6):482–501.

Becker E, Florin L, Pfizenmaier K, Kaufmann H. 2008. An XBP-1 dependent bottle-neck in production of IgG subtype antibodies in chemically defined serum-free Chinese hamster ovary (CHO) fed-batch processes. J Biotechnol 135:217–23.

Bieber MM, Twist CJ, Bhat NM, Teng NN. 2007. Effects of human monoclonal antibody 216 on B-progenitor acute lymphoblastic leukemia in vitro. Pediatr Blood Cancer 48(4):380–6.

Borys MC, Linzer DI, Papoutsakis ET. 1993. Culture pH affects expression rates and glycosylation of recombinant mouse placental lactogen proteins by Chinese hamster ovary (CHO) cells. Biotechnology (NY) 11(6):720–4.

Boyd PN, Lines AC, Patel AK. 1995. The effect of the removal of sialic acid, galactose and total carbohydrate on the functional activity of Campath-1H. Mol Immunol 32(17–18): 1311–18.

Bragonzi A, Distefano G, Buckberry LD, Acerbis G, Foglieni C, Lamotte D, Campi G, Marc A, Soria MR, Jenkins N, et al. 2000. A new Chinese hamster ovary cell line expressing alpha2,6-sialyltransferase used as universal host for the production of human-like sialylated recombinant glycoproteins. Biochim Biophys Acta 1474(3):273–82.

Brekke OH, Sandlie I. 2003. Therapeutic antibodies for human diseases at the dawn of the twenty-first century. Nat Rev Drug Discov 2(1):52–62.

Burton DR, Dwek RA. 2006. Immunology. Sugar determines antibody activity. Science 313(5787):627–8.

Butler M. 2006. Optimisation of the cellular metabolism of glycosylation for recombinant proteins produced by Mammalian cell systems. Cytotechnology 50(1–3):57–76.

Butler M, Perreault H. 2010. Approaches and methods for determining Glycosylation. In Encyclopedia of Industrial Biotechnology, Bioprocess, Bioseparation, and Cell Technology (Flickinger MC, Editor) Wiley, NY.

Cabrera G, Cremata JA, Valdes R, Garcia R, Gonzalez Y, Montesino R, Gomez H, Gonzalez M. 2005. Influence of culture conditions on the N-glycosylation of a monoclonal antibody specific for recombinant hepatitis B surface antigen. Biotechnol Appl Biochem 41:67–76.

Campbell MP, Royle L, Radcliffe CM, Dwek RA, Rudd PM. 2008. GlycoBase and autoGU: tools for HPLC-based glycan analysis. Bioinformatics 24(9):1214–16.

Cartron G, Dacheux L, Salles G, Solal-Celigny P, Bardos P, Colombat P, Watier H. 2002. Therapeutic activity of humanized anti-CD20 monoclonal antibody and polymorphism in IgG Fc receptor FcgammaRIIIa gene. Blood 99(3):754–8.

Castilho A, Strasser R, Stadlmann J, Grass J, Jez J, Gattinger P, Kunert R, Quendler H, Pabst M, Leonard R, et al. 2010. In planta protein sialylation through overexpression of the respective mammalian pathway. J Biol Chem 285(21):15923–30.

Chiba Y, Akeboshi H. 2009. Glycan engineering and production of 'humanized' glycoprotein in yeast cells. Biol Pharm Bull 32(5):786–95.

Chiba Y, Jigami Y. 2007. Production of humanized glycoproteins in bacteria and yeasts. Curr Opin Chem Biol 11(6):670–6.

Chiba Y, Suzuki M, Yoshida S, Yoshida A, Ikenaga H, Takeuchi M, Jigami Y, Ichishima E. 1998. Production of human compatible high mannose-type (Man5GlcNAc2) sugar chains in Saccharomyces cerevisiae. J Biol Chem 273(41):26298–304.

Choi BK, Bobrowicz P, Davidson RC, Hamilton SR, Kung DH, Li H, Miele RG, Nett JH, Wildt S, Gerngross TU. 2003. Use of combinatorial genetic libraries to humanize N-linked glycosylation in the yeast Pichia pastoris. Proc Natl Acad Sci USA 100(9):5022–7.

Chung CH, Mirakhur B, Chan E, Le QT, Berlin J, Morse M, Murphy BA, Satinover SM, Hosen J, Mauro D, et al. 2008. Cetuximab-induced anaphylaxis and IgE specific for galactose-alpha-1,3-galactose. N Engl J Med 358(11):1109–17.

Clarke A, Harmon B, DeFelippis MR. 2009. Analysis of 3-(acetylamino)-6-aminoacridine-derivatized oligosaccharides from recombinant monoclonal antibodies by liquid chromatography-mass spectrometry. Anal Biochem 390(2):209–11.

Cox KM, Sterling JD, Regan JT, Gasdaska JR, Frantz KK, Peele CG, Black A, Passmore D, Moldovan-Loomis C, Srinivasan M, et al. 2006. Glycan optimization of a human monoclonal antibody in the aquatic plant Lemna minor. Nat Biotechnol 24(12):1591–7.

Crispin M, Bowden TA, Coles CH, Harlos K, Aricescu AR, Harvey DJ, Stuart DI, Jones EY. 2009. Carbohydrate and domain architecture of an immature antibody glycoform exhibiting enhanced effector functions. J Mol Biol 387(5):1061–6.

Cruz HJ, Conradt HS, Dunker R, Peixoto CM, Cunha AE, Thomaz M, Burger C, Dias EM, Clemente J, Moreira JL, Rieke E, Carrondo MJ. 2002. Process development of a recombinant antibody/interleukin-2 fusion protein expressed in protein-free medium by BHK cells. J Biotechnol 96:169–83.

Dechant M, Beyer T, Schneider-Merck T, Weisner W, Peipp M, van de Winkel JG, Valerius T. 2007. Effector mechanisms of recombinant IgA antibodies against epidermal growth factor receptor. J Immunol 179(5):2936–43.

Deisenhofer J. 1981. Crystallographic refinement and atomic models of a human Fc fragment and its complex with fragment B of protein A from Staphylococcus aureus at 2.9- and 2.8-A resolution. Biochemistry 20(9):2361–70.

Dhume ST, Saddic GN, Anumula KR. 2008. Monitoring glycosylation of therapeutic glycoproteins for consistency by HPLC using highly fluorescent anthranilic acid (AA) tag. Methods Mol Biol 446:317–31.

Dillon TM, Bondarenko PV, Rehder DS, Pipes GD, Kleemann GR, Ricci MS. 2006. Optimization of a reversed-phase high-performance liquid chromatography/mass spectrometry method for characterizing recombinant antibody heterogeneity and stability. J Chromatogr A 1120:112–20.

Domann PJ, Pardos-Pardos AC, Fernandes DL, Spencer DI, Radcliffe CM, Royle L, Dwek RA, Rudd PM. 2007. Separation-based glycoprofiling approaches using fluorescent labels. Proteomics 7(Suppl 1):70–6.

Endo T, Wright A, Morrison SL, Kobata A. 1995. Glycosylation of the variable region of immunoglobulin G–site specific maturation of the sugar chains. Mol Immunol 32(13): 931–40.

Ferrara C, Brunker P, Suter T, Moser S, Puntener U, Umana P. 2006a. Modulation of therapeutic antibody effector functions by glycosylation engineering: influence of Golgi enzyme localization domain and co-expression of heterologous beta1, 4-N-acetylglucosaminyltransferase III and Golgi alpha-mannosidase II. Biotechnol Bioeng 93(5):851–61.

Ferrara C, Stuart F, Sondermann P, Brunker P, Umana P. 2006b. The carbohydrate at FcgammaRIIIa Asn-162. An element required for high affinity binding to non-fucosylated IgG glycoforms. J Biol Chem 281(8):5032–6.

Gala FA, Morrison SL. 2002. The role of constant region carbohydrate in the assembly and secretion of human IgD and IgA1. J Biol Chem 277(32):29005–11.

Galbraith DJ, Tait AS, Racher AJ, Birch JR, James DC. 2006. Control of culture environment for improved polyethylenimine-mediated transient production of recombinant monoclonal antibodies by CHO cells. Biotechnol Prog 22:753–62.

Galili U. 2004. Immune response, accommodation, and tolerance to transplantation carbohydrate antigens. Transplantation 78(8):1093–8.

Gasser B, Mattanovich D. 2007. Antibody production with yeasts and filamentous fungi: on the road to large scale? Biotechnol Lett 29(2):201–12.

Gaza-Bulseco G, Bulseco A, Chumsae C, Liu H. 2008. Characterization of the glycosylation state of a recombinant monoclonal antibody using weak cation exchange chromatography and mass spectrometry. J Chromatogr B Analyt Technol Biomed Life Sci 862(1–2):155–60.

Gaza-Bulseco G, Hickman K, Sinicropi-Yao S, Hurkmans K, Chumsae C, Liu H. 2009. Effect of the conserved oligosaccharides of recombinant monoclonal antibodies on the separation by protein A and protein G chromatography. J Chromatogr A 1216:2382–7.

Gennaro LA, Salas-Solano O. 2008. On-line CE-LIF-MS technology for the direct characterization of N-linked glycans from therapeutic antibodies. Anal Chem 80(10):3838–45.

Gerngross TU. 2004. Advances in the production of human therapeutic proteins in yeasts and filamentous fungi. Nat Biotechnol 22(11):1409–14.

Ghaderi D, Taylor RE, Padler-Karavani V, Diaz S, Varki A. 2010. Implications of the presence of N-glycolylneuraminic acid in recombinant therapeutic glycoproteins. Nat Biotechnol 28(8):863–7.

Ghirlando R, Lund J, Goodall M, Jefferis R. 1999. Glycosylation of human IgG-Fc: influences on structure revealed by differential scanning micro-calorimetry. Immunol Lett 68(1):47–52.

Godoy-Silva R, Chalmers JJ, Casnocha SA, Bass LA, Ma N. 2009. Physiological responses of CHO cells to repetitive hydrodynamic stress. Biotechnol Bioeng 103(6):1103–17.

Grey C, Edebrink P, Krook M, Jacobsson SP. 2009. Development of a high performance anion exchange chromatography analysis for mapping of oligosaccharides. J Chromatogr B Analyt Technol Biomed Life Sci 877(20–21):1827–32.

Guile GR, Rudd PM, Wing DR, Prime SB, Dwek RA. 1996. A rapid high-resolution high-performance liquid chromatographic method for separating glycan mixtures and analyzing oligosaccharide profiles. Anal Biochem 240(2):210–26.

Hamilton SR, Bobrowicz P, Bobrowicz B, Davidson RC, Li H, Mitchell T, Nett JH, Rausch S, Stadheim TA, Wischnewski H, et al. 2003. Production of complex human glycoproteins in yeast. Science 301(5637):1244–6.

Hang HC, Bertozzi CR. 2005. The chemistry and biology of mucin-type O-linked glycosylation. Bioorg Med Chem 13(17):5021–34.

Hansen R, Dickson AJ, Goodacre R, Stephens GM, Sellick CA. 2010. Rapid characterization of N-linked glycans from secreted and gel-purified monoclonal antibodies using MALDI-ToF mass spectrometry. Biotechnol Bioeng 107(5):902–8.

Harvey DJ, Crispin M, Scanlan C, Singer BB, Lucka L, Chang VT, Radcliffe CM, Thobhani S, Yuen CT, Rudd PM. 2008. Differentiation between isomeric triantennary N-linked glycans by negative ion tandem mass spectrometry and confirmation of glycans containing galactose attached to the bisecting (beta1-4-GlcNAc) residue in N-glycans from IgG. Rapid Commun Mass Spectrom 22(7):1047–52.

Hashim OH, Cushley W. 1987. Role of processing of N-linked oligosaccharides in control of immunoglobulin secretion from rat hybridomas. Mol Immunol 24(10):1087–96.

Hayter PM, Curling EM, Baines AJ, Jenkins N, Salmon I, Strange PG, Tong JM, Bull AT. 1992. Glucose-limited chemostat culture of Chinese hamster ovary cells producing recombinant human interferon-gamma. Biotechnol Bioeng 39(3):327–35.

Hermes PA, Castro CD. 2010. A fully defined, fed-batch, recombinant NS0 culture process for monoclonal antibody production. Biotechnol Prog 26:1411–1416 (May 2010 online advaned publication).

Hills AE, Patel A, Boyd P, James DC. 2001. Metabolic control of recombinant monoclonal antibody N-glycosylation in GS-NS0 cells. Biotechnol Bioeng 75:239–51.

Hodoniczky J, Zheng YZ, James DC. 2005. Control of recombinant monoclonal antibody effector functions by Fc N-glycan remodeling in vitro. Biotechnol Prog 21(6):1644–52.

Horvath B, Mun M, Laird MW. 2010. Characterization of a monoclonal antibody cell culture production process using a quality by design approach. Mol Biotechnol 45(3):203–6.

Houdebine LM. 2009. Production of pharmaceutical proteins by transgenic animals. Comp Immunol Microbiol Infect Dis 32(2):107–21.

Hsu TA, Takahashi N, Tsukamoto Y, Kato K, Shimada I, Masuda K, Whiteley EM, Fan JQ, Lee YC, Betenbaugh MJ. 1997. Differential N-glycan patterns of secreted and intracellular IgG produced in Trichoplusia ni cells. J Biol Chem 272(14):9062–70.

Huang L, Biolsi S, Bales KR, Kuchibhotla U. 2006. Impact of variable domain glycosylation on antibody clearance: an LC/MS characterization. Anal Biochem 349(2):197–207.

Huhn C, Selman MH, Ruhaak LR, Deelder AM, Wuhrer M. 2009. IgG glycosylation analysis. Proteomics 9:882–913.

Iida S, Misaka H, Inoue M, Shibata M, Nakano R, Yamane-Ohnuki N, Wakitani M, Yano K, Shitara K, Satoh M. 2006. Nonfucosylated therapeutic IgG1 antibody can evade the inhibitory effect of serum immunoglobulin G on antibody-dependent cellular cytotoxicity through its high binding to FcgammaRIIIa. Clin Cancer Res 12(9):2879–87.

Imai-Nishiya H, Mori K, Inoue M, Wakitani M, Iida S, Shitara K, Satoh M. 2007. Double knockdown of alpha1,6-fucosyltransferase (FUT8) and GDP-mannose 4,6-dehydratase (GMD) in antibody-producing cells: a new strategy for generating fully non-fucosylated therapeutic antibodies with enhanced ADCC. BMC Biotechnol 7:84.

Irie RF, Ollila DW, O'Day S, Morton DL. 2004. Phase I pilot clinical trial of human IgM monoclonal antibody to ganglioside GM3 in patients with metastatic melanoma. Cancer Immunol Immunother 53(2):110–17.

Jacquemin M, Radcliffe CM, Lavend'homme R, Wormald MR, Vanderelst L, Wallays G, Dewaele J, Collen D, Vermylen J, Dwek RA, et al. 2006. Variable region heavy chain glycosylation determines the anticoagulant activity of a factor VIII antibody. J Thromb Haemost 4(5): 1047–55.

Jarvis DL. 2003. Developing baculovirus-insect cell expression systems for humanized recombinant glycoprotein production. Virology 310(1):1–7.

Jefferis R. 2005a. Glycosylation of natural and recombinant antibody molecules. Adv Exp Med Biol 564:143–8.

Jefferis R. 2005b. Glycosylation of recombinant antibody therapeutics. Biotechnol Prog 21(1): 11–16.

Jefferis R. 2007. Antibody therapeutics: isotype and glycoform selection. Expert Opin Biol Ther 7(9):1401–13.

Jones D, Kroos N, Anema R, van Montfort B, Vooys A, van der Kraats S, van der Helm E, Smits S, Schouten J, Brouwer K, et al. 2003. High-level expression of recombinant IgG in the human cell line per.c6. Biotechnol Prog 19(1):163–8.

Jung ST, Reddy ST, Kang TH, Borrok MJ, Sandlie I, Tucker PW, Georgiou G. 2010. Aglycosylated IgG variants expressed in bacteria that selectively bind FcgammaRI potentiate tumor cell killing by monocyte-dendritic cells. Proc Natl Acad Sci USA 107(2):604–9.

Kanda Y, Imai-Nishiya H, Kuni-Kamochi R, Mori K, Inoue M, Kitajima-Miyama K, Okazaki A, Iida S, Shitara K, Satoh M. 2007a. Establishment of a GDP-mannose 4,6-dehydratase (GMD) knockout host cell line: a new strategy for generating completely non-fucosylated recombinant therapeutics. J Biotechnol 130(3):300–10.

Kanda Y, Yamada T, Mori K, Okazaki A, Inoue M, Kitajima-Miyama K, Kuni-Kamochi R, Nakano R, Yano K, Kakita S, et al. 2007b. Comparison of biological activity among nonfucosylated therapeutic IgG1 antibodies with three different N-linked Fc oligosaccharides: the high-mannose, hybrid, and complex types. Glycobiology 17(1):104–18.

Kanda Y, Yamane-Ohnuki N, Sakai N, Yamano K, Nakano R, Inoue M, Misaka H, Iida S, Wakitani M, Konno Y, et al. 2006. Comparison of cell lines for stable production of fucose-negative antibodies with enhanced ADCC. Biotechnol Bioeng 94(4):680–8.

Kaneko Y, Nimmerjahn F, Ravetch JV. 2006. Anti-inflammatory activity of immunoglobulin G resulting from Fc sialylation. Science 313(5787):670–3.

Kaneko Y, Sato R, Aoyagi H. 2010. Evaluation of Chinese hamster ovary cell stability during repeated batch culture for large-scale antibody production. J Biosci Bioeng 109(3): 274–80.

Kato T, Kajikawa M, Maenaka K, Park EY. 2010. Silkworm expression system as a platform technology in life science. Appl Microbiol Biotechnol 85(3):459–70.

Kawasaki N, Itoh S, Hashii N, Takakura D, Qin Y, Huang X, Yamaguchi T. 2009. The significance of glycosylation analysis in development of biopharmaceuticals. Biol Pharm Bull 32(5): 796–800.

Khetan A, Huang YM, Dolnikova J, Pederson NE, Wen D, Yusuf-Makagiansar H, Chen P, Ryll T. 2010. Control of misincorporation of serine for asparagine during antibody production using CHO cells. Biotechnol Bioeng 107(1):116–23.

Kim WD, Tokunaga M, Ozaki H, Ishibashi T, Honda K, Kajiura H, Fujiyama K, Asano R, Kumagai I, Omasa T, et al. 2010. Glycosylation pattern of humanized IgG-like bispecific antibody produced by recombinant CHO cells. Appl Microbiol Biotechnol 85(3):535–42.

Ko K, Brodzik R, Steplewski Z. 2009. Production of antibodies in plants: approaches and perspectives. Curr Top Microbiol Immunol 332:55–78.

Ko K, Tekoah Y, Rudd PM, Harvey DJ, Dwek RA, Spitsin S, Hanlon CA, Rupprecht C, Dietzschold B, Golovkin M, et al. 2003. Function and glycosylation of plant-derived antiviral monoclonal antibody. Proc Natl Acad Sci USA 100(13):8013–18.

Krapp S, Mimura Y, Jefferis R, Huber R, Sondermann P. 2003. Structural analysis of human IgG-Fc glycoforms reveals a correlation between glycosylation and structural integrity. J Mol Biol 325(5):979–89.

Kunkel JP, Jan DC, Jamieson JC, Butler M. 1998. Dissolved oxygen concentration in serum-free continuous culture affects N-linked glycosylation of a monoclonal antibody. J Biotechnol 62(1):55–71.

Kuster B, Wheeler SF, Hunter AP, Dwek RA, Harvey DJ. 1997. Sequencing of N-linked oligosaccharides directly from protein gels: in-gel deglycosylation followed by matrix-assisted laser desorption/ionization mass spectrometry and normal-phase high-performance liquid chromatography. Anal Biochem 250:82–101.

Lattova E, Chen VC, Varma S, Bezabeh T, Perreault H. 2007. Matrix-assisted laser desorption/ionization on-target method for the investigation of oligosaccharides and glycosylation sites in glycopeptides and glycoproteins. Rapid Commun Mass Spectrom 21(10): 1644–50.

Lattova E, Kapkova P, Krokhin O, Perreault H. 2006. Method for investigation of oligosaccharides from glycopeptides: direct determination of glycosylation sites in proteins. Anal Chem 78(9):2977–84.

Lattova E, Tomanek B, Bartusik D, Perreault H. 2010. N-glycomic changes in human breast carcinoma MCF-7 and T-lymphoblastoid cells after treatment with herceptin and herceptin/Lipoplex. J Proteome Res 9(3):1533–40.

Li H, Sethuraman N, Stadheim TA, Zha D, Prinz B, Ballew N, Bobrowicz P, Choi BK, Cook WJ, Cukan M, et al. 2006. Optimization of humanized IgGs in glycoengineered Pichia pastoris. Nat Biotechnol 24(2):210–15.

Lifely MR, Hale C, Boyce S, Keen MJ, Phillips J. 1995. Glycosylation and biological activity of CAMPATH-1H expressed in different cell lines and grown under different culture conditions. Glycobiology 5(8):813–22.

Lim A, Reed-Bogan A, Harmon BJ. 2008. Glycosylation profiling of a therapeutic recombinant monoclonal antibody with two N-linked glycosylation sites using liquid chromatography coupled to a hybrid quadrupole time-of-flight mass spectrometer. Anal Biochem 375(2): 163–72.

Liu H, Bulseco GG, Sun J. 2006. Effect of posttranslational modifications on the thermal stability of a recombinant monoclonal antibody. Immunol Lett 106(2):144–53.

Liu Y, Salas-Solano O, Gennaro LA. 2009. Investigation of sample preparation artifacts formed during the enzymatic release of N-linked glycans prior to analysis by capillary electrophoresis. Anal Chem 81(16):6823–9.

Lund J, Takahashi N, Pound JD, Goodall M, Jefferis R. 1996. Multiple interactions of IgG with its core oligosaccharide can modulate recognition by complement and human Fc gamma receptor I and influence the synthesis of its oligosaccharide chains. J Immunol 157(11):4963–9.

Mahassni SH, Klapper DG, Hiskey RG. 2009. Purification of a murine IgM monoclonal antibody. Hybridoma (Larchmt) 28(3):189–97.

Majid FA, Butler M, Al-Rubeai M. 2007. Glycosylation of an immunoglobulin produced from a murine hybridoma cell line: the effect of culture mode and the anti-apoptotic gene, bcl-2. Biotechnol Bioeng 97(1):156–69.

Marino K, Bones J, Kattla JJ, Rudd PM. 2010. A systematic approach to protein glycosylation analysis: a path through the maze. Nat Chem Biol 6(10):713–23.

Martinez T, Pace D, Brady L, Gerhart M, Balland A. 2007. Characterization of a novel modification on IgG2 light chain. Evidence for the presence of O-linked mannosylation. J Chromatogr A 1156(1–2):183–7.

Matsumiya S, Yamaguchi Y, Saito J, Nagano M, Sasakawa H, Otaki S, Satoh M, Shitara K, Kato K. 2007. Structural comparison of fucosylated and nonfucosylated Fc fragments of human immunoglobulin G1. J Mol Biol 368(3):767–79.

McCann KJ, Ottensmeier CH, Callard A, Radcliffe CM, Harvey DJ, Dwek RA, Rudd PM, Sutton BJ, Hobby P, Stevenson FK. 2008. Remarkable selective glycosylation of the immunoglobulin variable region in follicular lymphoma. Mol Immunol 45(6):1567–72.

Mehta A, Block TM. 2008. Fucosylated glycoproteins as markers of liver disease. Dis Markers 25(4–5):259–65.

Millward TA, Heitzmann M, Bill K, Langle U, Schumacher P, Forrer K. 2008. Effect of constant and variable domain glycosylation on pharmacokinetics of therapeutic antibodies in mice. Biologicals 36(1):41–7.

Mimura Y, Church S, Ghirlando R, Ashton PR, Dong S, Goodall M, Lund J, Jefferis R. 2000. The influence of glycosylation on the thermal stability and effector function expression of human IgG1-Fc: properties of a series of truncated glycoforms. Mol Immunol 37(12–13): 697–706.

Mimura Y, Sondermann P, Ghirlando R, Lund J, Young SP, Goodall M, Jefferis R. 2001. Role of oligosaccharide residues of IgG1-Fc in Fc gamma RIIb binding. J Biol Chem 276(49): 45539–47.

Miranda LR, Duval M, Doherty H, Seaman MS, Posner MR, Cavacini LA. 2007. The neutralization properties of a HIV-specific antibody are markedly altered by glycosylation events outside the antigen-binding domain. J Immunol 178(11):7132–8.

Mori K, Kuni-Kamochi R, Yamane-Ohnuki N, Wakitani M, Yamano K, Imai H, Kanda Y, Niwa R, Iida S, Uchida K, et al. 2004. Engineering Chinese hamster ovary cells to maximize effector function of produced antibodies using FUT8 siRNA. Biotechnol Bioeng 88(7):901–8.

Muthing J, Kemminer SE, Conradt HS, Sagi D, Nimtz M, Karst U, Peter-Katalinic J. 2003. Effects of buffering conditions and culture pH on production rates and glycosylation of clinical phase I antimelanoma mouse IgG3 monoclonal antibody R24. Biotechnol Bioeng 83:321–34.

Natsume A, Wakitani M, Yamane-Ohnuki N, Shoji-Hosaka E, Niwa R, Uchida K, Satoh M, Shitara K. 2005. Fucose removal from complex-type oligosaccharide enhances the antibody-dependent cellular cytotoxicity of single-gene-encoded antibody comprising a single-chain antibody linked the antibody constant region. J Immunol Methods 306(1–2):93–103.

Natsume A, Wakitani M, Yamane-Ohnuki N, Shoji-Hosaka E, Niwa R, Uchida K, Satoh M, Shitara K. 2006. Fucose removal from complex-type oligosaccharide enhances the antibody-dependent cellular cytotoxicity of single-gene-encoded bispecific antibody comprising of two single-chain antibodies linked to the antibody constant region. J Biochem 140(3):359–68.

Niwa R, Shoji-Hosaka E, Sakurada M, Shinkawa T, Uchida K, Nakamura K, Matsushima K, Ueda R, Hanai N, Shitara K. 2004. Defucosylated chimeric anti-CC chemokine receptor 4 IgG1 with enhanced antibody-dependent cellular cytotoxicity shows potent therapeutic activity to T-cell leukemia and lymphoma. Cancer Res 64(6):2127–33.

Nuttall J, Ma JK, Frigerio L. 2005. A functional antibody lacking N-linked glycans is efficiently folded, assembled and secreted by tobacco mesophyll protoplasts. Plant Biotechnol J 3(5): 497–504.

Okazaki A, Shoji-Hosaka E, Nakamura K, Wakitani M, Uchida K, Kakita S, Tsumoto K, Kumagai I, Shitara K. 2004. Fucose depletion from human IgG1 oligosaccharide enhances binding enthalpy and association rate between IgG1 and FcgammaRIIIa. J Mol Biol 336(5): 1239–49.

Orzaez D, Granell A, Blazquez MA. 2009. Manufacturing antibodies in the plant cell. Biotechnol J 4(12):1712–24.

Padlan E. 1991. Biological significance of carbohydrate chains on monoclonal antibodies. Proc Natl Acad Sci USA 80:6632–36.

Parekh RB, Dwek RA, Sutton BJ, Fernandes DL, Leung A, Stanworth D, Rademacher TW, Mizuochi T, Taniguchi T, Matsuta K, et al. 1985. Association of rheumatoid arthritis and primary osteoarthritis with changes in the glycosylation pattern of total serum IgG. Nature 316(6027):452–7.

Patel TP, Parekh RB, Moellering BJ, Prior CP. 1992. Different culture methods lead to differences in glycosylation of a murine IgG monoclonal antibody. Biochem J 285 (Pt 3):839–45.

Peipp M, Lammerts van Bueren JJ, Schneider-Merck T, Bleeker WW, Dechant M, Beyer T, Repp R, van Berkel PH, Vink T, van de Winkel JG, et al. 2008. Antibody fucosylation differentially impacts cytotoxicity mediated by NK and PMN effector cells. Blood 112(6):2390–9.

Prater BD, Connelly HM, Qin Q, Cockrill SL. 2009. High-throughput immunoglobulin G N-glycan characterization using rapid resolution reverse-phase chromatography tandem mass spectrometry. Anal Biochem 385:69–79.

Radaev S, Motyka S, Fridman WH, Sautes-Fridman C, Sun PD. 2001. The structure of a human type III Fcgamma receptor in complex with Fc. J Biol Chem 276(19):16469–77.

Radcliffe CM, Arnold JN, Suter DM, Wormald MR, Harvey DJ, Royle L, Mimura Y, Kimura Y, Sim RB, Inoges S, et al. 2007. Human follicular lymphoma cells contain oligomannose glycans in the antigen-binding site of the B-cell receptor. J Biol Chem 282(10):7405–15.

Raju TS, Briggs JB, Borge SM, Jones AJ. 2000. Species-specific variation in glycosylation of IgG: evidence for the species-specific sialylation and branch-specific galactosylation and importance for engineering recombinant glycoprotein therapeutics. Glycobiology 10(5):477–86.

Raju TS, Briggs JB, Chamow SM, Winkler ME, Jones AJ. 2001. Glycoengineering of therapeutic glycoproteins: in vitro galactosylation and sialylation of glycoproteins with terminal N-acetylglucosamine and galactose residues. Biochemistry 40(30):8868–76.

Raju TS, Scallon B. 2007. Fc glycans terminated with N-acetylglucosamine residues increase antibody resistance to papain. Biotechnol Prog 23(4):964–71.

Reid CQ, Tait A, Baldascini H, Mohindra A, Racher A, Bilsborough S, Smales CM, Hoare M. 2010. Rapid whole monoclonal antibody analysis by mass spectrometry: an ultra scale-down study of the effect of harvesting by centrifugation on the post-translational modification profile. Biotechnol Bioeng 107(1):85–95.

Restelli V, Wang MD, Huzel N, Ethier M, Perreault H, Butler M. 2006. The effect of dissolved oxygen on the production and the glycosylation profile of recombinant human erythropoietin produced from CHO cells. Biotechnol Bioeng 94(3):481–94.

Rook GA, Steele J, Brealey R, Whyte A, Isenberg D, Sumar N, Nelson JL, Bodman KB, Young A, Roitt IM, et al. 1991. Changes in IgG glycoform levels are associated with remission of arthritis during pregnancy. J Autoimmun 4(5):779–94.

Rothman RJ, Perussia B, Herlyn D, Warren L. 1989a. Antibody-dependent cytotoxicity mediated by natural killer cells is enhanced by castanospermine-induced alterations of IgG glycosylation. Mol Immunol 26(12):1113–23.

Rothman RJ, Warren L, Vliegenthart JF, Hard KJ. 1989b. Clonal analysis of the glycosylation of immunoglobulin G secreted by murine hybridomas. Biochemistry 28(3):1377–84.

Royle L, Campbell MP, Radcliffe CM, White DM, Harvey DJ, Abrahams JL, Kim YG, Henry GW, Shadick NA, Weinblatt ME, et al. 2008. HPLC-based analysis of serum N-glycans on a 96-well plate platform with dedicated database software. Anal Biochem 376(1):1–12.

Royle L, Radcliffe CM, Dwek R, Rudd P. 2007. Detailed structural analysis of N-glycans released from glycoproteins in SDS-PAGE gell bands using HPLC combined with exoglycosidase array digestions. In Glycobiology Protocols (Brockhausen I, Editor), "Methods in Molecular Biology" vol. 347. Humana Press, Totowa, NJ.

Rudd PM, Leatherbarrow RJ, Rademacher TW, Dwek RA. 1991. Diversification of the IgG molecule by oligosaccharides. Mol Immunol 28(12):1369–78.

Saint-Jore-Dupas C, Faye L, Gomord V. 2007. From planta to pharma with glycosylation in the toolbox. Trends Biotechnol 25(7):317–23.

Sandoval WN, Pham V, Ingle ES, Liu PS, Lill JR. 2007. Applications of microwave-assisted proteomics in biotechnology. Comb Chem High Throughput Screen 10:751–65.

Saldova R, Wormald MR, Dwek RA, Rudd PM. 2008. Glycosylation changes on serum glycoproteins in ovarian cancer may contribute to disease pathogenesis. Dis Markers 25(4–5):219–32.
Salfeld JG. 2007. Isotype selection in antibody engineering. Nat Biotechnol 25(12):1369–72.
Sarmay G, Lund J, Rozsnyay Z, Gergely J, Jefferis R. 1992. Mapping and comparison of the interaction sites on the Fc region of IgG responsible for triggering antibody dependent cellular cytotoxicity (ADCC) through different types of human Fc gamma receptor. Mol Immunol 29(5):633–9.
Scallon BJ, McCarthy S, Radewonuk J, Cai A, Naso M, Raju TS, Capocasale R. 2007a. Quantitative in vivo comparisons of the Fc gamma receptor-dependent agonist activities of different fucosylation variants of an immunoglobulin G antibody. Int Immunopharmacol 7(6):761–72.
Scallon BJ, Tam SH, McCarthy SG, Cai AN, Raju TS. 2007b. Higher levels of sialylated Fc glycans in immunoglobulin G molecules can adversely impact functionality. Mol Immunol 44(7): 1524–34.
Schachter H. 1986. Biosynthetic controls that determine the branching and microheterogeneity of protein-bound oligosaccharides. Adv Exp Med Biol 205:53–85.
Schahs M, Strasser R, Stadlmann J, Kunert R, Rademacher T, Steinkellner H. 2007. Production of a monoclonal antibody in plants with a humanized N-glycosylation pattern. Plant Biotechnol J 5(5):657–63.
Schellekens H. 2009. Biosimilar therapeutics-what do we need to consider? NDT Plus 2(Suppl 1):i27–i36.
Schillberg S, Fischer R, Emans N. 2003. Molecular farming of recombinant antibodies in plants. Cell Mol Life Sci 60:433–45.
Schmelzer AE, Miller WM. 2002. Hyperosmotic stress and elevated pCO2 alter monoclonal antibody charge distribution and monosaccharide content. Biotechnol Prog 18(2):346–53.
Serrato JA, Hernandez V, Estrada-Mondaca S, Palomares LA, Ramirez OT. 2007. Differences in the glycosylation profile of a monoclonal antibody produced by hybridomas cultured in serum-supplemented, serum-free or chemically defined media. Biotechnol Appl Biochem 47(Pt 2):113–24.
Sheeley DM, Merrill BM, Taylor LC. 1997. Characterization of monoclonal antibody glycosylation: comparison of expression systems and identification of terminal alpha-linked galactose. Anal Biochem 247(1):102–10.
Shen X, Hu GB, Jiang SJ, He FR, Xing W, Li L, Yang J, Zhu HF, Lei P, Shen GX. 2009. Engineering and characterization of a baculovirus-expressed mouse/human chimeric antibody against transferrin receptor. Protein Eng Des Sel 22(12):723–31.
Shields RL, Lai J, Keck R, O'Connell LY, Hong K, Meng YG, Weikert SH, Presta LG. 2002. Lack of fucose on human IgG1 N-linked oligosaccharide improves binding to human Fcgamma RIII and antibody-dependent cellular toxicity. J Biol Chem 277(30):26733–40.
Shinkawa T, Nakamura K, Yamane N, Shoji-Hosaka E, Kanda Y, Sakurada M, Uchida K, Anazawa H, Satoh M, Yamasaki M, et al. 2003. The absence of fucose but not the presence of galactose or bisecting N-acetylglucosamine of human IgG1 complex-type oligosaccharides shows the critical role of enhancing antibody-dependent cellular cytotoxicity. J Biol Chem 278(5): 3466–73.
Siberil S, de Romeuf C, Bihoreau N, Fernandez N, Meterreau JL, Regenman A, Nony E, Gaucher C, Glacet A, Jorieux S, et al. 2006. Selection of a human anti-RhD monoclonal antibody for therapeutic use: impact of IgG glycosylation on activating and inhibitory Fc gamma R functions. Clin Immunol 118(2–3):170–9.
Siberil S, Elluru S, Negi VS, Ephrem A, Misra N, Delignat S, Bayary J, Lacroix-Desmazes S, Kazatchkine MD, Kaveri SV. 2007. Intravenous immunoglobulin in autoimmune and inflammatory diseases: more than mere transfer of antibodies. Transfus Apher Sci 37(1):103–7.
Siemiatkoski J, Lyubarskaya Y, Houde D, Tep S, Mhatre R. 2006. A comparison of three techniques for quantitative carbohydrate analysis used in characterization of therapeutic antibodies. Carbohydr Res 341:410–19.

Sinha S, Pipes G, Topp EM, Bondarenko PV, Treuheit MJ, Gadgil HS. 2008. Comparison of LC and LC/MS methods for quantifying N-glycosylation in recombinant IgGs. J Am Soc Mass Spectrom 19(11):1643–54.

Sondermann P, Huber R, Oosthuizen V, Jacob U. 2000. The 3.2-A crystal structure of the human IgG1 Fc fragment-Fc gammaRIII complex. Nature 406(6793):267–73.

Song M, Park DY, Kim Y, Lee KJ, Lu Z, Ko K, Choo YK, Han YS, Ahn MH, Oh DB. 2010. Characterization of N-glycan structures and biofunction of anti-colorectal cancer monoclonal antibody CO17-1A produced in baculovirus-insect cell expression system. J Biosci Bioeng 110(2):135–40.

Srebalus Barnes CA, Lim A. 2007. Applications of mass spectrometry for the structural characterization of recombinant protein pharmaceuticals. Mass Spectrom Rev 26(3):370–88.

Sriraman R, Bardor M, Sack M, Vaquero C, Faye L, Fischer R, Finnern R, Lerouge P. 2004. Recombinant anti-hCG antibodies retained in the endoplasmic reticulum of transformed plants lack core-xylose and core-alpha(1,3)-fucose residues. Plant Biotechnol J 2(4):279–87.

Stadlmann J, Pabst M, Kolarich D, Kunert R, Altmann F. 2008. Analysis of immunoglobulin glycosylation by LC-ESI-MS of glycopeptides and oligosaccharides. Proteomics 8:2858–71.

Stadlmann J, Weber A, Pabst M, Anderle H, Kunert R, Ehrlich HJ, Peter Schwarz H, Altmann F. 2009. A close look at human IgG sialylation and subclass distribution after lectin fractionation. Proteomics 9(17):4143–53.

Stanley P, Schachter H, Taniguchi N. 2009. Chapter 8. Essentials of Glycobiology. 2nd edition. Varki A, Cummings RD, Esko JD, Freeze HH, Stanley P, Bertozzi CR, Hart GW, Etzler ME, editors. Cold Spring Harbor, NY: Cold Spring Harbor Laboratory Press.

Stork R, Zettlitz KA, Muller D, Rether M, Hanisch FG, Kontermann RE. 2008. N-glycosylation as novel strategy to improve pharmacokinetic properties of bispecific single-chain diabodies. J Biol Chem 283(12):7804–12.

Strasser R, Stadlmann J, Schahs M, Stiegler G, Quendler H, Mach L, Glossl J, Weterings K, Pabst M, Steinkellner H. 2008. Generation of glyco-engineered Nicotiana benthamiana for the production of monoclonal antibodies with a homogeneous human-like N-glycan structure. Plant Biotechnol J 6(4):392–402.

Takahashi M, Kuroki Y, Ohtsubo K, Taniguchi N. 2009. Core fucose and bisecting GlcNAc, the direct modifiers of the N-glycan core: their functions and target proteins. Carbohydr Res 344(12):1387–90.

Takegawa Y, Deguchi K, Nakagawa H, Nishimura S. 2005. Structural analysis of an N-glycan with "beta1-4 bisecting branch" from human serum IgG by negative-ion MSn spectral matching and exoglycosidase digestion. Anal Chem 77(18):6062–8.

Tao MH, Morrison SL. 1989. Studies of aglycosylated chimeric mouse-human IgG. Role of carbohydrate in the structure and effector functions mediated by the human IgG constant region. J Immunol 143(8):2595–601.

Taylor RJ, Chan SL, Wood A, Voskens CJ, Wolf JS, Lin W, Chapoval A, Schulze DH, Tian G, Strome SE. 2009. FcgammaRIIIa polymorphisms and cetuximab induced cytotoxicity in squamous cell carcinoma of the head and neck. Cancer Immunol Immunother 58(7):997–1006.

Tekoah Y, Ko K, Koprowski H, Harvey DJ, Wormald MR, Dwek RA, Rudd PM. 2004. Controlled glycosylation of therapeutic antibodies in plants. Arch Biochem Biophys 426(2):266–78.

Thobhani S, Yuen CT, Bailey MJ, Jones C. 2009. Identification and quantification of N-linked oligosaccharides released from glycoproteins: an inter-laboratory study. Glycobiology 19(3):201–11.

Thomas P, Smart TG. 2005. HEK293 cell line: a vehicle for the expression of recombinant proteins. J Pharmacol Toxicol Methods 51(3):187–200.

Tomiya N, Narang S, Lee YC, Betenbaugh MJ. 2004. Comparing N-glycan processing in mammalian cell lines to native and engineered lepidopteran insect cell lines. Glycoconj J 21(6):343–60.

Tornoe I, Titlestad IL, Kejling K, Erb K, Ditzel HJ, Jensenius JC. 1997. Pilot scale purification of human monoclonal IgM (COU-1) for clinical trials. J Immunol Methods 205(1):11–17.

Triguero A, Cabrera G, Cremata JA, Yuen CT, Wheeler J, Ramirez NI. 2005. Plant-derived mouse IgG monoclonal antibody fused to KDEL endoplasmic reticulum-retention signal is N-glycosylated homogeneously throughout the plant with mostly high-mannose-type N-glycans. Plant Biotechnol J 3(4):449–57.

Tsuchiya N, Endo T, Matsuta K, Yoshinoya S, Aikawa T, Kosuge E, Takeuchi F, Miyamoto T, Kobata A. 1989. Effects of galactose depletion from oligosaccharide chains on immunological activities of human IgG. J Rheumatol 16(3):285–90.

Umana P, Jean-Mairet J, Moudry R, Amstutz H, Bailey JE. 1999. Engineered glycoforms of an antineuroblastoma IgG1 with optimized antibody-dependent cellular cytotoxic activity. Nat Biotechnol 17(2):176–80.

Valliere-Douglass JF, Eakin CM, Wallace A, Ketchem RR, Wang W, Treuheit MJ, Balland A. 2010. Glutamine-linked and non-consensus asparagine-linked oligosaccharides present in human recombinant antibodies define novel protein glycosylation motifs. J Biol Chem 285(21):16012–22.

Valliere-Douglass JF, Kodama P, Mujacic M, Brady LJ, Wang W, Wallace A, Yan B, Reddy P, Treuheit MJ, Balland A. 2009. Asparagine-linked oligosaccharides present on a non-consensus amino acid sequence in the CH1 domain of human antibodies. J Biol Chem 284(47): 32493–506.

van Berkel PH, Gerritsen J, Perdok G, Valbjorn J, Vink T, van de Winkel JG, Parren PW. 2009. N-linked glycosylation is an important parameter for optimal selection of cell lines producing biopharmaceutical human IgG. Biotechnol Prog 25(1):244–51.

van Berkel PH, Gerritsen J, van Voskuilen E, Perdok G, Vink T, van de Winkel JG, Parren PW. 2010. Rapid production of recombinant human IgG With improved ADCC effector function in a transient expression system. Biotechnol Bioeng 105(2):350–7.

Vezina LP, Faye L, Lerouge P, D'Aoust MA, Marquet-Blouin E, Burel C, Lavoie PO, Bardor M, Gomord V. 2009. Transient co-expression for fast and high-yield production of antibodies with human-like N-glycans in plants. Plant Biotechnol J 7(5):442–55.

Wada Y, Azadi P, Costello CE, Dell A, Dwek RA, Geyer H, Geyer R, Kakehi K, Karlsson NG, Kato K, et al. 2007. Comparison of the methods for profiling glycoprotein glycans – HUPO Human Disease Glycomics/Proteome Initiative multi-institutional study. Glycobiology 17(4):411–22.

Wallick SC, Kabat EA, Morrison SL. 1988. Glycosylation of a VH residue of a monoclonal antibody against alpha (1–6) dextran increases its affinity for antigen. J Exp Med 168(3):1099–109.

Walsh G. 2010. Biopharmaceutical benchmarks 2010. Nat Biotechnol 28(9):917–24.

Ward M, Lin C, Victoria DC, Fox BP, Fox JA, Wong DL, Meerman HJ, Pucci JP, Fong RB, Heng MH, et al. 2004. Characterization of humanized antibodies secreted by Aspergillus niger. Appl Environ Microbiol 70(5):2567–76.

Warnock D, Bai X, Autote K, Gonzales J, Kinealy K, Yan B, Qian J, Stevenson T, Zopf D, Bayer RJ. 2005. In vitro galactosylation of human IgG at 1 kg scale using recombinant galactosyltransferase. Biotechnol Bioeng 92(7):831–42.

Wei Y, Li C, Huang W, Li B, Strome S, Wang LX. 2008. Glycoengineering of human IgG1-Fc through combined yeast expression and in vitro chemoenzymatic glycosylation. Biochemistry 47(39):10294–304.

Weng WK, Levy R. 2003. Two immunoglobulin G fragment C receptor polymorphisms independently predict response to rituximab in patients with follicular lymphoma. J Clin Oncol 21(21):3940–7.

Wright A, Morrison SL. 1994. Effect of altered CH2-associated carbohydrate structure on the functional properties and in vivo fate of chimeric mouse-human immunoglobulin G1. J Exp Med 180(3):1087–96.

Wright A, Morrison SL. 1998. Effect of C2-associated carbohydrate structure on Ig effector function: studies with chimeric mouse-human IgG1 antibodies in glycosylation mutants of Chinese hamster ovary cells. J Immunol 160(7):3393–402.

Wright A, Sato Y, Okada T, Chang K, Endo T, Morrison S. 2000. In vivo trafficking and catabolism of IgG1 antibodies with Fc associated carbohydrates of differing structure. Glycobiology 10(12):1347–55.

Wright A, Tao MH, Kabat EA, Morrison SL. 1991. Antibody variable region glycosylation: position effects on antigen binding and carbohydrate structure. EMBO J 10(10):2717–23.

Wu S, Luo J, O'Neil K, Kang J, Lacy ER, Canziani G, Baker A, Huang M, Tang Q, Raju TS, et al. 2010. Structure-based engineering of a monoclonal antibody for improved solubility. Protein Eng Des Sel 23(8):643–651.

Wuhrer M, Stam JC, van de Geijn FE, Koeleman CA, Verrips CT, Dolhain RJ, Hokke CH, Deelder AM. 2007. Glycosylation profiling of immunoglobulin G (IgG) subclasses from human serum. Proteomics 7(22):4070–81.

Yamada E, Tsukamoto Y, Sasaki R, Yagyu K, Takahashi N. 1997. Structural changes of immunoglobulin G oligosaccharides with age in healthy human serum. Glycoconj J 14(3): 401–5.

Yamaguchi Y, Nishimura M, Nagano M, Yagi H, Sasakawa H, Uchida K, Shitara K, Kato K. 2006. Glycoform-dependent conformational alteration of the Fc region of human immunoglobulin G1 as revealed by NMR spectroscopy. Biochim Biophys Acta 1760(4):693–700.

Yamane-Ohnuki N, Kinoshita S, Inoue-Urakubo M, Kusunoki M, Iida S, Nakano R, Wakitani M, Niwa R, Sakurada M, Uchida K, et al. 2004. Establishment of FUT8 knockout Chinese hamster ovary cells: an ideal host cell line for producing completely defucosylated antibodies with enhanced antibody-dependent cellular cytotoxicity. Biotechnol Bioeng 87(5):614–22.

Yamane-Ohnuki N, Satoh M. 2009. Production of therapeutic antibodies with controlled fucosylation. MAbs 1(3):230–6.

Yang M, Butler M. 2000. Effect of ammonia on the glycosylation of human recombinant erythropoietin in culture. Biotechnol Prog 16(5):751–9.

Ye J, Kober V, Tellers M, Naji Z, Salmon P, Markusen JF. 2009. High-level protein expression in scalable CHO transient transfection. Biotechnol Bioeng 103(3):542–51.

Zhao J, Kuroki M, Shibaguchi H, Wang L, Huo Q, Takami N, Tanaka T, Kinugasa T. 2008. Recombinant human monoclonal igA antibody against CEA to recruit neutrophils to CEA-expressing cells. Oncol Res 17(5):217–22.

Zhou Q, Shankara S, Roy A, Qiu H, Estes S, McVie-Wylie A, Culm-Merdek K, Park A, Pan C, Edmunds T. 2008. Development of a simple and rapid method for producing non-fucosylated oligomannose containing antibodies with increased effector function. Biotechnol Bioeng 99(3):652–65.

Zhu D, McCarthy H, Ottensmeier CH, Johnson P, Hamblin TJ, Stevenson FK. 2002. Acquisition of potential N-glycosylation sites in the immunoglobulin variable region by somatic mutation is a distinctive feature of follicular lymphoma. Blood 99(7):2562–8.

Zhu L, van de Lavoir MC, Albanese J, Beenhouwer DO, Cardarelli PM, Cuison S, Deng DF, Deshpande S, Diamond JH, Green L, et al. 2005. Production of human monoclonal antibody in eggs of chimeric chickens. Nat Biotechnol 23(9):1159–69.

Chapter 13
Quality Issues Arising from Post-translational Modification of Recombinant Antibodies

Raymond Tyther and Nigel Jenkins

Abstract This chapter explores the post translational modifications (PTMs) that can occur in recombinant monoclonal antibodies. The topic of glycosylation is covered in another chapter in this volume. These modifications can occur at each stage of bioprocessing i.e. cell expansion, fermentation, protein purification, formulation and long-term storage. The chapter focusses on the following PTMs: protein aggregation and misfolding, dimerization, oxidation (principally of methionine residues), and deamidation (principally of asparagine residues). It explores the mechanisms and possible causes of these PTMs, and also the assays used to track these changes. It is the responsibility of each manufacturer to define the limits of variation that exist for individual biopharmaceuticals through comprehensive analytics and submit these to the regulatory authorities. These limits, along with parameters defined during Quality by Design programs (QbD) using Process Analytical Technologies (PAT) are used to optimize the bioprocessing steps to minimize the presence of aberrant species.

13.1 Introduction

One of the more surprising revelations upon the completion of the Human Genome Project (Venter et al. 2001), was that the genetic code in itself did not fully account for the variety and complexity of human proteins (Dhingra et al. 2005). This is due to the significant amount of post-translational modifications (PTM) to which proteins are subjected. Some of this variety occurs as a consequence of normal cell processing, some as a result of stress and ageing, and yet more due to pathologies.

In the case of biopharmaceuticals, the scenario is further complicated because human proteins are routinely produced in non-human cell species, principally Chinese Hamster Ovary (CHO) cells (Jenkins 2007). Furthermore, these cell lines are selected for their accelerated growth and production levels, which permit product titers of the order of 1–3 g/L, but at these high production levels, protein quality issues can arise (Jenkins et al. 2009).

R. Tyther and N. Jenkins (✉)
National Institute for Bioprocessing Reasearch and Training, University College Dublin, Dublin, Ireland
e-mail: nigel.jenkins@nibrt.ie

M. Al-Rubeai (ed.), *Antibody Expression and Production*, Cell Engineering 7,
DOI 10.1007/978-94-007-1257-7_13, © Springer Science+Business Media B.V. 2011

The increased focus on product quality for major biopharmaceutical products such as recombinant monoclonal antibodies (MAbs) instead of simply product titer, has been driven by the Quality by Design (QbD) initiative (Rathore 2009). Manufacturing process changes can impact significantly on biopharmaceutical heterogeneity (Chirino and Mire-Sluis 2004), so the U.S. Food and Drug Administration is implementing QbD with the assistance biotechnology industry to minimize product heterogeneity which may compromise patient safety. For this to be achieved, the extent and nature of PTMs in therapeutic proteins must be defined and their influence on biological activity must be characterised. This gives rise to critical quality attributes (CQAs), which are defined as as physical, chemical, biological or microbiological properties or characteristics that need to be controlled (directly or indirectly) to ensure product quality (Lionberger et al. 2008).

Ranked in terms of a criticality continuum, the five main product quality attributes of concern for MAbs are as follows:

1. High Molecular Weight (HMW) Aggregates
2. Isoaspartate (deamidation of Asparagines in the CDR of IgG molecules)
3. Non-glycosylated IgG heavy chains (covered in another chapter in this volume)
4. Oxidized species (principally Methionine and Tryptophan)
5. Dimerization.

The PTM of primary concern is protein aggregation because of the profound effects in terms of product loss, drug efficacy, and patient immunoreactivity. Critical changes such as deamidation in the CDR region are also of particular importance because of the major implications for drug efficacy.

The genesis and consequences of aberrant glycosylation and glycation during bioprocessing will be discussed elsewhere in this volume, but we will discuss the nature the other major PTMs in the remainder of this chapter.

13.2 Aggregation

The presence of HMW aggregate species during biopharmaceutical manufacture or administration is of primary concern, because of the potential for aggregated species to elicit inappropriate autoimmune responses (Rosenberg 2006). In a study involving recombinant human interferon, Hermeling et al. demonstrated that preparations containing aggregates increased the immune response in the wild-type mice as compared to native forms (Hermeling et al. 2006). Similar drug-response profiles have been found during insulin treatment (Maislos et al. 1988), all of which indicate that aggregated or mis-folded versions of biopharmaceuticals evoke undesired autoimmune responses (Maas et al. 2007). There is no "silver bullet" solution to the aggregation problem, because aggregates can arise at every stage of production, be that intracellularly, in the cell media, during purification, or post-formulation. The unfolded protein response (UPR) is a natural mechanism for recycling unfolded or

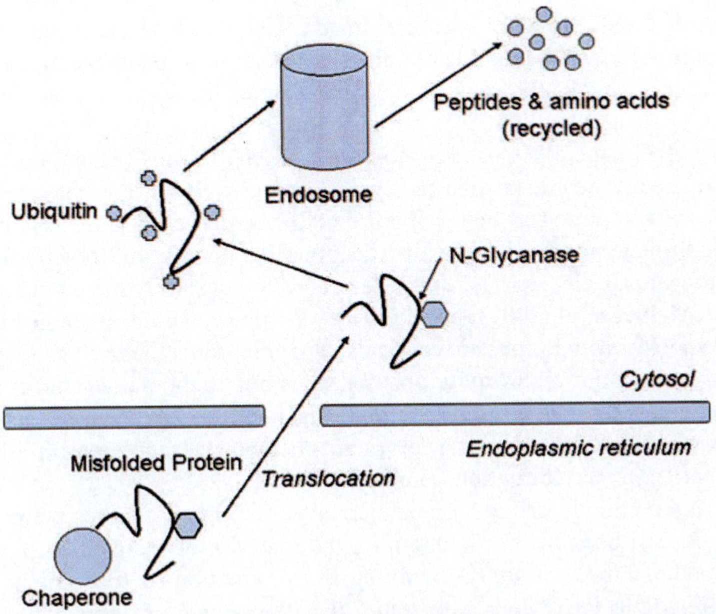

Fig. 13.1 The unfolded protein response (UPR) is a natural mechanism for recycling unfolded or misfolded proteins. Chaperones transport unfolded proteins from the endoplasmin reticulum to the cytosol. Glycan residues are removed, before the polypeptides are degraded to amino acids in the endosome

misfolded proteins (see Fig. 13.1). However this mechanism may become limited in cells producing large amounts of recombinant protein resulting in secretion of aggregated species (Jenkins 2007).

Aggregate species can be loosely divided into particulate and sub-particulate species (Cromwell et al. 2006). While consistent definitions for either do not exist, "particulate" species are largely considered to be covalently-bonded, visible particles which can be removed via a 0.22 μM filter. Sub-particulate or "soluble" aggregates particles are not visible to the naked eye, may be either covalently or non-covalently associated, and are not removed by 0.22 μM filtration. Covalently-bound aggregate species are the main focus of attention because they may be robust enough to persist through the various stages of bioprocessing. They are also more readily quantifiable than non-covalently associated HMW species such as dimers, whose presence may be concentration-dependent and which may only exist transiently. This transient binding may arise through weak non-covalent protein interaction (possibly through hydrophobic residues) and may be perturbed by pH or protein concentration changes. Such a transient association has been observed for a MAb against VEGF (Moore et al. 1999), and sometimes an increase in viscosity is indicative of reversible protein self association (Liu et al. 2005). However, covalently bound dimer species have also been detected in commercial antibodies such as Epratuzumab (Remmele et al. 2006).

Disulphide bond formation is crucial for the assembly and structural integrity of MAbs in particular, but disulphide bonding is also the most plausible mechanism for the formation of covalently associated HMW aggregates during biopharmaceutical production (Brych et al. 2010). Significant levels of free thiol content have been reported for IgG subclasses used as commercial MAb drugs (Zhang and Czupryn 2002), and specific residues such as Cys 22 and Cys 96 on the heavy chain of a MAb have been reported as being unpaired (Chaderjian et al. 2005; Harris 2005). Where incomplete intra-chain disulphide bonding occurs, the potential to form HMW aggregate species exists. Proteins can also be covalently linked via dityrosine formation (Mahler et al. 2009), but this has yet to be reported in studies involving aggregate species from biopharmaceuticals. The free thiols themselves may arise in CHO cells and similar high-production cell lines due to the native chaperones, and folding enzymes such as protein disulphide isomerase (Chakravarthi et al. 2006) being overwhelmed. The resulting thiol pair mismatches may promote misfolding and covalent aggregate formation (Harris 2005).

Whilst interventions at the intracellular site of protein synthesis are difficult, more success has been had in maintaining the integrity of secreted biopharmaceuticals in media. Firstly, a study involving the production of recombinant human antithrombin III in CHO cells, illustrated that the degree of aggregation is influenced by the expression level high-producing CHO clone and also the stage of the cell growth phase (Schroder and Friedl 1997). Through consideration of the merits of production levels against protein quality, and careful selection of the harvesting time, high aggregate levels can be curtailed. The addition of excipients to cell media can also maintain the correct redox equilibrium and prevent protein free thiol groups from cross-linking protein monomer species. This has been achieved through the addition of copper sulphate to culture to protect against excessive free thiol content (Chaderjian et al. 2005). In the case of G-CSF protein, the addition of polysorbate 80 to culture also prevented the formation of rhG-CSF protein aggregates (Bahrami et al. 2008).

The extensive cross-flow filtration and depth filtration steps included during downstream biopharmaceutical processing are adept at removing the large particulate aggregates (Shukla et al. 2007a; Zhou et al. 2008), but soluble species may escape capture by the filtration steps. Downstream polishing steps have been shown to be successful in capturing soluble aggregates and eliminating them prior to formulation (Chen et al. 2010; Shukla et al. 2007b). Modes of chromatography suitable for eliminating aggregate species include cation-exchange chromatography (Chen et al. 2010) and hydroxyapatite chromatography (Gagnon and Beam 2009) and preparative size exclusion chromatography. Aggregated species may still emerge post-formulation and during storage prior to administration. This may impinge on patient safety, so drug formulation composition is being optimized to ensure minimal drug aggregation ensues during storage. The precedent for formulation problems resulting in the development of neutralizing antibodies against interferon-alpha 2a has been described previously (Hochuli 1997), and a similar response to MAb drugs would drastically impair their effectiveness. Formulation additives that have demonstrated an ability to preserve biopharmaceutical integrity include histidine (Chen et al. 2003) and carbohydrates (Andya et al. 2003).

The standard established, FDA-approved method for analysis of biopharmaceutical aggregates is via size-exclusion HPLC (SEC-HPLC), which separates molecules based on size. However, non-specific column interactions may cause the proportion of aggregate to be under-estimated, and the addition of arginine has been shown to improve sample recovery and accuracy of HMW estimation (Arakawa et al. 2004). Analytical ultracentrifugation (AUC) in contrast is a matrix-free method of detecting HMW aggregate species (Pekar and Sukumar 2007), and is often used to validate SEC-HPLC. However, the low throughput of the device and the requirement for complex analysis and long experimental runs, make it unsuitable for routine batch analysis.

Another matrix-free method, field flow fractionation has been used to quantify HMW aggregate but it has yet to attract widespread use (Liu et al. 2006b). It shares the advantage of AUC in that column interaction does not disrupt HMW species, but it requires considerable optimization to achieve reproducible results (Silveira et al. 2006). Optical analysis of HMW species via dynamic light scattering (DLS) or multi-angle light scattering (MALLS) is also commonplace. The chief attraction of these modes of analysis is their ability to resolve HMW species at upper size scale of > 1,000 kDa. Such species can escape detection via SEC-HPLC (Mahler et al. 2005), but are conveniently detected via LS apparatus. Absolute molecular size can also be calculated by coupling SEC-HPLC to a MALLS detector (Oliva et al. 2004; Ye 2006). This is also a mean to overcoming the chief drawback of LS which is that resolution suffers below the 1,000 kDa threshold, which DLS cannot resolve dimer species from monomers.

13.3 Deamidation

Deamidation is considered the most commonplace PTM in proteins (Shire et al. 2004), and has traditionally been understood to be a form of protein damage. However, the pervasiveness and persistence of the modification has prompted speculation that deamidation events may act as molecular timers that regulate protein function and stability (Weintraub and Deverman 2007; Weintraub and Manson 2004). The deamidation event involves the rapid non-enzymatic conversion of asparagine residues to a cyclic imide intermediate. This intermediate hydrolyzes to create a mix of isoaspartic and aspartic residues at ratios of 3:1, thus isoaspartate is the most prevalent form of deamidation (Geigér and Clarke 1987). In terms of susceptibility to deamidation, asparagine residues found at the protein surface are more vulnerable, and each residue's microenvironment influences the rate and extent of deamidation (Sinha et al. 2009; Weintraub and Deverman 2007). *In silico* studies indicate that the presence of a carboxyl terminal glycine significantly enhances the rate of deamidation (Catak et al. 2008). Environmental factors such as temperature, buffer composition, ionic strength, and pH can all impact on the rate of deamidation, so there are obvious implications for upstream and downstream bioprocess conditions.

Nonetheless, not all asparagine residues are equally critical in biopharmaceuticals, and deamidation events in the complementarity-determining region (CDR) of immunoglobulin-type drugs are particularly undesirable (Liu 1992). PTMs in the CDR region potentially exert a direct effect on target binding and drug efficacy. Examples of such include deamidation that caused a 70% decrease in potency in a commercial recombinant IgG product (Harris et al. 2001), and asparagine deamidation in light chain complementarity determining region 1 (CDR1) of a humanized IgG1 monoclonal antibody (Vlasak et al. 2009). Deamidation has been linked to an enhanced immune response (Falini et al. 2008) and to promoting protein aggregation in certain proteins (Takata et al. 2008), but to date no such data exist for commercially-relevant MAb drugs.

Deamidation of biopharmaceuticals can be detected via a variety of methods involving both direct and indirect measurements. Conversion of the asparagines residue to isoaspartate alters the net charge of the proteins and creates acidic variants which can be detected via cation-exchange HPLC (Vlasak and Ionescu 2008). Hydrophobic interaction chromatography (HIC) has also been successfully employed to detect deamidated protein variants (Zhang et al. 2008). Alternative analytical methods include the detection of deamidation products such as ammonia (Tsai et al. 1993), isoasparate (Robinson et al. 1994), and alterations in Asp-N cleavage (Mimura et al. 1998). The most comprehensive routes to identification are Edman sequencing and LC/MS-MS which can identify site-specific deamidation events, and have been used to characterise deamidation sites in a variety of MAbs (Chelius et al. 2005; Harris et al. 2001; Liu et al. 2006a; Lyubarskaya et al. 2006).

13.4 Oxidation

As previously discussed in the section on aggregation, the correct redox equilibrium in terms of oxidized/reduced thiol groups is essential for maintaining correctly-folded protein products. Therefore the oxidation status of cysteine residues is of particular importance in relation to proteins with extensive disulphide-bonding such as MAbs.

Aside from cysteine residues, methionine residues are also susceptible to oxidation, and in the intracellular environment this is readily reversed through the presence of stereospecific methionine sulfoxide reductases (Levine et al. 2000). This mechanism does not protect secreted proteins however, so biopharmaceutical manufacturers introduce antioxidant excipients to help protect MAbs etc. against undesirable methionine oxidation (Lam et al. 1997a; Soenderkaer et al. 2004). Methionine oxidation is not just simply a source of product hetereogeneity, but also has the potential to impair drug efficacy so it must be carefully monitored. The therapeutically significant protease inhibitor alpha(1)-antitrypsin is inactivated by methionine oxidation at either methionine 358 and methionine 351 (Taggart et al.

2000). Methionine oxidation also has negative implications for the more commonplace MAb drug products, because oxidation of the conserved Met 33 and Met 209 of the human IgG1 Fc can reduce the physical and covalent stability of the protein (Liu et al. 2008). Interestingly, oxidation of these sites also enhanced the deamidation rate at the Asn 67 and Asn 96 sites, indicating that certain PTMs may induce a "domino effect" whereby changes at certain residues act as precursors for changes at unrelated residues. Methionine oxidation at Met 256 and Met 432 of a recombinant fully human monoclonal IgG1 antibody also reduced antibody affinity for Protein A and Protein G (Gaza-Bulseco et al. 2008; Pan et al. 2009), therefore there is a possibility of increased product loss during purification if the oxidized species elute earlier than the bulk of the MAb.

Proteins with oxidized methionine residues exhibit altered binding during HIC (Lam et al. 1997b), weak-cation IEX (Chumsae et al. 2007), and reverse phase HPLC (Kroon et al. 1992). Assignment of specific oxidation-sensitive methionine residues is possible via LC-MS (Houde et al. 2006).

While not oxidized as readily as methionine, examples of tryptophan oxidation have also been reported for biopharmaceuticals (Matamoros Fernandez et al. 2001). This is a potential area of concern because for the reported Trp residues that have been reported, several occur in the critical CDR region of MAbs (Matamoros Fernandez et al. 2001; Wei et al. 2007; Yang et al. 2007). Antibody fragmentation may also be a problem in long-term storage conditions (Lui et al. 2006a).

13.5 Conclusions

Drug product quality is governed by several factors during bioprocessing, purification and formulation. As illustrated in this chapter, multiple potential sources of product variation exist. Therefore it is the responsibility of each manufacturer to define the limits of variation that exist for individual biopharmaceuticals through comprehensive analytics, and optimize their protocols to minimize the presence of aberrant species.

References

Andya JD, Hsu CC, Shire SJ. 2003. Mechanisms of aggregate formation and carbohydrate excipient stabilization of lyophilized humanized monoclonal antibody formulations. AAPS PharmSci 5(2):26–32.

Arakawa T, Philo JS, Tsumoto K, Yumioka R, Ejima D. 2004. Elution of antibodies from a Protein-A column by aqueous arginine solutions. Protein Expr Purif 36(2):244–8.

Bahrami A, Shojaosadati SA, Khalilzadeh R, Mohammadian J, Vashghani Farahani E, Masoumian MR. 2008. Prevention of human G-CSF protein aggregation in recombinant Pichia pastoris fed-batch fermentation using additives. Biotechnol Appl Biochem 14:14.

Brych SR, Gokarn YR, Hultgen H, Stevenson RJ, Rajan R, Matsumura M. 2010. Characterization of antibody aggregation: role of buried, unpaired cysteines in particle formation. J Pharm Sci 99(2):764–81.

Catak S, Monard G, Aviyente V, Ruiz-Lopez MF. 2008. Computational study on nonenzymatic peptide bond cleavage at asparagine and aspartic acid. J Phys Chem A 112(37):8752–61. Epub 2008 Aug 20.

Chaderjian WB, Chin ET, Harris RJ, Etcheverry TM. 2005. Effect of copper sulfate on performance of a serum-free CHO cell culture process and the level of free thiol in the recombinant antibody expressed. Biotechnol Prog 21(2):550–53.

Chakravarthi S, Jessop CE, Bulleid NJ. 2006. The role of glutathione in disulphide bond formation and endoplasmic-reticulum-generated oxidative stress. EMBO Rep 7(3):271–5.

Chelius D, Rehder DS, Bondarenko PV. 2005. Identification and characterization of deamidation sites in the conserved regions of human immunoglobulin gamma antibodies. Anal Chem 77(18):6004–11.

Chen B, Bautista R, Yu K, Zapata GA, Mulkerrin MG, Chamow SM. 2003. Influence of histidine on the stability and physical properties of a fully human antibody in aqueous and solid forms. Pharm Res. 20(12):1952–60.

Chen S, Lau H, Brodsky Y, Kleemann GR, Latypov RF. 2010. The use of native cation-exchange chromatography to study aggregation and phase separation of monoclonal antibodies. Protein Sci 19(6):1191–204.

Chen J, Tetrault J, Zhang Y, Wasserman A, Conley G, Dileo M, Haimes E, Nixon AE, Ley A. 2010. The distinctive separation attributes of mixed-mode resins and their application in monoclonal antibody downstream purification process. J Chromatogr A 1217(2):216–24.

Chirino AJ, Mire-Sluis A. 2004. Characterizing biological products and assessing comparability following manufacturing changes. Nat Biotechnol 22(11):1383–91.

Chumsae C, Gaza-Bulseco G, Sun J, Liu H. 2007. Comparison of methionine oxidation in thermal stability and chemically stressed samples of a fully human monoclonal antibody. J Chromatogr B Analyt Technol Biomed Life Sci 850(1–2):285–94. Epub 2006 Dec 19.

Cromwell ME, Hilario E, Jacobson F. 2006. Protein aggregation and bioprocessing. Aaps J 8(3):E572–9.

Dhingra V, Gupta M, Andacht T, Fu ZF. 2005. New frontiers in proteomics research: a perspective. Int J Pharm 299(1–2):1–18.

Falini ML, Elli L, Caramanico R, Bardella MT, Terrani C, Roncoroni L, Doneda L, Forlani F. 2008. Immunoreactivity of antibodies against transglutaminase-deamidated gliadins in adult celiac disease. Dig Dis Sci 53(10):2697–701. Epub 2008 Feb 28.

Gagnon P, Beam K. 2009. Antibody aggregate removal by hydroxyapatite chromatography. Curr Pharm Biotechnol 10(4):440–6.

Gaza-Bulseco G, Faldu S, Hurkmans K, Chumsae C, Liu H. 2008. Effect of methionine oxidation of a recombinant monoclonal antibody on the binding affinity to protein A and protein G. J Chromatogr B Analyt Technol Biomed Life Sci 870(1):55–62. Epub 2008 Jun 5.

Geiger T, Clarke S. 1987. Deamidation, isomerization, and racemization at asparaginyl and aspartyl residues in peptides. Succinimide-linked reactions that contribute to protein degradation. J Biol Chem 262(2):785–94.

Harris RJ. 2005. Heterogeneity of recombinant antibodies: linking structure to function. Dev Biol (Basel) 122:117–27.

Harris RJ, Kabakoff B, Macchi FD, Shen FJ, Kwong M, Andya JD, Shire SJ, Bjork N, Totpal K, Chen AB. 2001. Identification of multiple sources of charge heterogeneity in a recombinant antibody. J Chromatogr B Biomed Sci Appl 752(2):233–45.

Hermeling S, Schellekens H, Maas C, Gebbink MF, Crommelin DJ, Jiskoot W. 2006. Antibody response to aggregated human interferon alpha2b in wild-type and transgenic immune tolerant mice depends on type and level of aggregation. J Pharm Sci 95(5):1084–96.

Hochuli E. 1997. Interferon immunogenicity: technical evaluation of interferon-alpha 2a. J Interferon Cytokine Res 17(Suppl 1):S15–21.

Houde D, Kauppinen P, Mhatre R, Lyubarskaya Y. 2006. Determination of protein oxidation by mass spectrometry and method transfer to quality control. J Chromatogr A 1123(2): 189–98.

Jenkins N. 2007. Modifications of therapeutic proteins: challenges and prospects. Cytotechnology 53:121–5.

Jenkins N, Meleady P, Tyther R, Murphy L. 2009. Strategies for analysing and improving the expression and quality of recombinant proteins made in mammalian cells. Biotechnol Appl Biochem 53(Pt 2):73–83.

Kroon DJ, Baldwin-Ferro A, Lalan P. 1992. Identification of sites of degradation in a therapeutic monoclonal antibody by peptide mapping. Pharm Res 9(11):1386–93.

Lam XM, Yang JY, Cleland JL. 1997a. Antioxidants for prevention of methionine oxidation in recombinant monoclonal antibody HER2. J Pharm Sci 86(11):1250–5.

Lam XM, Yang JY, Cleland JL. 1997b. Antioxidants for prevention of methionine oxidation in recombinant monoclonal antibody HER2. J Pharm Sci 86(11):1250–5.

Levine RL, Moskovitz J, Stadtman ER. 2000. Oxidation of methionine in proteins: roles in antioxidant defense and cellular regulation. IUBMB Life 50(4–5):301–7.

Lionberger RA, Lee SL, Lee L, Raw A, Yu LX. 2008. Quality by design: concepts for ANDAs. AAPS J 10(2):268–76.

Liu DT. 1992. Deamidation: a source of microheterogeneity in pharmaceutical proteins. Trends Biotechnol 10(10):364–9.

Liu J, Andya JD, Shire SJ. 2006b. A critical review of analytical ultracentrifugation and field flow fractionation methods for measuring protein aggregation. Aaps J 8(3):E580–9.

Liu H, Gaza-Bulseco G, Sun J. 2006a. Characterization of the stability of a fully human monoclonal IgG after prolonged incubation at elevated temperature. J Chromatogr B Analyt Technol Biomed Life Sci 837(1–2):35–43.

Liu J, Nguyen MD, Andya JD, Shire SJ. 2005. Reversible self-association increases the viscosity of a concentrated monoclonal antibody in aqueous solution. J Pharm Sci 94(9):1928–40.

Liu D, Ren D, Huang H, Dankberg J, Rosenfeld R, Cocco MJ, Li L, Brems DN, Remmele RL, Jr. 2008. Structure and stability changes of human IgG1 Fc as a consequence of methionine oxidation. Biochemistry 47(18):5088–100. Epub 2008 Apr 12.

Lyubarskaya Y, Houde D, Woodard J, Murphy D, Mhatre R. 2006. Analysis of recombinant monoclonal antibody isoforms by electrospray ionization mass spectrometry as a strategy for streamlining characterization of recombinant monoclonal antibody charge heterogeneity. Anal Biochem 348(1):24–39.

Maas C, Hermeling S, Bouma B, Jiskoot W, Gebbink MF. 2007. A role for protein misfolding in immunogenicity of biopharmaceuticals. J Biol Chem 282(4):2229–36. Epub 2006 Nov 29.

Mahler HC, Friess W, Grauschopf U, Kiese S. 2009. Protein aggregation: pathways, induction factors and analysis. J Pharm Sci 98(9):2909–34.

Mahler HC, Muller R, Friess W, Delille A, Matheus S. 2005. Induction and analysis of aggregates in a liquid IgG1-antibody formulation. Eur J Pharm Biopharm 59(3):407–17.

Maislos M, Bialer M, Mead PM, Robbins DC. 1988. Pharmacokinetic model of circulating covalent aggregates of insulin. Diabetes 37(8):1059–63.

Matamoros Fernandez LE, Kalume DE, Calvo L, Fernandez Mallo M, Vallin A, Roepstorff P. 2001. Characterization of a recombinant monoclonal antibody by mass spectrometry combined with liquid chromatography. J Chromatogr B Biomed Sci Appl 752(2):247–61.

Mimura Y, Nakamura K, Tanaka T, Fujimoto M. 1998. Evidence of intra- and extracellular modifications of monoclonal IgG polypeptide chains generating charge heterogeneity. Electrophoresis 19(5):767–75.

Moore JM, Patapoff TW, Cromwell ME. 1999. Kinetics and thermodynamics of dimer formation and dissociation for a recombinant humanized monoclonal antibody to vascular endothelial growth factor. Biochemistry 38(42):13960–7.

Oliva A, Llabres M, Farina JB. 2004. Applications of multi-angle laser light-scattering detection in the analysis of peptides and proteins. Curr Drug Discov Technol 1(3):229–42.

Pan H, Chen K, Chu L, Kinderman F, Apostol I, Huang G. 2009. Methionine oxidation in human IgG2 Fc decreases binding affinities to protein A and FcRn. Protein Sci 18(2):424–33.

Pekar A, Sukumar M. 2007. Quantitation of aggregates in therapeutic proteins using sedimentation velocity analytical ultracentrifugation: practical considerations that affect precision and accuracy. Anal Biochem 367(2):225–37.

Rathore AS. 2009. Roadmap for implementation of quality by design (QbD) for biotechnology products. Trends Biotechnol 27(9):546–53.

Remmele RL, Jr., Callahan WJ, Krishnan S, Zhou L, Bondarenko PV, Nichols AC, Kleemann GR, Pipes GD, Park S, Fodor S, et al. 2006. Active dimer of Epratuzumab provides insight into the complex nature of an antibody aggregate. J Pharm Sci 95(1):126–45.

Robinson DK, Chan CP, Yu Lp C, Tsai PK, Tung J, Seamans TC, Lenny AB, Lee DK, Irwin J, Silberklang M. 1994. Characterization of a recombinant antibody produced in the course of a high yield fed-batch process. Biotechnol Bioeng 44(6):727–35.

Rosenberg AS. 2006. Effects of protein aggregates: an immunologic perspective. Aaps J 8(3):E501–7.

Schroder M, Friedl P. 1997. Overexpression of recombinant human antithrombin III in Chinese hamster ovary cells results in malformation and decreased secretion of the recombinant protein. Biotechnol Bioeng 53:547–59.

Shire SJ, Shahrokh Z, Liu J. 2004. Challenges in the development of high protein concentration formulations. J Pharm Sci 93(6):1390–402.

Shukla AA, Gupta P, Han X. 2007a. Protein aggregation kinetics during Protein A chromatography. Case study for an Fc fusion protein. J Chromatogr A 1171(1–2):22–8. Epub 2007 Sep 22.

Shukla AA, Hubbard B, Tressel T, Guhan S, Low D. 2007b. Downstream processing of monoclonal antibodies – application of platform approaches. J Chromatogr B Analyt Technol Biomed Life Sci 848(1):28–39.

Silveira JR, Hughson AG, Caughey B. 2006. Fractionation of prion protein aggregates by asymmetrical flow field-flow fractionation. Methods Enzymol 412:21–33.

Sinha S, Zhang L, Duan S, Williams TD, Vlasak J, Ionescu R, Topp EM. 2009. Effect of protein structure on deamidation rate in the Fc fragment of an IgG1 monoclonal antibody. Protein Sci 18(8):1573–84.

Soenderkaer S, Carpenter JF, van de Weert M, Hansen LL, Flink J, Frokjaer S. 2004. Effects of sucrose on rFVIIa aggregation and methionine oxidation. Eur J Pharm Sci 21(5):597–606.

Taggart C, Cervantes-Laurean D, Kim G, McElvaney NG, Wehr N, Moss J, Levine RL. 2000. Oxidation of either methionine 351 or methionine 358 in alpha 1-antitrypsin causes loss of anti-neutrophil elastase activity. J Biol Chem 275(35):27258–65.

Takata T, Oxford JT, Demeler B, Lampi KJ. 2008. Deamidation destabilizes and triggers aggregation of a lens protein, betaA3-crystallin. Protein Sci 17(9):1565–75. Epub 2008 Jun 20.

Tsai PK, Bruner MW, Irwin JI, Ip CC, Oliver CN, Nelson RW, Volkin DB, Middaugh CR. 1993. Origin of the isoelectric heterogeneity of monoclonal immunoglobulin h1B4. Pharm Res 10(11):1580–6.

Venter JC, Adams MD, Myers EW, Li PW, Mural RJ, Sutton GG, Smith HO, Yandell M, Evans CA, Holt RA, et al. 2001. The sequence of the human genome. Science 291(5507):1304–51.

Vlasak J, Bussat MC, Wang S, Wagner-Rousset E, Schaefer M, Klinguer-Hamour C, Kirchmeier M, Corvaia N, Ionescu R, Beck A. 2009. Identification and characterization of asparagine deamidation in the light chain CDR1 of a humanized IgG1 antibody. Anal Biochem 392(2):145–54.

Vlasak J, Ionescu R. 2008. Heterogeneity of monoclonal antibodies revealed by charge-sensitive methods. Curr Pharm Biotechnol 9(6):468–81.

Wei Z, Feng J, Lin HY, Mullapudi S, Bishop E, Tous GI, Casas-Finet J, Hakki F, Strouse R, Schenerman MA. 2007. Identification of a single tryptophan residue as critical for binding activity in a humanized monoclonal antibody against respiratory syncytial virus. Anal Chem 79(7):2797–805.

Weintraub SJ, Deverman BE. 2007. Chronoregulation by asparagine deamidation. Sci STKE 2007(409):re7.

Weintraub SJ, Manson SR. 2004. Asparagine deamidation: a regulatory hourglass. Mech Ageing Dev 125(4):255–7.

Yang J, Wang S, Liu J, Raghani A. 2007. Determination of tryptophan oxidation of monoclonal antibody by reversed phase high performance liquid chromatography. J Chromatogr A 1156(1–2):174–82.

Ye H. 2006. Simultaneous determination of protein aggregation, degradation, and absolute molecular weight by size exclusion chromatography-multiangle laser light scattering. Anal Biochem 356(1):76–85. Epub 2006 Jun 9.

Zhang W, Czupryn MJ. 2002. Free sulfhydryl in recombinant monoclonal antibodies. Biotechnol Prog 18(3):509–13.

Zhang Y, Martinez T, Woodruff B, Goetze A, Bailey R, Pettit D, Balland A. 2008. Hydrophobic interaction chromatography of soluble interleukin I receptor type II to reveal chemical degradations resulting in loss of potency. Anal Chem 80(18):7022–8. Epub 2008 Aug 16.

Zhou JX, Solamo F, Hong T, Shearer M, Tressel T. 2008. Viral clearance using disposable systems in monoclonal antibody commercial downstream processing. Biotechnol Bioeng 100(3): 488–96.

Chapter 14
Recovery and Purification of Antibody

XueJun Han, Arthur Hewig, and Ganesh Vedantham

Abstract Monoclonal antibody drugs have become a large portion of protein therapeutics and many antibody molecules are being evaluated at various stages of clinical trials in the biopharmaceutical industry. This review article summarizes the state of the art antibody purification techniques. The main focus is chromatographic techniques that include protein A, ion exchange and HIC. For each technique, the mechanisms of antibody binding, factors affecting binding capacity, resin selection, as well as typical process parameters and chromatograms are discussed in detail. Cell culture clarification, filtration, alternative antibody purification techniques, and viral clearance strategies are discussed briefly. The goal of this article is to provide a broad coverage of antibody purification technology. Readers are referred to extensive articles for further in-depth reading.

14.1 Introduction

Monoclonal antibodies (mAbs) have unique properties that have made them the most prevalent therapeutics in the biopharmaceutical industry, presently accounting for a large proportion of recombinant protein drug candidates in clinical development (Walsh, 2004). Two of these unique properties are their specificity for disease targets and wide range of targets. In the last 2 decades, many mAbs have received marketing approval by the regulatory agencies for a variety of indications such as non-Hodgkin's lymphoma, rheumatoid arthritis, and colorectal cancer (Shukla and Kandula, 2009). Antibody therapies may involve frequent high doses. To meet the demand for some indications, several hundred kilograms of antibody product per year may be required. Recent advances in cell line selection, growth and production media, and feeding strategies have led to antibody expression levels as high as 5 g L^{-1} in a 12-day fed-batch process (Jagschies et al., 2006), with some reporting 10 g L^{-1} through longer cell culture duration (Luan et al., 2006). The combination of high titer and large bioreactors will result in >100 kg batch sizes. Consequently, purification costs are now greater than cell culture costs and process bottlenecks have moved to downstream (Gottschalk, 2008). The current focus for antibody

X. Han, A. Hewig, and G. Vedantham (✉)
Amgen Inc., Seattle, WA 98119, USA
e-mail: vedanthg@amgen.com

M. Al-Rubeai (ed.), *Antibody Expression and Production*, Cell Engineering 7,
DOI 10.1007/978-94-007-1257-7_14, © Springer Science+Business Media B.V. 2011

purification process development is to streamline process development activities, reduce manufacturing cost, and increase throughput in manufacturing facilities.

The primary considerations for downstream process development are product purity and process yield. The process needs to remove process related contaminants and product related impurities. Process related contaminants include cells, cell debris, host cell protein (HCP), DNA, endotoxin, leached Protein A, as well as chemical reagents from cell culture media such as methotrexate, growth promoters, antifoam, and buffer components that should not be present in the final bulk formulation. Product related impurities include high molecular weight (HMW) aggregates, clipped low molecular weight (LMW) species, and product variants (with slight differences in charge, confirmation, disulfide mispairing or glycosylation). In addition, the process should demonstrate capability for clearing potential adventitious agents to ensure products are safe for patients.

Efficient recovery and purification of mAbs from cell culture media is critical for the biopharmaceutical industry's success. Taking advantage of the structural and biochemical similarities in this product class, many companies have adopted a platform approach that is based on a common sequence of unit operations (Shukla et al., 2007, Kelley et al., 2009). A typical purification process for mAbs is shown in Fig. 14.1. Centrifugation and depth filtration are employed to harvest and clarify cell culture broth. Antibody is then captured, purified and concentrated by protein A affinity chromatography. After the protein A step the product purity is typically greater than 95% due to the superior selectivity of the protein A ligand. One or two additional chromatographic polishing steps are required to remove trace amount of contaminants to meet final drug substance purity specifications. Ion exchange or hydrophobic interaction chromatography are often chosen for the polishing steps to provide orthogonal modes of separation. The purification process also has other steps such as low pH viral inactivation, solvent/detergent inactivation, and viral filtration specifically designed to inactivate or remove viruses in the event of an

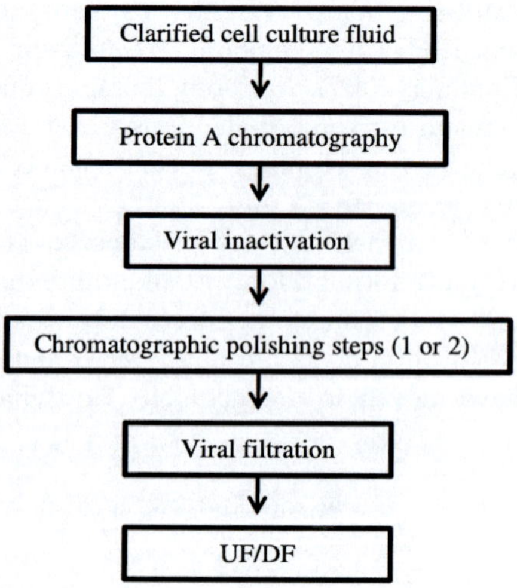

Fig. 14.1 A typical purification process for monoclonal antibodies

undetected contamination. At the end of the process, ultrafiltration/diafiltration (UF/DF) is implemented for concentrating protein and exchanging the solution environment, to prepare the antibody into formulated drug substance. This purification scheme is efficient, robust and is the standard for mAb processing in the biopharmaceutical industry. To increase process throughput and reduce process cost, the development of alternative antibody purification techniques is progressing (Low et al., 2007, Thommes and Etzel, 2007); however, none has come close to being a "platform" process. These techniques include membrane chromatography, high performance tangential flow filtration (HPTFF), impurity flocculation, antibody precipitation, crystallization, expanded bed adsorption, simulated moving bed chromatography, and liquid-liquid extractions using two-phase aqueous polymer systems.

This chapter provides concise information on downstream antibody purification. We hope that it will be useful for those who are interested in general information in the field. The main focus is on chromatography techniques and practical considerations. Non-chromatography modes of purification techniques are briefly discussed and readers are referred to references for details.

14.2 Harvest and Clarification

Mammalian cells are widely used in the production of mAbs because they have the ability to perform comprehensive post-translational modifications and to secrete glycoproteins that are correctly folded and contain complex antennary oligosaccharides with terminal sialic acid (Jefferis, 2005). A wide range of techniques have been evaluated for initial harvest and clarification at bench scale and only a few of them have been successfully applied at process scale purification processes. These techniques include homogenization, high-shear mixings, milling, precipitation/flocculation, centrifugation, microfiltration, depth filtration, aqueous two-phase extraction, expanded bed adsorption chromatography etc (Zhang and Van Cott, 2007, Kinolov et al., 2009). Even though different initial recovery and separation techniques are applied in different expression systems, further purification of antibody is generally carried out with same chromatographic techniques such as Protein A affinity, ion exchange (IEX), hydrophobic interaction (HIC) and Hydroxyapatite (HA) etc. Given the dominant status of mammalian cell expression system, this section will be focusing on the harvest and clarification techniques used in this system.

Since mAbs are produced extracelluarly in mammalian cell culture, there is no need to concentrate cells and disrupt them to release target protein. Mammalian cells are sensitive to breakage due to shear stress, and this can result in the release of proteases and other host cell proteins into the broth, which can affect product stability and/or purity. The first harvesting step for mammalian expression system is to remove the cells by centrifugation or microfiltration. Centrifugation cannot efficiently remove all particulates and the centrate is typically still turbid (Yigzaw et al., 2006). Additional clarification can be achieved by depth filtration. Sterile filtrations are used as a terminal harvest step to ensure sterility and the absence of particulates in the load material to prevent downstream capture step from fouling. Cell culture flocculation of impurities can create larger particles by clumping smaller ones together (Wang et al., 2009, Riske et al., 2007). This technique is typically added prior to centrifugation to improve depth filter capacity and sometimes even product purity. Expanded-bed adsorbent chromatography attempts to integrate solid-liquid separation, product capture and concentration (Sonnenfeld and Thommes, 2007). It has the potential to decrease process time and improve process throughput.

14.2.1 Centrifugation

Centrifugation is a widely used method to remove cells and cell debris, and takes advantage of the difference between the solid density and surrounding fluid density to drive particle settling. The greater the difference in density, the faster a particle will settle out. Centrifugal force accelerates the cell settling that would normally take much longer by simple sedimentation. Continuous flow disk-stack centrifuges (DSC) use a relatively generic set of processing parameters. Factors affecting solid-liquid separations include Sigma factor (the equivalent settling area of a centrifuge), density difference between the solid and liquid, angular velocity, viscosity of liquid, flow rate and residence time, as well as operating temperature (Russell et al., 2007). The percentage of solid in the feed stream and the shear sensitivity of the cells have a strong impact on the performance of DSC. For high percent solid feed stream, the flow rate should be low to allow for adequate solid-liquid separation. However, low flow rates lead to long residence time in the bowl, resulting in long cell exposure to shear. Therefore, shear sensitive material like mammalian cell culture is run at the higher end of acceptable flow rates to prevent cell breakage and at the same time still have adequate clarification. The discharge frequency, discharge type and the pre-discharge flush solution and volume are additional parameters to optimize (Shukla and Kandula, 2009). They can impact yield and possibly product quality. Performance of DSC is monitored through yield, centrate turbidity, and downstream filter capacity (e.g., depth filter or sterile filter). Since centrifugation cannot provide sufficient degree of solid removal, depth filtration is usually required for further clarification. Major DSC suppliers for process scale antibody purification include Westfalia and Alfa Laval.

14.2.2 Microfiltration

Microfiltration is another way to harvest antibody product from mammalian cell cultures. It can provide a particle-free harvest stream that needs minimal additional filtration (Van Reis and Zydney, 2007). Both flat sheet and hollow fiber designs have demonstrated some level of success (van Reis et al., 1991). The pore sizes of typical microfiltration membranes are from 0.2 to 0.45 μm. Newer asymmetric membranes with graded pore sizes have significantly improved the throughput of microfiltration processes and have minimized the effect of concentration polarization (Lee et al., 1995). A variety of alternative flow configurations have also been proposed to mitigate the effects of concentration polarization and fouling. Optimization of operating conditions and the post-use cleaning procedure can help to address this issue. Parameters including cross flow rate, trans-membrane pressure (TMP), flux, membrane area and membrane loading, cleaning solution and time need to be optimized for high yield, appropriate processing time and long membrane lifetime. Major microfiltration membrane suppliers for process scale antibody purification include Millipore, Pall, Sartorius and GE Healthcare. Recent trends towards higher density mammalian cell cultures, which tend to have much higher levels of cell debris, have created more challenges in the application of microfiltration for cell removal and clarification. Microfiltration is generally not a preferred harvest and clarification technique for mAb purification at process scale.

14.2.3 Depth Filtration

Depth filters use a porous medium that can retain particles throughout its matrix rather than just on its surface (Fiore et al., 1980). They trap particles in their tortuous flow channels to a level that size-based screening alone cannot achieve. These filters are usually applied in the situation where the feed stream contains large amount of particles. Being placed prior to a final sterile filter, they can dramatically increase sterile filter capacity by removing large insoluble contaminants from the feed stream. Without depth filtration, the sterile filter would clog relatively quickly and increased filter area would be required (van Reis and Zydney, 2001). Depth filters used in harvest and clarification process are usually composed of cellulose or polypropylene fibers with an appropriate filter aid such as diatomaceous earth and a binder. The role of the filter aid is to provide a large surface area (Smith, 1998). Some depth filters are charged, either because of the binder polymer or from additional charged polymers (Knight and Ostreicher, 1998). Sometimes a microfiltration membrane with an absolute pore size rating is integrated into the depth filter sheet as the bottommost layer. In addition to removing particulates, positively charged depth filters have been reported to also remove endotoxin, DNA, HCP, virus and prion (Yigzaw et al., 2006, Tipton et al., 2002). For process scale operations, depth filtration is usually placed after a centrifugation step for further clarification. Depth filtration can be used as the sole harvest and clarification step, but only at small scale. Depth filter screening involves testing a variety of depth filters with different chemistries, porosities, and charges. Filtrate turbidity, analyzed by a light-scattering or light-obstruction technique, and pressure drop across the filter are monitored for comparison. Since the goal for depth filtration is to improve filterability on sterile filters, a more effective way to evaluate depth filter performance is to test sterile filter capacity on the depth filtration filtrate. Major depth filter suppliers for process scale antibody purification include CUNO, Millipore, Pall and Sartorius.

14.2.4 Flocculation

Higher titer in mammalian cell culture is in part achieved by increasing cell density and cell culture duration, which typically leads to lower cell viability and higher percentage of solid. These high productivity bioreactors usually result in a large amount of particles and a wider distribution of particle size (Thommes and Gottschalk, 2009). This complex cell culture broth is more difficult to harvest and clarify. Flocculation has been evaluated for clumping smaller particles into larger ones which are easier to separate from the cell culture fluid. It is a process in which the suspended particles bind together when the attraction among them overcomes repulsion. The repulsion can be eliminated through the addition of inorganic electrolytes which shield the particle surface charges, or by the addition of polyelectrolytes which neutralize the particle surface charges (Coffman et al., 2006, Shpritzer et al., 2006). Flocculants such as caprylic acid, chitosan, polyethyleneimine have been used for clarification of mammalian cell culture broth (Wang et al., 2009, Riske et al., 2007). If the flocculants are toxic or their toxicity profile is unknown, the downstream purification process must show adequate clearance of them to ensure their levels in the bulk drug substance meet specifications. The combined use of calcium chloride and potassium phosphate has also been reported for flocculation. When the solutions of these two compounds are mixed, calcium phosphate precipitates and flocculates cellular debris (Coffman et al., 2006). Flocculation can also be induced by lowering the mammalian cell culture pH (Lydersen et al., 1994). In addition to simplifying the harvest process, flocculation has been reported to reduce soluble impurities such as HCP and DNA (Wang et al., 2009). Flocculation is often placed before centrifugation and/or depth filtration to improve depth filter capacity. Sometimes it is employed after protein A chromatography, mainly to reduce impurities, not for clarification purposes.

14.2.5 Expanded Bed Adsorption

Expanded bed adsorption (EBA) offers a unique mode of chromatography in which the mammalian cell culture can be applied directly to the column without clarification (Sonnenfeld and Thommes, 2007, Blank et al., 2001). EBA could potentially replace centrifugation, depth filtration, primary capture, and product concentration with a single column step. This could lead to shorter overall processing time and higher process throughput. By forcing the process stream to flow upward, the adsorbent expands, facilitating the passage of solids, and unbound contaminants, through the column while the antibody is selectively captured. EBA resins have larger diameter beads than conventional packed-bed resins, which enable the use of large pore adaptor screens for the column. A crucial factor for EBA performance is the distribution system design. The optimal choice of adsorption bead size and density, column hardware, and operating conditions ensure maintenance of bed expansion without loss of adsorbent in the column effluent. The loading and washing phases are operated in the expanded-bed mode. Product elution is typically done in a packed bed mode to minimize the volume of elution buffer and elution pool. Disadvantages of EBA technology include lower dynamic binding capacity of adsorbent and the limited column diameters and lengths (Spitali, 2009). It is also difficult to ensure uniform flow distribution from the bottom of the column as column diameters increase. Equipment cleaning and regeneration can be another issue (Shukla and Kandula, 2009). STREAMLINE products for EBA, manufactured by GE Healthcare, include STREAMLINE columns and systems.

14.3 Chromatography

14.3.1 Protein A Affinity Chromatography

A variety of chromatographic techniques have been employed in mAb purification processes. Protein A affinity chromatography is still the most effective technique for capture and purification of antibodies. Protein A binds mAbs with high affinity and specificity and will continue to be the workhorse for mAb purification in the foreseeable future (Vunnum et al., 2009). It can tolerate a wide range of loading conditions in terms of pH and conductivity and conditioning of the clarified cell culture fluid is not required. The yield for this step is very high, usually > 95%. Protein A chromatography is also relatively easy to develop. Even without optimization, this step typically delivers > 95% product purity starting directly from complex cell culture media (Gagnon, 1996). The large purification factor obtained from this step helps to simplify the entire downstream purification process. In general, only trace amount of contaminants (HMW aggregates, residual HCP, leached protein A, and DNA) remain to be removed after this unit operation. After one or two subsequent chromatography steps, the final bulk drug substance usually can meet purity specifications (Fahrner et al., 2001). There has been a considerable effort to identify low cost mimetics that can provide greater chemical stability and longer lifetime than Protein A, and yet to retain the selectivity and simplicity associated with Protein A resin. These alternative ligands have shown some selectivity, but the purification factors are significantly lower than those obtained with Protein A resin (Schwartz et al., 2001, Ghose et al., 2006).

The molecular weight of an intact protein A molecule is 54 kDa. A derivative of the protein A with the cell wall domain deleted is ~42 kDa. The five homologous antibody binding domains are named as E, D, A, B, C and they have approximately the same antibody binding ability (Hjelm et al., 1975). The interaction between antibody and protein A has been studied in detail by X-ray crystallography of the complex between an antibody Fc fragment and a 58 amino acids fragment spanning the B-domain of protein A (Deisenhofer, 1981). The three dimensional structure of the complex revealed two antiparallel α–helices on domain B interacting with both C_H2 and C_H3 domains of the Fc region (Gagnon, 1996). The interaction primarily consists of hydrophobic interactions along with some hydrogen bonding and two salt bridges (Li et al., 1998). The primary binding site for protein A on the Fc region is at the juncture of C_H2 and C_H3 domains. The histidyl residue in the center of the protein A binding site of antibody is highly conserved (Burton, 1985). This residue aligns face to face with a complementary and similarly conserved histidyl residue on protein A (Moks et al., 1986, Lindmark et al., 1977). At alkaline pH, these histidyl residues are uncharged and there are no restrictions on interfacial contact. The hydrophobicity of the uncharged imidazolium rings at the interface strengthens the association (Gagnon, 1996). At low pH, the histidyl residues are positively charged resulting in electrostatic repulsion between the two proteins. This repulsion is strong enough to elute the antibody off the protein A column. The variable region interactions in V_H3 antibodies can have an important effect on their binding affinity and their elution pH. The variable region interactions may even be stronger than Fc-mediated interactions in some cases (Ghose et al., 2005). For further in-depth reading, refer to the following review articles (Vunnum et al., 2009, Ghose et al., 2007).

Several commercially available protein A resins are listed in Table 14.1. These resins vary in the source of the protein A ligand (natural wild type vs recombinant), immobilization chemistries, and bead characteristics. Recombinant protein A is expressed in *Escherichia coli* and it lacks the cell wall associated region. Various constructs incorporate different features to support directional coupling of the ligand to the solid-phase supports. Differences in matrix composition, bead size, and pore size can give rise to differences in resin compressibility, chemical resistance, permeability, available surface area, and mass transfer properties, which can have significant impacts on the performance of the protein A column (McCue et al., 2003). The two leading manufacturers of industrial protein A chromatographic

Table 14.1 Some preparative protein A resins

Resin name	Ligand	Base matrix	Vendor
rProtein A Sepharose fast flow	Recombinant protein A (*E. coli*)	Cross-linked agarose	GE Healthcare
MabSelect	Recombinant protein A (*E. coli*)	Cross-linked agarose	GE Healthcare
MabSelect Xtra	Recombinant protein A (*E. coli*)	Cross-linked agarose	GE Healthcare
MabSelect SuRe	Alkali stable protein A derived (*E. coli*)	Cross-linked agarose	GE Healthcare
ProSep-vA high capacity	Native protein A	Controlled pore glass (1,000 Å)	EMD Millipore
ProSep-vA ultra	Native protein A	Controlled pore glass (700 Å)	EMD Millipore

media are GE Healthcare and Millipore. GE healthcare employs agarose with varying degrees of cross-linking as the base matrix. MabSelect uses recombinant protein A that is coupled to highly cross-linked agarose beads. MabSelect Xtra has a higher ligand density and a wider pore size than MabSelect, resulting in a higher binding capacity. MabSelect SuRe has been developed to withstand stronger alkaline conditions allowing the repeated use of 0.1–0.5 M NaOH for regeneration and sanitization. It has the same backbone as MabSelect but a genetically modified protein A ligand. Using protein engineering techniques, a number of asparagine residues were replaced in the Z domain (a functional analogue and energy-miminized version of the B domain) of protein A, and a new ligand was created as a tetramer of four identical modified Z domains (Braisted and Wells, 1996). The absence of D and E domains in the new resin also helps eliminating variable region interactions (Gulich et al., 2000). This is why the elution buffer pH for MabSelect SuRe can be higher than that used for MabSelect when mAbs have V_H3 domains. A higher elution pH can minimize low pH induced HMW aggregate formation in protein A capture step (Ghose et al., 2005). MabSelect or MabSelect SuRe are frequently the choice of resin for protein A affinity chromatography. Protein A resins from Millipore are based on controlled pore glass (CPG) matrix. They have good pressure-flow characteristics due to the rigid CPG backbone. However, CPG is hydrophobic as compared to agarose and thus exhibits higher levels of nonspecific interactions with impurities in the harvested cell culture fluid. Wash steps may be required for ProSep A resins to address this issue.

The objectives of the protein A affinity chromatography in antibody purification processes are product capture from harvest cell culture fluid (HCCF) and the removal of HCP, DNA and other process related impurities. It is also a volume reduction and product concentration step. The basic protocol for protein A chromatography is relatively straightforward: bind mAbs at neutral pH and elute them at acidic pH. Protein A steps usually deliver very high product purity because of high selectivity. The ease and simplicity of method development for protein A has been a key reason for its widespread adoption for antibody purification. A typical protein A process flow diagram is shown in Fig. 14.2 and a typical protein A chromatogram is shown in Fig. 14.3. The equilibration of the column is usually done under neutral pH conditions (6.0–7.5). The HCCF (at pH ~7.0) is loaded directly onto the column to ~90% of resin dynamic binding capacity (DBC). One to two wash steps are used to remove loosely bound HCP, DNA and other contaminants. Antibody product is eluted at low pH (3.4–3.8). The column is then stripped with an acid at an even lower pH (2.5–3.0). A wash step with equilibration buffer is used to neutralize the column before the column is regenerated with NaOH solution. Finally, the protein A column is stored in a solution that inhibits bacterial growth. For manufacturing scale purification, protein A process development is a compromise between throughput, cost, impurity removal, protein A leaching, flow characteristics, cleaning, and resin lifetime considerations. Protein A chromatography process development is discussed in detail below.

Most of the mAbs currently being used or investigated for therapeutic applications are human or humanized molecules belonging to IgG classes 1, 2 or 4, all

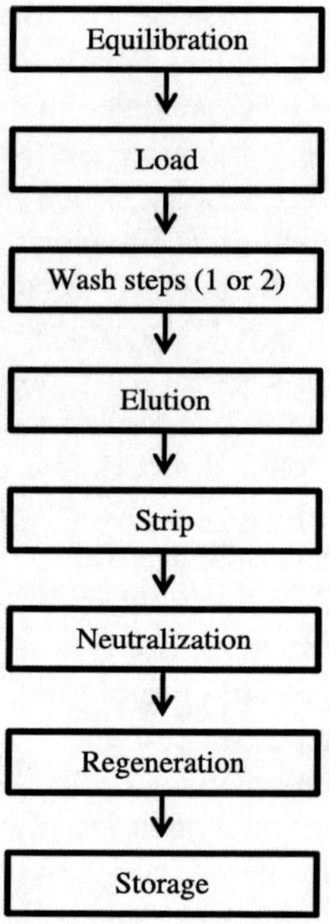

Fig. 14.2 A typical protein A process flow diagram

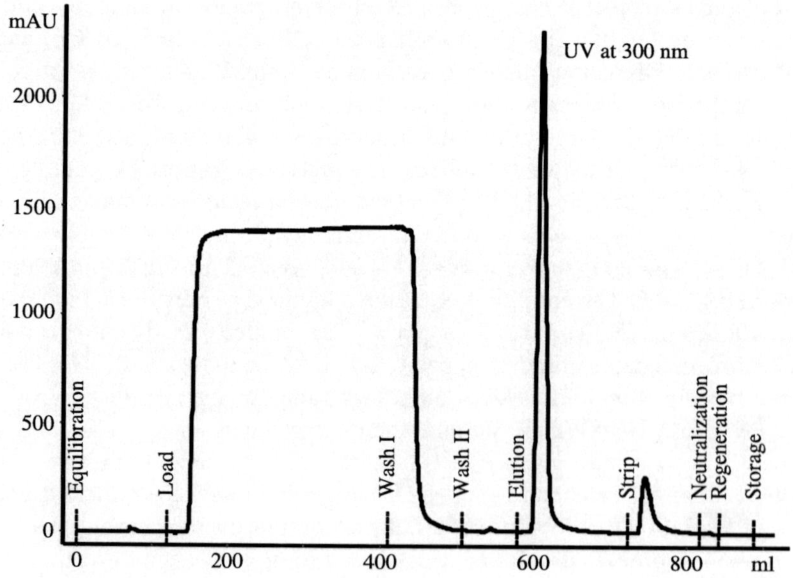

Fig. 14.3 A typical protein A chromatogram

of which bind strongly to protein A (Rohrbach et al., 2003). The direct capture of mAbs from HCCF is therefore common practice in the biopharmaceutical industry, and presents no significant technical issues due to the high titers and efficient resins currently available on the market. Salts can be added to the HCCF to promote antibody binding to protein A resins (Gagnon, 1996). The key parameter affecting protein A DBC is column residence time (Hahn et al., 2003). Usually the residence time is kept $\geq$ 6 min to maximize protein A binding capacity. The protein A column is typically loaded at > 30 mg mL^{-1} resin and can be higher for some mAbs. The primary disadvantage of protein A affinity chromatography is the high cost of the resin. Protein A media are almost an order of magnitude more expensive than traditional chromatographic media. However, with its efficient capture of antibody directly from HCCF and exceptional purification capabilities it is difficult to develop a better capture step. To reduce capital cost at process scale, companies usually pack a small protein A column and run it multiple cycles (3–10) to purify a single batch.

Despite the high specificity of a protein A chromatography step, HCP, DNA and other contaminants are still present at varying levels in the elution pool. Impurities can bind to antibodies and co-elute with them. Even though the polishing chromatographic steps are capable of removing the small amount of contaminants left in the protein A pool, it is desirable to minimize the impurity level in the protein A step itself to increase the overall robustness of the process. Intermediate wash steps can be applied for reducing pool impurities and minimizing pool turbidity. Several washing strategies have been evaluated. The first strategy is to optimize the wash buffer pH. For maximum removal of nonspecifically bound material, the wash buffer pH should be as low as possible but not so low as to initiate premature elution of the antibody. When an intermediate pH wash is impractical or insufficient to reduce impurities to acceptable levels, the second strategy is to evaluate the addition of salt, amino acids, detergents, or solvents to the wash buffer (Vunnum et al., 2009). Once the buffer additive is proven effective, the concentration can be optimized. Low pH (3.4–3.8) is the most commonly used method for eluting protein A columns. Even though the protein A ligand can tolerate low pH very well, mAbs tend to form HMW aggregates at low pH. HMW aggregates have the potential to compromise product safety and to complicate the development of the polishing steps. Several methods have been used to address the aggregation issue during protein A elution. The first method is to elute the column at as high a pH as possible while maintaining good step yield to minimize product exposure to low pH environments. MabSelect SuRe interacts with only the Fc portion of the antibodies (no Fab interaction), allowing the use of less acidic elution buffers. Addition of NaCl, ethylene glycol, urea, histidine, and imidazole to the elution buffer has been reported to moderate the elution pH on protein A column (Gagnon, 1996, Shukla et al., 2005). Proper mixing of the elution pool helps to dilute out the acidic tail of the elution peak, reducing the duration of product under very low pH conditions. The second method is to employ stabilizers such as arginine in the elution buffer to reduce aggregation (Arakawa et al., 2004). The third method is to apply low temperature operation for protein A column to reduce HMW formation (Shukla et al., 2005). Manipulating the pH transition between wash and elution phases is another viable strategy (Vunnum et al., 2009).

Another common issue associated with protein A chromatography is precipitation during low pH elution. Elution buffer type (e.g., acetate vs citrate) can have a significant impact on elution pool turbidity. Low concentrations of NaCl or Na_2SO_4 can be included in the elution buffer to counteract the tendency of some antibodies to precipitate at low ionic strength. If the turbidity is predominantly due to precipitation of contaminating HCP, an efficient wash buffer or flocculation prior to centrifugation may help to reduce protein A pool turbidity. Determining appropriate wash and elution conditions is one of the areas that require significant process development effort for protein A chromatography.

Efficient cleaning of protein A resin is crucial because the high cost of this resin makes it necessary to extensively cycle the protein A column. Protein A ligand has a high conformational stability and is remarkably resistant to physicochemical stress (Vunnum et al., 2009). It is stable in strong acidic conditions and it can refold after treatment with denaturing solutions such as urea and guanidine. After product elution at low pH, a protein A column is typically stripped at an even lower pH (2.5–3.0) to remove any tightly bound HCP, HMW aggregates, and other impurities. Protein A column regeneration is typically carried out with high concentrations of chaotropes (6 M urea or 6 M guanidine HCl). The lack of cysteine residues allows cleaning with reducing agents such as dithiothreitol (DTT). In some cases, a cleaning buffer containing both chaotropic and reducing reagents are more effective. Since protein A ligand cannot withstand strong alkaline conditions, only low concentrations of NaOH (typically < 0.1 M) can be used to clean the resin. Cycling studies with MabSelect have shown a 16% reduction in DBC after 61 cycles when regenerated with 50 mM NaOH (Hale et al., 1994). Addition of NaCl to the NaOH solution has a significant positive effect on the stability of protein A ligand and can help with resin cleaning. Regenerating MabSelect with 50 mM NaOH/100 mM NaCl resulted in no change in the DBC of the resin after 100 cycles, and only a 4% drop after 200 cycles, and 11% drop after 300 cycles. MabSelect SuRe resin with the engineered ligand is designed to improve ligand stability in alkaline conditions and can withstand NaOH concentrations up to 0.5 M for improved resin cleaning. Chaotropic solutions are costly and require special handling during disposal at process scale. NaOH is inexpensive, easy to dispose, and is a process reagent that does not require showing clearance throughout the rest of the process because it is not a potential safety concern. Therefore, NaOH solution is a preferred buffer to clean protein A resin, particularly for base stable MabSelect SuRe. Cleaning protein A resin with NaOH solution might not be required after every cycle. At process scale, a protein A column is usually stripped at low pH after every cycle and regenerated with NaOH solution once every batch. A lifetime of greater than 100 cycles for Protein A column is generally considered economically viable, and typically can be obtained with proper cleaning.

Protein A is known to cause immunogenic responses in human and has been proven toxic in a number of clinical trials (Terman and Bertram, 1985). Residual levels of leached protein A are often detected in protein A elution pools. The trace amount of leached protein A needs to be reduced to acceptable levels by the polishing steps. Leached protein A can come from breakdown of the support matrix or the immobilization linkage, and the proteolytic cleavage of protein A ligand (Gagnon, 1996). It is recommended that any protein A resin be selected from an established manufacturer with a history of good ligand stability. The flexible random coil sequences that link the protein A domains are susceptible to proteolytic cleavage. Addition of chelators (e.g., EDTA) to cell culture harvest or holding the cell culture load at lower temperature can minimize proteolytic degradation (Vunnum et al., 2009, Ghose et al., 2007).

14.3.2 Ion Exchange Chromatography

Various modes of chromatography including cation exchange (CEX), anion exchange (AEX), hydrophobic interaction chromatography (HIC), Hydroxyapatite (HA), and other multi-modal resin have been used as polishing steps in antibody purification processes. Ion exchange (IEX, including CEX and AEX) is the most widely used and well characterized chromatography mode (Staby et al., 2005, Stein and Kiesewetter, 2007). IEX chromatography is very useful for the removal of a wide range of impurities and is an extremely versatile chromatographic unit operation. High resolution with high yield can be obtained by choosing the optimal ion exchanger and the optimal operating conditions. In general IEX media have shown high loading capacities and good stability in strong alkaline cleaning conditions, both of which lead to manufacturing friendly processes. Usually IEX is employed as the first polishing step after protein A in mAb purification processes. Many antibody downstream processes have demonstrated that a 2-column purification scheme can meet drug substance purity and regulatory compliance guidelines, using just a protein A chromatography step followed by an IEX column step (Kelley, 2007, Kelley et al., 2008).

IEX is based on electrostatic interactions between surface charges on the antibody and charged functional groups on the resin. The net charge of an antibody is a function of the pH of its environment. At a pH below its isoelectric point (pI), the antibody is positively charged and at a pH above its pI the antibody is negatively charged. Antibodies are ampholytes, they contain both positive charges from the ionization of basic amino acid residues and negative charges from the acidic amino acid residues. The pI only reflects the overall surface charge of the protein – patches of either charge still exist on the antibody surface, and the antibody usually interacts through these charged patches rather than through an average charge over its entire surface. The composition of the binding site at one pH may be very different from its composition at a different pH (Shukla and Yigzaw, 2007).

Ion exchangers can be divided into two types: cation exchangers (negative charge on ligand) and anion exchangers (positive charge on ligand). Within each type, there are strong and weak ion exchangers. In general, strong ion exchangers can maintain their charge over a broader pH range than weak ion exchangers. Typically strong ion exchangers have functional groups with very low pK (e.g., <1 for sulfopropyl) or very high pK (e.g., >13 for Trimethyl ammoniumethyl). A large number of preparative resins from several commercial suppliers are available for IEX development. Theses media differ from each other in terms of ligand type (CEX or AEX, weak or strong), ligand density, base matrix, linker chemistry, pore size and particle size. All of these properties can contribute to the variations in resin capacity, selectivity, stability, and other separation characteristics. Some preparative IEX resins are listed in Table 14.2. Cross-linked agarose based IEX resins are available with a wide variety of functional ligands such as Q Sepharose FF, DEAE Sepharose FF, SP Sepharose FF, and CM Sepharose FF. A recent addition to this family of resins is the Sepharose XL series. In this series, functionalized long flexible dextran chains are covalently coupled to the agarose matrix to significantly improve resin capacity while maintaining good selectivity. IEX resins based on synthetic polymeric matrices are also widely employed. Fractogel and Fractoprep series provide a number of advantages

Table 14.2 Some preparative IEX resins

Resin name	Ligand	Base matrix	Vendor
CEX resins			
Fractogel COO^- (M)	Carboxymethyl	Methacrylate	EMD
Toyopearl CM-650 M	Carboxymethyl	Methacrylate	Tosoh
CM Sepharose fast flow	Carboxymethyl	Cross-linked agarose	GE Healthcare
Fractogel SO_3^- (M)	Sulfoisobutyl	Methacrylate	EMD
Fractoprep SO_3^-	Sulfoisobutyl	Hydrophilic synthetic vinyl polymer	EMD
SP Sepharose fast flow	Sulfopropyl	Cross-linked agarose	GE Healthcare
SP Sepharose XL	Sulfopropyl	Cross-linked agarose with bound dextram	GE Healthcare
Toyopearl SP-650 M	Sulfopropyl	Methacrylate	Tosoh
AEX resins			
DEAE Sepharose fast flow	Diethylaminoethyl	Cross-linked agarose	GE Healthcare
Fractogel DEAE (M)	Diethylaminoethyl	Methacrylate	EMD
Fractoprep DEAE	Diethylaminoethyl	Hydrophilic synthetic vinyl polymer	EMD
Toyopearl DEAE-650 M	Diethylaminoethyl	Methacrylate	Tosoh
Q Sepharose fast flow	Quaternary ammonium	Cross-linked agarose	GE Healthcare
Q Sepharose XL	Quaternary ammonium	Cross-linked agarose with bound dextran	GE Healthcare
Fractogel TMAE (M)	Trimethyl ammoniumethyl	Methacrylate	EMD
Fractogel TMAE HiCap (M)	Trimethyl ammoniumethyl	Methacrylate	EMD
Fractoprep TMAE	Trimethyl ammoniumethyl	Hydrophilic vinyl polymer	EMD
Toyopearl SuperQ-650 M	Quaternary ammonium	Methacrylate	Tosoh

such as high capacity, high flow rate, and high selectivity. The covalent surface modification using flexible linear polymers results in an increased spatial availability of the immobilized ion exchanger groups. Therefore, the binding capacity for antibody can be increased significantly over supports where the functional groups are anchored directly to the matrix with none or short linkers. Fractogel TMAE and Fractogel $SO3^-$ are frequently used in the biopharmaceutical industry. A group of recently introduced Toyopearl GigaCap IEX media is optimized specifically for the capture and purification of antibody. Toyopearl HW-65 resin is the polymeric base bead chemically modified to provide a high level of binding sites for antibody. For matrices that are too hydrophobic, a hydrophilic polymer is often applied as a coating to reduce nonspecific interactions.

The objectives of the IEX chromatography step are to remove trace amount of impurities (e.g., HCP, leached protein A, DNA, endotoxin, HMW species). IEX chromatography can also concentrate product if it is operated in a bind and elute mode and the resin has a high loading capacity. The basic concept for IEX chromatography (for a bind and elute mode) is very simple: binding antibody at low conductivity and eluting it at high conductivity. However, IEX process development is not straightforward. A variety of resins, pHs, buffers and salt concentrations need to be screened for process optimization. Resin screening for best selectivity and antibody DBC while maintaining high process yield should be the main focus. Loading capacity is mainly influenced by IEX resin, loading pH and salt concentration. Wash buffer is essential for the efficient removal of bound contaminants and it should be screened with different buffer species at different pHs and ionic strengths. Salt gradient elution can be steep or shallow and this has a significant impact to IEX resolution and pool purity. Since most mAbs have a higher pI (6.5–9.0), CEX chromatography is usually operated in a bind and elute mode at lower pH (5–6) and AEX chromatography is operated in a flow-through mode at higher pH (7–8).

A typical CEX process flow diagram (in a bind and elute mode) is shown in Fig. 14.4 and a typical CEX chromatogram (in a bind and elute mode) is shown in Fig. 14.5. The equilibration of the column is usually done at low conductivity. Feed stream is also loaded at low conductivity and loading is typically up to ~90% of DBC. After one or two wash steps to remove loosely bound HCP, DNA and other contaminants, antibody product is eluted with a salt gradient of increasing salt concentration. The column is striped using a buffer with high salt concentration, sanitized with 0.5–1.0 M NaOH solution and stored in 0.1–0.2 M NaOH solution. IEX process development is discussed in detail below. CEX (in a bind and elute mode) development will be the focus and the AEX (in a flow-through mode) development will be discussed briefly.

Resin screening is usually the first step in process development. Screening IEX resins with different properties over a wide pH range is recommended for identifying resins with suitable selectivity and binding capacity. Due to the large number of possible combinations of operating parameters (e.g., resins, pHs, buffers, gradient), initial IEX resin screening should be done with high throughput screening (HTS) technique to rapidly narrow down the number of runs for further screening. After the initial screening with HTS, a manageable number of conditions are then screened

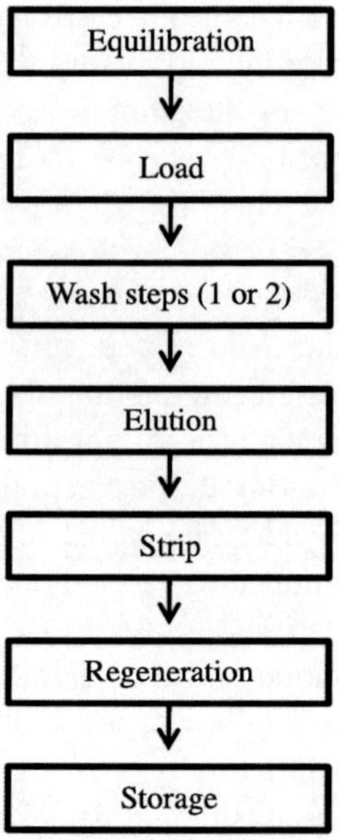

Fig. 14.4 A typical CEX process flow diagram (in a bind and elute mode)

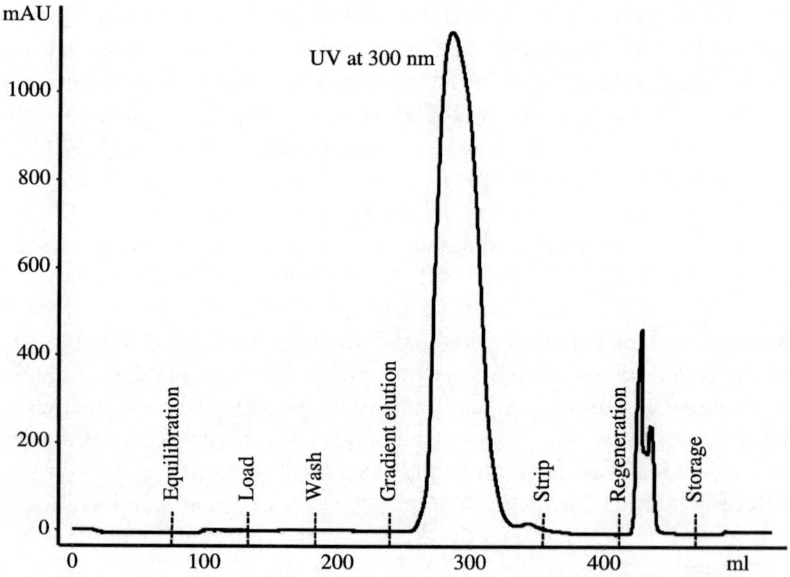

Fig. 14.5 A typical CEX chromatogram (in a bind and elute mode with gradient elution)

for resin selectivity with a packed column using a linear salt gradient at moderate antibody loading. Fractions are collected and analyzed with relevant in-process assays for purity and yield. Fractioning the product elution peak can help to identify impurity profiles. In most cases, the contaminants elute either at the beginning or the trailing end of the peak. A plot of cumulative key impurity level vs. cumulative product yield can help to compare resin selectivity. In addition to selectivity, a variety of other performance attributes (e.g., pressure flow characteristics, resin lifetime, cleaning, sanitization, resin lot to lot reproducibility) should be considered during resin selection. For details of resin screening process and case studies, readers are referred to review articles (Shukla and Han, 2007, Shukla and Yigzaw, 2007).

Ion exchange resins have demonstrated relatively high binding capacities (up to 100 mg mL^{-1} resin) for monoclonal antibodies (Stein and Kiesewetter, 2007). Newer resins with novel chemistries are developed for even higher capacity to meet increased purification throughput demand from higher titer cell culture processes. Binding capacities are usually limited by the presence of other binding contaminants or by the elevated salt concentrations in the feed stream. Loading at low conductivities are preferred for maximizing IEX binding capacity. If the feed stream has high conductivity, dilution or even buffer exchange into a low salt buffer is often required. Intuitively, one would expect maximum CEX capacity to be obtained under very low conductivity. However, recent studies have shown that binding capacity for antibodies on CEX resins can increase with increasing conductivity at the lower range of ionic strength conditions (Harinarayan et al., 2006). This atypical dependence of DBC on conductivity is explained by a protein exclusion mechanism. At very low conductivity, proteins initially bound to the outer pore regions can electrostatically and sterically hinder subsequent protein molecules from diffusing into the pores. Increasing the ionic strength shields the charges on the protein and attenuates the exclusion between proteins bound at the pore and new incoming molecules, thereby resulting in a higher binding capacity. The increased ionic strength also allows for proteins to further penetrate the resin bead by decreasing the binding strength, and can help to minimize steric exclusion in the resin pores. As the conductivity further increases, the traditional trend of decreased capacity reasserts itself due to reduced electrostatic interactions between the ligand and the antibody molecules. This observation should be kept in mind when optimizing IEX resin binding capacity.

Acetate, MES, citrate and phosphate are used for CEX chromatography at lower pH while Tris, HEPES and borate are applied for AEX chromatography at higher pH. Buffering species being employed should not bind to the IEX resins, otherwise pH fluctuation can be an issue when the bound buffer species are displaced from the stationary phase. Sufficient buffer capacity should be maintained when weak IEX resins are used. For example, the bound H^+ (e.g. to the COO^- group) under low salt conditions can be displaced during NaCl gradient elution and this could lead to a transient pH decrease if the buffer capacity is not high enough (Ghose et al., 2002). When Tris buffer is used, IEX operating temperature should be controlled in a tight range because the pH of Tris buffer is very sensitive to temperature. Buffer disposal and impurities in buffer reagents are additional considerations for process scale antibody purification.

Another important area in IEX process development is defining the peak collection criteria during product elution. Peak collection is typically based on UV signal of the column effluent. Triggering peak collection is usually set at a low absorbance value at 280 nm. This value should not be too low, otherwise peak collection may initiate prematurely due to baseline drifting. Terminating peak collection is usually based on a percentage of the maximum height of the elution peak. This percentage is set according to the impurity distribution profile at the end of the peak and target yield for this step. Since antibody concentration at elution peak can be very high, one must ensure that the UV detector is not saturated during elution. This can be done by using a small path length flow cell or by moving away from 280 nm to another wavelength (typically 300 nm) where the extinction coefficient is lower. Well defined peak collection criteria are critical for ensuring good and consistent product quality.

Due to the high pI (6.5–9.0) of most mAbs, AEX is often used in a flow-through mode under basic pH conditions and low conductivity. In a flow-through mode operation, positively charged basic antibodies flow through the column while negatively charged HCP, DNA and endotoxin bind to the resin. Since the contaminants after protein A capture step are in trace amount, AEX flow-through step generally has very high loading capacity. Recently, Kelley et al. reported principles of weak-partitioning chromatography (WPC) for antibody purification by AEX using isocratic conditions (Kelley et al., 2008). Appropriate pH and counterion concentration are chosen to allow significant amount of product to bind to the resin. In the same load conditions, impurities bind even stronger and this leads to good flow through pool purity. To minimize product yield loss due to product binding, very high loading and short isocratic washes at the end of load stage are usually required. Another advantage of very high loading is smaller percentage of product pool dilution.

CEX has been shown to clear HCP, leached protein A and HMW aggregates. It is more effective than AEX in reducing HMW aggregates. AEX is useful for DNA, HCP and endotoxin removal. AEX is superior to CEX for DNA and endotoxin removal. DNA molecules or fragments and endotoxins typically bear strong negative charges and thus bind strongly to AEX columns.

14.3.3 Hydrophobic Interaction Chromatography

Hydrophobic interaction chromatography (HIC) is often employed when IEX cannot achieve the required level of product purity. Since HIC and IEX separations are based on completely different mechanisms they can be used as orthogonal polishing steps in antibody purification. HIC is typically the mode of choice when trying to remove significant amounts of HMW aggregates (Lu et al., 2009). Removal of antibody aggregates is important because the aggregates have the potential to enhance immune responses that can cause adverse clinical effect (Rosenberg, 2006). Aggregates are typically more hydrophobic than monomers and are more retained on HIC resins. HIC development involves screening a variety of resins, pHs, buffers and lyotropic salts for high DBC, best selectivity, and high process yield.

Fig. 14.6 Hofmeister series

Increasing salting out effect ←

Anions: PO_4^{3-}, SO_4^{2-}, CH_3COO^-, Cl^-, Br^-, NO_3^-, ClO_4^-, I^-, SCN^-

Cations: NH_4^+, Rb^+, K^+, Na^+, Li^+, Mg^{2+}, Ca^{2+}, Ba^{2+}

→ Increasing salting in effect

HIC is based on the interaction between hydrophobic (aliphatic and aromatic) ligands on the resin and hydrophobic patches on the antibody surface. Antibody retention in HIC depends not only on the stationary phase but also on characteristics of the mobile phase, such as the type and concentration of the salt, pH, temperature and additives. The influence of different salts follows the Hofmeister series (see Fig. 14.6). The salts at the beginning of the series promote hydrophobic interactions while the salts at the end of the series decrease the strength of hydrophobic interactions. A decrease in pH or increase in temperature typically promotes hydrophobic interactions (Queiroz et al., 2001). Mobile phase additives including water soluble alcohols (e.g., ethyleneglycol), detergents (e.g., Triton X-100), and aqueous solutions of chaotropic salts have significant effect to hydrophobic interactions and they can be used to strip very tightly bound proteins off the HIC column.

Both HIC and reversed-phase chromatography (RPC) exploit hydrophobic interactions for protein purification. However, the differences between them are worth mentioning. Separations on HIC matrices are usually done in aqueous salt solutions which generally are nondenaturing. Separations on RPC media are usually done in mixtures of aqueous and organic solvents which are often denaturing. HIC depends on protein surface hydrophobic groups and is carried out under conditions which maintain the biological activity of the protein molecule. RPC depends on the native hydrophobicity of the protein and is carried out under conditions which expose nearly all hydrophobic groups of the protein to the matrix (Kennedy, 1990).

A selection of commercially available HIC resins for preparative chromatography are listed in Table 14.3. The resins shown in the table are listed in their generally accepted order of increasing hydrophobicity. These media differ on the type of ligand, ligand density, base matrix, and linker chemistry. These properties play significant roles in determining hydrophobicity, which affects resin capacity, selectivity, yield, and cleaning. The media also have different pore sizes and particle sizes, which affect resin capacity, resolution, and column operating pressures. Cross-linked agarose based HIC resins are available with a variety of functional ligands such as Butyl Sepharose 4 fast flow, Octyl Sepharose 4 fast flow and Phenyl Sepharose 6 fast flow (high sub and low sub). The Phenyl Sepharose resin (high sub) is frequently applied in a polishing step at process scale. Toyopearl HIC resins with methacrylate base matrix provide different surface chemistry and hydrophobicity. They enable optimal separations at the extremes of hydrophobic spectrum. Highly retentive Toyopearl Hexyl can be used to separate mildly hydrophobic proteins while less retentive Toyopearl Ether is recommended for the purification of very hydrophobic proteins.

Table 14.3 Some preparative HIC resins

Resin name	Ligand	Base matrix	Vendor
Toyopearl Ether-650 M	Ether	Methacrylate	Tosoh
Phenyl Sepharose 6 fast flow (low sub)	Phenyl	Cross-linked agarose	GE Healthcare
Toyopearl Phenyl-650 M	Phenyl	Methacrylate	Tosoh
Butyl Sepharose 4 fast flow	Butyl	Cross-linked agarose	GE Healthcare
Octyl Sepharose 4 fast flow	Octyl	Cross-linked agarose	GE Healthcare
Toyopearl Butyl-650 M	Butyl	Methacrylate	Tosoh
Phenyl Sepharose 6 fast flow (high sub)	Phenyl	Cross-linked agarose	GE Healthcare
Toyopearl Hexyl-650 C	Hexyl	Methacrylate	Tosoh

HIC can be operated in either flow-through or bind and elute modes, with flow through being the most preferable option when appropriate. A comparison of HIC operated in flow-through mode versus bind and elute mode is shown in Table 14.4. When operated in flow-through mode the removal of impurities such as HMW aggregates are often on par with a bind and elute step. There are situations where bind and elute can deliver improved purification but often this level of purification is not required. HIC in flow-through mode is especially successful if the majority of impurities is more hydrophobic than the product and thus retained on the column while antibody product flows through, which is generally the case with HMW aggregates. Like an AEX flow-through step, HIC flow-through can be operated in a WPC mode by choosing optimal resin, salt type and concentration, as well as pH. In this mode the antibody weakly binds to the resin while impurities bind stronger, resulting in excellent contaminant removal. If placed after an IEX bind and elute step the salt concentration is typically high enough for efficient impurity removal. HIC step can often be developed at the pH of the previous IEX step, and the only development required is resin selection and column load factor. By screening resins of increasing hydrophobicity and examining the resulting impurity removal and step yield, the optimal resin choice can be made. Subsequent investigation of product quality and yield as a function of column load factor can then be used to define the column's operating space. High loading can be used to increase step yield and decrease the percentage of product pool dilution caused by the collection of flow-through peak tail. Fig. 14.7 shows a typical HIC process flow diagram (in a flow-through mode) and Fig. 14.8 shows a typical HIC chromatogram (in a flow-through mode). If sufficient removal of impurities cannot be achieved then

Table 14.4 A comparison of HIC in flow-through mode versus in bind and elute mode

	Flow-through	Bind and elute
Load conditioning	Typically none	Development required
Load factor (mg mL^{-1} resin)	>100	<50
Yield (%)	>95	85–90
Sensitivity to temperature	Low	Moderate

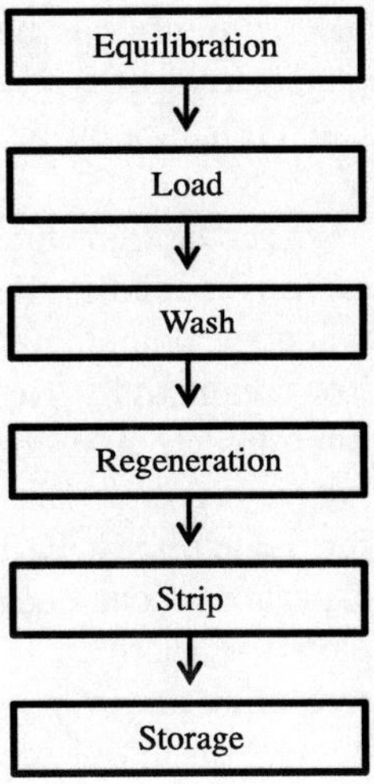

Fig. 14.7 A typical HIC process flow diagram (in a flow through mode)

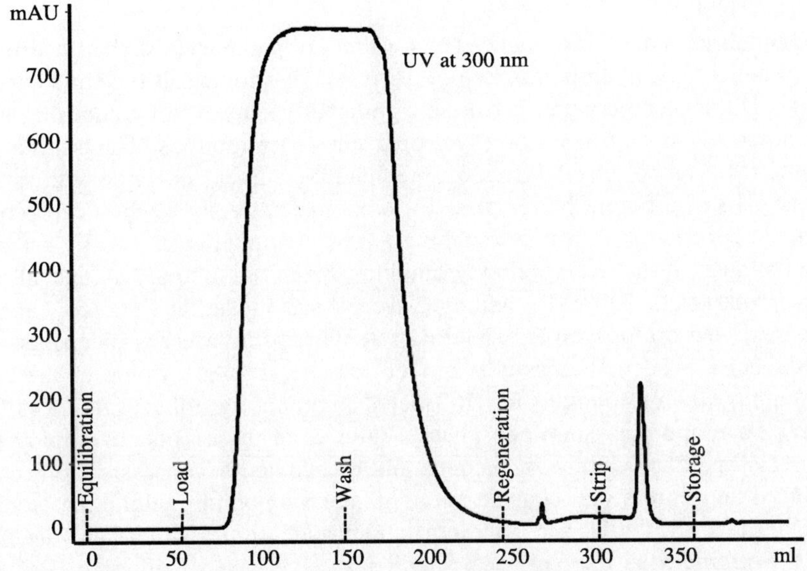

Fig. 14.8 A typical HIC chromatogram (in a flow through mode)

increasing the salt concentration in the load may be necessary. It is important to generate antibody precipitation curves, as described in the following HIC bind and elute section, to minimize aggregation and precipitation problems when conditioning the HIC load.

The first step in the development of a HIC bind and elute chromatography step is to generate antibody precipitation curves. This is done by measuring turbidity of the antibody solution in the presence of different salt concentrations at a defined pH. Turbidity can be monitored by light scattering or absorbance at 410 nm. The goal is to identify the highest salt concentration that can be used for loading without causing antibody precipitation. After antibody precipitation curves are generated, antibody retention on HIC resins can be evaluated. For a HIC bind and elute step resin screening is the most critical part of development. During resin screening antibody is loaded to the HIC resins at the highest salt concentration possible (a little below the point that starts to cause precipitation) and then eluted with a buffer at very low conductivity. Resins that do not bind the antibody (early breakthrough) or that bind the antibody too tightly (low recovery from column) should be avoided. The goal is to identify the HIC resins that demonstrate good retention and good recovery for the antibody. Following the antibody retention study, linear gradient selectivity experiments can be run under preparative loading conditions to generate selectivity plots as described in the IEX resin selection section. The goal is to compare the resolution of different HIC resins. As in IEX resin screening, HTS technique should be used in initial screening to narrow down the conditions to be screened by the column format. It is recommended to select the most hydrophobic resin that does not denature the antibody and still allows elution at low salt concentrations (Shukla and Yigzaw, 2007).

HIC binding capacity is influenced by a variety of parameters such as hydrophobicity of antibodies, hydrophobicity of HIC resins, lyotropic salt type and concentrations, pH, and temperature. It can be significantly improved by choosing more hydrophobic resins and increasing lyotropic salt concentrations. The optimal salt concentrations can be selected from the precipitation curves generated with various lyotropic salts at different buffer pHs. To maximize antibody binding capacity on HIC, a salt concentration just below the point of precipitation is usually selected. Another strategy to increase antibody binding capacity is to apply a dual salt system (Senczuk et al., 2009). By adding a second salt in the feed stream, capacity can be increased up to twofold. The dual salt synergistic effect is independent of salt type and has been successfully applied to several mAbs. Some of the commonly employed lyotropic salts for HIC include ammonium sulfate, sodium sulfate, sodium citrate, and potassium phosphate. While all of these salts are employed at bench scale operations, some of them should be avoided at process scale because of disposal and corrosivity considerations. Since ammonium sulfate and sodium phosphate have disposal issues, sodium citrate and sodium sulfate are the most suitable lyotropic salts for large scale applications. When conditioning HIC load with high salt stock solution, efficient mixing is important to avoid local high salt concentration that can potentially cause antibody precipitation.

In a typical HIC (in a bind and elute mode) process, the equilibration of the column is done with high salt concentration buffers. Feed streams are also loaded at high salt concentrations to the column at ~90% of DBC. The column is then washed with equilibration buffer to remove any weakly bound impurities. Sometimes a wash buffer at lower salt concentrations can be employed for better impurity removal. pH is another variable that can be optimized for the wash buffer since protein conformation changes significantly with pH. Antibody elution is achieved under low salt conditions. Product pool collection is based on impurity distribution across the peak and the target yield for the step. After product elution, the HIC column is cleaned with 0.5–1.0 M NaOH followed by water wash. HIC column can be stored in 0.2 M NaOH when not used.

HIC has been shown to clear HMW aggregates, HCP, DNA and leached protein A. It is very effective for removing HMW aggregates and is often chosen as a polishing step when HMW removal is the primary concern. In flow-through modes, it is not effective for leached protein A and DNA removal because these impurities have low hydrophobicity and they flow through with antibody. HIC in bind and elute modes support more efficient DNA and leached Protein A clearance.

14.3.4 Membrane Chromatography

Due to their high selectivity and binding capacity, packed-bed chromatography remains the choice for capture steps. However, the packed bed systems are facing growing substitution pressure from membrane based chromatography for polishing steps. Membrane chromatography operated in flow-through modes offers an alternative to conventional packed-bed chromatography (Knudsen et al., 2001). Membrane chromatography uses microfiltration pore size membranes (e.g., regenerated cellulose, polyethylene, and polyethersulfone) that contain functional ligands attached to the inner pore surface throughout the membrane structure to provide highly selective separations through adsorption interactions. Disposable membrane adsorbers can provide several benefits: no cleaning validation cost, no column hardware, and no column packing which allows for direct use straight from the manufacturers (Fraud et al., 2009). Ion exchange, hydrophobic interaction, and affinity membranes have been developed for membrane chromatography (Thommes and Kula, 1995, Zeng and Ruckenstein, 1999). Millipore, Sartorius and Pall are three major membrane adsorber manufacturers. The principles used in method development for conventional chromatography media are applicable for membrane chromatography development.

Membrane chromatography can be operated at very high flow rates (Thommes and Gottschalk, 2009, Gottschalk, 2008). This is the key advantage of membrane chromatography over conventional packed-bed chromatography. In contrast to bead porous material, whose internal surface is mainly present as dead-ended pores, membrane adsorbers usually have through-pores. Since ligands are immobilized in the membrane pores and the convective flow brings the solute molecules very close to the ligands, only film diffusion may limit the mass transport for membrane adsorbers (see Fig. 14.9). However, for bead matrices the dominant limitations to mass transport arise from both pore and film diffusion. Film diffusion is usually orders of magnitude faster than pore diffusion. The presence of convective transport reduces mass transfer resistance and thus binding kinetics are the dominant factor in the adsorption process. By eliminating the need for pore diffusion the dynamic binding capacities for membrane adsorbers are typically independent of the flow rate over a fairly large range (Knudsen et al., 2001, Fraud et al., 2009). The typical residence time for membrane adsorbers is only a few seconds, much shorter than the time,

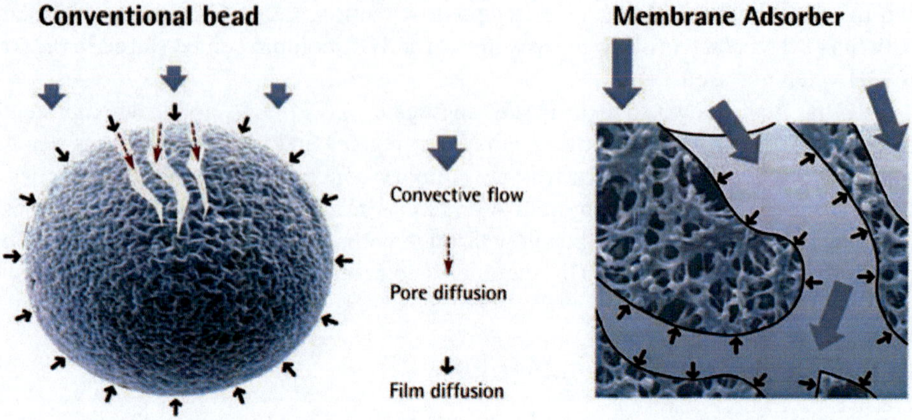

Fig. 14.9 Comparison of solute transport in conventional beads and membrane adsorbers

typically a few minutes, usually needed for conventional chromatography media. High flow rates used in membrane chromatography lead to short processing time and high throughput.

Membrane adsorbers typically have low binding capacities (Boi, 2007). For preparative separations of proteins they are not suitable for bind and elute mode operations. However, they can be ideally employed in flow-through modes as polishing steps for removal of small amount of impurities (Zhou and Tressel, 2006, Zhou et al., 2006). In flow-through applications, buffer conditions are set to enable binding of impurities while allowing product to flow through the membrane. In this case, membrane devices have sufficient capacity. The use of small membrane units with minimal buffer consumption and space requirement has been proved to be well suited for this particular application.

Membrane adsorbers have much larger pores compared to conventional chromatographic media. This can be particularly important for purification of large biomolecules and viruses that can have significant diffusion limitations in resins. Membrane chromatography has been applied in bind and elute applications for large molecules such as DNA, RNA, and viruses (Lajmi et al., 2007). Most bead-based media have pore sizes that exclude very large molecules from entering the pores. The dynamic binding capacity for these media is therefore limited to the outer surface of the beads. Membrane chromatography provides competitive DBC when used for purification of large biomolecules or viruses.

Improvements in upstream production have boosted productivity in the biopharmaceutical industry, but this is leading to bottlenecks in downstream processing as current technology platforms reach their limits of throughput and scalability. This fact makes membrane adsorbers more attractive for very large scale antibody purification. With membrane devices up to 5 L and load factors approaching 20 kg mAb L^{-1} membrane, membrane adsorbers can polish 100 kg batches of antibodies and can thus be operated in high titer processes (Gottschalk, 2008). On the other hand, conventional packed-bed column operations would suffer from the rising costs of resins, buffers, as well as the scale-related packing issues.

The following are a few examples of membrane adsorber applications in the last several years. Zhou et al. evaluated Sartobind Q devices in a flow-through mode as a polishing step. The membrane devices demonstrated HCP, DNA, and leached protein A removal that was comparable to Butyl-650 M and Q Sepharose FF resins. Comparable membrane performance was also observed at 2 kL scale using 70 and 180 mL Sartobind Q capsules (Zhou et al., 2006). A cost analysis was completed to compare Sartobind Q membrane with Q Sepharose FF resin. Even though many assumptions (such as membrane and resin binding capacity, resin lifetime, labor cost, buffers cost, cleaning and validation cost etc.) had to be made, they concluded that Q membrane is a viable alternative to Q resin as a polishing step in a flow-through mode for process scale antibody production (Zhou and Tressel, 2006). Hydrophobic type membrane adsorber Sartobind Phenyl was reported to have a >10 mg mL^{-1} binding capacity for mAb. Flow rates from 5 to 20 mL min^{-1} had no impact to the antibody binding capacity. The Phenyl device was used in a bind and elute mode for model protein separations (Fraud et al., 2009). Different types of membrane adsorbers were also employed in series for antibody purification (Giovannoni et al., 2009). In this study, a Sartobind Q – Sartobind S disposable system tackles both positively and negatively charged impurities in a single step. Mathematical models of membrane chromatography have been described in multiple articles (Thommes and Kula, 1995, Boi, 2007).

Among membrane adsorbers evaluated for applications in antibody purification, AEX type membranes in flow-through modes as polishing steps showed the most promising results in terms of binding capacity and the removal of impurities (Zhou and Tressel, 2006). Typically these membranes perform well only at low conductivities (< 6 ms cm^{-1}). Therefore product pool from a CEX bind and elute chromatography step cannot be directly loaded to these membranes, due to high conductivity. A large dilution or buffer exchange step is required to lower conductivity before CEX pool can be processed by the traditional AEX type membrane adsorbers. More recently, membranes have been developed that can tolerate much higher conductivities (up to 40 ms cm^{-1}), specifically the Millipore ChromaSorb and Sartorius STIC. They have shown good HCP clearance from CEX eluates without any dilution (Amgen internal data). A comparison of ChromaSorb and STIC membrane adsorbers are shown in Table 14.5.

At the current binding capacities, it is difficult for membrane adsorbers to effectively compete with conventional chromatography media on bind and elute applications for small protein molecules. At their current capacities they are often limited to polishing steps operated in a flow-through mode. Membrane cleaning and reuse is another area to be studied with the goal to reduce overall manufacturing cost. With research activities leading to improved membrane adsorbers, broad applications of membrane chromatography at process scales will come in the near future. Readers are referred to review articles for in-depth reading (Gottschalk, 2008, Van Reis and Zydney, 2007).

Table 14.5 Properties of two salt tolerant membrane adsorbers

Membrane adsorber properties	ChromaSorb	Sartobind STIC
Membrane type	Anion exchanger	Anion exchanger
Ligand	Poly (allylamine hydrochloride)	Primary amine
Membrane material	Ultra high molecular weight polyethylene	Stabilized and reinforced cellulose
Capacity for bovine serum albumin	≥ 75 mg/mL on average at 0 mM NaCl, ≥ 20 mg/mL on average at 280 mM NaCl	50 mg/mL in 20 mM Tris/HCl buffer, pH 7.5, 150 mM NaCl
Membrane pore size (μm)	0.65	> 3
Number of layers	8	15
Maximum pressure (psi)	30	58
Recommended flow rate (CV/min)	12.5	up to 30
Salt tolerance	High	High

14.3.5 Mixed-Mode Chromatography

Mixed-mode chromatography employs a combination of interaction mechanisms for antibody purification. It can provide unique selectivity for impurity removal and complement standard purification tools such as protein A affinity, IEX, and HIC methods. At a certain level, every class of chromatography media exploits multiple modes of interactions (Gagnon, 2009). For examples, size-exclusion chromatography (SEC) media often have charged groups that may affect SEC performance significantly. Hydrophobicity of base matrix can change the selectivity and capacity of ion exchange media. Primary interactions for Hydroxyapatite (HA) include both phosphoryl CEX and calcium metal affinity. To successfully apply mixed-mode resins at process scale antibody purification, we must understand the mechanisms of separation and the toxicity profiles of ligands in case of ligand leaching during purification. Some commercially available mixed-mode resins are from GE Healthcare and Pall. This section will focus on two of them: Capto Adhere from GE Healthcare and MEP HyperCel from Pall.

The ligand for Capto adhere is N-benzyl-N-methyl ethanolamine. The interactions between ligand and antibody include both anion exchange and hydrophobic properties. Operating pH, conductivity and loading have the largest impact to Capto Adhere performance whether it is used in bind and elute modes or flow-through modes (Gagnon, 2009, Ghose et al., 2009). In a bind and elute mode application, modifying the pH (to induce electrostatic repulsion) or adding chaotropic reagents can be necessary to facilitate antibody elution (Yang et al., 2007). GE Healthcare recommends operating the resin in a flow-through mode. Based on design of experiments (DOE) performed with several antibodies, some general trends have been identified: (1) Best yield is obtained at higher loading, lower pH, and higher conductivity; (2) Best aggregate clearance is obtained at higher pH, lower loading, and lower conductivity. Aggregate clearance is often less affected by conductivity than leached protein A and HCP clearance; (3) Best leached protein A and HCP clearance is obtained at higher pH, lower conductivity, and lower loading. At pH 6.75, Capto adhere showed about 5.8 logs clearance of Minute Virus of Mouse (MVM) at conductivities ranging from 10 to 30 ms cm^{-1}, whereas clearance of Murine Leukemia Virus (XMuLV) declines from 4.5 to 3.6 logs over the same range. The high purity obtained after capture on protein A affinity chromatography and the multimodal functionality of Capto Adhere make it possible to design a two-step antibody purification process applying MabSelect SuRe followed by Capto Adhere.

The ligand for MEP HyperCell is 2-mercaptoethyl pyridine. At neutral pH, MEP acts primarily as a HIC ligand, and the antibody binds to the resin through hydrophobic interactions. At lower pH, the nitrogen in the ligand pyridyl ring becomes positively charged (Burton and Harding, 1998). Decreasing pH also increases the positive charge on antibody amino residues, inducing electrostatic repulsion between the ligand and the antibody. This overcomes the remaining hydrophobic interactions and results in antibody elution. This type of mixed-mode chromatography is called hydrophobic charge induction chromatography (HCIC). Initially MEP HyperCell was evaluated as a capture step for antibody purification (Schwartz et al., 2001). However, recent studies have revealed that its selectivity as a capture step is much lower than that from protein A resin (Ghose et al., 2006). Readers are referred to review articles for detailed discussions on these two mixed-mode resins (Gagnon, 2009, Ghose et al., 2009).

14.3.6 Hydroxyapatite Chromatography

Hydroxyapatite (HA) chromatography has been employed in antibody purification processes to mainly remove leached protein A, endotoxin, and DNA. HA is the ligand and also the matrix (Gagnon, 1996). Its formula is $Ca_{10}(PO_4)_6(OH)_2$. There are two types of binding sites on HA. One is comprised of positively charged calcium and is called the C site. The other one contains negatively charged phosphate and is called the P site. Amino groups are attracted to P sites and repelled by C sites. The situation is reversed for carboxyl groups (Kawasaki, 1991). Carboxyl groups bind to the C sites by metal chelate interaction that is much stronger than normal electrostatic interaction. Basic proteins bind to HA through electrostatic interaction between the positively charged amino groups on the protein and the negatively charged P sites on HA. Elution of basic proteins can be achieved by increasing salt concentration. Acidic proteins bind to HA through metal chelate interaction between carboxyl groups on protein and the C sites on HA. Elution of acidic proteins cannot be accomplished by increasing salt concentration. Phosphate ions that have stronger affinity for C sites have proven to be effective for eluting acidic proteins from the matrix (Gorbunoff, 1984a, b, Gorbunoff and Timasheff, 1984). Due to the unique interactions between protein and the HA resin, high NaCl concentration can be used to selectively wash basic proteins off the column and then phosphate buffer can be applied to elute acidic proteins. Ceramic HA from Biorad is one of the few stationary phases available for process scale chromatography. Because ceramic HA resins are incompressible, a HA column can be operated at high flow rates. Obtaining a uniform packed-bed can be a challenge and special precautions need to be taken to prevent rapid settling of the high density beads and particle fracturing. Since the HA matrix contains Ca^{2+}, it is sensitive to chelators. Even low concentration of chelators such as ethylene diamine tetraacetic acid (EDTA) or citrate can dissolve the matrix over time. The matrix is unstable at low pH conditions but can tolerate strong alkaline conditions very well. Yellow-reddish brown discoloration is sometimes observed at the top of an HA column, indicating an uncontrolled source of metal contamination. The strip buffer for HA is usually a high concentration of sodium phosphate and the storage buffer typically contains a low concentration of phosphate to maintain matrix stability. HA chromatography is a very selective polishing step and can effectively remove leached protein A, endotoxin, DNA, and model viruses.

14.4 Ultrafiltration/Diafiltration and Sterile Filtration

Ultrafiltration/diafiltration (UF/DF) is employed for protein concentration and buffer exchange (Van Reis and Zydney, 2007). Protein retention in UF/DF is achieved by using a small pore size membrane (e.g., 30 kD nominal molecular weight limit membrane). UF/DF is usually applied at the end of antibody purification processes to prepare final bulk drug substance. Sometimes it is also applied between different unit operations when process intermediates need to be conditioned for the following step. UF/DF membranes are cast from a variety of polymers including polysulfone, polyethersulfone, and regenerated cellulose. Synthetic polymers have high thermal stability and chemical resistance and harsh cleaning conditions can be used to regenerate the membranes. Regenerated cellulose membranes are more hydrophilic, reducing both protein adsorption and fouling. New composite regenerated cellulose membranes have excellent mechanical strength and cause very little fouling. They also provide higher flux and better retention characteristics than other membranes. As a guideline, concentration of solutes during diafiltration can be predicted with the equation $C = C_0 e^{-SN}$ where C is the solute concentration, C_0 is the initial concentration of the solute, S is the sieving coefficient for the solute, and N is the number of diavolumes (buffer volume divided by retentate volume). Clearance of solutes may not follow the theoretically predicted value due to many reasons such as preferential binding to the product. The filtrate flux in UF/DF is typically governed by concentration polarization. High protein concentrations can reduce the membrane permeability through irreversible fouling or through the formation of a protein gel or cake on the membrane surface. UF/DF process development is accomplished by optimization of feed flow rate, trans-membrane pressure (TMP), and flux.

Sterile filtration is widely used in antibody purification processes (Van Reis and Zydney, 2007, Shukla and Kandula, 2009). The dual roles of sterile filtration include removing particulates to protect downstream units from fouling and sterilization for bioburden control. Sterile filters operate in normal flow filtration mode. Bacteria, cell debris, and insoluble aggregates are retained by the membrane based on size exclusion. The capacity of a sterile filter is determined by the fouling characteristics of the feed solution. Fouling can occur on the upper surface of the membrane, both by pore blockage and by the formation of a cake, and also within the membrane pore structure. Fouling causes a decay in flow rate in constant pressure mode and it increases the pressure in constant flux mode. The pore sizes for sterile filters typically are 0.1 and 0.2 μm. Sterile filters are made from a variety of base polymers including polyethersulfone, polyvinylidene fluoride, nylon, and polypropylene. Many of these base membranes have surface coatings or are cast as polymeric alloys to reduce protein adsorption and fouling. Membrane manufactures can cast a number of different membrane structures. The structural characteristics of the membrane have significant impact to permeability, retention capability, membrane flux, and membrane capacity. Newer composite and multi-layer membranes have dramatically increased membrane capacity compared to the isotropic membranes used in early bioprocesses. Sterile membrane capacity is typically measured by V_{max} that can be used to compare various membrane filters during filter screening.

14.5 Viral Clearance

The majority of monoclonal antibodies for therapeutic purposes are produced in mammalian cell culture (e.g., Chinese hamster ovary cells). Viral contamination can come from the original source of the cell lines and adventitious introduction of virus during the antibody production process (Chu and Robinson, 2001, Miesegaes et al., 2010). To ensure patient safety, three complementary approaches are widely used in the biopharmaceutical industry (Kundu and Reindel, 2007): (1) selecting and testing cell lines and raw materials for the absence of undesirable viruses; (2) evaluating the capability of the production processes for clearing infectious viruses; (3) testing the products at appropriate steps of production for the absence of contaminating infectious viruses. This section will focus on evaluating the capability of the purification process for clearing infectious viruses.

The goal for viral clearance studies is to demonstrate that the purification process is capable of clearing significantly more virus than what may be potentially present in the unprocessed bulk. If a viral clearance study has demonstrated good clearance of viruses representing different virus groups and characteristics, then there is a high probability that any potential virus contamination would be cleared by the purification process. A typical panel of viruses used for viral clearance studies include Xenotropic Murine Leukemia Virus (XMuLV), Minute Virus of Mice (MVM), Psenorabies Virus (PRV), and Reovirus 3 (Reo 3). Since cell lines from rodents are known to contain endogenous retroviral particles or retroviral-like particles, XMuLV is often chosen as a model for murine retrovirus. The other three model viruses are selected so that the panel of viruses represents a wide range of properties: (1) DNA and RNA genomes with single and double stranded; (2) lipid-enveloped and nonenveloped; (3) large and small sizes; (4) low to high resistance to chemical reagents. For biological license application, antibody purification processes are typically required to show sufficient clearance of the four viruses. However, for early stage clinical trials (e.g., first in human study), generally only two viruses, XMuLV and MMV, are chosen to show clearance. In a viral clearance study, model viruses are deliberately spiked into the load material. Viruses in load and product pool are analyzed by infectivity or Q-PCR assays. The virus clearance is expressed by the $\log_{10}$ ratio of the virus in load to virus in product pool. Viral clearance studies are always performed with scale-down models of the manufacturing process steps. It's not practical to perform viral clearance at large scale because it's not appropriate to spike virus into GMP material that will be used by patients. Another reason is that the amount of virus required for large scale viral clearance study would be prohibitively large and expensive.

In theory, all mAb purification techniques described in this chapter can remove and or inactivate viruses to a certain level, but not all of them will be chosen for evaluating viral clearance because it's difficult to have a representative scale-down model for some unit operations such as precipitation, centrifugation and depth filtration. Typically, in a mAb purification process, chromatography steps including protein A, IEX, HIC, and membrane chromatography (Zhou et al., 2006, Wang et al., 2009, Miesegaes et al., 2010) are evaluated for viral clearance. Protein A clears viruses by both removal and low pH inactivation. To assess viral clearance only by removal, it is necessary to use Q-PCR rather than an infectivity assay. All other chromatography steps only remove viruses, and often use an infectivity assay to measure viral clearance. Whenever possible it is desirable to use an infectivity assay to quantify viral clearance, this way the measurement is on the removal of infectious virus and not just virus fragments that would be detected by Q-PCR methods. There are other steps in the purification process, that are specifically dedicated to viral clearance, including low pH viral inactivation, solvent/detergent inactivation, and nanofiltration. Low pH and solvent/detergent inactivate enveloped viruses by disrupting the virus envelope (Bronson, 2007). Low pH viral inactivation is typically placed after the protein A column since the protein A elution pool is already at a relatively low pH. This step is typically performed at ambient temperature at pH 3.7 for 60 min. and usually complete inactivation of enveloped viruses can be achieved. Monoclonal antibodies tend to endure exposure to low pH conditions with no ill effects. However, some antibodies form HMW aggregates during low pH hold. In this case a solvent/detergent inactivation can be used instead. Nanofiltration removes retroviruses (large size, 80–110 nm) and parvoviruses viruses (small size, ~20 nm) based on size-exclusion mechanism. It is typically placed at the end of the purification process, after the final chromatography step. Viral inactivation and viral filtration are always evaluated for viral clearance because their mechanisms for viral

Table 14.6 Typical viral clearance data for 4 model viruses

Unit operation	XMuLV	MMV	PRV	Reo-3
Protein A	1.0–3.5	0.0–2.5	0.0–3.0	0.0–4.0
Low pH viral inactivation	>4.0 to >6.5	N/A	>4.0 to >6.0	N/A
CEX	2.0–6.0	0.0–2.0	3.0–5.0	2.0–6.0
AEX	2.0–6.0	0.0–6.0	3.0–6.0	3.0–6.0
HIC	0.0–4.0	0.0–2.5	1.0–4.0	2.0–5.0
Viral filtration	>4.0 to >6.0	>4.0 to >6.0	>4.0 to >6.0	>4.0 to >6.0

inactivation or removal are best understood and they are considered the most robust viral clearance strategies. The sum of viral clearance from these orthogonal steps shows the total viral clearance achieved by the entire purification process. The typical viral clearance ranges are summarized in Table 14.6. To learn more about viral clearance, readers are referred to two review articles (Kundu and Reindel, 2007, Bronson, 2007).

14.6 Conclusions

A typical mAb purification process usually includes cell culture harvest and clarification, protein A affinity chromatography, one or two more orthogonal chromatographic polishing steps, TFF and sterile filtration, as well as dedicated viral clearance steps (viral inactivation and viral filtration). Preparative chromatography is the fundamental unit operation for biopharmaceutical downstream processing. Protein A affinity, IEX and HIC will continue to be the workhorse in the foreseeable future. A platform approach for mAb purification has been widely adopted in the biopharmaceutical industry. The efficiency, robustness and scalability of this standardized process can enable short development time and multiproduct harmonization at manufacturing scales. The recent increases in cell culture titers are likely to continue with increased cellular productivity, higher cell density, and extended cell culture duration time. The combination of higher titer and larger bioreactors has sparked debate about downstream purification bottlenecks that limit a production plant's capacity. To increase process throughput and reduce manufacturing cost, the development of alternative antibody purification techniques is gaining speed. These techniques include membrane chromatography, high performance tangential flow filtration (HPTFF), flocculation/precipitation, crystallization, expanded bed adsorption, simulated moving bed chromatography, and liquid-liquid extractions using two-phase aqueous polymer systems.

Acknowledgements The authors are grateful to Yinges Yigzaw at Amgen for his feedbacks during the preparation of the chapter. We also would like to thank Priyanka Gupta at Sartorius Stedim North America Inc. for the permission to use Fig. 14.9.

References

Arakawa, T., Philo, J. S., Tsumoto, K., Yumioka, R. & Ejima, D. (2004) Elution of antibodies from a Protein-A column by aqueous arginine solutions. *Protein Expr Purif*, 36, 244–248.

Blank, G. S., Zapata, G., Fahrner, R., Milton, M., Yedinak, C., Knudsen, H. & Schmelzer, C. (2001) Expanded bed adsorption in the purification of monoclonal antibodies: a comparison of process alternatives. *Bioseparation*, 10, 65–71.

Boi, C. (2007) Membrane adsorbers as purification tools for monoclonal antibody purification. *J Chromatogr B Analyt Technol Biomed Life Sci*, 848, 19–27.

Braisted, A. C. & Wells, J. A. (1996) Minimizing a binding domain from protein A. *Proc Natl Acad Sci USA*, 93, 5688–5692.

Bronson, K. (2007) Advances in viral clearance. In Shukla, A. A., Etzel, M. R., Gadam, S., (Eds.) *Process scale bioseparations for the biopharmaceutical industry*. Boca Raton, FL, CRC Press.

Burton, D. R. (1985) Immunoglobulin-G – functional sites. *Mol Immunol*, 22, 161–206.

Burton, S. C. & Harding, D. R. (1998) Hydrophobic charge induction chromatography: salt independent protein adsorption and facile elution with aqueous buffers. *J Chromatogr A*, 814, 71–81.

Chu, L. & Robinson, D. K. (2001) Industrial choices for protein production by large-scale cell culture. *Curr Opin Biotechnol*, 12, 180–187.

Coffman, J., Shpritzer, R. & Vicik, S. (2006) Flocculation of antibody-producing mammalian cells with precipitating solutions of soluble cations and anions. *Recovery of biological products XII.* Litchfield, AZ.

Deisenhofer, J. (1981) Crystallographic refinement and atomic models of a human Fc fragment and its complex with fragment B of protein A from Staphylococcus aureus at 2.9- and 2.8-A resolution. *Biochemistry*, 20, 2361–2370.

Fahrner, R. L., Knudsen, H. L., Basey, C. D., Galan, W., Feuerhelm, D., Vanderlaan, M. & Blank, G. S. (2001) Industrial purification of pharmaceutical antibodies: development, operation, and validation of chromatography processes. *Biotechnol Genet Eng Rev*, 18, 301–327.

Fiore, J. V., Olson, W. P. & Holst, S. L. (1980) Depth filtration. In Curling, J. M., (Ed.) *Methods of plasma protein fractionation*. New York City, NY, Academic Press.

Fraud, N., Faber, R., Kiss, C., Demmer, W., Hoerl, H. -H. & Fischer-Fruehholz, S. (2009) Hydrophobic-interaction membrane chromatography for large-scale purification of biopharmaceuticals. *BioProcess Int*, 7, 30–35.

Gagnon, P. (1996) *Purification tools for monoclonal antibodies*. Tucson, AZ, Validated Biosystems.

Gagnon, P. (2009) Purification of monoclonal antibodies by mixed-mode chromatography. In Gottschalk, U., (Ed.) *Process scale purification of antibodies*. Hoboken, NJ, Wiley

Ghose, S., Allen, M., Hubbard, B., Brooks, C. & Cramer, S. M. (2005) Antibody variable region interactions with Protein A: implications for the development of generic purification processes. *Biotechnol Bioeng*, 92, 665–673.

Ghose, S., Hubbard, B. & Cramer, S. M. (2006) Evaluation and comparison of alternatives to Protein A chromatography – mimetic and hydrophobic charge induction chromatographic stationary phases. *J Chromatogr A*, 1122, 144–152.

Ghose, S., Jin, M., Liu, J. & Hickey, J. (2009) Integrated polishing steps for monoclonal antibody purification. In Gottschalk, U., (Ed.) *Process scale purification of antibodies*. Hoboken, NJ, Wiley.

Ghose, S., Mcnerney, T. M. & Hubbard, B. (2002) pH Transitions in ion-exchange systems: role in the development of a cation-exchange process for a recombinant protein. *Biotechnol Prog*, 18, 530–537.

Ghose, S., Mcnerney, T. & Hubbard, B. (2007) Protein A affinity chromatography for capture and purification of monoclonal antibodies and Fc-Fusion Proteins: practical considerations for process development. In Shukla, A. A., Etzel, M. R., Gadam, S., (Eds.) *Process scale bioseparations for the biopharmaceutical industry.* Boca Raton, FL, CRC Press.

Giovannoni, L., Ventani, M. & Gottschalk, U. (2009) Antibody purification using membrane adsorbers. *Biopharm Int*, 28–32.

Gorbunoff, M. J. (1984a) The interaction of proteins with hydroxyapatite. 1. Role of protein charge and structure. *Anal Biochem*, 136, 425–432.

Gorbunoff, M. J. (1984b) The interaction of proteins with hydroxyapatite. 2. Role of acidic and basic groups. *Anal Biochem*, 136, 433–439.

Gorbunoff, M. J. & Timasheff, S. N. (1984) The interaction of proteins with hydroxyapatite. 3. Mechanism. *Anal Biochem*, 136, 440–445.

Gottschalk, U. (2008) Bioseparation in antibody manufacturing: the good, the bad and the ugly. *Biotechnol Prog*, 24, 496–503.

Gulich, S., Uhlen, M. & Hober, S. (2000) Protein engineering of an IgG-binding domain allows milder elution conditions during affinity chromatography. *J Biotechnol*, 76, 233–244.

Hahn, R., Schlegel, R. & Jungbauer, A. (2003) Comparison of protein A affinity sorbents. *J Chromatogr B Analyt Technol Biomed Life Sci*, 790, 35–51.

Hale, G., Drumm, A., Harrison, P. & Phillips, J. (1994) Repeated cleaning of protein-a affinity column with sodium-hydroxide. *J Immunol Methods*, 171, 15–21.

Harinarayan, C., Mueller, J., Ljunglof, A., Fahrner, R., Van Alstine, J. & Van Reis, R. (2006) An exclusion mechanism in ion exchange chromatography. *Biotechnol Bioeng*, 95, 775–787.

Hjelm, H., Sjodahl, J. & Sjoquist, J. (1975) Immunologically active and structurally similar fragments of protein-a from staphylococcus-aureus. *Eur J Biochem*, 57, 395–403.

Jagschies, G., Gronberg, A., Bjorkman, T., Lacki, K. & Johansson, H. J. (2006) Technical and economical evaluation of downstream processing options for monoclonal antibody (MAb) production. *Biopharm Int* June Suppl., 10–19.

Jefferis, R. (2005) Glycosylation of recombinant antibody therapeutics. *Biotechnol Prog*, 21, 11–16.

Kawasaki, T. (1991) Hydroxyapatite as a liquid-chromatographic packing. *J Chromatogr*, 544, 147–184.

Kelley, B. (2007) Very large scale monoclonal antibody purification: the case for conventional unit operations. *Biotechnol Prog*, 23, 995–1008.

Kelley, B., Blank, G. & Lee, A. (2009) Downstream processing of monoclonal antibodies: current practices and future opportunities. In Gottschalk, U., (Ed.) *Process scale purification of antibodies*. Hoboken, NJ, John Wiley & Sons, Inc.

Kelley, B. D., Tobler, S. A., Brown, P., Coffman, J. L., Godavarti, R., Iskra, T., Switzer, M. & Vunnum, S. (2008) Weak partitioning chromatography for anion exchange purification of monoclonal antibodies. *Biotechnol Bioeng*, 101, 553–566.

Kennedy, R. M. (1990) Hydrophobic chromatography. *Methods Enzymol*, 182, 339–357.

Kinolov, Z. L., Regan, J. T., Dickey, L. F. & Woodard, S. L. (2009) Purification of antibodies from transgenic plants. In Gottschalk, U., (Ed.) *Process scale purification of antibodies*. Hoboken, NJ, Wiley.

Knight, R. & Ostreicher, E. (1998) Charge-modified filter media. In Meltzer, T. H., Jornitz, M. W., (Eds.) *Filtration in the biopharmaceutical industry*. New York, NY, Marcel Dekker.

Knudsen, H. L., Fahrner, R. L., Xu, Y., Norling, L. A. & Blank, G. S. (2001) Membrane ion-exchange chromatography for process-scale antibody purification. *J Chromatogr A*, 907, 145–154.

Kundu, A. & Reindel, K. (2007) Evaluation of viral clearance in purification processes. In Shukla, A. A., Etzel, M. R., Gadam, S., (Eds.) *Process scale bioseparations for the biopharmaceutical industry*. Boca Raton, FL, CRC Press.

Lajmi, A., Kutner, R. & Reiser, J. (2007) A membrane chromatography application: a rapid, high capacity gene therapy vector purification tool. In Shukla, A. A., Etzel, M. R., Gadam, S., (Eds.) *Process scale bioseparations for the biopharmaceutical industry*. Boca Raton, FL, CRC Press.

Lee, S. S., Burt, A., Russotti, G. & Buckland, B. (1995) Microfiltration of recombinant yeast-cells using a rotating-disk dynamic filtration system. *Biotechnol Bioeng*, 48, 386–400.

Li, R., Dowd, V., Stewart, D. J., Burton, S. J. & Lowe, C. R. (1998) Design, synthesis, and application of a Protein A mimetic. *Nat Biotech*, 16, 190–195.

Lindmark, R., Movitz, J. & Sjoquist, J. (1977) Extracellular protein-a from a methicillin-resistant strain of staphylococcus-Aureus. *Eur J Biochem*, 74, 623–628.

Low, D., O'leary, R. & Pujar, N. S. (2007) Future of antibody purification. *J Chromatogr B Analyt Technol Biomed Life Sci*, 848, 48–63.

Lu, Y., Williamson, B. & Gillespie, R. (2009) Recent advancement in application of hydrophobic interaction chromatography for aggregate removal in industrial purification process. *Curr Pharm Biotechnol*, 10, 427–433.

Luan, Y. T., Wang, W., Nolan, R. & Drapeau, D. (2006) Defined medium development for high yielding mammalian cell culture processes. *IBC & Bioprocess International conference*. San Francisco, CA.

Lydersen, B. K., Brehm-Gibson, T. & Murel, A. (1994) Acid precipitation of mammalian cell fermentation broth. *Ann NY Acad Sci*, 745, 222–231.

Mccue, J. T., Kemp, G., Low, D. & Quinones-Garcia, I. (2003) Evaluation of protein-A chromatography media. *J Chromatogr A*, 989, 139–153.

Miesegaes, G., Lute, S. & Brorson, K. (2010) Analysis of viral clearance unit operations for monoclonal antibodies. *Biotechnol Bioeng*, 106, 238–246.

Moks, T., Abrahmsen, L., Nilsson, B., Hellman, U., Sjoquist, J. & Uhlen, M. (1986) Staphylococcal protein-a consists of 5 Igg-binding domains. *Eur J Biochem*, 156, 637–643.

Queiroz, J. A., Tomaz, C. T. & Cabral, J. M. (2001) Hydrophobic interaction chromatography of proteins. *J Biotechnol*, 87, 143–159.

Riske, F., Schroeder, J., Belliveau, J., Kang, X., Kutzko, J. & Menon, M. K. (2007) The use of chitosan as a flocculant in mammalian cell culture dramatically improves clarification throughput without adversely impacting monoclonal antibody recovery. *J Biotechnol*, 128, 813–823.

Rohrbach, P., Broders, O., Toleikis, L. & Dubel, S. (2003) Therapeutic antibodies and antibody fusion proteins. *Biotechnol Genet Eng Rev*, 20, 137–163.

Rosenberg, A. S. (2006) Effects of protein aggregates: an immunologic perspective. *AAPS J*, 8, E501–E507.

Russell, E., Wang, A. & Rathore, A. S. (2007) Harvest of a therapeutic protein product form high cell density fermentation broths: principles and case study. In Shukla, A. A., Etzel, M. R., Gadam, S., (Eds.) *Process scale bioseparations for the biopharmaceutical industry*. Boca Raton, FL, CRC Press.

Schwartz, W., Judd, D., Wysocki, M., Guerrier, L., Birck-Wilson, E. & Boschetti, E. (2001) Comparison of hydrophobic charge induction chromatography with affinity chromatography on protein A for harvest and purification of antibodies. *J Chromatogr A*, 908, 251–263.

Senczuk, A. M., Klinke, R., Arakawa, T., Vedantham, G. & Yigzaw, Y. (2009) Hydrophobic interaction chromatography in dual salt system increases protein binding capacity. *Biotechnol Bioeng*, 103, 930–935.

Shpritzer, R., Vicik, S., Orlando, S., Acharya, H. & Coffman, J. (2006) Calcium phosphate flocculation of antibody-producing mammalian cells at pilot scale. *232*nd *American chemical society national meeting*. San Francisco, CA.

Shukla, A. A. & Han, X. S. (2007) Screening chromatographic stationary phases. In Shukla, A. A., Etzel, M. R., Gadam, S., (Eds.) *Process scale bioseparations for the biopharmaceutical industry*. Boca Raton, FL, CRC Press.

Shukla, A. A., Hinckley, P. J., Gupta, P., Yigzaw, Y. & Hubbard, B. (2005) Strategies to address aggregation during protein A chromatography. *BioProcess Int*, 3, 36–45.

Shukla, A. A., Hubbard, B., Tressel, T., Guhan, S. & Low, D. (2007) Downstream processing of monoclonal antibodies–application of platform approaches. *J Chromatogr B Analyt Technol Biomed Life Sci*, 848, 28–39.

Shukla, A. A. & Kandula, J. R. (2009) Harvest and recovery of monoclonal antibodies: cell removal and clarification. In Gottschalk, U., (Ed.) *Process scale purification of antibodies*. Hoboken, NJ, Wiley.

Shukla, A. A. & Yigzaw, Y. (2007) Modes of preparative chromatography. In Shukla, A. A., Etzel, M. R., Gadam, S., (Eds.) *Process scale bioseparations for the biopharmaceutical industry*. Boca Raton, FL, CRC Press.

Smith, G. R. S. (1998) Filter aid filtration. In Meltzer, T. H., Jornitz, M. W., (Eds.) *Filtration in the biopharmaceutical industry*. New York City, NY, Marcel Dekker.

Sonnenfeld, A. & Thommes, J. (2007) Expanded bed adsorption for capture from crude solution. In Shukla, A. A., Etzel, M. R., Gadam, S., (Eds.) *Process scale bioseparations for the biopharmaceutical industry*. Boca Raton, FL, CRC Press.

Spitali, M. (2009) Downstream processing of monoclonal antibody fragments. In Gottschalk, U., (Ed.) *Process scale purification of antibodies*. Hoboken, NJ, John Wiley & Sons, Inc.

Staby, A., Sand, M. B., Hansen, R. G., Jacobsen, J. H., Andersen, L. A., Gerstenberg, M., Bruus, U. K. & Jensen, I. H. (2005) Comparison of chromatographic ion-exchange resins IV. Strong and weak cation-exchange resins and heparin resins. *J Chromatogr A*, 1069, 65–77.

Stein, A. & Kiesewetter, A. (2007) Cation exchange chromatography in antibody purification: pH screening for optimised binding and HCP removal. *J Chromatogr B Analyt Technol Biomed Life Sci*, 848, 151–158.

Terman, D. S. & Bertram, J. H. (1985) Antitumor effects of immobilized protein A and staphylococcal products: linkage between toxicity and efficacy, and identification of potential tumoricidal reagents. *Eur J Cancer Clin Oncol*, 21, 1115–1122.

Thommes, J. & Etzel, M. (2007) Alternatives to chromatographic separations. *Biotechnol Prog*, 23, 42–45.

Thommes, J. & Gottschalk, U. (2009) Alternatives to packed-bed chromatography for antibody extraction and purification. In Gottschalk, U., (Ed.) *Process scale purification of antibodies*. Hoboken, NJ, John Wiley & Sons, Inc.

Thommes, J. & Kula, M. R. (1995) Membrane chromatography – an integrative concept in the downstream processing of proteins. *Biotechnol Prog*, 11, 357–367.

Tipton, B., Boose, J. A., Larsen, W., Beck, J. & O'brien, T. (2002) Retrovirus and parvovirus clearance from an affinity column product using adsorptive depth filtration. *Biopharm-Appl T Bio*, 15, 43–50.

Van Reis, R., Leonard, L. C., Hsu, C. C. & Builder, S. E. (1991) Industrial scale harvest of proteins from mammalian cell culture by tangential flow filtration. *Biotechnol Bioeng*, 38, 413–422.

Van Reis, R. & Zydney, A. (2001) Membrane separations in biotechnology. *Curr Opin Biotechnol*, 12, 208–211.

Van Reis, R. & Zydney, A. (2007) Bioprocess membrane technology. *J Memb Sci*, 297, 16–50.

Vunnum, S., Vedantham, G. & Hubbard, B. (2009) Protein A-based affinity chromatography. In Gottschalk, U., (Ed.) *Process scale purification of antibodies*. Hoboken, NJ, John Wiley & Sons, Inc.

Walsh, G. (2004) Modern antibody-base therapeutics. *BioPharm Int*, 17, 18–25.

Wang, J., Diehl, T., Aguiar, D., Dai, X. P. & Arunakumari, A. (2009) Precipitation of process-derived impurities in non-protein A purification schemes for antibodies. *Biopharm Int*, 32, 4–10.

Yang, T., Malmquist, G., Johansson, B. L., Maloisel, J. L. & Cramer, S. (2007) Evaluation of multi-modal high salt binding ion exchange materials. *J Chromatogr A*, 1157, 171–177.

Yigzaw, Y., Piper, R., Tran, M. & Shukla, A. A. (2006) Exploitation of the adsorptive properties of depth filters for host cell protein removal during monoclonal antibody purification. *Biotechnol Prog*, 22, 288–296.

Zeng, X. & Ruckenstein, E. (1999) Membrane chromatography: preparation and applications to protein separation. *Biotechnol Prog*, 15, 1003–1019.

Zhang, C. & Van Cott, K. E. (2007) Product recovery from transgenic sources. In Shukla, A. A., Etzel, M. R., Gadam, S., (Eds.) *Process scale bioseparations for the biopharmaceutical industry*. Boca Raton, FL, CRC Press.

Zhou, J. X. & Tressel, T. (2006) Basic concepts in Q membrane chromatography for large-scale antibody production. *Biotechnol Prog*, 22, 341–349.

Zhou, J. X., Tressel, T., Gottschalk, U., Solamo, F., Pastor, A., Dermawan, S., Hong, T., Reif, O., Mora, J., Hutchison, F. & Murphy, M. (2006) New Q membrane scale-down model for process-scale antibody purification. *J Chromatogr A*, 1134, 66–73.

Subject Index

M. Al-Rubeai (ed.), *Antibody Expression and Production*, Cell Engineering 7, DOI 10.1007/978-94-007-1257-7, © Springer Science+Business Media B.V. 2011